Wastewater to Resource Recovery

Wastewater to Resource Recovery

Applying the Circular Economy Toward Sustainable Development

Edited by

Arvind Kumar Mungray
Department of Chemical Engineering
S. V. National Institute of Technology (SVNIT)
Surat, Gujarat, India

Alka A. Mungray
Department of Chemical Engineering
S. V. National Institute of Technology (SVNIT)
Surat, Gujarat, India

Ambika Arkatkar
Department of Biotechnology
Veer Narmad South Gujarat University (VNSGU)
Surat, Gujarat, India

Mohammad Ali Abdelkareem
Sustainable & Renewable Energy Engineering Department
College of Engineering, University of Sharjah
Sharjah, UAE; and Faculty of Engineering
Minia University, Egypt

WILEY

Published by John Wiley & Sons, Inc., Hoboken, New Jersey.

Library of Congress Cataloging-in-Publication Data has been applied for:

Hardback: 9781394274284
ePDF: 9781394274307
epub: 9781394274291
oBook: 9781394274314

Cover Design: Wiley
Cover Image: © Surasak Suwanmake/Getty Images

Dedicated to my parents, wife, and lovely children, Dakshit and Prisha.

—Arvind Kumar Mungray

Contents

List of Contributors *xi*
Preface *xxi*
Acknowledgments *xxiii*

1 **The Centralized Systems and Strategies for Wastewater Treatment: Overview** *1*
Srishti Mishra, Anil Dhanda, Brajesh K. Dubey, and Makarand M. Ghangrekar

2 **Linear and Circular Economy Strategies for Waste Treatment: Fundamentals** *17*
Abhishek, Anag Katiyar, Ishwar Sharan, Tezna Singh, Hanshika, Parul, Ashish Yadav, and Vimal Chandra Srivastava

3 **Infrastructure Requirement and Regulatory Framework for Wastewater Management System** *73*
Utkarsh Dubey, Divyani Pathak, Sanjay Kumar Gupta, and Bhaskar Singh

4 **Sewage Composition and Treatment: Economic Burden and Future Implications** *95*
Soumyajit Chandra, Soumya Pandit, and Abdurrasheed Abubakar Yunus

5 **Conventional Physical and Chemical Treatment Technologies and Economic Implications** *119*
Mohamed Sameh El-harty, Ahmed Salah Elkomy, and Nabila Shehata

6 **Conventional Biological Treatment Technologies and Economic Implications** *145*
Sachin Chunchu, Soubhagya Nayak, Nayan Babulal Gadling, Sudipa Bhadra, Soujanya Jonnalagadda, Venkateswara Rao Polisetty, and Surajbhan Sevda

7 **Composition of Industrial Wastewater and Its Conventional Treatment Facility** *167*
Heba A. Younes and Nabila Shehata

8 **Legal Framework and Industrial Insight for Treatment of Waste** *189*
Saji Simon, Bibin K. Suresh, Anusree P. C., Anantha Singh T. S., George K. Varghese, and Nidheesh P. V.

9 **Industrial Wastewater Treatment: Environmental and Economic Implications** *213*
Mona Ebaid, Ghada Abdallah, Nabila Shehata, and Michael Kornaros

10 **Case Study for Wastewater Treatment** *235*
Nabila Shehata and Rania Osama

11 **Sewage a Wasting Resource: Need to Rethink** *253*
Olfat Saied, Tabbi Wilberforce, and Nabila Shehata

12 **Wastewater as a Source of Different Forms of Energy: Resource Recovery, Circular Economy and/or Cleaner Production Perspectives from the Mining and Food Processing Industry** *269*
Quinnetta Blessing Clement, Abayomi B. C. Grant, Abubakar M. Bilyaminu, Hector García Hernandez, Alka A. Mungray, Arvind Kumar Mungray, Eldon R. Rene, and Anina James

13 **Utilization of Industrial Wastewater as an Input for Resource Recovery** *295*
Ahmed Salah Elkomy and Nabila Shehata

14 **Decentralized Systems for Resource Recovery: Water Recovery Strategies from Wastewater** *313*
Upexa Patel, Ambika Arkatkar, Pankaj Pardeshi, Alka A. Mungray, and Arvind Kumar Mungray

15 **Conversion Strategies for Utilization of Wastewater as a Fertilizer** *337*
Sudeeptha Girijan, K. Hasim Suhaib, M. Devasena, and Praveena Gangadharan

16 **Contribution of Modern Biological Sewage Resource Recovery Technologies in Circular Economy of Water** *369*
Minh-Ky Nguyen, Quoc-Minh Truong, The-Anh Luu, Van-Giang Le, Hoang-Lam Nguyen, Minh-Thuan Pham, and D. Duc Nguyen

17 Adopting Conventional and Modern Water Treatment Technologies to Achieve Circular Economy by Resource Recovery 385
Hebatullah H. Farghal, Asmaa Ali, Nabila Shehata, and Mayyada El-Sayed

18 Resource Recovery from Urine and Fecal Material at On-site Treatment and Its Economic Evaluation: A Case Study 401
Zavin Gajera, Monali Chhatbar, Ambika Arkatkar, Alka A. Mungray, and Arvind Kumar Mungray

19 Technologies for Treatment of Wastewater: Scale-up Plant 427
Nabila Shehata

20 Membrane Technology Enhanced by Nanostructures for Wastewater Circular Economy 455
Alaa El Din Mahmoud, Nourhan S. Sultan, Nikolay G. Chechenin, Mayar M. Osman, E. M. El-Maghraby, and E. M. Elsehly

21 Economic Benefits of Source Separation of Wastewater 495
Anushree R. Gahlowt, Ambika Arkatkar, and Anand Kishore Kola

22 Integration of Different Technologies to Enhance the Efficiency of Wastewater Treatment 517
Nabila Shehata and Davidson Egirani

23 Utilization of Artificial Intelligence (AI) for Modernization of Wastewater Treatment and Resource Recovery 533
Priyanka Ray Choudhury, Muskan Syed, Alakto Choudhury, Sheetal Kumari, Abhishek Sengupta, Priyanka Narad, Smriti Agarwal, and Manoj Chandra Garg

24 Sustainability and Interdependency of Multi-step Scale Up of Wastewater Treatment Technologies 561
Ambika Arkatkar, Dipak A. Jadhav, Mohammed Hussein, Hai Yen Nguyen, Tasnim Eisa, Ahmed Bahaa, Arvind Kumar Mungray, and Kyu-Jung Chae

25 Government Strategies to Economically Support the Cleaning of Water Bodies in India 575
Abdul Gani, Athar Hussain, and Shray Pathak

Index 597

List of Contributors

Ghada Abdallah
Environmental Science and Industrial Development Department, Faculty of Postgraduate Studies for Advanced Sciences
Beni-Suef University
Beni-Suef
Egypt

Abhishek
Department of Chemical Engineering
Indian Institute of Technology
Roorkee
Roorkee
Uttarakhand
India

Smriti Agarwal
Motilal Nehru National Institute of Technology Allahabad
Prayagraj
Uttar Pradesh
India

Asmaa Ali
Department of Chemistry,
School of Sciences and Engineering
The American University in Cairo
New Cairo
Egypt

Environmental Science and Industrial Development Department, Faculty of Postgraduate Studies for Advanced Sciences (PSAS)
Beni-Suef University
Beni Suef
Egypt

Anusree P. C.
Department of Civil Engineering,
National Institute of Technology
Calicut
Kozhikode
Kerala
India

Ambika Arkatkar
Department of Biotechnology
Veer Narmad South Gujarat University
Surat
Gujarat
India

Ahmed Bahaa
Department of Environmental Engineering
College of Ocean Science and Engineering, Korea Maritime and Ocean University
Busan
Republic of Korea

Interdisciplinary Major of Ocean Renewable Energy Engineering
Korea Maritime and Ocean University
Busan
Republic of Korea

Sudipa Bhadra
Department of Biotechnology
National Institute of Technology Warangal
Warangal
Telangana
India

Abubakar M. Bilyaminu
Department of Water Supply, Sanitation and Environmental Engineering
IHE Delft Institute for Water Education
Delft
The Netherlands

Kyu-Jung Chae
Department of Environmental Engineering
College of Ocean Science and Engineering, Korea Maritime and Ocean University
Busan
Republic of Korea

Interdisciplinary Major of Ocean Renewable Energy Engineering
Korea Maritime and Ocean University
Busan
Republic of Korea

Soumyajit Chandra
Department of Life Sciences
Sharda School of Basic Sciences and Research, Sharda University
Greater Noida
Uttar Pradesh
India

Nikolay G. Chechenin
Skobeltsyn Institute of Nuclear Physics, Lomonosov Moscow State University
Moscow
Russia

Monali Chhatbar
Department of Chemical Engineering
Sardar Vallabhbhai National Institute of Technology
Surat
Gujarat
India

Alakto Choudhury
Amity Institute of Biotechnology
Amity University
Noida
Uttar Pradesh
India

Priyanka Ray Choudhury
Amity Institute of Biotechnology
Amity University
Noida
Uttar Pradesh
India

Department of Civil Engineering
Indian Institute of Technology
Kharagpur
Kharagpur
West Bengal
India

Sachin Chunchu
Department of Civil Engineering
National Institute of Technology
Warangal
Warangal
Telangana
India

Quinnetta Blessing Clement
Department of Water Supply, Sanitation and Environmental Engineering
IHE Delft Institute for Water Education
Delft
The Netherlands

M. Devasena
Global Sanitation Center of Excellence
IIT Palakkad
Palakkad
Kerala
India

Anil Dhanda
Department of Civil Engineering
Indian Institute of Technology
Kharagpur
Kharagpur
West Bengal
India

Brajesh K. Dubey
School of Water Resources
Indian Institute of Technology
Kharagpur
Kharagpur
West Bengal
India

Utkarsh Dubey
Department of Environmental Sciences
Central University of Jharkhand
Ranchi
Jharkhand
India

Mona Ebaid
Material Science and Nanotechnology Department, Faculty of Postgraduate Studies for Advanced Sciences
Beni-Suef University
Beni-Suef
Egypt

Davidson Egirani
Faculty of Science
Niger Delta University
Wilberforce Island
Nigeria

Tasnim Eisa
Department of Environmental Engineering
College of Ocean Science and Engineering, Korea Maritime and Ocean University
Busan
Republic of Korea

Interdisciplinary Major of Ocean Renewable Energy Engineering
Korea Maritime and Ocean University
Busan
Republic of Korea

Mohamed Sameh El-harty
Materials Science and Nanotechnology Department, Faculty of Postgraduate Studies for Advanced Sciences
Beni-Suef University
Beni-Suef
Egypt

Ahmed Salah Elkomy
Materials Science and Nanotechnology Department, Faculty of Postgraduate Studies for Advanced Sciences
Beni-Suef University
Beni-Suef
Egypt

E. M. El-Maghraby
Physics Department, Faculty of Science
Damanhour University
Damanhour
Egypt

Mayyada El-Sayed
Department of Chemistry, School of Sciences and Engineering
The American University in Cairo
New Cairo
Egypt

E. M. Elsehly
Physics Department, Faculty of Science
Damanhour University
Damanhour
Egypt

Hebatullah H. Farghal
Department of Chemistry, School of Sciences and Engineering
The American University in Cairo
New Cairo
Egypt

Nayan Babulal Gadling
Department of Biotechnology
National Institute of Technology
Warangal
Warangal
Telangana
India

Anushree R. Gahlowt
Department of Biotechnology
Veer Narmad South Gujarat University
Surat
Gujarat
India

Zavin Gajera
Department of Chemical Engineering
Sardar Vallabhbhai National Institute of Technology
Surat
Gujarat
India

Praveena Gangadharan
Global Sanitation Center of Excellence
IIT Palakkad
Palakkad
Kerala
India

Department of Civil Engineering
IIT Palakkad
Palakkad
Kerala
India

Abdul Gani
Department of Civil Engineering
Netaji Subhas University of Technology
New Delhi
India

Manoj Chandra Garg
Amity Institute of Environmental Sciences
Amity University
Noida
Uttar Pradesh
India

Makarand M. Ghangrekar
Department of Civil Engineering
Indian Institute of Technology Kharagpur
Kharagpur
West Bengal
India

National Institute of Technology Puducherry
Karaikal
Puducherry
India

Sudeeptha Girijan
Global Sanitation Center of Excellence
IIT Palakkad
Palakkad
Kerala
India

Abayomi B. C. Grant
Environmental & Safeguard Officer,
Monrovia Metropolitan Climate Resilience Project (MMCRP),
Environmental Protection Agency of Liberia (EPA)
Monrovia
Liberia

Sanjay Kumar Gupta
Department of Civil Engineering
Indian Institute of Technology Delhi
New Delhi
India

Hanshika
Department of Chemistry
Indian Institute of Technology Roorkee
Roorkee
Uttarakhand
India

Ashish Yadav
Department of Chemical Engineering
Indian Institute of Technology Roorkee
Roorkee
Uttarakhand
India

Hector García Hernandez
Department of Water Supply, Sanitation and Environmental Engineering
IHE Delft Institute for Water Education
Delft
The Netherlands

Athar Hussain
Department of Civil Engineering
Netaji Subhas University of Technology
New Delhi
India

Mohammed Hussein
Department of Environmental Engineering
College of Ocean Science and Engineering, Korea Maritime and Ocean University
Busan
Republic of Korea

Interdisciplinary Major of Ocean Renewable Energy Engineering
Korea Maritime and Ocean University
Busan
Republic of Korea

Dipak A. Jadhav
School of Civil and Environmental Sciences, Faculty of Science and Technology
JSPM University Pune
Maharashtra
India

Anina James
J & K Pocket
Delhi
India

Soujanya Jonnalagadda
Department of Civil Engineering
National Institute of Technology Warangal
Warangal
Telangana
India

Anag Katiyar
Department of Chemical Engineering
Indian Institute of Technology Roorkee
Roorkee
Uttarakhand
India

Anand Kishore Kola
Department of Chemical Engineering
National Institute of Technology Warangal
Warangal
Telangana
India

Michael Kornaros
Laboratory of Biochemical Engineering & Environmental Technology (LBEET), Department of Chemical Engineering
University of Patras
Patras
Greece

Sheetal Kumari
Amity Institute of Environmental Sciences
Amity University
Noida
Uttar Pradesh
India

School of Eco-environment
Harbin Institute of Technology
Shenzhen
China

Van-Giang Le
Central Institute for Natural Resources and Environmental Studies
Vietnam National University (CRES-VNU)
Hanoi
Vietnam

The-Anh Luu
Central Institute for Natural Resources and Environmental Studies
Vietnam National University (CRES-VNU)
Hanoi
Vietnam

Alaa El Din Mahmoud
Environmental Sciences Department,
Faculty of Science
Alexandria University
Alexandria
Egypt

Green Technology Group,
Faculty of Science
Alexandria University
Alexandria
Egypt

Srishti Mishra
School of Water Resources
Indian Institute of Technology
Kharagpur
Kharagpur
West Bengal
India

Alka A. Mungray
Department of Chemical Engineering
Sardar Vallabhbhai National Institute
of Technology
Surat
Gujarat
India

Arvind Kumar Mungray
Department of Chemical Engineering
Sardar Vallabhbhai National Institute
of Technology
Surat
Gujarat
India

Priyanka Narad
Indian Council of Medical Research
Ansari Nagar
New Delhi
India

Soubhagya Nayak
Department of Biotechnology
National Institute of Technology
Warangal
Warangal
Telangana
India

D. Duc Nguyen
Department of Civil & Energy System
Engineering
Kyonggi University
Suwon
South Korea

Hoang-Lam Nguyen
Department of Civil Engineering
McGill University
Montreal
Canada

Minh-Ky Nguyen
Faculty of Environment and Natural
Resources
Nong Lam University
Ho Chi Minh City
Vietnam

Nidheesh P. V.
National Environmental Engineering
Research Institute
Nagpur
Maharashtra
India

Rania Osama
Department of Civil Engineering,
Faculty of Engineering
Minia University
Minia
Egypt

Mayar M. Osman
Physics Department, Faculty of Science
Damanhour University
Damanhour
Egypt

Soumya Pandit
Department of Life Sciences
Sharda School of Basic Sciences and Research, Sharda University
Greater Noida
Uttar Pradesh
India

Pankaj Pardeshi
Technical & Applied Chemistry Department
Veermata Jijabai Technological Institute (VJTI)
Mumbai
India

Parul
Department of Chemical Engineering
Indian Institute of Technology Roorkee
Roorkee
Uttarakhand
India

Upexa Patel
Department of Biotechnology
Veer Narmad South Gujarat University
Surat
Gujarat
India

Divyani Pathak
Department of Environmental Sciences
Central University of Jharkhand
Ranchi
Jharkhand
India

Shray Pathak
Department of Civil Engineering
Indian Institute of Technology Ropar
Roopnagar
Punjab
India

Minh-Thuan Pham
Center for Environmental Toxin and Emerging-Contaminant Research
Cheng Shiu University
Kaohsiung
Taiwan

Venkateswara Rao Polisetty
Department of Civil Engineering
National Institute of Technology Warangal
Warangal
Telangana
India

Eldon R. Rene
Department of Water Supply, Sanitation and Environmental Engineering
IHE Delft Institute for Water Education
Delft
The Netherlands

Olfat Saied
Environmental Science and Industrial Development Department, Faculty of Postgraduate Studies for Advanced Sciences (PSAS)
Beni-Suef University
Beni Suef
Egypt

Abhishek Sengupta
Amity Institute of Biotechnology
Amity University
Noida
Uttar Pradesh
India

Surajbhan Sevda
Department of Biotechnology
National Institute of Technology
Warangal
Warangal
Telangana
India

Ishwar Sharan
Department of Chemical Engineering
Indian Institute of Technology
Roorkee
Roorkee
Uttarakhand
India

Nabila Shehata
Environmental Science and Industrial Development Department, Faculty of Postgraduate Studies for Advanced Sciences (PSAS)
Beni-Suef University
Beni Suef
Egypt

Saji Simon
Department of Civil Engineering,
National Institute of Technology
Calicut
Kozhikode
Kerala
India

Anantha Singh T. S.
Department of Civil Engineering,
National Institute of Technology
Calicut
Kozhikode
Kerala
India

Bhaskar Singh
Department of Environmental Sciences
Central University of Jharkhand
Ranchi
Jharkhand
India

Tezna Singh
Department of Chemical Engineering
Indian Institute of Technology
Roorkee
Roorkee
Uttarakhand
India

Vimal Chandra Srivastava
Department of Chemical Engineering
Indian Institute of Technology
Roorkee
Roorkee
Uttarakhand
India

K. Hasim Suhaib
Department of Civil Engineering
IIT Palakkad
Palakkad
Kerala
India

Nourhan S. Sultan
Biotechnology and Biomolecular Chemistry Department, Faculty of Science
Cairo University
Giza
Egypt

Bibin K. Suresh
Department of Civil Engineering, National Institute of Technology Calicut
Kozhikode
Kerala
India

Muskan Syed
Amity Institute of Biotechnology
Amity University
Noida
Uttar Pradesh
India

Quoc-Minh Truong
Faculty of Management Science
Thu Dau Mot University
Binh Duong
Vietnam

George K. Varghese
Department of Civil Engineering, National Institute of Technology Calicut
Kozhikode
Kerala
India

Tabbi Wilberforce
Faculty of Natural, Mathematical and Engineering Sciences
King's College London
Strand
London
UK

Hai Yen Nguyen
Department of Environmental Engineering
College of Ocean Science and Engineering, Korea Maritime and Ocean University
Busan
Republic of Korea

Interdisciplinary Major of Ocean Renewable Energy Engineering
Korea Maritime and Ocean University
Busan
Republic of Korea

Heba A. Younes
Environmental Science and Industrial Development Department, Faculty of Postgraduate Studies for Advanced Sciences (PSAS)
Beni-Suef University
Beni Suef
Egypt

Abdurrasheed Abubakar Yunus
Department of Life Sciences
Sharda School of Basic Sciences and Research
Sharda University
Greater Noida
Uttar Pradesh
India

Preface

Background

This book is a collective effort to understand the economic viability of the treatment processes used for wastewater treatment. The book is divided into three sections that deal with the strategies for treatment, resource recovery, and relative analysis of treatment with water pollution and solid waste management. The three conventional methods of treatment, namely, physical, chemical, and biological, are considered. The wastewater resources are divided into sewage and industrial sector. Through this book, we would like to present a coverage of the economic sustainability of the treatment process, as a single-step process and also as a multistep combinational process.

The key points in each chapter are listed as follows:

Chapters 1, 2, 3, and 4 give an introduction to the present scenario of the centralized treatment facility, compare the linear and circular economic ideation of treatment, present-day regulatory guidelines for wastewater management, and the effect of wastewater composition on economic sustainability of the treatment approach. Chapters 5 and 6 discuss the economics behind the physical, chemical, and biological treatment technologies. Chapters 7, 8, and 9 are dedicated to understand the industrial wastewater treatment. Chapter 10 specifically addresses the case studies of wastewater treatment.

Resource recovery from wastewater is directly related to the economic impact of the treatment process. Chapters 11–18 focus on the resource recovery from sewage as well as industrial wastewater using a decentralized setup, modern technologies, a combinational approach of conventional and modern technology, and source separation of wastewater.

It is important to take the optimized lab-based technologies to field application. Chapter 19 discusses the scale-up of treatment technologies. Chapter 20 discusses advancements in membrane technology, and Chapter 21 emphasizes the source separation of wastewater. The near future is all about integration of different technologies and artificial intelligence (AI); thus Chapters 22–24 elaborate the interdependency and integration of different treatment processes and application of AI in wastewater treatment. The last chapter, Chapter 25 of this book, sheds light on the application of the governmental approach and support for cleaning water bodies in India.

Benefits to Audience

The compiled book with 25 chapters will give the readers information regarding the possible factors that can be considered while planning the treatment facilities. Economic implications of the chosen process, possible resource recovery, and utilization of the generated waste combine to evaluate the circular approach and sustainability of the process. The overview regarding the available treatment technologies and the possible combinations that will lead to lessen the financial implication of the treatment plan are discussed through case studies. This book content can be referred by the students and researchers while selecting and planning the wastewater treatment.

Dr. Arvind Kumar Mungray
Dr. Alka A. Mungray
Dr. Ambika Arkatkar
Dr. Mohammad Ali Abdelkareem

Acknowledgments

First of all, I want to thank my wife, Dr. Alka, for her unwavering support throughout the entire publication process of this book. Her mental and technical support boosted the outcome of this book. As my wife and one of the editors of this book, she has always been a motivation during the challenging period of the publication journey. She is full of novel ideas and worked thoroughly from the inception of the idea toward writing this book, editing and proofreading, image selection, graphics, etc. Thank you for being my constant source of encouragement. I would also like to express my gratitude to my children, Dakshit and Prisha, for being so supportive during writing work; they kept their demands to a minimum, so I could concentrate on my work.

I am very much thankful to Dr. Ambika Arkatker, one of the editors of this book, for having a lot of energy and taking on all challenges and maintaining the flow of work. She knows how to complete the work on time and coordinated with the contributors and solved their queries. Whenever required, she drafted and finalized the chapters and displayed her dedication and hard work in completing the book on time.

I am really thankful to Dr. Mohammad Ali Abdelkareem, one of the editors of this book, for keeping an eye on the work progress, arranging many overseas contributors, giving many suggestions wherever I got stuck. I am fortunate enough that Dr. Abdelkareem accepted my proposal and involved in this project, and now we have completed this valuable book. I am deeply grateful to Dr. Abdelkareem for his invaluable collaboration and unwavering support throughout this book project.

I cannot forget my gratitude toward the contributors from around the world who accepted my invitation and produced the chapters as per the requirements of the book. Thank you for your hard work and for dealing with my onslaught of comments and particularities through the publication process. I sincerely appreciate the help and support provided by my PhD students, Ms. Monali Chhatbar, Ms. Parini Surti, and Mr. Zavin Gajera, for the compilation of the data along with their research work.

Dr. Arvind Kumar Mungray

1

The Centralized Systems and Strategies for Wastewater Treatment: Overview

Srishti Mishra[1], *Anil Dhanda*[2], *Brajesh K. Dubey*[1,2], *and Makarand M. Ghangrekar*[2,3*]

[1]*School of Water Resources, Indian Institute of Technology Kharagpur, Kharagpur, West Bengal, India*
[2]*Department of Civil Engineering, Indian Institute of Technology Kharagpur, Kharagpur, West Bengal, India*
[3]*National Institute of Technology Puducherry, Karaikal, Puducherry, India*

*Corresponding author: ghangrekar@civil.iitkgp.ac.in

1.1 Introduction

The water used for everyday purposes such as drinking, bathing, cooking, and industrial processes ultimately becomes wastewater. Treating this wastewater is crucial to effectively eliminate pollutants, uphold public health, and comply with environmental regulations before discharging it into natural water bodies or potential reuse. Wastewater treatment (WWT) not only removes pollutants but also provides reclaimed water, which can be used to address water scarcity. With growing environmental concerns about water contamination, quality degradation, and scarcity of freshwater, WWT and reuse are becoming a priority. According to the UN World Water Development Report, there is an estimated annual growth of 1% in overall global water demand, leading to a projected 20–30% increase in water demand by 2050 (Singh et al., 2024). Declining groundwater levels and deteriorating water quality are impacting the water supply and have left more than 30% of the groundwater systems in distress (Richey et al., 2015). India is also facing a water stress situation due to an imbalance between supply and demand for water. As per the central pollution control board, only 37% of wastewater is being adequately treated, leaving 63% of the generated wastewater untreated (Doki et al., 2024).

Wastewater to Resource Recovery: Applying the Circular Economy Toward Sustainable Development, First Edition. Edited by Arvind Kumar Mungray et al. Published 2026 by John Wiley & Sons, Inc.

Currently, India ranks thirteenth for overall water stress and has three times the population of the other 17 most extremely stressed countries combined together, representing a significant challenge (Varma et al., 2022). Moreover, the list of contaminants of concern has been expanding, with newly detected pollutants at alarming levels, such as perfluorinated compounds, pharmaceuticals, personal care products, cyanotoxins, detergents, and nanomaterials (Boretti and Rosa, 2019). Hence, recent developments in sustainable methods of WWT have come into the picture, extending the conventional centralized WWT methods to decentralized and semi-centralized approaches with the introduction of newly advanced technologies targeting emerging contaminants. The centralized WWT approach involves large-scale treatment plants that depend on extensive pipe networks, while decentralized strategies are more localized and require fewer pipelines.

The strategies for WWT involve physical, biological, and chemical methods for removing contaminants to meet the acceptable limit of target pollutants for easy disposal or reuse purposes. Different physical strategies include screening, which eradicates large and floating solid materials; comminution, for crushing solids into smaller pieces as an alternative to screening; skimming, which removes grease and oil-like substances; and sedimentation, which is performed for removal of heavier suspended solids through settling. The secondary treatment is used to oxidize mainly organic matter present in colloidal or dissolved form. Biological strategies employed in secondary WWT are grouped into aerobic and anaerobic, described as organic matter decomposition by microorganisms in the presence or absence of oxygen, respectively. Aerobic biological WWT, such as the activated sludge process (ASP), trickling filter (TF), and membrane bioreactor, produces water, biomass, and carbon dioxide, whereas methane, carbon dioxide, and biomass are released in anaerobic treatment methods. The chemical strategies adopted for offering tertiary WWT involve the use of chemicals to treat the dissolved and colloidal pollutants, such as coagulation, ozonation, photolysis, photocatalysis, advanced oxidation process (AOP), electrochemical treatment, electro-Fenton, etc. (Samsami et al., 2020). The various chemical methods are broadly categorized into two types: the conventional oxidation process, which is not that effective in refractory organic compounds removal, and the AOP, which utilizes the hydroxyl radicals to treat inflexible organics present in wastewater (Ahmed et al., 2021).

This chapter provides an overview of the centralized approaches utilized for WWT, as well as the factors that influence the method of selection. Different unit operations and unit processes involved are also briefly presented. Additionally, the chapter offers a comparative analysis of the approaches, by highlighting the various advantages and disadvantages, and suitability considerations that influence method selection preferences. Finally, it provides comparative economics between centralized and decentralized WWT systems.

1.2 History of Centralized System

In previous centuries, waste from water closets was disposed of manually, while wastewater from bathrooms and kitchens was directed into open drains. This led to the spread of disease-causing germs through manual waste handling and unpleasant odors from open drains. To address these issues, a water carriage system was introduced for transporting wastewater through closed conduits using gravity flow (Ghangrekar, 2022). The closed conduits transport the wastewater from the point of its generation to disposal, effectively eliminating bad odors and health hazards associated with open disposal. However, the accumulation of untreated wastewater often causes pollution of freshwater bodies and health problems.

Wastewater produced from different sources releases organic and inorganic pollutants such as salts, metals, and chemicals in the form of suspended and sediment solids that polluted the water bodies in which they were discharged. This is because a high concentration of organic pollutants, when discharged into a freshwater body, reduces dissolved oxygen (DO) and exerts biochemical oxygen demand (BOD), leading to further depletion of DO due to aerobic oxidation of organic waste harming the aquatic life. The effluents released from the industrial production units, such as detergents, paints, and insecticides, posed another challenge, making the water unfit for human consumption. Pathogens entering through sewage discharge or industries like slaughterhouses caused the spread of waterborne diseases and caused deaths worldwide. As a result, WWT (the centralized approach) came into the picture to ensure proper sewage treatment prior to disposal, safeguarding human health and the natural environment.

1.3 WWT System: Centralized Approach

With the advancement of technology, infrastructure development, and growing demand for systematic and effective management of wastewater from residential, industrial, and commercial facilities, the implementation of centralized WWT has extended to a greater scale. In a centralized system, sewers gather the wastewater released and transfer it to a centrally established facility for treatment, which is usually located away from the city. The sewer network delivering this wastewater could be of separate, combined, and partially separate types for stormwater and wastewater. These sewer networks follow design guidelines for peak flow, dimension, and type of material used and are placed as a gravity flow network. They are avoided to flow full and restricted to either two-third or three-fourth full at the maximum design discharge (Ghangrekar, 2022). The quantity of sewage

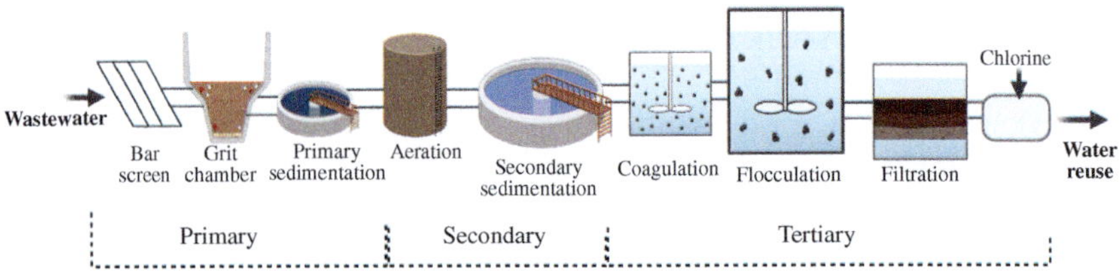

Figure 1.1 Centralized system for wastewater treatment.

is usually estimated to be 75–80% of the water supplied. Fluctuation of the flow of sewage due to seasonal and hourly variation is also taken into consideration while designing the sewer lines. Minimum velocity is to be maintained to avoid settling of suspended solids into the pipe network. The wastewater, after reaching the centralized facility, undergoes a series of treatment units categorized into preliminary/primary, secondary, and tertiary sections. The choice of unit operations and unit processes employed depends on the characteristics of the sewage to be treated and the permitted pollutant level required for the treated effluent. The conventional centralized WWT comprises primary treatment involving screening and grit removal, followed by primary sedimentation tank (PST) and biological treatment methods (secondary treatment) either employing ASP or TF. The secondary treated wastewater then undergoes tertiary treatment for water reuse, as illustrated in Figure 1.1.

1.3.1 Preliminary/Primary Treatment

Screening and grit removal—A screen of parallel bars or rods, or wire mesh is placed ahead of pumping stations or first unit operations of the WWT plant to remove floating and coarser materials to protect the pumps, pipes, and other equipment. They are designed with a cross-sectional area of two to three times more than the incoming sewer and adequate length to prevent eddies formation at the screen, avoiding declogging of materials through screen openings (Ghangrekar, 2022). The bed slope of the channel is kept to maintain the self-cleansing velocity to prevent the deposition of heavier materials. The screen is designed for handling peak flow conditions with the incorporation of at least two screens (each designed for peak flow) to allow divergence arrangement at times of maintenance. Based on the opening size, screens are classified into coarse (75–150 mm for the pumping station, 20–50 mm as a first-stage screen at the sewage treatment plant), medium, and fine screens (0.035–6 mm) to be used as a second-stage screening or for polishing secondary treated effluent (Ghangrekar, 2022). On the basis of the cleaning method, screens are categorized as manually and mechanically cleaned screens. After the removal of debris, the wastewater flows to the grit chamber.

A grit chamber is placed after screening and PST to remove suspended inorganic materials like sand and grit to prevent abrasion into the conduits and other equipment of the WWT system, reducing the need for maintenance. This unit removes particles of specific gravity around 2.65 with the smallest particle size of 0.20 mm diameter (sometimes even up to 0.15 mm diameter) (Ghangrekar, 2022). The design of grit chambers can be classified into four categories: rectangular horizontal flow, square, aerated, and vortex flow type grit chambers.

1.3.2 Secondary Treatment

Biological treatment methods—Biological processes of aerobic, anoxic, and anaerobic types are provided as secondary treatment units for microbial degradation of organic matter present in wastewater. The aerobic systems usually adopted are of two types: the suspended growth process (e.g. ASP), where microbes are present in suspended form, and the attached growth process, where microorganisms are attached to the media material such as a TF.

An ASP comprises an aeration tank, a secondary sedimentation tank (SST), and a sludge recirculation/excess sludge removal system. Stabilization of organic matter by active aerobic bacteria in the aeration tank takes place, followed by bacterial sludge settlement in the sedimentation tank after biochemical reactions. In the ASP, the ratio of MLVSS/MLSS ranges from 0.7 to 0.85 (Ghangrekar, 2022). The ASP successfully reduces organic matter quantified in terms of chemical oxygen demand (COD) and BOD. The other type of secondary treatment unit used for domestic and industrial WWT involving the microbial community is a rotating biological contactor. In this, a biofilm process involves disks installed on the shaft axis of the contactor. This disk is partially submerged in wastewater, which allows the microbial film to come in contact with wastewater and then perform microbial metabolism, thus digesting the organic matter. This unit is advantageous in terms of process stability, low maintenance, better sludge settling, low power consumption, high activated sludge, and high specific surface area (Singh et al., 2024).

Similar to aerobic processes, anaerobic reactors, such as up-flow anaerobic sludge blanket (UASB) reactors and anaerobic filters, are used as secondary treatment units that achieve more efficient and cost-effective degradation of wastewater. They convert 70–90% of organic matter into biogas, compared to 50–60% of organic matter getting converted into biomass in aerobic systems (Ghangrekar, 2022). Additionally, anaerobic processes produce a better-concentrated sludge than aerobic methods and eliminate the need for external energy for aeration. The choice between aerobic and anaerobic processes for either a centralized or decentralized strategy is determined by the characteristics of the wastewater.

Although anaerobic processes consume less energy, they are not suitable for wastewater with a low BOD of less than 230 mg/L (Saadatinavaz et al., 2024). This can pose challenges for domestic sewage treatment, as these processes may produce methane-saturated or supersaturated effluent (in dissolved form). Methane has nearly 30 times the global warming potential of carbon dioxide (Saadatinavaz et al., 2024). To achieve better overall quality of treated wastewater, a common practice is to follow anaerobic treatment with aerobic treatment to offer polishing the final effluent quality. Examples of anaerobic secondary treatment units include UASB reactors, anaerobic fluidized bed bioreactors, as well as anaerobic fixed bed reactors. Further, the sludge obtained from the treatment of wastewater (primary and secondary clarifier) undergoes treatment through adequate sludge management, which involves sludge thickening, conditioning, dewatering, stabilization, and drying before being disposed of or used as fertilizer. In this regard, the anaerobic treatment has an advantage since the excess sludge generated doesn't need further digestion, whereas from most of the aerobic processes, the excess sludge generated needs further treatment to improve its dewatering characteristics.

1.3.3 Tertiary Treatment

Although primary and secondary treatment remove most of the pollutants from wastewater, tertiary treatment, the final stage of treatment, is undertaken that prepares wastewater for reuse or discharge into surface water bodies. It is a combination of physical unit operations and chemical processes that opt for the removal of remaining pollutants after secondary treatment, including the removal of harmful emerging contaminants as per the need. Commonly adopted tertiary methods of treatment involve coagulation and flocculation, which allows floc formation of the turbid particles present in secondary treated effluent; filtration (removal of finer suspended solids); and disinfection (using technologies like ultraviolet light, ozone sterilization, and chlorination), as shown in Figure 1.1. Adsorption, membrane filtration, electrodialysis, ion exchange, and pressure filters are some of the other examples of tertiary treatment units.

The adequately planned centralized WWT removes various pollutants, including colloids, organic matter, and pathogens. Treated wastewater also prevents the possibility of eutrophication of water bodies where effluents are discharged. Different studies worldwide illustrate the successful implementation of the centralized WWT approach. A centralized WWT system located in Kuching, Malaysia, successfully treated over 400 000 m^3 of wastewater annually from 2017 to 2020 and processed over 2000 tonnes of sludge during this period. Notably, it met the discharge criteria for Standard A of the Environmental Quality Act of 1974 for most parameters, with further improvements observed in the following years (Kuok et al., 2022). A centralized WWT strategy ensures the treatment of large volumes of wastewater at high rates. It also guarantees compliance with environmental regulations through strict discharge quality standards for effluent. Although the centralized WWT approach is widely used and offers some environmental benefits, it does face certain limitations.

1.4 Limitations and Effects or Impacts of Centralized Systems

Centralized systems are preferable for maintaining high-quality treated wastewater standards because of the complex treatments applied. However, the complexity of different units of the system poses the challenge of installation in overcrowded cities. Also, with the rapid population growth and rise in water scarcity issues, the call for water reuse has gained importance. The disposal of treated wastewater as per established quality standards is no longer the sole objective of WWT plants, and centralized systems may turn out to be economically and environmentally unsustainable when additional purposes are considered (Pasciucco et al., 2022; Salgot and Folch, 2018). Reclaimed water reuse, though of lower quality, has gained attention as it meets various non-potable and even potable water demands, highlighting the advantage of other WWT strategies like decentralized over centralized ones.

Another limitation of a centralized WWT facility is the treatment of only a section of a larger population area while the rest of the section remains deprived, causing improper sanitation and waste management issues in that area (Varma et al., 2022). Moreover, the centralized WWT system when implemented to serve a geographically large area observes topographical impact, as laying a gravity flow sewer network is tedious, while installation of intermediate pumping stations at suitable intervals in undulating terrain makes the system expensive and complex. Approximately, the cost of wastewater collection accounts for 60% of the overall management cost and 70% of the annual operational budget for pumping and maintenance. Further, the use of an extensively large conveyance network burdens this budget in a centralized system (Saadatinavaz et al., 2024). Additionally, rainwater runoff problems caused during wet weather conditions lead to infiltration, hence the rise of water quantity into the sewers by flowing through the manholes, roofs, and faulty joints of the sewer network. Moreover, centralized WWT may receive effluents from small-scale industries located within the city, thus affecting the treated water quality and hindering the reuse and resource recovery aspect. For ensuring proper operation of the system installed, consistent and corrective maintenance is crucial; neglect of proper checkups may cause adverse effects on the system functioning by exposing them to environmental stresses such as pH and toxic shock loading (Saadatinavaz et al., 2024). Hence, further making the centralized approach more vulnerable and less reliable.

The WWT scheme should serve as a reliable supplementary source of water, particularly in areas with high water demand; hence, the choice of treatment technology employed plays an important role. Utilizing treated water also requires transportation from the treatment plant to residential or agricultural areas. In this regard, centralized WWT involves an additional financial burden for transportation, whether by truck or pumping. In contrast, other strategies like decentralized systems generate treated water in close proximity, allowing for more efficient use of treated water in horticulture, gardening, flushing, and other applications (Saadatinavaz et al., 2024).

1.5 Other Strategies for WWT

The WWT faces significant social and economic challenges that hinder its full potential. Limited funding agencies and public acceptance of treated wastewater reuse restrict the widespread adoption of reclamation methods. However, advancements in the WWT technologies have emerged despite these obstacles. Some of the pressing challenges like the inadequacy of localized implementation of the centralized WWT strategy and the increase in demand for target contaminant removal, along with increasing emerging contaminant concentrations in domestic wastewater, have drawn attention to newly introduced technologies. Therefore, the importance of decentralized WWT systems over centralized ones has grown. The decentralized systems effectively bridge the gap between on-site and centralized treatment approaches. A decentralized WWT system is designed to treat wastewater locally and at the community level (small/medium scale) rather than the centralized system that is prevalent on a large scale. They provide treatment for both domestic and industrial wastewater with low COD/BOD ratio and flows ranging from 1 to 1000 m^3/day, entailing the treatment, release, and reuse of wastewater, close to the point of origin (Varma et al., 2022). A decentralized approach allows for the change and upgrade of the facility through component addition and reorganization without affecting the overall operation of the existing system. Some of the methods opted for the treatment of wastewater in either centralized or decentralized form are as follows:

1.5.1 Septic Tank

One of the simplest decentralized methods implemented in rural and urban areas involves the septic systems where wastewater gets collected in a tank followed by the settlement of solids and discharge of liquid effluent using a drain field for soil filtration. Anaerobic bacteria present within the tank decompose solids and liquids through a process called “septic” digestion, which leads to the production of heat and biogas.

1.5.2 Membrane Bioreactor

The other technologies include membrane bioreactors, which combine biological treatment with membrane filtration, effectively eradicating the contaminants. Microorganisms decompose the organic pollutants in the bioreactor, and their separation from the treated water occurs in the membrane module. Membrane bioreactors are advantageous in terms of producing high-quality effluent, lesser sludge production, and smaller space requirements than the conventional methods. This is also effective in the treatment of landfill leachates, which often have high concentrations of organic and inorganic compounds. However, due to the expensive nature of this system, it is less preferred by countries like India.

1.5.3 Constructed Wetland

A constructed wetland is a system that mimics the natural wetland for treating wastewater with microbial activity, plant, and soil working together. It involves sedimentation of solids as wastewater flows into the system, followed by removal of large particles through plant roots, degradation of nutrients and pollutants by plants and bacteria, killing of pathogens with UV radiation, and plant secretion of antibiotics. The advantages of constructed wetlands involve cost efficiency, as construction and maintenance are cheaper, require less energy, and also assist in maintaining groundwater and surface water levels. However, the requirement of a large area for its implementation prevents extensive application, especially in areas of low land availability.

1.5.4 Up-flow Anaerobic Sludge Blanket Reactor

The UASB reactor is a single-tank process used in the anaerobic treatment of wastewater for high organic content removal. It is implemented as a unit in both centralized and decentralized approaches. A suspended sludge blanket with bacteria disintegrates the organic content through anaerobic digestion, resulting in biogas formation. Baffles at the top aid the escape of gases, preventing the outflow of the sludge blanket. Application of UASB reactor is advantageous in terms of high BOD reduction, and low sludge production, which prevents the need for frequent desludging, nutrient-rich effluent, production of biogas, and less area requirement. However, its limitations include additional treatment requirements for the effluent and inadaptability in the colder regions.

1.5.5 Biological Filtration System (Biofilter)

Filters such as slow sand filters, rapid sand filters, and granular active carbon filters are tertiary treatment processes. Biofilters (aerobic/anaerobic) have microorganisms attached to filter media called as biomat, which consume the pollutants from the wastewater. The advantage of biofilter implementation is that they can help to eliminate the clarification stage, which saves space, requires initial low investment cost, and is compact in design, which controls the odor. However, the operation includes challenges of uneven biomass distribution, clogging of the filtration media, and increase in operation cost when one layer is placed on top of the other for fulfilling large surface area requirement. Additionally, periodic replacement of the filter material causes problems. Thus, adequate maintenance is necessary during system operation.

1.5.6 Sequencing Batch Reactor

Sequencing batch reactors (SBRs) are a type of ASP suitable for the treatment of intermittent or low-flow situations of wastewater. It performs in an alternating mode of operation in a single-volume basin as per the need of the final clarifier. A repeated time-based sequence is followed for biological operation (fill-aeration), solid/liquid separation (fill-settle), and treated effluent removal (decant). It is also advantageous over the conventional ASP in terms of growth control of the filamentous sludge bulking. Other benefits include shock load toleration for organic and hydraulic load variation, eliminating the need for a secondary clarifier. Additionally, it provides nutrient removal without the addition of chemicals. It also provides the benefit of providing an equalization basin and clarifier in the same basin, avoiding the need for separate equalization.

1.5.7 Anaerobic Baffled Reactor

The anaerobic baffled reactor (ABR) is a wastewater system that consists of a series of baffles for treating wastewater through a sludge blanket. They have proved effective for medium- to small-size communities, as they consist of multiple chambers and perform degradation of organic matter anaerobically. It is a modified version of the septic tank, and it is often used in decentralized systems. Its advantages include less sludge production and resistance to organic and hydraulic shock loads, whereas, the limitations of ABR involve inefficiency in colder climate regions and regular removal of sludge from the settling compartment.

1.5.8 Composting Toilets

The composting toilet is a system that converts human waste into compost, and it is also called a waterless or dry toilet. They hold and process waste material to capture the nutrients like nitrogen and phosphorus. They are aerobic processing systems and use no or little water to function and do not produce black water. The different types of composting toilets comprise slow composting toilets, vermifilter toilets, and active composters.

Other decentralized system technologies comprise bioelectrochemical systems like microbial fuel cells (Dhanda et al., 2023, 2024) and microbial desalination cells (Mishra et al., 2024). However, the decentralized system units alone are not effective, but rather when implemented in combination as per the influent wastewater characteristics. A study from Japan documented 2500 successful cases of decentralized systems for a cluster of residential buildings, schools, and hospitals treating and reusing their own wastewater (Yamagata et al., 2003). Similarly, the Solitaire residential complex in New York and the Metropolitan Government facilities in Tokyo reported wastewater reuse for cooling systems and flushing (Bernal et al., 2021). Besides fully centralized and decentralized WWT approaches, semi-centralized treatment for wastewater and resource recovery are also implemented. A case study of the same for the Qingdao Horticultural EXPO 2014 separated blackwater and greywater at source and treated into two separate systems, which exhibited removal of organics and nutrients, meeting effluent discharge standards, accompanied by 7–10 m^3 of biogas production per cubic meter of raw sludge (Tang et al., 2020).

1.6 Comparative Economics of Centralized and Decentralized Systems

Decentralized WWT systems are both cost-effective due to small-scale implementation, efficient at a local community level, and also offer great potential for establishing a long-term environmental sanitation system. Research undertaken by the United States Environmental Protection Agency remarked decentralized WWT as more cost-effective than the centralized for areas of low-population density and variation in site conditions. Also, the smaller conveyance leads to savings in terms of the sewer pipe network. They can accommodate local resources and energy use, making it feasible for adoption at the local community level. One of the advantages of a decentralized system over a centralized one involves site-specific treatment and progressive expansion capabilities (Capodaglio et al., 2017). Decentralized systems are designed for relatively small clusters of dwellings or a single household. If there is a failure or damage to the system, it will only impact the particular cluster and require less monetary investment to repair compared to a centralized system.

A life cycle cost assessment (LCCA) of the sludge treatment scenarios of centralized and decentralized WWT systems from a case study of Bangkok, Thailand, concluded that the decentralized treatment had a lower total environmental cost and total financial cost (dewatering and fertilizer) than the centralized treatment (Prateep Na Talang et al., 2022). A comparative cost analysis of centralized and cluster-type decentralized WWT of Alibag, India, showed that decentralized systems when configured with large cluster subsystems are less expensive than centralized, which require higher capital input but lower operation and maintenance costs. The costs of decentralized WWT systems vary depending on the layout and the degree of decentralization. A cost-benefit study of centralized and decentralized strategies for Surabaya (Indonesia) stated centralized WWT featured a high net present value cost and low cost-benefit ratio while the decentralized approach was economically feasible for serving the same population of three million (Prihandrijanti et al., 2008). In terms of economies of scale, the centralized approach benefits from lower overall costs and efforts, while the decentralized approach offers greater multifunctionality and flexibility. Hence, when deciding whether to use a centralized or decentralized approach for WWT, a thorough evaluation of factors like geographical features of the area, population density, land availability, technical accessibility, finances, energy efficiency, flexibility, regulatory requirements, and environmental impact is pivotal.

1.7 Summary

The WWT, utilizing various approaches and advanced technologies, significantly reduces the dependence of the human population on freshwater for producing goods and services known as the human water footprint. Proper treatment of wastewater addresses the growing demand for freshwater, reduces the need for wastewater disposal facilities, prevents pollutants from being discharged into water bodies, and lowers the risks associated with untreated wastewater. The adequately treated wastewater as per effluent standards is safer to dispose of and can additionally be used for recharging aquifers and groundwater. Numerous benefits accompanying WWT led to the emergence of centralized WWT strategies. Traditional centralized methods are now transitioning to preferred decentralized or hybrid approaches that focus on nature-based processes, making WWT more environmentally friendly, economically viable, and socially acceptable. Choosing the right technology for optimal treatment requires expertise, experience, and feasibility studies.

Important criteria for selecting a suitable WWT approach include cost minimization, eliminating mistakes during the planning and design stages, and making decisions based on prevailing conditions and user requirements, all while promoting sustainable development goals (SDGs) 6. Proper evaluation of area and energy requirements to reduce the footprint of the chosen system is also essential. Higher centralization levels support higher energy recovery potential and lower urban and agricultural reuse (Bernal et al., 2021). Maximizing wastewater recycling and reuse with optimized system design can significantly contribute to the sustainable development and circular economy of both society and the nation.

Review Questions

1 What are the different strategies for WWT?
2 What are the significances of limitations of primary, secondary, and tertiary treatment?
3 What is the origin of centralized WWT systems?
4 What is the difference between aerobic, anoxic, and anaerobic type of treatment and their significance, accordingly type of technologies based on above?
5 What are the limitations of centralized WWT systems?
6 Compare ASP and UASB for centralized WWT system.
7 Mention the advantages and limitations of composite toilets.
8 What would be the strategies for choosing centralized and decentralized WWT systems?
9 Comment on the societal acceptance of recovered water from centralized and decentralized wastewater systems.
10 What are the challenges in current WWT systems?

References

Ahmed, S.F., Mofijur, M., Nuzhat, S. et al. (2021). Recent developments in physical, biological, chemical, and hybrid treatment techniques for removing emerging contaminants from wastewater. *Journal of Hazardous Materials*, **416**, 125912. https://doi.org/10.1016/j.jhazmat.2021.125912

Bernal, D., Restrepo, I., and Grueso-Casquete, S. (2021). Key criteria for considering decentralization in municipal wastewater management. *Heliyon*, **7**, e06375. https://doi.org/10.1016/j.heliyon.2021.e06375

Boretti, A. and Rosa, L. (2019). Reassessing the projections of the world water development report. *NPJ Clean Water*, **2**, 15. https://doi.org/10.1038/s41545-019-0039-9

Capodaglio, A.G., Callegari, A., Cecconet, D. et al. (2017). Sustainability of decentralized wastewater treatment technologies. *Water Practice & Technology*, **12**, 463–477. https://doi.org/10.2166/wpt.2017.055

Dhanda, A., Tripathi, A., Raj, R. et al. (2024). Amalgamation of constructed wetland and microbial fuel cell for cost-effective and low carbon emitting treatment of industrial wastewater. In: *Resource Recovery from Industrial Wastewater Through Microbial Electrochemical Technologies*, 193–206. IWA Publishing. https://doi.org/10.2166/9781789063813_0193

Dhanda, A., Yadav, S., Raj, R. et al. (2023). Fuel cells and biofuel cells. In: *Microbial Electrochemical Technologies*, 19–52. https://doi.org/https://doi.org/10.1002/9783527839001.ch1

Doki, M.M., Dhanda, A., Yadav, S. et al. (2024). An overview of the advances in biological and hybrid wastewater treatment technologies in developing countries with experience from developed countries. In: *Emerging Technologies in Biological and Hybrid Wastewater Treatment: Lessons from Developed to Enhancing Practices in Developing Countries*, 3–19. Cham: Springer. https://doi.org/10.1007/978-3-031-74515-7_1

Ghangrekar, M.M. (2022). *Wastewater to Water*. Singapore: Springer Nature Singapore. https://doi.org/10.1007/978-981-19-4048-4

Kuok, K.K., Chiu, P.C., Rahman, M.R. et al. (2022). Effectiveness of centralized wastewater treatment plant in removing emerging contaminants: a case study at Kuching, Malaysia. *Journal of Water Resource and Protection*, **14**, 650–663. https://doi.org/10.4236/jwarp.2022.149034

Mishra, S., Dhanda, A., Dubey, B.K. et al. (2024). Enhancing electrokinetics and desalination efficiency through catalysts and electrode modifications in microbial desalination cells. *Journal of Environmental Management*, **366**, 121719. https://doi.org/10.1016/j.jenvman.2024.121719

Pasciucco, F., Pecorini, I., and Iannelli, R. (2022). Planning the centralization level in wastewater collection and treatment: a review of assessment methods. *Journal of Cleaner Production*, **375**, 134092. https://doi.org/10.1016/j.jclepro.2022.134092

Prateep Na Talang, R., Sirivithayapakorn, S., and Polruang, S. (2022). Life cycle impact assessment and life cycle cost assessment for centralized and decentralized wastewater treatment plants in Thailand. *Scientific Reports*, **12**, 14540. https://doi.org/10.1038/s41598-022-18852-y

Prihandrijanti, M., Malisie, A., and Otterpohl, R. (2008). Cost–benefit analysis for centralized and decentralized wastewater treatment system (Case Study in Surabaya-Indonesia). In: *Efficient Management of Wastewater*, 259–268. Berlin, Heidelberg: Springer. https://doi.org/10.1007/978-3-540-74492-4_22

Richey, A.S., Thomas, B.F., Lo, M.H. et al. (2015). Quantifying renewable groundwater stress with GRACE. *Water Resources Research*, **51**, 5217–5237. https://doi.org/10.1002/2015WR017349

Saadatinavaz, F., Alomari, M.A., Ali, M. et al. (2024). Striking a balance: decentralized and centralized wastewater treatment systems for advancing sustainable development goal 6. *Advanced Energy and Sustainability Research*, **5**(10), 2400097. https://doi.org/10.1002/aesr.202400097

Salgot, M. and Folch, M. (2018). Wastewater treatment and water reuse. *Current Opinion in Environmental Science & Health*, **2**, 64–74. https://doi.org/10.1016/j.coesh.2018.03.005

Samsami, S., Mohamadizaniani, M., Sarrafzadeh, M.-H. et al. (2020). Recent advances in the treatment of dye-containing wastewater from textile industries: overview and perspectives. *Process Safety and Environmental Protection*, **143**, 138–163. https://doi.org/10.1016/j.psep.2020.05.034

Singh, D., Singh, D., Mishra, V. et al. (2024). Strategies for biological treatment of waste water: a critical review. *Journal of Cleaner Production*, **454**, 142266. https://doi.org/10.1016/j.jclepro.2024.142266

Tang, X., Wen, Y., He, Y. et al. (2020). Full-scale semi-centralized wastewater treatment facilities for resource recovery: operation, problems and resolutions. *Water Science and Technology*, **82**, 303–314. https://doi.org/10.2166/wst.2020.169

Varma, V.G., Jha, S., Raju H.K. et al. (2022). A review on decentralized wastewater treatment systems in India. *Chemosphere*, **300**, 134462. https://doi.org/10.1016/j.chemosphere.2022.134462

Yamagata, H., Ogoshi, M., Suzuki, Y. et al. (2003). On-site water recycling systems in Japan. *Water Science and Technology: Water Supply*, **3**, 149–154.IWA Publishing. https://doi.org/10.2166/ws.2003.0020

2

Linear and Circular Economy Strategies for Waste Treatment: Fundamentals

*Abhishek[1], Anag Katiyar[1], Ishwar Sharan[1], Tezna Singh[1], Hanshika[2], Parul[1], Ashish Yadav[1], and Vimal Chandra Srivastava[1]**

[1]*Department of Chemical Engineering, Indian Institute of Technology Roorkee, Roorkee, Uttarakhand, India*
[2]*Department of Chemistry, Indian Institute of Technology Roorkee, Roorkee, Uttarakhand, India*

*Corresponding authors: vimalcsr@yahoo.co.in; vimal.srivastava@ch.iitr.ac.in (VCS)

2.1 Introduction

2.1.1 Importance of Waste Management

Environmental protection is the main goal of waste management in developed countries, and it greatly outweighs all other goals. The objectives of waste management, however, are much more complicated for a rising nation like India. In addition to protecting the environment, other important goals include increasing production, generating jobs, recovering precious resources, and meeting the welfare needs of a large population. As a result, there are several factors to consider when setting waste management priorities in these situations.

Despite being the seventh largest country in the world, India currently lacks clear regulations and policy frameworks for waste management services. Consequently, the performance of municipal authorities in this sector has been subpar. The daily operations of these entities are governed by a few laws found in the various municipal statutes, but their effectiveness has been limited because they are not strictly enforced (Gupta et al., 1998). To guarantee efficient waste management in India, the Ministry of Environment, Forest and Climate Change (MoEF&CC) issued the Municipal Solid Waste (MSW) (Management and Handling) Rules 2000.

Wastewater to Resource Recovery: Applying the Circular Economy Toward Sustainable Development, First Edition. Edited by Arvind Kumar Mungray et al. Published 2026 by John Wiley & Sons, Inc.

In 2015, new, updated draft regulations were published. Enforcing these laws and building the necessary infrastructure for MSW collection, storage, segregation, transportation, processing, and disposal are under the purview of municipalities. Compared to other Indian cities, waste management has improved in Chandigarh, which was the first to implement Solid Waste Management (SWM) systematically (Rana et al., 2015).

Waste management directly affects the health of the population, the stability of the economy, and the sustainability of the environment, and, therefore, it is a core aspect of contemporary civilization. It is forecasted that by 2050, there will be over 9 billion people on Earth, making it much harder to manage waste. The generation of waste has rapidly increased through unprecedented urbanization and an upsurge in consumption habits. Hence, there is a strong need for effective waste treatment systems to manage this increasing volume as well as curb its negative effects on the environment and society at large. Such effects include depleting limited natural resources, contributing to climate change through greenhouse gas emissions, and contaminating air and water.

Waste generation rates are influenced by a number of factors, including city/region, culture, business activity, population density, and economic standing. The amount of solid trash produced per person in each state and union territory is shown in Figure 2.1. Per capita, Delhi produces the most solid garbage, followed by Lakshadweep and Mizoram. State-wise treated solid waste (%) 2020–2021 is shown in Figure 2.2. Arunachal Pradesh has the lowest percentage of processed garbage, while Chhattisgarh has the most.

The complexity of waste management underscores how critical it is to achieve Sustainable Development Goals (SDGs). Good systems are not only about sanitation but also about resource conservation and minimizing ecological footprints. For instance, improper waste management endangers biodiversity and food

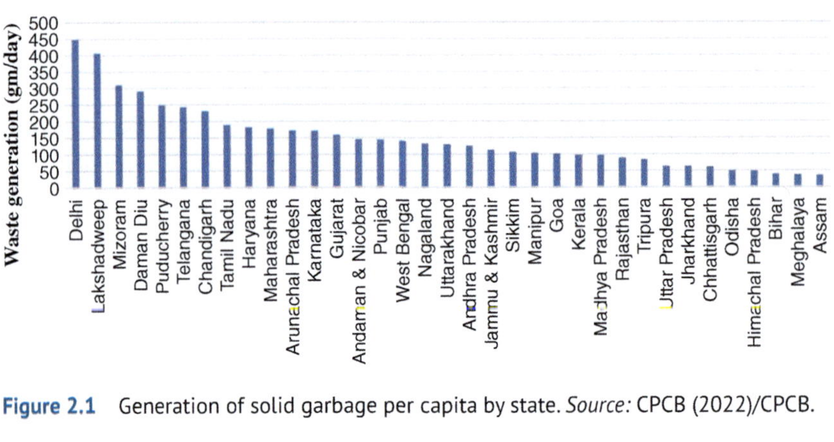

Figure 2.1 Generation of solid garbage per capita by state. *Source:* CPCB (2022)/CPCB.

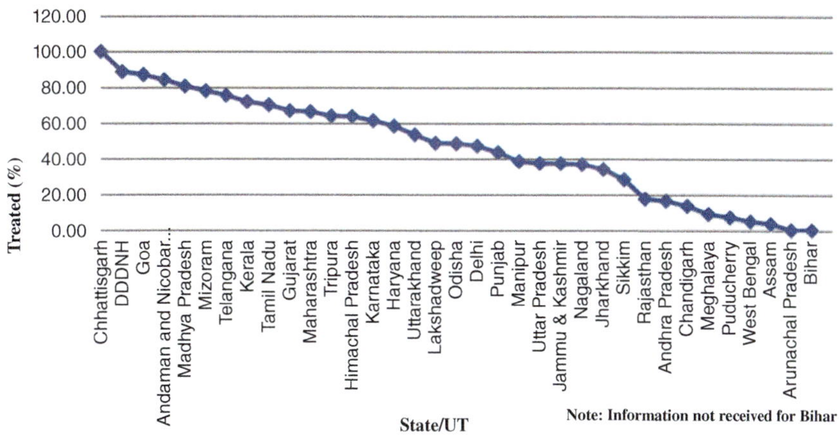

Figure 2.2 State-wise treated solid waste (%) 2020–2021. *Source:* CPCB (2022)/CPCB.

security through marine pollution. Moreover, low-income areas bear a disproportionate share of the socioeconomic consequences of inadequate waste disposal, which worsens global inequality. Thus, dealing with these problems requires an integrated approach that incorporates economic policies, technical development, and community participation to build robust and flexible waste management plans.

2.1.2 Historical Perspective of Waste Treatment

Throughout history, waste management techniques have changed to meet the demands of society and technological breakthroughs. Early societies disposed of rubbish using crude techniques like cremation and open dumping. However, these methods frequently have detrimental effects on the environment and human health. The fast growth of metropolitan areas during the Industrial Revolution made more structured waste management systems necessary, which was a watershed moment. Wastewater treatment facilities and landfills were commonplace in several nations by the 20th century. Despite these developments, the linear character of conventional waste management, which prioritizes disposal above resource recovery, has shown itself untenable in the face of growing garbage production.

The impact of public health theories and societal shifts over time is reflected in the historical development of waste treatment. Based on the Miasmatic Theory, which connected illnesses to unpleasant scents, waste management in the ancient and medieval eras was crude. It included methods like open dumping and burying organic garbage. Urbanization made waste problems worse during the Renaissance and early Industrial Revolution, which led to the implementation of crude municipal structures and public health improvements. The 19th century saw the development of germ theory, which changed the focus to microbes as disease agents. This led to the development of organized municipal systems, waste segregation, and sanitation technologies like incineration. By the 20th century, improvements in toxicology had exposed the risks associated with chemical waste, leading to resource recovery methods, sanitary landfills, and more stringent restrictions (Sandhu, 2014). By introducing ideas like waste hierarchy and the polluter-pays principle, the environmental movement of the 1970s placed even more emphasis on sustainable waste management. However, many developing countries continue to struggle with filthy landfills and informal disposal techniques, showing continuing global imbalances, even as rich countries have embraced sophisticated garbage systems.

India's waste management laws have evolved historically in response to growing environmental and public health concerns. At first, waste management was seen as a sanitary problem linked to infectious diseases, and informal methods predominated. Urbanization and industrialization increased trash production in the 1970s and 1980s, which prompted attention to pollution and health issues. Laws like the Water Act of 1974 and the Environment Protection Act of 1986 established the groundwork for regulatory oversight. Hazardous waste management became crucial after the Bhopal Gas Disaster in 1984, and regulations such as the Hazardous Waste Rules (1989) were created. Urban SWM gained more attention in the 1990s as a result of the Bajaj Committee's recycling and segregation recommendations and incidents like the Surat plague. The Municipal Solid Trash Rules (2000) signaled a change to scientific trash disposal as the 2000s got underway. Sector-specific legislation also emerged during this time, such as those pertaining to plastic (2011) and electronic waste (E-waste) (2011), which were later improved in 2016 (Deeksha, 2022).

Notwithstanding these developments, issues still exist, such as poor infrastructure, insufficient enforcement, and a lack of stakeholder responsibility. With a focus on source segregation and technology use, the 2016 Solid Waste Management Rules expanded their application to non-municipal regions. Implementation gaps and exclusion from the informal sector, however, continue to be major obstacles, underscoring the need for more stringent enforcement and decentralized solutions.

2.1.3 Types of Waste and Their Treatment Needs

Waste can be categorized into several primary categories, each of which presents unique difficulties and necessitates a specific treatment approach to reduce its impact on the environment and human health effectively:

1. **Municipal Waste:** Waste from homes and businesses, such as food scraps, packaging, and other commonplace discards, falls under the category of

municipal waste. Effective collection, classification, and recycling systems are essential to managing municipal garbage. Additionally, in order to lessen the strain on urban disposal sites, landfill diversion through energy recovery and composting techniques has become increasingly crucial.

2. **Industrial Waste:** Industrial and manufacturing waste is frequently complicated, containing a combination of nonhazardous and hazardous components. Specialized treatments are required for substances such as hazardous chemicals, solvents, and heavy metals. For industrial waste to be managed ethically, advanced techniques, including thermal treatment and chemical stabilization, safe disposal, and waste minimization at the source, are essential.
3. **Electronic Waste:** Due to the rapid advancement of technology, the number of discarded devices has increased, including laptops, smartphones, and appliances. Along with dangerous substances like lead and mercury, e-waste also contains valuable elements like copper, silver, and gold. Safe disassembly, the recovery of valuable metals, and the appropriate disposal of hazardous materials are all necessary for efficient e-waste treatment. Addressing this expanding problem requires enacting extended producer responsibility (EPR) laws and raising consumer awareness.
4. **Agricultural Waste:** Crop leftovers, animal excrement, and pesticide containers are examples of trash that comes from farming operations. Agricultural waste can be effectively handled by environmentally friendly techniques like composting, to produce organic fertilizers and anaerobic digestion to produce biogas. These techniques improve soil health and sustainable energy production in addition to reducing trash.
5. **Wastewater:** Wastewater management, which includes industrial effluents, home sewage, and agricultural runoff, is essential to environmental preservation. Advanced biological, chemical, and physical treatment methods are used to eliminate contaminants and recover resources like clean water, nutrients, and energy. Water reuse and ecosystem conservation are supported by innovations like membrane filtration and artificial wetlands, which improve wastewater treatment substantially.

2.1.4 Global Waste Generation Trends and Challenges

The World Bank predicts that worldwide garbage creation will rise by 70% by 2050 if current trends continue. This spike is being driven by growing urbanization, economic expansion, and changes in purchasing patterns. While high-income countries produce large per capita trash, middle- and low-income countries confront increasing issues as their urban populations rise. Key difficulties include insufficient infrastructure, insufficient financing, and low public knowledge, all of which impede effective garbage management.

Urbanization, economic progress, and population expansion are all closely linked to the sharp rise in waste output that is occurring globally. Global MSW generation was estimated at 2.01 billion tons in 2016, with a 70% increase expected by 2050 to 3.4 billion tons. Waste generation and consumption patterns, as well as urbanization rates and income levels, are all strongly correlated. Roughly 34% of the world's rubbish is produced in high-income nations, which make up 16% of the world's population. However, although making up 9% of the world's population, low-income nations only produce 5% of the waste produced worldwide. Income level also affects the type of trash; high-income countries produce more inorganic waste, like metals and plastics, while low-income countries produce more organic waste, including food scraps and agricultural waste (Singh et al., 2014). A breakdown of global waste treatment practices, such as landfilling, recycling, and incineration, shows the prevalence of open dumping in low-income regions, as reported by the World Bank Group. Disparities in collection and treatment infrastructure worsen waste management issues. Low-income countries collect about 39% of waste, while high-income countries collect more than 96%. The bulk of waste in low-income neighborhoods is thrown in open landfills, which pose substantial health and environmental risks such as greenhouse gas emissions, poisoned soil, and contaminated water. Methane emissions, which have a 72-fold greater potential to cause global warming than CO_2 over 20 years, are greatly increased by open dumping, a prevalent garbage disposal technique in developing nations (Sharma and Jain, 2020).

2.1.5 Overview of Economic Perspectives on Waste Management

The economics of managing waste is complex, with costs, benefits, and environmental externalities all interdependent. Traditional linear approaches, which prioritize disposal, are frequently economically unsustainable due to escalating landfill and incinerator prices, as well as the degradation of natural resources. Circular economic models, on the other hand, emphasize waste minimization and resource recovery, making them a more sustainable and economically viable option. For example, recycling and composting not only reduce trash but also produce cash through the sale of recovered materials and organic fertilizers. Similarly, modern technologies like anaerobic digestion and chemical recycling may transform trash into energy, fuels, and valuable compounds, opening up new economic prospects. High efficiency and the creation of new internal capabilities can come from abandoning a linear manufacturing paradigm. This can lower expenses, boost output, promote brand names, lessen risks, develop new goods, and satisfy consumer demands and legal requirements for sustainability (Cazcarro et al., 2016; Voukkali et al., 2023).

Government laws and regulations influence the economic landscape of waste management. Tax credits, subsidies, and grants can encourage innovation and investment in sustainable waste treatment systems. Furthermore, public–private partnerships are widely seen as a way to close financing shortfalls and increase operational efficiencies (Gupta et al., 2016).

Efficient waste management systems demand significant capital investment, but insufficient financial resources frequently impede their adoption. Developing countries invest only a small portion of their finances in waste management, with a primary concentration on residential trash. Inadequate financing leads to antiquated infrastructure, poorly maintained collection vehicles, and insufficient waste treatment facilities. The private sector may play an important role in bridging this gap, but such collaborations require appropriate incentives and policy frameworks (Sharma et al., 2024).

Furthermore, waste management provides major economic prospects, such as employment development and money generation via recycling and Waste-to-Energy (W2E) efforts. For example, industries that recycle and repurpose materials not only minimize waste but also supply raw manufacturing resources, thereby contributing to economic growth. Furthermore, good garbage management can increase property prices and attract tourists, resulting in further economic benefits. However, problems remain, including inadequate regulation enforcement, low public knowledge, and informal garbage-collecting methods that limit their effectiveness and scalability (Jagun et al., 2023).

2.1.6 Bridging the Gap to Wastewater Treatment

Safe drinking water is critical to human health, but waterborne diseases such as diarrhea continue to kill millions of people each year, particularly children under the age of five. Arsenic contamination, which frequently results from natural geological sources or human activity such as mining, has exposed millions to dangerous levels of arsenic. Efforts to reduce arsenic contamination encounter major hurdles, including expensive costs and technical complications in lowering arsenic to safe levels. These issues are exacerbated in resource-constrained environments, where solutions must also be compatible with local socioeconomic realities and community acceptance. To address these challenges, an IUPAC project was undertaken to investigate the origins of arsenic contamination, examine on-site testing kits, estimate health hazards, and propose acceptable remediation solutions. Several mitigating strategies have been investigated, including chemical treatments, bioremediation, membrane technologies, and alternate water sources such as rainwater collecting. However, their effectiveness is determined by aspects such as cost, efficiency, and local applicability. The paper recommends utilizing

Multi-Criteria Analysis (MCA) as a decision-making method for evaluating and choosing arsenic remediation technologies. MCA considers technical, economic, and social issues, allowing stakeholders to make educated decisions that strike a balance between practicality and community demands. Finally, the study emphasizes the need to combine scientific knowledge with local realities in order to establish successful and long-term solutions for arsenic pollution reduction (Hemda and Huw, 2008).

While waste management covers a wide range of commodities, wastewater treatment has distinct economic and environmental issues. Unlike solid waste, wastewater requires ongoing and technically advanced operations to remove contaminants and recover resources like water, energy, and nutrients. Wastewater treatment has economic worth beyond cost reductions because it provides ecosystem services such as water purification and habitat conservation.

In the move from a linear to a Circular Economy (CE), wastewater treatment can serve as a paradigm for sustainable practices. Wastewater can be turned into an economic asset by using technologies such as membrane filtering, anaerobic digestion, and nutrient recovery. This chapter goes into these topics, providing a thorough examination of the economic aspects of waste and wastewater management.

2.2 Linear Economy Strategies for Waste Treatment

The linear economy, often known as the take-make-waste economy, uses resources to generate items that are then discarded. Linear economies use products and materials in one direction, from raw material to waste, and rarely maximize their potential. Pollution destroys natural systems and causes climate change and biodiversity loss. The Industrial Revolution established this linear economy, which had many benefits. The first mass-produced commodities were made. Industrialized countries became civilizations of abundance, the world population expanded, and millions were lifted out of poverty as raw materials and energy looked inexhaustible. This global development required exploiting scarce resources and deteriorating natural capital to create economic value. Over time, we have learned more methods to accomplish this and produce more advanced goods. But the presiding model remains. This strategy harms the environment and wastes essential materials (Debrah et al., 2022; Hatley et al., 2024; Hemidat et al., 2022; Mattson et al., 2024; Ribarova et al., 2024; Sundar et al., 2023; Voukkali et al., 2023).

Environmental pollution and resource depletion are significant obstacles to sustainability in the contemporary world, attributed to urbanization and industrialization in the past decade. Heightened human demands precipitate the overexploitation of natural resources, culminating in augmented environmental pollution encompassing water, air, and soil contamination—diminishing quality of life and constraining economic development. The existing linear economic model produces a lot of waste and frequently results in excessive use of natural resources. A relentless population surge and unsustainable consumption patterns generate excessive trash, exacerbating both direct and indirect environmental issues. By 2050, human needs will require the resources of three Earths to meet if human consumption continues on its current trend. The amount of rubbish produced worldwide is expected to rise in the ensuing decades. Among other problems, increasing waste production combined with insufficient waste management systems causes climate change, soil, water, and air pollution, and biodiversity loss (Voukkali et al., 2023).

Since the Industrial Revolution, society has operated inside a linear economy. Our consumption and "single-use" habits have transformed the globe into a "take, make, dispose" environment. This describes a unidirectional production model: natural resources supply our industrial inputs utilized to manufacture mass-produced commodities purchased and discarded after a single usage. This linear economic model of mass production and consumption is straining the physical limitations of the world. It is, therefore, not sustainable, and the shift to a CE is increasingly inevitable. Although their efforts significantly contribute to establishing a CE, they may enhance their focus on strategies enabling companies to shift from a linear production model to a CE framework. Our economy's reliance on natural capital is being depleted by its operations. Degrading soils, polluted oceans, dwindling biodiversity, depleted freshwater supplies, and deforested areas are all examples of this. For us to succeed in the future, the shift to a CE will be essential.

2.2.1 Significance of the Linear Economy in Waste Management

1. **E-waste Management:** Technology and population growth have increased the use of electrical and electronic items, increasing e-waste. E-waste happens when a product's structure and state are no longer suitable for its intended application. Hazardous, poisonous, rare, and precious metals are included. E-waste generation is quickly increasing, with an estimated 74.7 Mt by 2030. Formal e-waste management and treatment procedures are required. Incorporating CE concepts into waste management is regarded as efficient. However, the transition from linear to CE is still difficult. To improve e-waste management in the United Kingdom, the WEEE Directive 2002/96/EC and the RoHS Directive were passed.

   Despite legislation, the United Kingdom collected 499 Kt of e-waste properly and legitimately for recycling in 2019, 17% of the total. Large household appliances, which generate most of the waste, were recycled 176 Kt, 11% of the total. The remaining e-waste is used for disposal, incineration, or export to developing nations like China, India, and Nigeria. Improperly managing e-waste can harm the environment and humans. E-waste that enters landfills without proper carding pollutes the soil, water, and air, causing environmental damage and serious health problems. E-waste burns, generating hazardous substances such as carbon dioxide (CO_2) and sulfur dioxide, which destroy the ozone layer, contribute to global warming, and cause skin problems. Official e-waste recycling benefits both the environment and the economy. Recycled e-waste can provide gold and silver. Retrieving these metals will minimize the requirement for virgin raw materials, supply chain risk, and ore mining pressure, benefiting the environment. Reusing things beyond their life cycle adds value to the CE and reduces waste. CE reduces waste and resource consumption, preserves resources in the economy to generate additional value, and extends the useful life of materials and products. Recently, a number of sectors and governments have pushed CE due to its many advantages (Sundar et al., 2023).

2. **Plastic Waste Management:** The linear approach ignores several environmental issues, perpetuating plastic consumption and environmental damage to disposal. The linear plastic economy generates a lot of waste. According to studies, just 10% of plastics were recycled between 1950 and 2015, with the remaining 60% disposed of in landfills or the environment. Although landfills absorb buried greenhouse gases, their linear management technique worsens their environmental impact by releasing dioxins, furans, and other harmful combustion by-products. Plastic manufacturing relies mainly on petroleum, which conflicts with energy needs. Microplastics operate as sinks for hydrophobic organic pollutants. Therefore, landfill leachate may pollute soil and water. Implementing a CE is the most effective strategy to reduce dependency on finite fossil fuels, alleviate environmental concerns, and recycle plastic rubbish once it has served its purpose. Concerns about the environmental impact of microplastics have led to the replacement of the linear plastic consumption and disposal model with a CE paradigm that emphasizes resource efficiency, waste reduction, and sustainable behaviors. Plastic depolymerization is critical in this scenario. Depolymerization decreases the environmental impact of plastics by converting complex polymers into monomers or valuable chemical feedstocks.

The landfill- and incineration-based plastic waste management system is unsustainable. These processes produce microplastics through aging and release them into the environment. Microplastics are too small to cure directly and may produce harmful nanoplastics. Since primary and secondary recycling lead to secondary contamination or downcycling, which restricts reclaimed plastic to low-quality uses, creative and sustainable approaches to plastic waste management and value-adding are required to address the global plastic waste problem. Therefore, encouraging proactive recycling of plastic into useful chemicals is crucial for sustainability, as it replaces unsustainable methods (Shi et al., 2024).

3. **Domestic Wastewater Management:** Simultaneously, linear technological solutions for full-scale applications and laboratory- or pilot-scale developments are established. One of the earliest decentralized solutions is a septic tank. Its design, control, and operation are complex and lead to operational, social, and environmental problems despite its extensive use. Improvements to septic tank performance are being researched. The analysis showed recent septic tank optimization research focuses on building strategies, including zoning, baffles, novel materials, and technical advances. These improvements include linking the septic tank with a zeolite filter, using an up-flow regime, inoculating with microalgae, lighting, and electrolytic stimulation. Removal mechanism studies have also shown how bacterivorous protozoa remove dangerous indicator bacteria and how Cu-based nanoparticles may be hazardous. Uncontrolled discharge or soil or sand infiltration systems discharge septic tank effluents into the Earth.

   Most agree that soil/sand infiltration systems improve effluent quality. Outflow nutrient cycle studies also try to understand nutrient transformations for system optimization and environmental sustainability. Multiple septic tanks enhance the quality of water after the third tank. Septic tank intake wastewater's variable quantity and quality with household water usage patterns makes qualitative analysis difficult. Septic tank pollutant reduction percentages are difficult to calculate due to this heterogeneity. Sludge accumulation and septic tank emptying are other sectors' hot subjects, complicating system management and maintenance. Final treatment with a zeolite filter eliminates nitrogen, total phosphorus, and organic debris. Aquifer loading can be decreased, and nitrogen removal in on-site treatment systems can be increased via passive engineering. Drip irrigation, shallow trenches, and single-pass sand filters with bioreactors can all be used to denitrify or increase plant nitrogen intake in the carbon-rich root zone. Built wetlands, lagoons, and stabilized ponds are examples of aquatic-based natural treatment techniques that, while less complex than centralized systems, can remove toxins in diverse ways based on the local climate. Since septic tanks and other basic solutions have been in use for many years, linear decentralized systems are technologically advanced. Many cutting-edge solutions have also been in use for a comparable amount of time. Therefore, linear systems may be regarded as mature as they have been explored more fully (Ribarova et al., 2024).

4. **Municipal Waste Management:** MSW increased worldwide demand for energy, waste management, and industrial sustainability. The United States, China, India, Brazil, and Russia produce the most MSW. Mismanagement of MSW can harm the environment, safety, health, and finances. MSW volume and complexity are rising, making waste management more difficult for current and future societies. Increased fossil energy usage, pollution, and global warming worsen this natural resource depletion. Related threats include natural resource depletion.

   In conclusion, converting MSW into energy or other products can reduce pollution, fossil energy usage, and resource depletion. For social sustainability, energy supply, conversion, and use efficiency must be improved. Modern technology can convert nonrecyclable MSW into power, heat, biofuel, and biogas. Anaerobic digestion, incineration, pyrolysis, gasification, and hydrothermal processing can convert MSW into valuable chemicals and fuels, whereas composting and landfilling are standard waste management methods. MSW has a high creation rate and can reduce resource depletion and pollution if appropriately handled. MSW recycling is crucial to circularizing the linear economy (Hoang et al., 2022).
5. **Hospital Waste Management:** The World Health Organization (WHO) defines this as medical waste generated during the diagnosis, treatment, and immunization of humans and animals. The WHO divides medical waste into seven primary categories: infectious waste, pathological waste, sharps, chemicals, pharmaceuticals, genotoxic waste, and radioactive waste. The U.S. Environmental Protection Agency (EPA) defines hospital trash as "all waste materials produced at healthcare facilities, including hospitals, clinics, physicians' offices, dental practices, blood banks, and veterinary hospitals/clinics, along with medical research facilities and laboratories." Improper hospital waste management exposes patients, healthcare personnel, the general public, the environment, and the ecosystem to both infectious and noninfectious diseases by spreading germs. Vulnerability and infections are just two of the concerns that can result from poor medical waste collection and disposal. In order to solve the critical waste management issue, the European Circular Economy Action Plan of 2015 reaffirmed the waste hierarchy. It underlined the importance of recycling, reusing, and ongoing diversion from landfills—inadequate medical waste management pollutes the environment and endangers public health. Medical waste management requires careful planning and oversight at all phases, including collection, segregation, treatment, recycling, and disposal. This fosters the circular approach (Govindan et al., 2022).

The advancements in technology and the future of resource allocation indicate a necessity for a comprehensive reevaluation of the construction and provision of products and services. Transitioning to a circular model necessitates altering the supply logic of our linear economy. Products should be engineered to use reclaimed secondary materials for economical end-of-life recycling to complete the manufacturing cycle efficiently. The CE necessitates the design of components for reuse in goods that may be updated and refurbished rather than discarded and replaced to enhance resource efficiency. The problem also resides in how a corporation interacts with its clients and their involvement during and after the product's utilization.

2.3 CE Strategies for Waste Treatment

A sustainable economic framework called the CE aids in the resolution of some major global issues, such as waste accumulation, environmental deterioration, and resource depletion. The core principle of Economic Circularity is to reduce waste and create additional value for items by reusing them in a variety of ways once their life cycle is over (Trigkas et al., 2020). As per the authors (Kirchherr et al., 2018), "When a product reaches the end of its life, its value is preserved for as long as possible, waste and resource consumption is kept to a minimum, and resources are retained in the economy to be utilized repeatedly to generate additional value," is how CE is described. This approach is highly relevant for waste treatment, as it diverts the concentration from disposal to resource recovery and sustainable waste management, ensuring that waste is not an endpoint but a valuable input. This "take, make, dispose" model of conventional linear economy leads to excessive resource exploitation and waste production. The CE, on the other hand, places more emphasis on the "make, use, reuse" strategy. It creates closed-loop systems where materials constantly circulate through the economy by converting waste streams into resource flows (Vennila and Karthikeyan, 2024). Waste treatment plays a pivotal part in CE, transforming waste into resources and limiting environmental burdens while promoting social as well as economic benefits (Sondh et al., 2024).

Waste management is a serious problem, also stated by the United Nations (UN), and addressing it can solve many other current global challenges. Effective waste management is either directly or indirectly related to the SDGs of the UN (Sharma et al., 2021). For instance, the issue of the global climate crisis can be palliated by robust waste management practices. If current practices persist, projected estimated emissions from landfilled waste will reach approximately 2.4 billion tonnes of CO_2 equivalents annually by 2050 (Sondh et al., 2024). Environmental justice is given top priority in a waste disposal management system that is driven by CE, which also offers the basis for addressing these problems comprehensively.

Additionally, the three fundamental verticals of sustainability—social, economic, and environmental—are in line with CE policies (Merli et al., 2018; Mies and Gold, 2021). It provides an opportunity to stabilize and grow the economy, improve natural resource management, and protect human rights. Alarmingly, between 400000 and 1 million deaths each year are connected to diseases caused by unmanaged garbage, including diarrhea, malaria, heart disease, and cancer (Mari et al., 2019). These difficulties show how urgently CE-based waste management techniques are needed to safeguard the nature and health of the people.

2.3.1 Principles of CE in Waste Treatment

CE in waste management provides a detailed conceptual understanding for creating such systems that prioritize sustainability and resource utility efficiency. These ideas are a foundation for developing waste treatment techniques for a green economy (Ellen, 2020).

1. **Eliminating Waste and Pollution:** We extract raw materials and natural deposits from the Earth, manufacture products, and dispose of them as waste. A significant portion of this garbage is disposed of in landfills, water bodies, or incinerators and is consequently lost (Ellen, 2020). This arrangement is unsustainable in the long term due to the restricted resources of our planet. CE emphasizes designing out waste by rethinking processes and products. This involves reducing material use, adopting cleaner production methods, and preventing pollution at its source (Korhonen et al., 2018).
2. **Moving Materials and Products:** The central theme of CE is extending the useful life of materials. This principle encourages 10Rs, as shown in Figure 2.3, reuse, repair, remanufacture, and recycle to maximize resource utilization and minimize disposal (Ho et al., 2024).
3. **Regenerating Natural Systems:** Waste streams are treated as valuable resources. Practices like composting, vermifiltration, and biochar production not only reduce waste but also add value to the system and enhance soil health and biodiversity (Arora and Saraswat, 2021; Carvalho et al., 2022; De Corato, 2020).

These principles form the basis for the practical strategies discussed in the following section.

2.3.2 Implementation Strategies for CE

CE strategies for waste treatment focus on optimizing resource efficiency and minimizing waste through innovative technologies and sustainable practices. These tactics seek to increase material reuse and recycling, decrease environmental effects, and prolong product life cycles.

Circular Economy	R1 (Reuse and Reject)	Smarter Use and Production	Making a product about a bid by abandoning its function or supplying a radically different product. We have to choose as to why not buy anything we don't need
	R2 (Rethink)		Intensify product use (for example, by sharing, multifunctional products, or refiling pot/glass rather than plastics)
	R3 (Reduce)		The amount of raw materials and materials in the product or in their use
Rise Circular	R4 (Reuse)	Extended Lifetime of Products and Parts	Reuse by another consumer of discarded product in the same function by another user
	R5 (Repair)		Maintenance of broken products for use in their old function (restore old products and bring them up to date)
	R6 (Refurbish)		Modernize old product (up-cycling)
	R7 (Remanufacture)		Use parts of discarded products in new products with the same function
	R8 (Repurpose)		Using discarded products or parts of it in new products with the same function
	R9 (Recycle)	Useful Application of Materials	Process materials to the same (high-quality) or medium (low-quality) quality
Linear Economy	R10 (Recover)		Incineration of material with energy recovery

Figure 2.3 The 10Rs concept for waste management based on the circular economy (CE). *Source:* Zorpas (2020)/with permission of Elsevier.

1. **Waste Valorization**
   The goal of value extraction is to transform rubbish into resources or products that may be unutilized.
   a. **Organic Waste:** Composting can be used to create nutrient-rich manure from organic waste, including food scraps and agricultural wastes. Significant environmental advantages of composting include lower greenhouse gas emissions, longer landfill life, less pollution, the creation of valuable compost, and resource-saving (Manea et al., 2024).
   b. **Industrial Waste:** Industrial by-products such as coal bottom ash, wheat straw ashes, fly ash from thermal power plants, sugarcane bagasse ashes, palm oil fuel ashes, and powdered Ground Granulated Blast-Furnace Slag (GGBFS) are used to stabilize soil and manufacture building materials

like cement and concrete (Oyejobi et al., 2024; Singh et al., 2017). The results show that these industrial wastes can enhance the durability and mechanical qualities of concrete while drastically lowering the carbon footprint associated with its manufacturing (Oyejobi et al., 2024). In India, fly ash from thermal power plants is widely used in manufacturing eco-friendly bricks and road construction (Rastogi and Kumar Paul, 2020).

   c. **Wastewater:** Textile wastewater, distillery wastewater, etc., have high organic content and great resource potential. Technologies like vermifilters treat wastewater, and by-products, like treated water and manure, can be utilized for fertigation, closing the loop (CE) (Arora and Saraswat, 2021; Vergine et al., 2016). Nutrient recovery such as nitrogen and phosphorus via membranes, bioelectronic processes, and biofertilizer production also aids in overcoming environmental pollution (Robles et al., 2020).

2. **Recycling and Upcycling Technologies**
   The recycling techniques in the CE incorporate advanced technologies to improve efficiency and material recovery.
   a. **Mechanical Recycling:** Classical methods for recycling materials such as plastics, paper, and metals involve mechanical operations such as shredding and melting. These procedures preserve the quality of materials, enabling their reuse in manufacturing (Lieder and Rashid, 2016).
   b. **Chemical Recycling:** This technologically advanced method breaks down complicated materials, like mixed plastics, into their base elements. In methods for managing plastic waste, chemical and solvent-based recycling works effectively as an addition to mechanical recycling, which is better for the environment (Klotz et al., 2024). Options that preserve chemical structures, like depolymerization and solvolysis, provide the most substantial reductions in CO_2 emissions. Chemical recycling can generate food-grade plastic for packaging, which is significant for the shift toward circular packaging (Geert, 2019). It makes difficult-to-recycle materials feasible by facilitating the extraction of premium raw materials from polluted waste streams (Klotz et al., 2024). Companies like BASF are investing in chemical recycling technologies for CE.
3. **Energy Recovery**
   Reducing dependency on fossil fuels, decreasing trash disposed of in landfills, and using non-recyclable garbage to generate energy are the main goals of energy initiatives.
   a. **Waste-to-Energy:** W2E can tackle large-scale waste-energy challenges while simultaneously promoting the CE. Figure 2.4 depicts the first step in optimizing resource recovery efficiency: waste segregation. Properly segregated dry waste is incinerated to generate electricity and heat

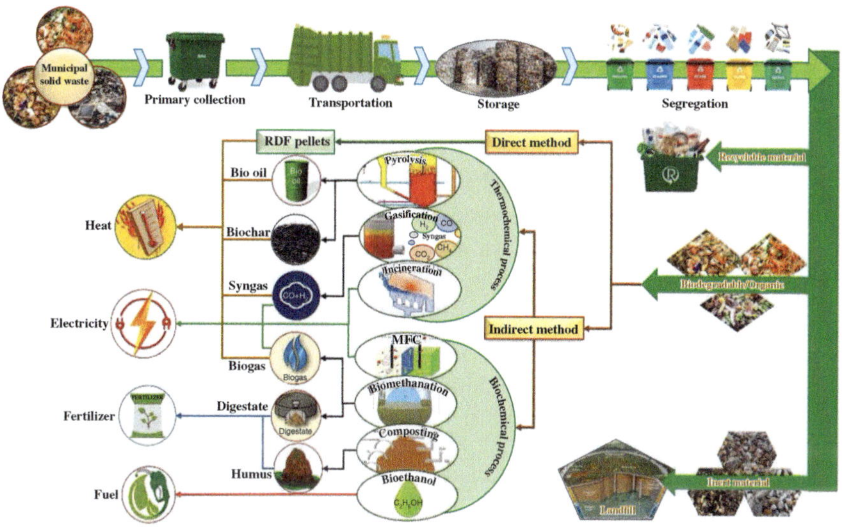

Figure 2.4 Current technologies for waste treatment for the CE. *Source:* Hoang et al. (2022)/with permission of Elsevier.

(Hoang et al., 2022; You et al., 2016). Additionally, it lowers energy efficiency by 25–30% and waste volume by 70–80% (Hoang et al., 2022). The anaerobic digestion process utilizes wastewater and organic waste to produce biogas. Also, digestate can be used for fertilization or nutrient recovery, which reduces waste by 60–70% (Nwokolo et al., 2020). According to the study, when sewage sludge and poultry processing wastes are co-digested, 7.63 GWh of power is produced annually on average, or 2.54 GWh (Masłoń et al., 2020).

b. **Advanced Thermal Methods:** Pyrolysis, gas reforming, and gasification are innovative methods that convert and reduce waste volume and also produce by-products such as hydrogen fuel, charcoal, chemicals, bio-oil or liquid fuels, etc. (Casanovas et al., 2015; Hoang et al., 2022; You et al., 2016). These methods are particularly effective for nonrecyclable plastics and other high-energy waste (Shah et al., 2022). In Japan, pyrolysis plants and plastic waste are converted into synthetic crude oil, contributing to the CE goals of the country (Kabeyi et al., 2023).

4. **Industrial Symbiosis**

Industrial symbiosis is recognized as a strategy to enhance materials and energy efficiency within industrial systems and achieve CE objectives (Martin et al., 2015). By using wastes or by-products from one of the operations as raw materials for the other, industrial symbiosis encourages collaborations across industries. This method establishes a resource-sharing network, improving efficiency and minimizing overall waste. There are many examples of successful industrial symbiosis, especially in industries like mining, construction, chemicals, metals, forestry, and paper (Domenech et al., 2019). The Kalundborg Eco-Industrial Park demonstrates industrial symbiosis effectively. Excess heat generated by a power plant is employed in industrial applications, and the by-products, such as fly ash, are reused in cement manufacturing (Chertow, 2008). This system illustrates the potential for inter-industry collaboration to yield both environmental and economic advantages.

5. **Digital Innovations**

   Waste may be efficiently eliminated, repurposed, and recovered with the use of digital and mechanical applications brought about by Industrial Revolution 4.0 (Cheah et al., 2022; Maiurova et al., 2022).

   a. **IoT-based Waste Management:** Technology from IoT has the potential to monitor trash cans in real time, optimize the garbage collection route, and lower operating expenses (Fatimah et al., 2020; Wang et al., 2021). Food waste generated was reduced by about 33% per month in the Gangnam district due to the application of user-friendly smart garbage bins (Hong et al., 2014).

   b. **Blockchain Technology:** This technology is vital for precise tracking of the product's lifecycle, ensuring transparent transactions among all stakeholders, and reducing the amount of waste (Bułkowska et al., 2023). Circularize—a startup uses blockchain to verify recycled product content, supporting CE practices.

   c. **Additive Manufacturing:** 3D printing allows the creation of complex product layers by layer, utilizing computerized technology like selective laser sintering and direct metal laser sintering, which is generally difficult to manufacture. It can print on-site and reduce human error and material wastage, reducing the waste produced during fabrication. Transportation requirements also reduced results in reducing emissions and energy use (Ashima et al., 2021; Cheah et al., 2022).

2.3.3 Benefits

The CE provides environmental, economic, and social advantages in waste treatment. Environmentally, CE uses W2E, biological treatment, recycling, and thermochemical treatment using renewable resources to reduce dependency on landfills. Economically, it lowers the cost of waste management practices, creates alternate revenue streams from the recovered materials, and promotes the creation of jobs in green sectors (Ghisellini et al., 2016). Socially, through waste treatment by lowering pollution and creating cleaner surroundings, CE enhances the quality of life, raises community involvement in trash reduction initiatives, and raises public knowledge of sustainability issues (Murray et al., 2015).

2.3.4 Challenges in Implementing Circular Strategies

1. **Financial Barriers:** One of the barriers is the consideration of cost efficiency, which is critical in sectors like the food supply chain and textile industry. Implementation of CE practices is hampered by financial limitations, such as a lack of funding and resources (Mehmood et al., 2021; Pannila et al., 2022; Saha et al., 2021).
2. **Technological Limitations:** The limitations of technological advancements and human resources are a significant challenge, particularly in industries. This includes insufficient infrastructure and innovation capabilities, which are important for CE (Mehmood et al., 2021; Saha et al., 2021).
3. **Institutional and Regulatory Barriers:** Institutional challenges, including inadequate regulatory frameworks and governance, are prevalent. These barriers are compounded by the need for international standards and government interventions to stimulate CE initiatives (Cantú et al., 2021; Mehmood et al., 2021).
4. **Cultural and Behavioral Challenges:** There is often a reluctance from management and indifference from end-users toward sustainability, which impedes CE implementation. This is particularly noted in small and medium-sized enterprises (SMEs) and emerging economies, where aligning business strategies with local market conditions is crucial (Cantú et al., 2021; Saha et al., 2021).
5. **Supply Chain Redesign:** Redesigning supply chains to accommodate CE principles is complex and involves numerous challenges, such as integrating different levels of the supply chain and ensuring vertical and transverse integration, which leads to the need for robust collaboration among stakeholders, advanced technological solutions, regulatory alignment, and the adoption of environment-friendly practices to balance economic, environmental, and social objectives effectively (Bressanelli et al., 2018).
6. **Lack of Collaboration and Knowledge Sharing:** Effective CE implementation requires collaborative efforts and knowledge sharing across the value chain among stakeholders, industry, academia, and government, which is often lacking. This includes marketization of waste recycling and sustainability management (Saha et al., 2021).

2.4 Economic Viability of Different Strategies for Waste Treatment

2.4.1 Cost Analysis of Traditional Versus Circular Approaches

The current global garbage generation boom is ascribed to increased living standards, population growth, and technological advancements (Cudjoe et al., 2021). As items become more complex, the composition of MSW also becomes more complicated.

Annually, about two billion tonnes of MSW are generated. This waste would travel more than a round-trip ticket to the Moon if it were piled into regular shipping containers and placed end-to-end, circling the Earth's equator 25 times. Every minute, the equivalent of one garbage truck's load of plastic is released into the ocean. Globally, the costs of social garbage and its (mis)management totaled \$361 billion in 2020. MSW is expected to cost \$640.3 billion worldwide in 2050 if improperly managed. Adopting the CE model is essential to mitigate escalating waste management expenses, wherein costs in 2050 will align with current expenditures. It is better to reduce trash by preventing its formation or to utilize things that would otherwise become waste rather than to create rubbish and then recycle it. Kitakyushu, Japan, exemplifies effective waste reduction, generating merely 0.42 kg of municipal solid garbage per capita a day, in contrast to the estimated global average of 0.75 kg per capita daily. The expenses associated with developing and maintaining waste treatment facilities may affect the selection of cities for final disposal methods. For example, a municipality adjusting to one million people may need to spend about \$10 million to build a landfill; the expense of a composting facility can range from several million dollars for simple systems to around \$10 million for sophisticated mechanized facilities, and the capital costs of an incinerator (with heat and energy recovery) are approximately \$600 per annual tonne, calculated by dividing the total capital expenditure over the plant's lifespan by its total annual capacity. Recent installations in Mexico, Poland, Singapore, and the United States have shown capital costs of about \$600 per tonne (Kaza et al., 2018).

Disposal expenses fluctuate significantly. In several nations, the disposal of waste is conducted informally and not recorded officially. In high-income nations, disposal costs are more accurately recorded, generally ranging from \$50 to \$100 per tonne. The expenses associated with recycling in high-income nations frequently parallel those of landfilling. Recycling can become more economical when landfills are subjected to taxation or constrained capacity, resulting in elevated market pricing for landfilling. Due to a lack of information on construction and tipping expenses, it is difficult to estimate the cost of open dumping. However, dumping raises the risk of huge disaster-related costs and causes large losses in land value. Dumpsite closures may result in significant costs. Inadequate waste treatment through dumping or uncontrolled incineration results in environmental expenses owing to water and air pollution as well as adverse health effects, together with disasters, land costs, and landfill closures. The long-term financial consequences may be enormous (Kaza et al., 2018).

As discussed in the linear and circular economy sections, the traditional waste treatment approach includes landfilling, dumping and open waste burning, and incineration with low energy recovery. It aims only to dispose of waste. At the same time, the circular one focuses on the maximum possible recovery of resources by recycling, composting, and converting to energy in a sustainable and economically feasible manner.

Adopting a zero-waste approach and moving toward a CE represents the sole method to ensure an economical, sustainable, and secure future (Bhatt et al., 2022).

For every area, year, and situation, estimates of waste generation and treatment techniques have been made by providing average expenses for each of these approaches, encompassing collecting, recycling, and processing (utilizing landfilling or W2E technology) expenditures per tonne of waste across nations with varying income levels, using 2020 as a reference point. The expenses have been aggregated via weighted averages for the countries within each area to create regional and worldwide cost estimates, data published in the *UNEP 2024 report* (Bhatt et al., 2022).

"Uncontrolled waste" is defined as waste that is either not collected and must be disposed of by the generator in an open environment or collected and subsequently incinerated at its location. In contrast, "controlled waste" is characterized as waste that is collected and further disposed of or recycled in a regulated manner. Analysis indicates that 38% of the MSW created worldwide in 2020 was uncontrolled (Bhatt et al., 2022). The direct charges do not cover the whole picture of waste management expenses. The wide-ranging externalities of poorly managed waste, including pollution, biodiversity loss, and climate change, represent a large cost and societal burden. The environmental implications of dumpsites extend beyond groundwater supplies; they can affect soil quality and human health. All dumpsites, regardless of size, are usually associated with odorous gas and leachate plumes that contaminate water resources and the environment. Research indicated that leachates generated at the dumpsite seldom persist at the discharge point (Mary et al., 2022) as they are transported through the porous medium in the subsurface. The main causes of water pollution are toxins seeping through the soil at waste disposal sites and river water percolating (Alao, 2023).

2.4.2 Potential for Recovered Resources

Cudjoe et al. (2021) calculated the amount of diesel, electricity, and greenhouse gas emissions that are reduced by recycling. According to the main conclusions, the global electricity savings of 132.4 thousand tonnes of oil equivalent (toe) per day in 2012 increased to 182.3 thousand toe per day by 2025. Power conservation averaged 63.6% between 2012 and 2025. Recycling reduced greenhouse gas emissions by 1.4 million kilo tons CO_2 eq per day in 2012; by 2025, that amount will have reached 1.9 million kilo tons CO_2 eq per day. Between 2012 and 2025, the total amount of diesel fuel saved grew by 37.7%, from 511 146 to 703 887 million liters/day. According to the sensitivity analysis, the environmental and energy benefits of recyclable materials will increase in 2025 if the waste collection rate, the percentage of plastic and paper waste, and the recycling rate all increase. In terms of solid waste recycling, the results of this study may provide scientific direction for attaining the SDGs. Returnable plastic packaging can yield significant environmental advantages over single-use plastic packaging, potentially decreasing emissions and water consumption by 35–70%. Rich countries work hard to recover resources from the source, emphasizing recycling and using organic waste effectively. Conveyor belt systems and the use of magnetic forces and optical lasers for waste separation are examples of automation in recycling operations (Kaza et al., 2018).

2.4.3 Challenges in Economic Implementation

Since the Stone Age, humans have recycled agricultural waste and the Romans reprocessed metals and glass 2000 years ago (Freestone, 2015). Notwithstanding this extensive history, just 19% of MSW, encompassing metals, glass, paper, cardboard, certain plastics, and biodegradable garbage converted into biogas by anaerobic digestion and composting, is currently recycled.

353 million metric tons (Mt) of plastic waste were produced worldwide between 2000 and 2019, with landfilling accounting for 44% of post-consumer plastic and open dumping or incineration for 22% of it. Recycling is only a small portion (9%) of the total (Velis and Cook, 2021).

According to D'Angelo and Meccariello (2021), over 12 billion tons of plastic garbage are expected to end up in landfills and the environment by 2050. Uncontrolled MSW discharge will increase until 2028, reaching 730 million metric tons per year. On the other hand, even under the best of circumstances, the construction of new infrastructure such as recycling facilities, designed landfills, incineration plants, and biological treatment plants is only expected to grow by 2% annually, which is insufficient to meet the present and future needs for MSW treatment facilities (Ferronato et al., 2024). Budget deficits for waste services are common in developing nations, notwithstanding the substantial allocation of SWM costs within municipal budgets. Coordination is necessary to maintain consistency among various levels of government. Numerous governments encounter challenges due to overlapping tasks among agencies or deficiencies in responsibilities, as SWM activities frequently span multiple ministries. For many communities, financing waste management systems is a major obstacle. It is essential to recover costs to avoid relying on internal or external subsidies. Operations expenses and waste management investment costs are typically funded using different strategies. Recycling markets may be hampered by insufficient regulation governing the collection and recycling of particular waste streams, alongside an absence of national rules or regulations for incorporating recycled materials in new goods. Without such regulations, the private sector faces difficulties in operating enterprises and sectors reliant on and generating demand for secondary resources. Capital expenditures are often financed through grants or subsidies from the national government, international donors, or collaborations with private enterprises because infrastructure and equipment improvements come with high costs. Additional major challenges encompass a disparity between the institutional prerequisites of financers and the capacity of municipalities to furnish official assurances of the project feasibility, as well as the lack of knowledge among national and subnational entities regarding finance accessibility (Kaza et al., 2018).

2.4.4 Role of Policy and Regulation in Economic Viability

A popular adage in the waste management industry says that waste management comprises 20% of technology and 80% human factors. All individuals on the Earth produce municipal waste and are responsible for personal decisions regarding its management (Bhatt et al., 2022).

Research on national waste regulation indicates that formal national laws or guidelines govern SWM in 86% of countries and economies. In lower-income nations, waste disposal is typically managed by local villages or towns. SWM legislation varies from comprehensive environmental regulations to specific programs. Peru's General Law on Solid Waste Management encompasses all facets of SWM, from generation to final disposal. Certain waste management initiatives that mitigate greenhouse gas emissions such as the installation of landfill gas capture systems or the composting of organic waste utilize carbon financing.

1. **National Strategies:** A national strategy spanning 5–10 years that delineates the current waste conditions and establishes sector-specific objectives for recycling, financial viability, public awareness, and the promotion of a green economy, reducing greenhouse gas emissions, and rehabilitating

contaminated sites is an example of advanced waste governance. Sometimes the national strategy includes extra guidelines for people and organizations or legally binding regulations. National governments may offer financial support or technical assistance, including cost-sharing or plan evaluation, to attain national objectives (Kaza et al., 2018).

2. **Regional Strategic Frameworks:** Due to the regional nature of waste management, it is more prevalent for municipalities to construct a regional Strategic framework centered on SWM than for nations to establish a national strategy. Frameworks define the community's objectives for SWM and include implementation strategies (Kaza et al., 2018).

2.4.5 Public–Private Partnerships in Waste Management

To attain technical proficiency, efficiency, and financial investment in waste management systems, partnerships with the business sector are commonly sought. Over 30% of waste management services, encompassing primary collection, treatment, and disposal, are provided via public–private partnerships despite the difficulties in their design and execution. Private operators can improve waste management systems' financial stability and efficiency under specific circumstances. Alongside general solid waste legislation, more than 50% of the studied nations possess formal regulations directing public bodies in their interactions with the private sector. Regulation is nonexistent in 21% of countries, while data is unavailable for 25%. About half of all waste services investments worldwide come from local governments, with the remaining 20% coming from national governments and the other 10–25% derived from the private sector, based upon the service provided (Kaza et al., 2018).

The private sector gathers user charges for their services to offset costs. This phenomenon predominantly occurs in underdeveloped nations when the sector of waste management is split into formal and informal components. The formal sector comprises public and private firms that receive legal and financial assistance from the government. In contrast, the informal sector operates beyond official oversight and, thus, is ineligible for subsidies like pensions or medical insurance. In the informal sector of Waste Management in developing countries, recycling informal garbage is a common way for urban poor people to make money. Around 1% of the city's population, which is more than 15 million people, works in the waste industry on an informal basis (Martin, 2005). In China's urban areas, for instance, between 3.3 million and 5.6 million people recycle informally. The informal sector has the potential to alleviate poverty, provide employment, enhance local industrial competitiveness, and decrease municipal expenditure on social services and SWM if it is properly encouraged and organized (Martin, 2005).

2.5 Future of Wastewater Treatment Based on Economic Value

2.5.1 Existing Challenges and Future Prospects of Nanotechnology Incorporated Wastewater Treatment

The combined effect of industrialization, urbanization, climate change, and population growth has exacerbated the shortage of water. Many regions across the globe have witnessed water scarcity, severe water pollution, and detrimental effects on the ecological environment due to increasing population growth, industrialization, and climate change. Anthropogenic activities have resulted in the contamination of water resources, which has become a global concern. The alarming levels of contaminants pose various threats to the environment and human health. As global water resources are facing increasing pressure, wastewater treatment and the reuse of treated water have become necessary. The use of treated wastewater proves to be one of the viable steps to address the concept of CE in the field of wastewater treatment. Reusing treated water and treating wastewater are creative ways to augment water supplies by purifying and reusing wastewater for use in industry, agriculture, and even drinking (Hurtado and Berbel, 2024; Singh et al., 2020; Wu et al., 2023).

Various nanomaterials, which include metal and metal oxide-based nanoparticles, polymeric and carbon-based nanomaterials, and nanocomposites, are increasingly being used for wastewater treatment to effectively remove pollutants and minimize risks to the environment and water bodies (Attri et al., 2022; Jain et al., 2021; Pandit et al., 2025; Singh et al., 2020). Adsorption, nanofiltration, photocatalysis, and other nanotechnology-based techniques are some of the methods utilized to remediate wastewater. Given the market evaluation of nanoparticles in wastewater treatment, the study shows that the application of nanomaterials in wastewater treatment is a profitable industry because of the rising need for cutting-edge water treatment technology (Jain et al., 2021; Pandit et al., 2025). In a study, Singh et al. (2020) used carbon soot nanoparticles coated with nickel/platinum nanofilms and iron-coated carbon soot particles for the degradation of methylene blue dye. Attri et al. (2022) degraded 4-aminopyridine with silver-doped cuprous oxide nanoparticles.

A report by Wilson (2023) provides an overview of the global and regional market for the technologies that are used in water and wastewater treatment and analyzes global market trends. In the report, 2022 was considered the base year, and a prediction was made for the period from 2023 to 2028. It contains a forecast for 2028 along with the anticipated Compound Annual Growth Rate (CAGR) for the years 2023–2028 and an estimate of market data from

2022 to 2023. In addition to quantifying the global market for water and wastewater technologies by end-user, process, product, and region, the study also examines the market's present and prospects. It is anticipated that the water and wastewater treatment technology industry will reach $303.1 billion globally. By 2028, the market is estimated to increase to $515.8 billion, with a CAGR of 11.2% between 2023 and 2028.

The cost incurred in treating wastewater with nanoparticles ranges from $10 to $500 per gram, depending on the quality and type of the nanoparticles used. Expenses incurred for synthesis and functionalization also vary significantly depending on the method used. Infrastructure, equipment, operating and maintenance costs, and regulatory compliance procedures contribute substantially. As a result, even with their benefits, nanoparticles in wastewater treatment require a thorough evaluation of the financial and environmental implications. Activated carbon in wastewater treatment facilities may have yearly operating expenses of up to $0.1 million. Because of the rise in micropollutants, its lifespan can be as little as 6 months, necessitating costly full replacements. An expensive adsorbent activated carbon can cost up to $1320 per ton (Pandit et al., 2025).

The amount of CO_2 released depends on the type of methodologies used to treat wastewater. The catalytic activity of nanomaterials introduces CO_2 generation, which is a greenhouse gas and a global concern; hence, it necessitates effective techniques and pathways, such as photosynthetic bacteria, advanced nanomaterials, and tri-reforming of methane, that can either utilize or mitigate emitted CO_2. The use of nanomaterials also raises the concern of ecotoxicity and leaching into the environment, which must be mitigated to make the process eco-friendly (Jain et al., 2021; Pandey and Biswas, 2024; Pandit et al., 2025).

The lack of a standardized and universal approach in this field is a significant gap despite significant research efforts being made toward the use of integrated nano-biotechnology for water and wastewater treatment. This emphasizes the necessity of thorough research to address and improve the present comprehension of this complex problem. Literature lacks detailed scientific reports and technical specifications for Wastewater Treatment Plants (WWTPs) based on nanotechnology. The estimation of initial and operating costs must carefully consider factors such as treatment process kinetics, energy, labor, and nanomaterials. Environmental issues such as CO_2 emission during treatment must be carefully examined and mitigated while treating wastewater with nanotechnology (Pandit et al., 2025).

2.5.2 Current Status and Practices Related to CE in Wastewater Treatment

2.5.2.1 Nutrient Recovery

In the modern world, effective treatment of wastewater poses a significant challenge. It is necessary to develop new economical and sustainable technology or upgrade the current ones (Tamta et al., 2020). In the context of sustainable technology, a CE can be a suitable option; however, one of the biggest challenges for the future CE is recovering different nutrients from the different wastewater matrices and recycling them in the form of soil fertilizers. Uncontrolled and relentless discharge of wastewater, limited accessibility to fertilizer in developing nations, and higher cost of fertilizer are the driving forces behind nutrient recovery and their recycling. Nutrient recovery adds to the environmental and economic value, thereby fostering the CE (Saliu and Oladoja, 2021; Zhang et al., 2024).

WWTPs' nutrient recycling process benefits the environment by lowering the need for fossil fuel-based traditional fertilizers, which in turn uses less water and energy. Sewage sludge (biosolids), raw wastewater, and semi-treated wastewater streams can all be utilized to recover nutrients (Neczaj and Grosser, 2018). Numerous studies have been successfully carried out to recover different nutrients from various sources of industrial wastewater, including semiconductor, food processing, tannery, textile, yeast, dye, tanning, fertilizers and agrochemical, rare earth, coking, slaughterhouse, and meatpacking industries. The total nitrogen concentration from these industrial effluents ranges from 100 to 4535 mg/L, whereas the total phosphorous content varies between 2.5 and 3490 mg/L (Foletto et al., 2013; Ryu and Lee, 2016; Saliu and Oladoja, 2021).

Phosphorus can be recovered from burned sludge ashes, wastewater, and sewage sludge. Phosphorus is typically recycled in WWTPs primarily through struvite crystallization procedures. In 2018, about 2000 metric tons of phosphorus per year (Mt P/year) were recovered in Europe (Neczaj and Grosser, 2018). Another well-known and abundant source of nutrients is reclaimed water. The presence of nutrients like potassium, phosphorus, and nitrogen means that they can either completely or partially replace traditional fertilizers (Hurtado and Berbel, 2024).

Huang et al. (2012) reported ammonia removal of 88% from anaerobic digester effluents of nylon wastewater using a chemical precipitation technique. In a study, Lee et al. (2003) eliminated phosphate and ammonia from wastewater that had undergone biological treatment and was derived from the effluent of an activated sludge technique that was treating swine wastewater.

Since the societal myth around pee has raised ethical and discriminatory concerns about its adoption in different countries, the idea of using human urine as a fertilizer substitute may be a pipe dream or a practical reality. However, if a comprehensive approach is implemented through the cooperation of the stakeholders—plant biologists, soil microbiologists, agronomists, and environmental scientists/engineers—using human urine as fertilizer is possible (Saliu and Oladoja, 2021). After nutrient recovery, human urine leaves behind a significant amount of organic matter, requiring additional treatment for environmental protection. Human urine is a highly saline solution that is full of nutrients that plants can consume. Although there is a wealth of literature on the recovery of nutrients from human urine, there is very little study on how to manage the remaining liquid phase (Akdag et al., 2024). Separating urine from the main wastewater stream can help recover nutrients. Urine contains around 70–80% of nitrogen and 50% of phosphorus, which can theoretically be collected at a rate of 70% by employing the urine-collecting systems in the toilets. This technology has not been widely adopted due to major technological issues and public acceptance (Neczaj and Grosser, 2018). In a study, Akdag et al. (2024) eliminated most of the nutrients from urine, including 30% of organic matter, 82% of ammonium, and 100% of phosphorus.

2.5.2.2 Sewage Sludge: Energy, Agriculture, and Construction Applications

Based on demographic projections for the world and European Union countries, it is anticipated that the amount of sewage sludge produced by municipal wastewater treatment facilities will increase (Manara and Zabaniotou, 2012). The oldest approach to sewage sludge (biosolids) is land application, which uses wastewater treatment by-product in the form of fertilizer, which is put on the soil's surface or injected into it. Anaerobic or aerobic digestion, composting, drying, and chemical treatment (mainly alkaline treatment) are the methods that WWTP can use to treat biosolids prior to land application. The construction sector can also make use of sewage sludge, and sewage sludge ash can be used to make building materials like bricks and tiles. It can also be used as a raw material to make concrete, cement, and other products (Neczaj and Grosser, 2018). Ghodke et al. (2021) employed a WWTP's sewage sludge and pyrolyzed it at different temperatures. A maximum of 22.4% bio-oil, 18.9% pyrolysis gas, and 58.7% biochar were reported to be produced at 500 °C. In one study, sewage sludge was used to evaluate a continuous rapid microwave-assisted pyrolysis system. The results showed that the maximum bio-oil production of 41.39 weight percent was obtained at a pyrolysis temperature of 550 °C; nevertheless, 500 °C was the ideal pyrolysis temperature for the maximum overall energy recovery.

The High Heating Value (HHV) of gas (22.5 MJ/Nm^3) and bio-oil (20.61 MJ/kg) indicated that sewage sludge has the potential to be a biomass resource in the future (Zhou et al., 2018).

Thermochemical valorization techniques appear to have a lot of potential for recovering energy from sewage sludge. Energy recovery via thermochemical valorization procedures appears to be one of the key alternatives among different sludge treatment methods. The main objective of thermochemical procedures is to maximize the energy content of sewage sludge while reducing environmental impacts (Manara and Zabaniotou, 2012). In Europe and other nations, sewage sludge is widely used for land applications. In 2015, approximately 968 000 Mt of sewage sludge was employed in agriculture. Considering the case of Europe, sewage sludge's application in agriculture is widely practiced in France and Germany. The primary obstacles related to this kind of biosolids disposal include health and safety concerns, odor, and public acceptance (Neczaj and Grosser, 2018).

2.5.2.3 Nutrient Recovery by Aqua-species

Aquatic plants have the potential to absorb various nutrients from sewage to produce high-value products directly. Sewage can be converted into useful nutrients such as nitrogen, phosphorus, potassium, manganese, zinc, nickel, iron, and boron (Zhang et al., 2024). Aqua species such as microalgae, crops, wetland plants, and duckweed can also recover nutrients from wastewater, which can then be used in the form of fertilizers or animal feeds. Recovering nutrients using aqua species is an environmentally benign technique owing to its low energy requirement and the advantage of combining wastewater treatment and nutrient recycling (Neczaj and Grosser, 2018). In a study by Zhang et al. (2024), the ability to recover nutrients from municipal sewage was assessed for six aquatic plants: *Alternanthera philoxeroides, Pistia stratiotes, Salvinia natans, Ludwigia adscendens, Eichhornia crassipe*, and *Myriophyllum aquaticum*. Since *Alternanthera philoxeroides* could react to municipal sewage with different levels of phosphorus and nitrogen, it was said to be the most promising of the six aquatic plants for extracting these elements from municipal sewage. It was reported that 34.7% nitrogen and 66.0% phosphorus had been removed on the eighth day.

2.5.2.4 Wastewater Treatment by Constructed Wetlands

Constructed wetlands are engineered systems that are designed and constructed to treat wastewater using wetland vegetation, soils, and their associated microbial assemblages. They are designed in such a controlled environment so that they can give advantages of the similar processes that take place in natural wetlands (Vymazal, 2014). Due to their low construction, operational, and maintenance costs, these systems are currently being employed at the field scale to deal with wastewater treatment (Tamta et al., 2020). Constructed wetland application primarily focuses on treating water, not necessarily on nutrient recycling for secondary use (in agriculture) (Neczaj and Grosser, 2018). Some of these are specifically designed to treat wastewater while others are implemented for a range of alternative purposes including the utilization of treated wastewater as a water source for agricultural usage and creating/restoring wetland habitats for wildlife usage. (Vymazal, 2014).

An innovative approach was used by Tamta et al. (2020) to reduce the land footprint by treating high-strength household wastewater and producing power at the same time, utilizing artificial wetlands and microbial fuel cells stacked vertically in a 12-cell stack. At 0.08 kΩ external resistance, 98.5% chemical oxygen demand (COD) reduction was accomplished in this investigation. The highest removal efficiencies of nitrate (NO_3^-) and ammonium (NH_4^+) were 90.4% and 86.9%, respectively, at external resistances of 0.08 and 15 kΩ. A current density of 48.56 mA/m^3 was attained by the serially coupled wetland–microbial fuel cells with an external resistance of 0.33 kΩ.

2.5.2.5 Treated/Reclaimed Water

The use of treated wastewater for irrigation and construction purposes has been well-known for many years, and it has the potential to replace agricultural demand while also reducing local water stress. It is quite economical to use reclaimed water as a resource for irrigation. It can also be used for toilet flushing for industrial and construction purposes (Hurtado and Berbel, 2024; Neczaj and Grosser, 2018).

Reclaimed water, if combined with nutrient management, proves to be a highly effective approach to increasing agricultural productivity, reducing environmental impact, and promoting the sustainability of water resources (Hurtado and Berbel, 2024). Reclaimed water utilized for irrigation has the potential to improve soil fertility and encourage plant development. Because it contains a range of nutrients, such as potassium, phosphate, and nitrogen, it can lower the need for fertilizer and lower the cost of producing crops (Lyu et al., 2016).

Utilizing recycled water for irrigation has several benefits for nutrient management, including a decrease in fertilizer usage. Rich in nutrients, reclaimed water can either fully or partially replace traditional fertilizers used in crop production. It usually consists of nitrogen, phosphorus, and potassium—remaining elements from human and industrial activity. Reducing the use of commercial fertilizers, which are typically costly and subject to price swings, can result in significant financial savings for farmers (Hurtado and Berbel, 2024).

2.5.3 Economic Analysis of Wastewater Treatment

The economic viability of WWTP in wastewater treatment indicates that all of the advantages of this procedure exceed all of its expenses, proving that wastewater treatment is advantageous from both an environmental and financial perspective (Molinos-Senante et al., 2010). Chen and Wang (2009) developed a net benefit value (NBV) model for evaluating the cost-benefit of a decentralized wastewater treatment and reuse (DESAR) system after conducting a study on the subject.

$$\text{NBV} = \sum B_i - \sum C_i$$

where B_i is the value of benefit item i and C_i is the value of cost item i. For a particular system, if the calculating result is NBV > 0, the project is considered economically viable; if the calculating result is NBV < 0, the project is considered economically not viable. The higher the NBV, the more advantageous the project.

2.6 Case Studies

2.6.1 Designing a Sustainable Bioenergy Generation Network Based on Managing MSW in the Face of Uncertainty

Effective MSW treatment is essential due to the significant rate of MSW generation (Figure 2.5). It lessens the effects on the surroundings by lowering pollution and resource depletion simultaneously (Seferlis et al., 2021). MSW recycling is crucial to shift the paradigm of the linear economy to a circular one. It examines and separates MSW into energy technologies both directly and indirectly using the CE viewpoint (Hoang et al., 2021). MSW is burned for heat recovery using the direct approach. The indirect approach is made possible by the range of valorized products, including digestate, humus, bio-oil, charcoal, and syngas, which include thermochemical and biochemical processes more appealing despite being more complex. The contributions of waste-to-energy production are examined, and the benefits of MSW valuation, management of hazardous materials, and the creation of models for the CE are considered. It discusses current challenges, future prospects, and research gaps. The findings aim to maintain a healthy environment, support adopting the CE, and promote MSW reuse and value-adding. Discussing waste management perspectives expands and extends the range of technologies within the broader plan for creating a circular bioeconomy (Zsófia Závodi and Jiri, 2011).

2.6.1.1 Key Concerns, Sources, and MSW Composition

MSW is gathered from various locations, including factories, industries, homes, workplaces, schools, marketplaces, and retail establishments. MSW may contain a range of inorganic and organic compounds, such as polymers

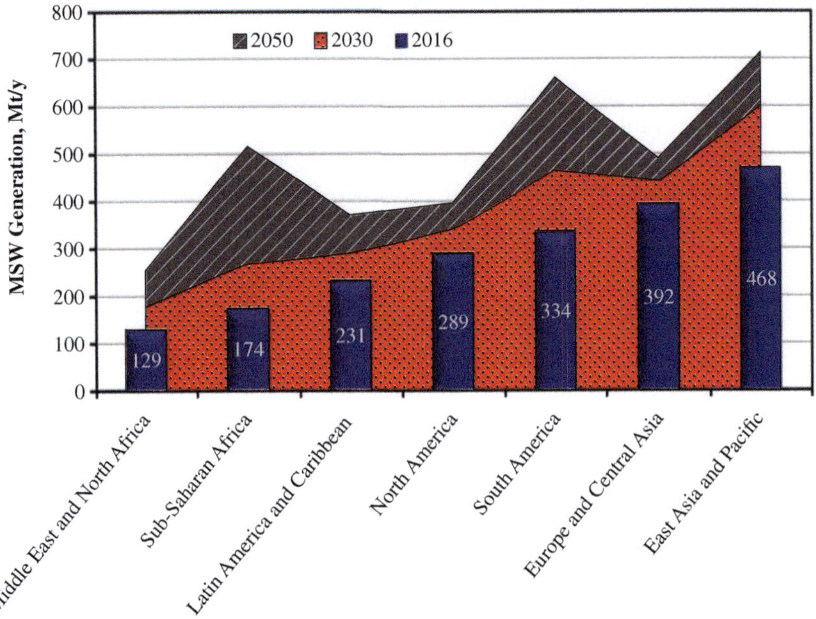

Figure 2.5 The annual increase in MSW for certain nations and regions worldwide. *Source:* Adapted from Kaza et al. (2018).

and nonrenewable resources, or a combination (Gundupalli et al., 2017). The components of MSW can be divided into seven main divisions (Figure 2.6): inerts, glass, metals, paper and board, plastics, organics, and textiles. The rest is miscellaneous (De Morais Lima et al., 2019). It is essential to comprehend the quantity and properties of the MSW gathered to streamline the handling procedure and optimize the energy recovery using the best W2E techniques. The physicochemical characteristics and calorific value of MSW are essential for producing a lot of energy and leaving no residue after treatment. According to ASTM D5231, the pie chart displays the disbursement of each component by respective shares (Rezaei et al., 2018).

2.6.1.2 Technologies for Turning MSW into Energy

By decreasing biodegradables to avoid CH_4 emissions, converting non-recyclable trash into heat or power, and reducing landfill volume, W2E techniques like incineration, pyrolysis, and anaerobic digestion efficiently process MSW. These techniques provide a sustainable solution for both business and residential waste. Table 2.1 describes different methods and their advantages and disadvantages.

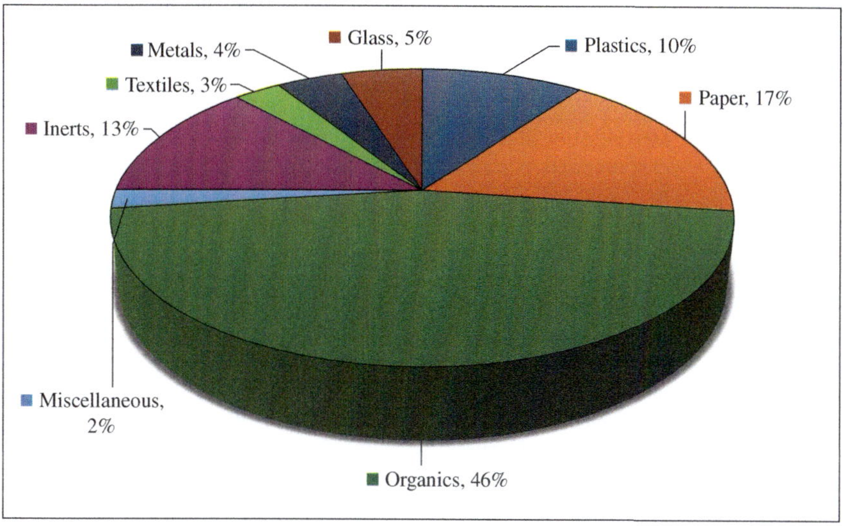

Figure 2.6 MSW composition worldwide. *Source:* Adapted from Sharma and Jain (2020).

Table 2.1 Benefits and drawbacks of different waste treatment methods.

Methods	Advantage	Disadvantage	Reference
Direct W2E process	• High-heating-value generated Refuse-derived fuel (RDF) that functions as a homogeneous fuel. • There is minimal by-product generation, which lowers pollution levels. • Greater efficiency because less extra air is needed. • Being utilized as additional fuel in coal-fired power plants. • The ability to extract non-combustible MSW makes handling it simple.	• Higher cost for the pretreatment process. • High cost for equipment maintenance. • Higher dangerous level.	(Malav et al., 2020; Paulraj et al., 2019; Moya et al., 2017)

(Continued)

Table 2.1 (Continued)

Methods	Advantage	Disadvantage	Reference
Incineration	• Minimization of contaminants, particularly for biomedical MSW. • Lowering the transportation costs and reducing the volume of MSW by 80–90%. • Considerable decrease in air pollution and the area used for disposing of MSW. • The capacity to recuperate heat for alternative uses, like home heating. • The capacity to eradicate bacteria and viruses due to its high-temperature functioning. • Construction sites could make use of generated ash. • Capacity to function in any weather conditions. • Ability to control odor and noise.	• High cost for installation. • Personnel needed for routine maintenance and operation. • Environmental contamination caused by ash, heavy metals, flue gases, and particles from the incineration process.	(Bhat et al., 2018; Moya et al., 2017; Shareefdeen et al., 2015)
Gasification	• The high net energy of syngas, up to 10 MJ/Nm^3. • High capacity for land conservation and potential use in generating heat or electricity. • Availability of high-temperature working range. • Mitigation of pollution in the environment.	• High startup costs and the need for specialized labor and technical personnel to run the process. • Energy recovery issues brought on by excessive MSW moisture.	(Malav et al., 2020; Piazzi et al., 2020; Safarian et al., 2020)
Pyrolysis	• Lower temperature in contrast to incineration. • Reduction of vol/wt of the MSW. • High energy recovery rate and minimal process area requirements. • Variety of things produced and how they are used.	• High startup costs and the need for specialized labor and technical personnel to run the process. • Difficulty in destroying hazardous organic compounds. • Excessive moisture from MSW raises issues with energy recovery.	(Chhabra et al., 2021; Yang et al., 2018)

Methods	Advantage	Disadvantage	Reference
Composting	• Serving as an organic input and soil conditioner in agriculture. • Reduction of the burden for landfills. • Acting as organic input in agriculture.	• High cost of transportation. • Low nutrient value when used as fertilizers. • Acting as an infectious agent's intermediary. • Requirement of a large area.	(Briassoulis et al., 2021; Zhou et al., 2020)
Anaerobic digestion	• Mitigation in the rate of volatile-solid. • End products that are stable, odorless, and have a high fertilizer value. • Easy functioning in the manufacture of gaseous fuel. • Low GHG and low capital cost.	• High pre-aeration power costs. • Poor sludge dewaterability. • Extremely sensitive to temperature fluctuations. • It causes odors in the case of ineffective handling.	(Bhat et al., 2018; Moya et al., 2017)
Landfill	• Boilers and power plants might consume 40–60% of the methane generated by landfills.	• Unpleasant odor from gas produced from landfill. • Causing fire and explosion risk.	(Bhat et al., 2018; IEA, 2008)
Bioethanol/ biodiesel production	• There could be no conflict between food security and MSW-derived bioethanol or biodiesel. • Reduction of GHG and climate change.	• Expensive synthesis and processing techniques.	(Kalyani and Pandey, 2014; Nair et al., 2017; Salati et al., 2013)
Microbial fuel cell	• Different MSWs are used to generate power without emitting any net CO_2. • Direct energy conversion from chemical to electrical, resulting in reduced energy loss. • More reliable and safer operation.	• High cost of microbial fuel cell materials. • Low electricity power.	(Budihardjo et al., 2021; Chandrasekhar et al., 2018; Gebreslassie et al., 2021; Karluvalı et al., 2015; Nawaz et al., 2020)

Source: Adapted from Kaza et al. (2018).

2.6.1.3 Views on Waste Management for a Circular Bioeconomy Energy Generation Plan

Source efficiency and CE models are crucial to turn MSW into high-value products. Sustainable and integrated MSW management techniques minimize expenses and their negative effects on the environment while streamlining operations. For biomass valorization, Solid-State Fermentation (SSF) is more cost-effective than Submerged Fermentation (SmF) (Cerda et al., 2017). Limited resources, socioeconomic considerations, and unsustainable disposal methods are some issues with MSW management in underdeveloped nations, underscoring the need for specialized approaches. W2E operations can be improved by valuing through pyrolysis, gasification, and energy recovery; nevertheless, pollution and public health issues must be addressed (Ferronato et al., 2018). Integrating waste models—like enhanced landfill mining—offers sustainable alternatives by treating garbage as an intermediary resource for potential future value. Policy backing, incentives, and public acceptability are essential to develop successful W2E solutions and promote circular business models (Okot-Okumu and Nyenje, 2011). Infrastructure for advanced garbage sorting is required to improve recyclable recovery and lower contamination (Yoshida, 2020). Carbon capture from waste treatment and novel bio-based processes presents opportunities to reduce greenhouse gas emissions and produce high-value products. Reliable market channels are necessary to reuse by-products, such as digestate and compost, to optimize the financial and ecological advantages of MSW treatment.

2.6.1.4 Future Directions in the Field

Environmental pollution and problems with energy supply are addressed by W2E technologies that produce energy from the organic part of MSW. Direct methods, such as incineration, use combustion to produce heat but less energy. Gasification and pyrolysis are examples of indirect methods that transform waste into energy intermediates and provide a diversity of outputs, including syngas and bio-oil. Anaerobic digestion, fermentation, and other biological processes are constrained by nonorganic waste and necessitate sophisticated sorting equipment, which is still in its infancy. W2E adoption is hampered by high sorting, equipment, and shipping expenses. Lean operations, integrating energy systems, and optimizing by-product recovery are all successful tactics. Encouraging W2E in the CE while reducing hazards and pollutants requires thorough waste utilization, safety procedures, and societal preparedness.

The sustainability of W2E technologies for MSW can be assessed in technology, economics, environment, and socio-political domains. Integrated waste management systems, sorting, and pre-processing plants enhance resource recovery and recycling rates. Waste minimization and household sorting through municipal programs, public awareness, and citizen engagement are crucial. Financial incentives and continued research are needed to advance W2E technologies. MSW treatment addresses environmental impacts while generating renewable energy with lower GHG emissions. Still, energy valorization should remain a "final resort" after exploring other processing routes like chemical and material recovery. CE principles prioritize product design for reuse, repurposing, and material recovery, though these processes also require energy and emit GHGs. Insights from global waste management practices and literature can improve understanding of trade-offs and waste prevention strategies.

2.6.2 Designing a Green Reverse Logistics Network for Medical Waste Management: A Case-based Approach to the CE Transition

Inadequate management of medical waste endangers the lives of Community members and pollutes the environment. Effective medical waste management requires proper planning and monitoring at each stage, including collection, separation, treatment, recycling, and disposal, as this promotes a circular approach (Kenny and Priyadarshini, 2021). For the first time, vehicles transporting infectious trash at treatment facilities are subjected to waiting time management using the queuing theory idea (Aydemir-Karadag, 2022). Furthermore, the suggested model employs a stochastic scenario-based strategy to handle waste generation uncertainty while concurrently addressing the locations of the centers and the routing of diverse vehicles (Windfeld and Brooks, 2015). The proposed model uses an upgraded augmented epsilon-constraint technique (AUGMECON2) to lower overall cost and population risk (Koo and Jeong, 2015). A case study from the province of Alborz is used to evaluate the efficacy of the suggested model and solution method. It contains information from six hospitals, three potential disposal sites, four potential recycling sites, three prospective treatment facilities, and three designated collecting stations (Makajic-Nikolic et al., 2016).

Medical waste generated to diagnose, cure, and immunize humans and animals is what the WHO refers to as this. According to the WHO, medical waste can be classified into seven major categories: "pathological waste," "infection waste," "chemicals," "pharmaceuticals," "sharps," "radioactive waste," and "genotoxic waste" (Rolewicz-Kalińska, 2016).

According to the U.S. EPA's 2013 definition of hospital garbage, it includes all waste products produced at healthcare institutions, including clinics, hospitals, dentistry offices, doctor's offices, veterinary clinics, blood banks, and medical research labs (Rolewicz-Kalińska, 2016). By optimizing logistical networks, the circular transition reduces waste and expenses while facilitating medical waste management driven by the CE (Alshraideh and Abu Qdais, 2016).

This field has attracted some researchers since the 2019 coronavirus (COVID-19) outbreak. In light of this, Kargar et al. (2020b) proposed a unique multi-objective mathematical model to design a recovery logistics system to handle medical waste during the COVID-19 outbreak (Valizadeh et al., 2021). The government protection strategy in COVID-19 hazardous infectious waste collection was examined using a two-level planning model.

2.6.2.1 Proposed Model

This part introduces the bi-objective MILP CE transition model for the management of medical waste that incorporates aspects of the queuing theory and concerns regarding location routing. This multi-item, multi-period network comprises hospitals and clinics (h), multiple levels of collection centers (c), treatment facilities (t), recycling units (r), and disposal sites (d) (Khalili Nasr et al., 2021). In order to determine the Pareto optimal solution or the precise Pareto set, this work applies the AUGMECON2 to the bi-objective model. Interestingly, AUGMECON2 is an improved epsilon-constraint method that avoids the creation of unwanted or ineffective solutions by employing an acceleration technique (Govindan et al., 2022).

2.6.2.2 Validation of Model

This part assesses the effectiveness of the suggested CE transition mathematical model and solution methodology utilizing actual data. In Alborz province, information is used for this purpose from six hospitals: Alborz, Bahonar, Madani, Kosar, Rejaei, and Kamali; three possible collection centers: Khor amdasht, Meshkindasht, and Malekabad; three possible treatment centers: Ramjin, Zakiabad, and Pashand; four possible recycling centers: Nazarabad, Hashtgerd, Saeidabad, and Baharestan; and three possible disposal centers: Halqeh Darreh, Aliabad, and Dangizak. Among the infectious wastes examined in this study are glass, metal, and plastic. The time horizon is one week, and there are four phases because the trash is collected from hospitals four times a week.

2.6.2.3 Result and Discussion

Mavrotas and Florios (2013) introduced the AUGMECON2 approach to solve the proposed bi-objective model. They use the lexicographic approach to form a payoff table (Basirati et al., 2019).

In the proposed paradigm, the first objective function is considered the objective function 1, while the objective function 2 is relegated to the list of constraints (Mavrotas, 2009). Therefore, the enhanced augmented epsilon-constraint approach (AUGMECON2) is employed in the suggested model. Several different equations given by Govindan et al. (2021) are used to solve the problem.

When utilizing a CPLEX solver to run the suggested model in GAMS software, ten grid points are taken into consideration. First, the lexicographic approach was used to calculate the values of ε_2 by determining the upper and lower bounds of objective function 2. Table 2.2 provides the ε_2 values and the ideal value for every iteration, and Figure 2.7 shows their graphical representation. It should be mentioned that 10^{-6} is the value of eps.

Table 2.2 Results obtained from AUGMECON2.

Iteration (Grid point)	ε_2	Optimal value of Z_1	Optimal value of Z_2
1	1 283 432 + 0 × 68 098 = 1 283 432	1 987 011	1 283 432
2	1 283 432 + 1 × 68 098 = 1 351 530	1 682 732	1 350 988
3	1 283 432 + 2 × 68 098 = 1 419 628	1 494 830	1 419 626
4	1 283 432 + 3 × 68 098 = 1 487 726	1 300 642	1 470 831
5	1 283 432 + 4 × 68 098 = 1 555 824	1 163 272	1 543 108
6	1 283 432 + 5 × 68 098 = 1 623 922	1 069 814	1 595 862
7	1 283 432 + 6 × 68 098 = 1 692 020	971 193	1 664 367
8	1 283 432 + 7 × 68 098 = 1 760 118	923 674	1 748 001
9	1 283 432 + 8 × 68 098 = 1 828 216	886 909	1 820 633
10	1 283 432 + 9 × 68 098 = 1 896 314	812 263	1 896 314

Source: Adapted from Mavrotas and Florios (2013).

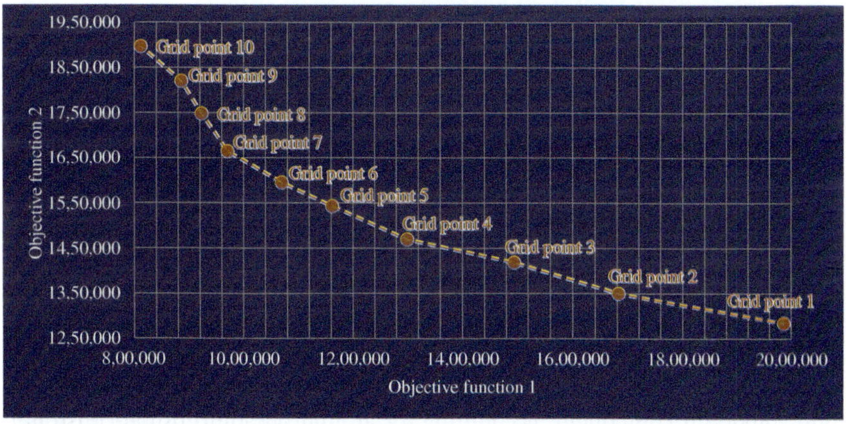

Figure 2.7 Pareto frontier obtained from AUGMECON2. *Source:* Adapted from Mavrotas and Florios (2013).

2.6.2.4 Managerial Insights

The literature on medical waste management frequently uses mathematical programming models to handle problems such as controlling vehicle fuel use and minimizing population exposure, as demonstrated by the works of Kargar et al. (2020a) and Tirkolaee et al. (2021). This work combines these elements with queuing theory-based truck waiting time management to reduce infectious and fuel-related pollution. Queuing theory and mathematical programming are combined to build a dependable, affordable, and environmentally friendly network that supports the CE transition. Using the epsilon-constraint method provides decision-makers with the best options for budgets and policies. The model's adaptable framework can be used for hazardous waste management, industry, and municipalities (Baradaran et al., 2020; Govindan et al., 2021).

The demand for medical services has grown due to the world's expanding population, resulting in a rise in medical waste production and complicated management issues. This study incorporates location-routing issues and queuing theory under uncertainty into its proposal for a green reverse logistics network for medical waste disposal. A bi-objective MILP CE transition model uses a scenario-based strategy for uncertain waste generation to minimize overall costs and population exposure. The model was tested in the province of Alborz using data from six hospitals, resulting in a $1 069 814 cost and 1 595 862 infection exposures. Decision-makers chose the effective solutions produced by the AUGMECON2 approach based on financial limitations. Future research should concentrate on creating metaheuristic algorithms to address large-scale NP-hard issues.

2.6.3 Wastewater Treatment Economic Feasibility Study: A Cost-benefit Evaluation

The economic feasibility of 22 WWTPs was evaluated using a cost-benefit analysis (Molinos-Senante et al., 2010). Twenty-two WWTPs in Valencia, on the Mediterranean coast of Spain, were assessed in this study. The area wastewater treatment authority supplied statistics for 2007. Each WWTP handles between one million and eight million m^3/year. Four undesirable outputs: suspended solids, COD, nitrogen, phosphorus, and the desired output of treated water define the wastewater treatment process. The following resources are needed for the treatment: electricity, personnel, chemicals, waste disposal, and upkeep.

Each of the 22 WWTPs' operational cost distribution is shown in Figure 2.8 as a weighted average depending on the volume of treated wastewater. Staff was the most important component, accounting for around one-third of total expenses (32%). While energy and maintenance make up 18% and 21% of the total cost, respectively, waste management and reagent costs make up nearly equal shares at 15% and 14%.

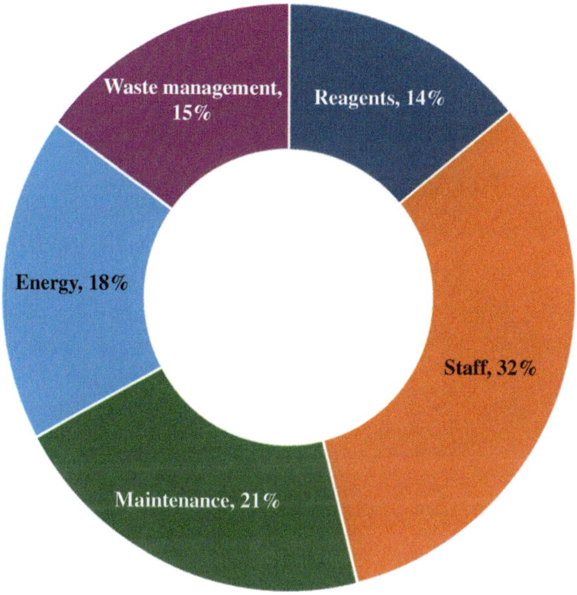

Figure 2.8 Each of the 22 WWTPs' operating costs is distributed in weighted averages according to the volume of treated wastewater. *Source:* Molinos-Senante et al. (2010)/ with permission of Elsevier.

The environmental benefits gained after removing pollutants are depicted in Figure 2.9. Nitrogen removal accounts for 66% of total profit; it has the most associated environmental benefits. With 21% of the overall benefit, phosphorus is the second most significant pollutant. The environmental gain that COD accounts for is only 12.4% because even if a significant amount of organic matter (COD) is eliminated, its shadow price is relatively low. Similarly, suspended solids (SS), which, even after eliminating significant amounts from wastewater, contribute only 0.2% to environmental benefits due to their extremely low shadow price.

The overall environmental benefit resulting after wastewater treatment was calculated by considering the volume of pollutants eliminated during the treatment process (kg/year), the volume of the treated wastewater (m^3/year), and the shadow price (€/kg) of each pollutant (SS, N, P, and COD). The 22 WWTPs provided varying environmental benefits with a minimum value of 0.0099 €/m^3 and a maximum value of 1.0039 €/m^3. The weighted average of environmental benefits based on the treated volume of wastewater was reported to be 0.3609 €/m^3.

The removal of SS from wastewater adds the least to the overall environmental advantages, according to the analysis and findings of the 22 WWTPs situated in the Valencia region on the Mediterranean coast of Spain. On the other hand, the greatest environmental advantages were linked to the removal of fertilizers from wastewater.

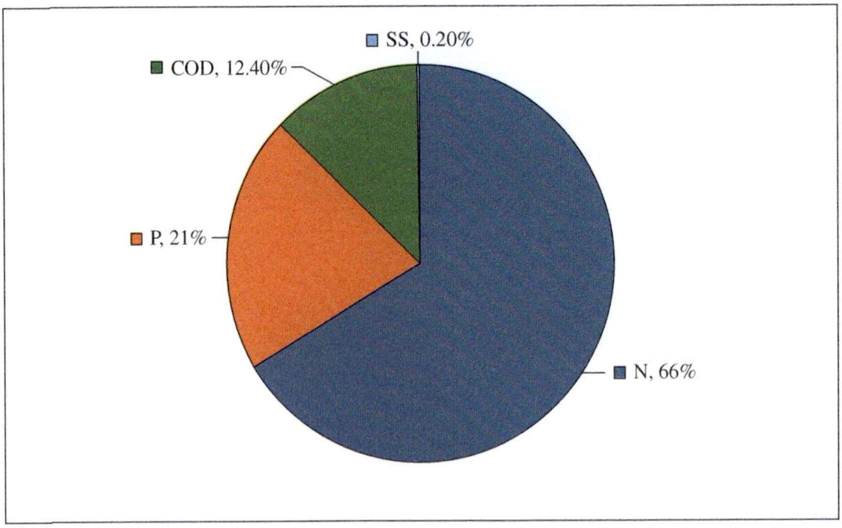

Figure 2.9 Benefits to the environment from removing different wastewater contaminants. *Source:* Molinos-Senante et al. (2010)/with permission of Elsevier.

The cost-benefit analysis supplied information on the treatment process's economic feasibility. The average net profit for each of the three possible scenarios involving the sale of regenerated water was determined. Even though the WWTPs' operational costs and environmental benefits from treatment differ greatly, the average net profit for the three scenarios was positive. This demonstrates that even if reclaimed water was not sold, the operation of WWTPs under study was economically viable.

2.7 Summary

In order to meet the growing environmental, economic, and social concerns, the study highlights the urgent need for a paradigm shift in waste treatment from linear to circular economic solutions. Although historically common, linear techniques frequently put disposal ahead of resource recovery, which results in resource depletion, environmental damage, and increased expenses. CE techniques, on the other hand, minimize waste and pollution by focusing on sustainability through resource recovery, recycling, and reuse. These models combine concepts like industrial symbiosis and digital innovation with cutting-edge technology like waste valorization, sophisticated recycling techniques, and energy recovery.

Circular methods have many advantages for the environment and the economy, but they also have drawbacks, such as high implementation costs, limited technology, and a lack of governmental support. Strong policy frameworks, public–private partnerships, and international cooperation are needed to remove these obstacles. According to the chapter's conclusion, implementing the concepts of CE is crucial for attaining long-term environmental and economic sustainability as well as for reducing the effects of the fast generation of garbage. In addition to ensuring effective resource use, a circular strategy supports global sustainability objectives and gets society ready for a more resilient and sustainable future.

Review Questions

1. Why effective waste management is necessary for developing and developed countries?
2. What are the different types of waste, their associated issues, and treatment strategies?
3. Define Linear Economy model for waste treatment and its significance and limitations.
4. Define CE model for waste management, its significance, and limitations.
5. Brief about current technologies for waste treatment for CE model.
6. What are the various challenges in implementing CE and their probable solutions?
7. What are the future/advanced technologies for wastewater treatment and their scope toward CE?
8. What are the possible uses of Sewage sludge and their limitations?
9. What is the Net Benefit Value (NBV) toward economic analysis of wastewater?
10. Mention various W2E technologies, their advantages, and disadvantages.

References

Akdag, Y., Altinbas, M., and Beler-Baykal, B. (2024). Nutrient recovery from source-separated urine via sorption and a comparative investigation on the improvement of the residual liquid quality. *Journal of Environmental Management*, **370**, 122507. https://doi.org/10.1016/j.jenvman.2024.122507

Alao, J.O. (2023). Impacts of open dumpsite leachates on soil and groundwater quality. *Groundwater for Sustainable Development*, **20**, 100877. https://doi.org/10.1016/j.gsd.2022.100877

Alshraideh, H. and Abu Qdais, H. (2016). Stochastic modeling and optimization of medical waste collection in Northern Jordan. *Journal of Material Cycles and Waste Management*, **19**(2), 743–753. https://doi.org/10.1007/s10163-016-0474-3

Arora, S. and Saraswat, S. (2021). Vermifiltration as a natural, sustainable and green technology for environmental remediation: a new paradigm for wastewater treatment process. *Current Research in Green and Sustainable Chemistry*, **4**(6), 100061. https://doi.org/10.1016/j.crgsc.2021.100061

Ashima, R., Haleem, A., Bahl, S. et al. (2021). Automation and manufacturing of smart materials in additive manufacturing technologies using Internet of Things towards the adoption of industry 4.0. *Materials Today: Proceedings*, **45**, 5081–5088. https://doi.org/10.1016/j.matpr.2021.01.583

Attri, P., Garg, S., and Ratan, J.K. (2022). Degradation of 4-aminopyridine using bio-mediated Ag-doped Cu_2O nanoparticles under visible light. *Nanotechnology for Environmental Engineering*, **7**(2), 415–423. https://doi.org/10.1007/s41204-021-00182-5

Aydemir-Karadag, A. (2022). Bi-objective adaptive large neighborhood search algorithm for the healthcare waste periodic location inventory routing problem. *Arabian Journal for Science and Engineering*, **47**(3), 3861–3876. https://doi.org/10.1007/s13369-021-06106-4

Baradaran, V., Nikzamir, M., and Panahi, Y. (2020). Designing a logistic network for hospital waste management: a benders decomposition algorithm. *Environmental Engineering and Management Journal*, **19**(11), 1937–1956. https://doi.org/10.30638/eemj.2020.184

Basirati, M., Akbari Jokar, M.R., and Hassannayebi, E. (2019). Bi-objective optimization approaches to many-to-many hub location routing with distance balancing and hard time window. *Neural Computing and Applications*, **32**(17), 13267–13288. https://doi.org/10.1007/s00521-019-04666-z

Bhat, R.A., Dar, S.A., Dar, D.A. et al. (2018). Municipal solid waste generation and current scenario of its management in India. *International Journal of Advance Research in Science and Engineering*, **7**, 419–431.

Bhatt, M., Kumar, S., Soneja, M. et al. (2022). *Emissions Gap Report 2024* (0034-6233 (Print), 2084-9834 (Electronic), 0034-6233 (Linking)). Retrieved from https://www.unep.org/resources/emissions-gap-report-2024

Bressanelli, G., Perona, M., and Saccani, N. (2018). Challenges in supply chain redesign for the Circular Economy: a literature review and a multiple case study. *International Journal of Production Research*, **57**(23), 7395–7422. https://doi.org/10.1080/00207543.2018.1542176

Briassoulis, D., Pikasi, A., and Hiskakis, M. (2021). Organic recycling of post-consumer /industrial bio-based plastics through industrial aerobic composting and anaerobic digestion – Techno-economic sustainability criteria and indicators.

Polymer Degradation and Stability, **190**, 109642. https://doi.org/10.1016/j.polymdegradstab.2021.109642

Budihardjo, M.A., Syafrudin, Effendi, A.J. et al. (2021). Waste valorization using solid-phase microbial fuel cells (SMFCs): recent trends and status. *Journal of Environmental Management*, **277**, 111417. https://doi.org/10.1016/j.jenvman.2020.111417

Bułkowska, K., Zielińska, M., and Bułkowski, M. (2023). Implementation of blockchain technology in waste management. *Energies*, **16**(23), 7742. https://doi.org/10.3390/en16237742

Cantú, A., Aguiñaga, E., and Scheel, C. (2021). Learning from failure and success: the challenges for circular economy implementation in SMEs in an emerging economy. *Sustainability*, **13**(3), 1529. https://doi.org/10.3390/su13031529

Carvalho, J., Nascimento, L., Soares, M. et al. (2022). Life Cycle Assessment (LCA) of biochar production from a circular economy perspective. *Processes*, **10**(12), 2684. https://doi.org/10.3390/pr10122684

Casanovas, A., Galvis, A., and Llorca, J. (2015). Catalytic steam reforming of olive mill wastewater for hydrogen production. *International Journal of Hydrogen Energy*, **40**(24), 7539–7545. https://doi.org/10.1016/j.ijhydene.2014.11.083

Cazcarro, I., López-Morales, C.A., and Duchin, F. (2016). The global economic costs of the need to treat polluted water. *Economic Systems Research*, **28**(3), 295–314. https://doi.org/10.1080/09535314.2016.1161600

Cerda, A., Gea, T., Vargas-Garcia, M.C. et al. (2017). Towards a competitive solid state fermentation: cellulases production from coffee husk by sequential batch operation and role of microbial diversity. *Science of The Total Environment*, **589**, 56–65. https://doi.org/10.1016/j.scitotenv.2017.02.184

Malav, L.C., Yadav, K.K., Gupta, N. et al. (2020). A review on municipal solid waste as a renewable source for waste-to-energy project in India: current practices, challenges, and future opportunities. *Journal of Cleaner Production*, **277**, 123227. https://doi.org/10.1016/j.jclepro.2020.123227

Chandrasekhar, K., Kadier, A., Kumar, G. et al. (2018). Challenges in microbial fuel cell and future scope. In: *Microbial Fuel Cell: A Bioelectrochemical System that Converts Waste to Watts* (ed. D. Das), 483–499. Cham: Springer International Publishing.

Cheah, C.G., Chia, W.Y., Lai, S.F. et al. (2022). Innovation designs of industry 4.0 based solid waste management: machinery and digital circular economy. *Environmental Research*, **213**, 113619. https://doi.org/10.1016/j.envres.2022.113619

Chen, R. and Wang, X.C. (2009). Cost-benefit evaluation of a decentralized water system for wastewater reuse and environmental protection. *Water Science & Technology*, **59**(8), 1515–1522. https://doi.org/10.2166/wst.2009.156

Chertow, M.R. (2008). "Uncovering" industrial symbiosis. *Journal of Industrial Ecology*, **11**(1), 11–30. https://doi.org/10.1162/jiec.2007.1110

Chhabra, V., Parashar, A., Shastri, Y. et al. (2021). Techno-economic and life cycle assessment of pyrolysis of unsegregated urban municipal solid waste in India. *Industrial & Engineering Chemistry Research*, **60**(3), 1473–1482. https://doi.org/10.1021/acs.iecr.0c04746

CPCB. (2022). *Annual Report on Solid Waste Management (2020–21)*. Retrieved from Delhi: https://cpcb.nic.in/uploads/MSW/MSW_AnnualReport_2020-21.pdf

Cudjoe, D., Zhu, B., Nketiah, E. et al. (2021). The potential energy and environmental benefits of global recyclable resources. *Science of The Total Environment*, **798**, 149258. https://doi.org/10.1016/j.scitotenv.2021.149258

D'Angelo, S. and Meccariello, R. (2021). Microplastics: a threat for male fertility. *International Journal of Environmental Research and Public Health*, **18**(5), 2392. https://doi.org/10.3390/ijerph18052392

De Corato, U. (2020). Agricultural waste recycling in horticultural intensive farming systems by on-farm composting and compost-based tea application improves soil quality and plant health: a review under the perspective of a circular economy. *Science of The Total Environment*, **738**, 139840. https://doi.org/10.1016/j.scitotenv.2020.139840

De Morais Lima, P., Olivo, F., Paulo, P.L. et al. (2019). Life cycle assessment of prospective MSW management based on integrated management planning in Campo Grande, Brazil. *Waste Management*, **90**, 59–71. https://doi.org/10.1016/j.wasman.2019.04.035

Debrah, J.K., Teye, G.K., and Dinis, M.A.P. (2022). Barriers and challenges to waste management hindering the circular economy in sub-Saharan Africa. *Urban Science*, **6**(3), 57. https://doi.org/10.3390/urbansci6030057

Deeksha, D. (2022). Journey of waste management regulations in India: a historical account. *International Journal of Basic and Applied Sciences*, **11**, 96–101. https://doi.org/10.13140/RG.2.2.14106.11200

Domenech, T., Bleischwitz, R., Doranova, A. et al. (2019). Mapping industrial symbiosis development in Europe_ typologies of networks, characteristics, performance and contribution to the circular economy. *Resources, Conservation and Recycling*, **141**, 76–98. https://doi.org/10.1016/j.resconrec.2018.09.016

Ellen, M. (2020). *What is a circular economy*? Retrieved from https://www.ellenmacarthurfoundation.org/topics/circular-economy-introduction/overview

Fatimah, Y.A., Govindan, K., Murniningsih, R. et al. (2020). Industry 4.0 based sustainable circular economy approach for smart waste management system to achieve sustainable development goals: a case study of Indonesia. *Journal of Cleaner Production*, **269**, 122263. https://doi.org/10.1016/j.jclepro.2020.122263

Ferronato, N., Gorritty Portillo, M.A., Guisbert Lizarazu, E.G. et al. (2018). The municipal solid waste management of La Paz (Bolivia): challenges and opportunities for a sustainable development. *Waste Management & Research*, **36**(3), 288–299. https://doi.org/10.1177/0734242X18755893

Ferronato, N., Maalouf, A., Mertenat, A. et al. (2024). A review of plastic waste circular actions in seven developing countries to achieve sustainable development goals. *Waste Management & Research*, **42**(6), 436–458. https://doi.org/10.1177/0734242X231188664

Foletto, E.L., Santos, W.R.B.d., Mazutti, M.A. et al. (2013). Production of struvite from beverage waste as phosphorus source. *Materials Research*, **16**(1), 242–245. https://doi.org/10.1590/s1516-14392012005000152

Freestone, I.C. (2015). The recycling and reuse of roman glass: analytical approaches. *Journal of Glass Studies*, **57**, 29–40.

Gebreslassie, T.R., Nguyen, P.K.T., Yoon, H.H. et al. (2021). Co-production of hydrogen and electricity from macroalgae by simultaneous dark fermentation and microbial fuel cell. *Bioresource Technology*, **336**, 125269. https://doi.org/10.1016/j.biortech.2021.125269

Geert, B. (2019). *Chemical recycling and its CO_2 reduction potential (Online pdf)*. from CE Delft chrome-extension://efaidnbmnnnibpcajpcglclefindmkaj/https://cedelft.eu/wp-content/uploads/sites/2/2021/04/Chemical-recycling-and-its-CO2-reduction-potential.pdf

Ghisellini, P., Cialani, C., and Ulgiati, S. (2016). A review on circular economy: the expected transition to a balanced interplay of environmental and economic systems. *Journal of Cleaner Production*, **114**, 11–32. https://doi.org/10.1016/j.jclepro.2015.09.007

Ghodke, P.K., Sharma, A.K., Pandey, J.K. et al. (2021). Pyrolysis of sewage sludge for sustainable biofuels and value-added biochar production. *Journal of Environmental Management*, **298**, 113450. https://doi.org/10.1016/j.jenvman.2021.113450

Govindan, K., Nasr, A.K., Mostafazadeh, P. et al. (2021). Medical waste management during coronavirus disease 2019 (COVID-19) outbreak: a mathematical programming model. *Computers & Industrial Engineering*, **162**, 107668. https://doi.org/10.1016/j.cie.2021.107668

Govindan, K., Nosrati-Abarghooee, S., Nasiri, M.M. et al. (2022). Green reverse logistics network design for medical waste management: a circular economy transition through case approach. *Journal of Environmental Management*, **322**, 115888. https://doi.org/10.1016/j.jenvman.2022.115888

Gundupalli, S.P., Hait, S., and Thakur, A. (2017). A review on automated sorting of source-separated municipal solid waste for recycling. *Waste Management*, **60**, 56–74. https://doi.org/10.1016/j.wasman.2016.09.015

Gupta, S., Mohan, K., Prasad, R. et al. (1998). Solid waste management in India: options and opportunities. *Resources, Conservation and Recycling*, **24**(2), 137–154. https://doi.org/10.1016/s0921-3449(98)00033-0

Gupta, V., Goel, S., and Rupa, T.G. (2016). Good governance and solid waste management: an overview of legislative regulations in India. *Journal of Business and Management Studies*, **2**(1), 1–9.

Hatley, G.A., Pahlow, M., Bello-Mendoza, R. et al. (2024). Identifying leverage points using material flow analysis to circularise resources from urban wastewater and organic waste. *Journal of Environmental Management*, **369**, 122215. https://doi.org/10.1016/j.jenvman.2024.122215

Hemda, G. and Huw, J. (2008). Bridging the gap between knowledge and practice. *Chemistry International – Newsmagazine for IUPAC*, **30**(4), 7–12. https://doi.org/10.1515/ci.2008.30.4.7

Hemidat, S., Achouri, O., El Fels, L. et al. (2022). Solid waste management in the context of a circular economy in the MENA region. *Sustainability*, **14**(1), 480. https://doi.org/10.3390/su14010480

Ho, O., Iyer-Raniga, U., Sadykova, C. et al. (2024). A conceptual model for integrating circular economy in the built environment: an analysis of literature and local-based case studies. *Journal of Cleaner Production*, **449**, 141516. https://doi.org/10.1016/j.jclepro.2024.141516

Hoang, A.T., Pham, V.V., and Nguyen, X.P. (2021). Integrating renewable sources into energy system for smart city as a sagacious strategy towards clean and sustainable process. *Journal of Cleaner Production*, **305**, 127161. https://doi.org/10.1016/j.jclepro.2021.127161

Hoang, A.T., Varbanov, P.S., Nižetić, S. et al. (2022). Perspective review on municipal solid waste-to-energy route: characteristics, management strategy, and role in circular economy. *Journal of Cleaner Production*, **359**, 131897. https://doi.org/10.1016/j.jclepro.2022.131897

Hong, I., Park, S., Lee, B. et al. (2014). IoT-based smart garbage system for efficient food waste management. *Scientific World Journal*, **2014**, 646953. https://doi.org/10.1155/2014/646953

Huang, H., Song, Q., Wang, W. et al. (2012). Treatment of anaerobic digester effluents of nylon wastewater through chemical precipitation and a sequencing batch reactor process. *Journal of Environmental Management*, **101**, 68–74. https://doi.org/10.1016/j.jenvman.2011.12.035

Hurtado, A.R. and Berbel, J. (2024). A cost–benefit analysis of reclaimed water and desalinated seawater for irrigation in Axarquia, Southern Spain. *Land*, **13**(12), 2156. https://doi.org/10.3390/land13122156

IEA. (2008). *Turning a Liability into an Asset: Landfill Methane Recovery in India*. Retrieved from Paris: https://www.iea.org/reports/turning-a-liability-into-an-asset-landfill-methane-recovery-in-india

Jagun, Z.T., Daud, D., Ajayi, O.M. et al. (2023). Waste management practices in developing countries: a socio-economic perspective. *Environmental Science and Pollution Research*, **30**(55), 116644–116655. https://doi.org/10.1007/s11356-022-21990-5

Jain, K., Patel, A.S., Pardhi, V.P. et al. (2021). Nanotechnology in wastewater management: a new paradigm towards wastewater treatment. *Molecules*, **26**(6), 1797. https://doi.org/10.3390/molecules26061797

Kabeyi, M.J.B., Olanrewaju, O.A., and Hustad, J.E. (2023). Review and design overview of plastic waste-to-pyrolysis oil conversion with implications on the energy transition. *Journal of Energy*, **2023**, 1–25. https://doi.org/10.1155/2023/1821129

Kalyani, K.A. and Pandey, K.K. (2014). Waste to energy status in India: a short review. *Renewable and Sustainable Energy Reviews*, **31**, 113–120. https://doi.org/10.1016/j.rser.2013.11.020

Kargar, S., Paydar, M.M., and Safaei, A.S. (2020a). A reverse supply chain for medical waste: a case study in Babol healthcare sector. *Waste Management*, **113**, 197–209. https://doi.org/10.1016/j.wasman.2020.05.052

Kargar, S., Pourmehdi, M., and Paydar, M. M. (2020b). Reverse logistics network design for medical waste management in the epidemic outbreak of the novel coronavirus (COVID-19). *Science of The Total Environment*, **746**, 141183. https://doi.org/10.1016/j.scitotenv.2020.141183

Karluvalı, A., Köroğlu, E.O., Manav, N. et al. (2015). Electricity generation from organic fraction of municipal solid wastes in tubular microbial fuel cell. *Separation and Purification Technology*, **156**, 502–511. https://doi.org/10.1016/j.seppur.2015.10.042

Kaza, S., Yao, L.C., Bhada-Tata, P. et al. (2018). A global snapshot of solid waste management to 2050. In: *What a Waste 2.0*. Washington, DC: Urban Development.

Kenny, C. and Priyadarshini, A. (2021). Review of current healthcare waste management methods and their effect on global health. *Healthcare (Basel)*, **9**(3), 284. https://doi.org/10.3390/healthcare9030284

Khalili Nasr, A., Tavana, M., Alavi, B. et al. (2021). A novel fuzzy multi-objective circular supplier selection and order allocation model for sustainable closed-loop supply chains. *Journal of Cleaner Production*, **287**, 124994. https://doi.org/10.1016/j.jclepro.2020.124994

Kirchherr, J., Piscicelli, L., Bour, R. et al. (2018). Barriers to the circular economy: evidence from the European Union (EU). *Ecological Economics*, **150**, 264–272. https://doi.org/10.1016/j.ecolecon.2018.04.028

Klotz, M., Oberschelp, C., Salah, C. et al. (2024). The role of chemical and solvent-based recycling within a sustainable circular economy for plastics. *Science of The Total Environment*, **906**, 167586. https://doi.org/10.1016/j.scitotenv.2023.167586

Koo, J.K. and Jeong, S.I. (2015). Sustainability and shared smart and mutual – green growth (SSaM-GG) in Korean medical waste management. *Waste Management & Research*, **33**(5), 410–418. https://doi.org/10.1177/0734242X15574561

Korhonen, J., Nuur, C., Feldmann, A. et al. (2018). Circular economy as an essentially contested concept. *Journal of Cleaner Production*, **175**, 544–552. https://doi.org/10.1016/j.jclepro.2017.12.111

Lee, S.I., Weon, S.Y., Lee, C.W. et al. (2003). Removal of nitrogen and phosphate from wastewater by addition of bittern. *Chemosphere*, **51**(4), 265–271. https://doi.org/10.1016/S0045-6535(02)00807-X

Lieder, M. and Rashid, A. (2016). Towards circular economy implementation: a comprehensive review in context of manufacturing industry. *Journal of Cleaner Production*, **115**, 36–51. https://doi.org/10.1016/j.jclepro.2015.12.042

Lyu, S., Chen, W., Zhang, W. et al. (2016). Wastewater reclamation and reuse in China: opportunities and challenges. *Journal of Environmental Sciences (China)*, **39**, 86–96. https://doi.org/10.1016/j.jes.2015.11.012

Maiurova, A., Kurniawan, T.A., Kustikova, M. et al. (2022). Promoting digital transformation in waste collection service and waste recycling in Moscow (Russia): applying a circular economy paradigm to mitigate climate change impacts on the environment. *Journal of Cleaner Production*, **354**, 131604. https://doi.org/10.1016/j.jclepro.2022.131604

Makajic-Nikolic, D., Petrovic, N., Belic, A. et al. (2016). The fault tree analysis of infectious medical waste management. *Journal of Cleaner Production*, **113**, 365–373. https://doi.org/10.1016/j.jclepro.2015.11.022

Manara, P. and Zabaniotou, A. (2012). Towards sewage sludge based biofuels via thermochemical conversion – a review. *Renewable and Sustainable Energy Reviews*, **16**(5), 2566–2582. https://doi.org/10.1016/j.rser.2012.01.074

Manea, E.E., Bumbac, C., Dinu, L.R. et al. (2024). Composting as a sustainable solution for organic solid waste management: current practices and potential improvements. *Sustainability*, **16**(15), 6329. https://doi.org/10.3390/su16156329

Mari, W., Rich, G., Joanne, G. et al. (2019). *No Time to Waste: Tackling the plastic pollution crisis before it's too late*. Retrieved from chrome-extension://efaidnbmnnnibpcajpcglclefindmkaj/https://wasteaid.org/wp-content/uploads/2019/05/NO_TIM2.pdf

Martin, M. (2005). *Waste Picker Cooperatives in Developing Countries*. Retrieved from Membership-Based Organizations of the Poor, Ahmedabad, India.: https://www.wiego.org/wp-content/uploads/2019/09/Medina-wastepickers.pdf

Martin, M., Svensson, N., and Eklund, M. (2015). Who gets the benefits? An approach for assessing the environmental performance of industrial symbiosis. *Journal of Cleaner Production*, **98**, 263–271. https://doi.org/10.1016/j.jclepro.2013.06.024

Mary, E.T., Adekanye, O.O., Abdullahi, D. et al. (2022). Effect of an active open dumpsite on the Earth's subsurface and groundwater resource. *Asian Journal of Physical and Chemical Sciences*, **10**, 15–24. https://doi.org/10.9734/ajopacs/2022/v10i230152

Masłoń, A., Czarnota, J., Szaja, A. et al. (2020). The enhancement of energy efficiency in a wastewater treatment plant through sustainable biogas use: case study from Poland. *Energies*, **13**(22), 6056. https://doi.org/10.3390/en13226056

Mattson, K.R., Pettersen, J.B., and Brattebø, H. (2024). Incineration economy: waste management policy failing the circular economy transition in Norway. *Resources, Conservation and Recycling*, **210**, 107838. https://doi.org/10.1016/j.resconrec.2024.107838

Mavrotas, G. (2009). Effective implementation of the ε-constraint method in multi-objective mathematical programming problems. *Applied Mathematics and Computation*, **213**(2), 455–465. https://doi.org/10.1016/j.amc.2009.03.037

Mavrotas, G. and Florios, K. (2013). An improved version of the augmented ε-constraint method (AUGMECON2) for finding the exact pareto set in multi-objective integer programming problems. *Applied Mathematics and Computation*, **219**(18), 9652–9669. https://doi.org/10.1016/j.amc.2013.03.002

Mehmood, A., Ahmed, S., Viza, E. et al. (2021). Drivers and barriers towards circular economy in agri-food supply chain: a review. *Business Strategy & Development*, **4**(4), 465–481. https://doi.org/10.1002/bsd2.171

Merli, R., Preziosi, M., and Acampora, A. (2018). How do scholars approach the circular economy? A systematic literature review. *Journal of Cleaner Production*, **178**, 703–722. https://doi.org/10.1016/j.jclepro.2017.12.112

Mies, A. and Gold, S. (2021). Mapping the social dimension of the circular economy. *Journal of Cleaner Production*, **321**, 128960. https://doi.org/10.1016/j.jclepro.2021.128960

Molinos-Senante, M., Hernandez-Sancho, F., and Sala-Garrido, R. (2010). Economic feasibility study for wastewater treatment: a cost-benefit analysis. *Science of The Total Environment*, **408**(20), 4396–4402. https://doi.org/10.1016/j.scitotenv.2010.07.014

Moya, D., Aldás, C., López, G. et al. (2017). Municipal solid waste as a valuable renewable energy resource: a worldwide opportunity of energy recovery by using Waste-To-Energy Technologies. *Energy Procedia*, **134**, 286–295. https://doi.org/10.1016/j.egypro.2017.09.618

Murray, A., Skene, K., and Haynes, K. (2015). The circular economy: an interdisciplinary exploration of the concept and application in a global context. *Journal of Business Ethics*, **140**(3), 369–380. https://doi.org/10.1007/s10551-015-2693-2

Nair, R.B., Lennartsson, P.R., and Taherzadeh, M.J. (2017). 8 – bioethanol production from agricultural and municipal wastes. In *Current Developments in Biotechnology and Bioengineering-Solid Waste Management*, 157–190. https://doi.org/10.1016/B978-0-444-63664-5.00008-3

Nawaz, A., Hafeez, A., Abbas, S.Z. et al. (2020). A state of the art review on electron transfer mechanisms, characteristics, applications and recent advancements in microbial fuel cells technology. *Green Chemistry Letters and Reviews*, **13**(4), 365–381. https://doi.org/10.1080/17518253.2020.1854871

Neczaj, E. and Grosser, A. (2018). Circular economy in wastewater treatment plant–challenges and barriers. *Proceedings*, **2**(11), 614. https://doi.org/10.3390/proceedings2110614

Nwokolo, N., Mukumba, P., Obileke, K. et al. (2020). Waste to energy: a focus on the impact of substrate type in biogas production. *Processes*, **8**(10), 1–21. https://doi.org/10.3390/pr8101224

Okot-Okumu, J. and Nyenje, R. (2011). Municipal solid waste management under decentralisation in Uganda. *Habitat International*, **35**(4), 537–543. https://doi.org/10.1016/j.habitatint.2011.03.003

Oyejobi, D.O., Firoozi, A.A., Fernández, D.B. et al. (2024). Integrating circular economy principles into concrete technology: enhancing sustainability through industrial waste utilization. *Results in Engineering*, **24**(12), 102846. https://doi.org/10.1016/j.rineng.2024.102846

Pandey, A. and Biswas, P. (2024). Tri-reforming of methane over Ni/ZrO_2 catalyst derived from Zr-Mof for the production of synthesis gas. *Environmental Science and Pollution Research*, **31**(24), 35069–35082. https://doi.org/10.21203/rs.3.rs-3968447/v1

Pandit, S., Yadav, N., Sharma, P. et al. (2025). Life cycle assessment and techno-economic analysis of nanotechnology-based wastewater treatment: status, challenges and future prospectives. *Journal of the Taiwan Institute of Chemical Engineers*, **166**, 105567 https://doi.org/10.1016/j.jtice.2024.105567

Pannila, N., Jayalath, M.M., Thibbotuwawa, A. et al. (2022). Challenges in applying circular economy concepts to food supply chains. *Sustainability*, **14**(24), 16536. https://doi.org/10.3390/su142416536

Paulraj, C.R.K.J., Bernard, M.A., Raju, J. et al. (2019). Sustainable waste management through waste to energy technologies in India – opportunities and environmental impacts. *International Journal of Renewable Energy Research*, **9**(v9i1), 309–342. https://doi.org/10.20508/ijrer.v9i1.9017.g7586

Piazzi, S., Zhang, X., Patuzzi, F. et al. (2020). Techno-economic assessment of turning gasification-based waste char into energy: a case study in South-Tyrol. *Waste Management*, **105**, 550–559. https://doi.org/10.1016/j.wasman.2020.02.038

Rana, R., Ganguly, R., and Gupta, A.K. (2015). An assessment of solid waste management system in Chandigarh City, India. *Electronic Journal of Geotechnical Engineering*, **20**(6), 1547–1572.

Rastogi, A. and Kumar Paul, V. (2020). A critical review of the potential for fly ash utilisation in construction-specific applications in India. *Environmental Research, Engineering and Management*, **76**(2), 65–75. https://doi.org/10.5755/j01.erem.76.2.25166

Rezaei, M., Ghobadian, B., Samadi, S.H. et al. (2018). Electric power generation from municipal solid waste: a techno-economical assessment under different scenarios in Iran. *Energy*, **152**, 46–56. https://doi.org/10.1016/j.energy.2017.10.109

Ribarova, I., Vasilaki, V., and Katsou, E. (2024). Review of linear and circular approaches to on-site domestic wastewater treatment: analysis of research achievements, trends and distance to target. *Journal of Environmental Management*, **367**, 121951. https://doi.org/10.1016/j.jenvman.2024.121951

Robles, A., Aguado, D., Barat, R. et al. (2020). New frontiers from removal to recycling of nitrogen and phosphorus from wastewater in the Circular Economy.

Bioresource Technology, **300**, 122673. https://doi.org/10.1016/j.biortech.2019.122673

Rolewicz-Kalińska, A. (2016). Logistic constraints as a part of a sustainable medical waste management system. *Transportation Research Procedia*, **16**, 473–482. https://doi.org/10.1016/j.trpro.2016.11.044

Ryu, H.-D. and Lee, S.-I. (2016). Struvite recovery from swine wastewater and its assessment as a fertilizer. *Environmental Engineering Research*, **21**(1), 29–35. https://doi.org/10.4491/eer.2015.066

Safarian, S., Unnthorsson, R., and Richter, C. (2020). Techno-economic and environmental assessment of power supply chain by using waste biomass gasification in Iceland. *Biophysical Economics and Sustainability*, **5**(2), 1–13. https://doi.org/10.1007/s41247-020-00073-4

Saha, K., Dey, P.K., and Papagiannaki, E. (2021). Implementing circular economy in the textile and clothing industry. *Business Strategy and the Environment*, **30**(4), 1497–1530. https://doi.org/10.1002/bse.2670

Salati, S., Scaglia, B., di Gregorio, A. et al. (2013). Mechanical biological treatment of organic fraction of MSW affected dissolved organic matter evolution in simulated landfill. *Bioresource Technology*, **142**, 115–120. https://doi.org/10.1016/j.biortech.2013.05.049

Saliu, T.D. and Oladoja, N.A. (2021). Nutrient recovery from wastewater and reuse in agriculture: a review. *Environmental Chemistry Letters*, **19**(3), 2299–2316. https://doi.org/10.1007/s10311-020-01159-7

Sandhu, K. (2014). Historical trajectory of waste management; an analysis using the health belief model. *Management of Environmental Quality: An International Journal*, **25**(5), 615–630. https://doi.org/10.1108/meq-09-2013-0098

Seferlis, P., Varbanov, P.S., Papadopoulos, A.I. et al. (2021). Sustainable design, integration, and operation for energy high-performance process systems. *Energy*, **224**, 120158. https://doi.org/10.1016/j.energy.2021.120158

Shah, H.H., Amin, M., Iqbal, A. et al. (2022). A review on gasification and pyrolysis of waste plastics. *Frontiers in Chemistry*, **10**, 960894. https://doi.org/10.3389/fchem.2022.960894

Shareefdeen, Z., Elkamel, A., and Tse, S. (2015). Review of current technologies used in municipal solid waste-to-energy facilities in Canada. *Clean Technologies and Environmental Policy*, **17**(7), 1837–1846. https://doi.org/10.1007/s10098-015-0904-2

Sharma, H.B., Vanapalli, K.R., Samal, B. et al. (2021). Circular economy approach in solid waste management system to achieve UN-SDGs: solutions for post-COVID recovery. *Science of the Total Environment*, **800**, 149605. https://doi.org/10.1016/j.scitotenv.2021.149605

Sharma, K.D. and Jain, S. (2020). Municipal solid waste generation, composition, and management: the global scenario. *Social Responsibility Journal*, **16**(6), 917–948. https://doi.org/10.1108/srj-06-2019-0210

Sharma, P., Bano, A., Singh, S.P. et al. (2024). Sustainable organic waste management and future directions for environmental protection and techno-economic perspectives. *Current Pollution Reports*, **10**(3), 459–477. https://doi.org/10.1007/s40726-024-00317-7

Shi, X., Chen, Z., Wei, W. et al. (2024). Perspectives on sustainable plastic treatment: a shift from linear to circular economy. *TrAC Trends in Analytical Chemistry*, **173**, 117631 https://doi.org/10.1016/j.trac.2024.117631

Singh, A.K., Bhuyan, T., Maity, S. et al. (2020). Magnetically actuated carbon soot nanoparticle-based catalytic CARBOts coated with Ni/Pt nanofilms for water detoxification and oil-spill recovery. *ACS Applied Nano Materials*, **3**(4), 3459–3470. https://doi.org/10.1021/acsanm.0c00199

Singh, J., Laurenti, R., Sinha, R. et al. (2014). Progress and challenges to the global waste management system. *Waste Management & Research*, **32**(9), 800–812. https://doi.org/10.1177/0734242X14537868

Singh, M.L., Sharma, R., and Abhishek. (2017). Soil stabilization using industrial waste (wheat husk and sugarcane straw ash). *International Research Journal of Engineering and Technology*, **4**(9), 589–596.

Sondh, S., Upadhyay, D.S., Patel, S. et al. (2024). Strategic approach towards sustainability by promoting circular economy-based municipal solid waste management system- a review. *Sustainable Chemistry and Pharmacy*, **37**, 101337. https://doi.org/10.1016/j.scp.2023.101337

Sundar, D., Mathiyazhagan, K., Agarwal, V. et al. (2023). From linear to a circular economy in the e-waste management sector: experience from the transition barriers in the United Kingdom. *Business Strategy and the Environment*, **32**(7), 4282–4298. https://doi.org/10.1002/bse.3365

Tamta, P., Rani, N., and Yadav, A. K. (2020). Enhanced wastewater treatment and electricity generation using stacked constructed wetland–microbial fuel cells. *Environmental Chemistry Letters*, **18**(3), 871–879. https://doi.org/10.1007/s10311-020-00966-2

Tirkolaee, E.B., Abbasian, P., and Weber, G.W. (2021). Sustainable fuzzy multi-trip location-routing problem for medical waste management during the COVID-19 outbreak. *Science of The Total Environment*, **756**, 143607. https://doi.org/10.1016/j.scitotenv.2020.143607

Trigkas, M., Karagouni, G., Mpyrou, K. et al. (2020). Circular economy. The Greek industry leaders' way towards a transformational shift. *Resources, Conservation and Recycling*, **163**, 105092. https://doi.org/10.1016/j.resconrec.2020.105092

Valizadeh, J., Hafezalkotob, A., Seyed Alizadeh, S.M. et al. (2021). Hazardous infectious waste collection and government aid distribution during COVID-19: a robust mathematical leader-follower model approach. *Sustainable Cities and Society*, **69**, 102814. https://doi.org/10.1016/j.scs.2021.102814

Velis, C.A. and Cook, E. (2021). Mismanagement of plastic waste through open burning with emphasis on the global south: a systematic review of risks to occupational and public health. *Environmental Science & Technology*, **55**(11), 7186–7207. https://doi.org/10.1021/acs.est.0c08536

Vennila, S. and Karthikeyan, K. (2024). An EOQ model for circularity index with waste minimization and reduced emission in electrical and electronic equipment. *Contemporary Mathematics*, **5**(3), 3002–3037. https://doi.org/10.37256/cm.5320244381

Vergine, P., Lonigro, A., Salerno, C. et al. (2016). Nutrient recovery and crop yield enhancement in irrigation with reclaimed wastewater: a case study. *Urban Water Journal*, **14**(3), 325–330. https://doi.org/10.1080/1573062x.2016.1141224

Voukkali, I., Papamichael, I., Loizia, P. et al. (2023). Waste metrics in the framework of circular economy. *Waste Management & Research*, **41**(12), 1741–1753. https://doi.org/10.1177/0734242X231190794

Vymazal, J. (2014). Constructed wetlands for treatment of industrial wastewaters: a review. *Ecological Engineering*, **73**, 724–751. https://doi.org/10.1016/j.ecoleng.2014.09.034

Wang, B., Farooque, M., Zhong, R.Y. et al. (2021). Internet of Things (IoT)-enabled accountability in source separation of household waste for a circular economy in China. *Journal of Cleaner Production*, **300**, 126773. https://doi.org/10.1016/j.jclepro.2021.126773

Wilson. (2023). *Water and Wastewater Treatment Technologies: Global Markets* (ENV008B). Retrieved from Boston: https://www.bccresearch.com/market-research/environment/wastewater-treatment-tech-env008b.html?srsltid=AfmBOopZUSOtulyvDtszivcRWgJGcCSP0X10SafJiJICqj5RKyV8Hgld. Accessed on December 2024.

Windfeld, E.S. and Brooks, M.S. (2015). Medical waste management – a review. *Journal of Environmental Management*, **163**, 98–108. https://doi.org/10.1016/j.jenvman.2015.08.013

Wu, C., Liu, W., and Deng, H. (2023). Urbanization and the emerging water crisis: identifying water scarcity and environmental risk with multiple applications in Urban agglomerations in western China. *Sustainability*, **15**(17), 12977. https://doi.org/10.3390/su151712977

Yang, Y., Wang, J., Chong, K. et al. (2018). A techno-economic analysis of energy recovery from organic fraction of municipal solid waste (MSW) by an integrated intermediate pyrolysis and combined heat and power (CHP) plant. *Energy Conversion and Management*, **174**, 406–416. https://doi.org/10.1016/j.enconman.2018.08.033

Yoshida, M. (2020). Social development and the environment—A view from solid waste management. In: *International Development and the Environment: Social Consensus and Cooperative Measures for Sustainability* (eds. S. Hori, Y. Takamura, T. Fujita et al.), 27–43. Singapore: Springer Singapore.

You, S., Wang, W., Dai, Y. et al. (2016). Comparison of the co-gasification of sewage sludge and food wastes and cost-benefit analysis of gasification- and incineration-based waste treatment schemes. *Bioresource Technology*, **218**, 595–605. https://doi.org/10.1016/j.biortech.2016.07.017

Zhang, N., Jin, W., Zhou, X. et al. (2024). Nutrients recovery during sewage treatment by aquatic plants: a comprehensive evaluation. *Process Safety and Environmental Protection*, **185**, 143–152. https://doi.org/10.1016/j.psep.2024.02.083

Zhou, J., Liu, S., Zhou, N. et al. (2018). Development and application of a continuous fast microwave pyrolysis system for sewage sludge utilization. *Bioresource Technology*, **256**, 295–301. https://doi.org/10.1016/j.biortech.2018.02.034

Zhou, X., Yang, J., Xu, S. et al. (2020). Rapid in-situ composting of household food waste. *Process Safety and Environmental Protection*, **141**, 259–266. https://doi.org/10.1016/j.psep.2020.05.039

Zorpas, A.A. (2020). Strategy development in the framework of waste management. *Science of The Total Environment*, **716**, 137088. https://doi.org/10.1016/j.scitotenv.2020.137088

Zsófia Závodi, F. and Jiri, K. (2011). Municipal solid waste as alternative fuel–minimising emissions and effluents. *Chemical Engineering Transactions*, **25**, 31–38. https://doi.org/10.3303/CET1125006

3

Infrastructure Requirement and Regulatory Framework for Wastewater Management System

Utkarsh Dubey[1#], *Divyani Pathak*[1#], *Sanjay Kumar Gupta*[2*], *and Bhaskar Singh*[1*]

[1]*Department of Environmental Sciences, Central University of Jharkhand, Ranchi, Jharkhand, India*
[2]*Department of Civil Engineering, Indian Institute of Technology Delhi, New Delhi, Delhi, India*

[*]Corresponding authors: bhaskar.singh@cuj.ac.in; sanjuenv@gmail.com
[#]These authors contributed equally to the manuscript

3.1 Introduction

In the contemporary context, global society is witnessing a significant intensification of competition for water resources. Concurrently, the risks of water contamination and environmental degradation continue to rise significantly (Zainudeen, 2007). As water resource demands continue to grow, sustainable and efficient treatment solutions have become increasingly critical (Wysowska and Kicińska, 2022). For instance, the expansion of urban populations and increase in industrial activities necessitate structured interventions, including the upgradation of existing facilities and the construction of new treatment plants (Wang et al., 2022). Furthermore, the challenges presented by climate change and deteriorating infrastructure have rendered the task of securing an adequate and easily accessible water supply progressively more complex (Falinski et al., 2020). Despite the establishment of 139 sewage treatment plants (STPs) by 2023, which possess the capability to handle 2514 Million Liter per Day (MLD) sewage, there are still notable disparities between wastewater production and treatment capacity, resulting in untreated wastewater releases (Singh and Farswan, 2023). Therefore, it is imperative to address the treatment of wastewater in order to uphold environmental conservation and safeguard public health from potential waterborne diseases and pollution stemming from untreated waste (Al-Taai, 2023).

Wastewater to Resource Recovery: Applying the Circular Economy Toward Sustainable Development, First Edition. Edited by Arvind Kumar Mungray et al. Published 2026 by John Wiley & Sons, Inc.

The water and wastewater infrastructure encompasses the facilities essential for transferring, refining, preserving, dispensing, and eliminating drinking water and sewage, along with regulating water levels and quality (Alcaraz and Zeadally, 2015). The fundamental water and wastewater treatment infrastructure is essential for safeguarding public health and economic prosperity and ensuring environmental well-being, thereby drastically influencing and shaping the quality of life (EPA, 2017; Jacobsen, 2018). Infrastructure development for wastewater treatment is considerably shaped by diverse regulatory frameworks across different jurisdictions, highlighting the necessity of obtaining requisite licenses and complying with government oversight (Arroyo and Vega, 2015). Effective wastewater management requires the integration of advanced technologies that correspond with evolving regulatory standards, particularly those addressing persistent and bio-accumulative pollutants (Monteith and Sterne, 2001). Moreover, local resource managers must devise watershed management strategies that consider the ecological consequences of wastewater discharge, which are frequently exacerbated by antiquated systems (Capps, 2019).

The journey toward effective and sustainable wastewater treatment systems is filled with challenges, but there's a lot of optimism about addressing these issues constructively (Cullis et al., 2018; Singh et al., 2021). Many wastewater treatment facilities and their pipelines have been operational for decades throughout the industrialized world, but their functionality has deteriorated due to wear and strain (Furlong et al., 2019). System breakdowns, environmental contamination, and public health hazards can arise from a backlog of necessary repairs and replacements brought on by improper maintenance and a lack of investment in system upgrades (Cullis et al., 2018). This situation, while concerning, presents a valuable opportunity to rejuvenate our infrastructure and enhance public health by preventing potential system failures and environmental contamination. Moreover, the growing urban populations and expanding economic activities are significantly increasing the demand for wastewater treatment services. Therefore, the amount of wastewater that has to be treated has increased dramatically, exceeding the design capacity of many current facilities and needing significant infrastructure upgrades to keep up (Wang et al., 2018; Hukka and Katko, 2015). This surge calls for innovative infrastructure upgrades that can accommodate our modern needs. One more thing to consider is that the energy-intensive nature of wastewater treatment processes raises concerns due to the sector's substantial fossil fuel consumption and the resulting greenhouse gas emissions, which contribute to environmental sustainability challenges (Wang et al., 2018; Gallagher and Gill, 2021). However, such challenges also open the door for advancements in sustainable practices and technologies. Thus, by embracing a coordinated strategy for regulation and infrastructure development, we can create a resilient and sustainable wastewater management framework that benefits our communities and the environment (Smail, 2024).

The journey of developing wastewater treatment infrastructure is certainly challenging. However, there are exciting strategies available to conquer these obstacles. Prescott et al. (2021) emphasize that the effective deployment of decentralized sanitation methods, such as constructed wetlands, not only hinges on technical solutions but also thrives on sociocultural appropriateness and robust community engagement. Embracing ecological sanitation techniques that encourage resource recovery and environmental sustainability is more important than ever, especially in light of the shortcomings of traditional sanitation systems (Werkneh and Gebru, 2022). To make a meaningful transition toward multi-benefit wastewater infrastructure, it is important to address the social and institutional obstacles through collaborative efforts with stakeholder engagement in defining common objectives (Harris-Lovett et al., 2018). Lastly, engaging stakeholders in collaborative planning approaches can aid in anticipating future needs and regulatory changes, guaranteeing that wastewater infrastructure is resilient and sustainable (Neth et al., 2022).

Resilience, sustainability, and efficiency have increasingly emerged as critical considerations in the construction of wastewater treatment infrastructure, especially in regions that are vulnerable to flooding. The implementation of strategies for asset restoration can substantially minimize interruptions (Karamouz et al., 2022). A comprehensive and systematic assessment of treatment facilities can determine the best locations for new sludge processing plants, underscoring the importance of establishing environmentally sustainable sludge management systems (Chertes et al., 2022). Furthermore, immobilized microorganisms have demonstrated significant effectiveness in the removal of pollutants from industrial wastewater, thereby ensuring compliance with discharge regulations (Sabliy, 2023). Lastly, integrating green infrastructure with traditional systems has proven effective in reducing nutrient concentrations in reclaimed water and advancing sustainable water management practices (Arrington et al., 2023).

3.2 Treatment System for Wastewater and Its Management

Numerous cutting-edge systems are utilized and employed in treating and managing wastewater, aiming to improve purification efficiency, reliability, and dependability. Wastewater treatment plants employ a combination of technologies and procedures to remove pollutants through physical, chemical, and biological methods (Srivastava and Singh, 2022). A comprehensive wastewater treatment system may incorporate components such as coagulation–flocculation assemblies, slurry separators, and aeration mechanisms to eliminate biological and nutrient contaminants (Muhammad et al., 2020). However, the escalation in micropollutants within wastewater streams has introduced significant challenges for existing wastewater treatment processes (Akula et al., 2021). An effective wastewater treatment is guaranteed through the utilization of advanced systems, including sequencing batch reactors (SBR) and membrane bioreactors (MBR). These technologies significantly enhance the nitrification and denitrification processes, thereby ensuring optimal treatment outcomes (Akula et al., 2021).

Wastewater treatment systems are essential for effective management of wastewater generated by domestic, industrial, and agricultural activities. These systems utilize a variety of technologies to remove the contaminants, thereby ensuring that treated water can be safely discharged or repurposed for reuse. So, the principal types of wastewater treatment systems encompass membrane technologies, biological processes, chemical and physical treatments, and advanced systems like bio-electrochemical systems. Urban wastewater treatment usually includes multiple stages, namely preliminary, primary, secondary, and occasionally tertiary treatments. Each stage is designed to systematically and gradually eliminate suspended solids, colloidal substances, and dissolved contaminants, thereby reducing environmental harm (Fadel et al., 2024). Biological treatment methods, which utilize natural processes to degrade pollutants, are also integral to a comprehensive wastewater management strategy. This highlights the importance of integrating various treatment technologies to enhance purification efficiency (Gadirova et al., 2024). Moreover, decentralized wastewater treatment systems (DEWATSs) are increasingly acknowledged and recognized as successful options for rural communities, enabling the recycling of treated water for non-drinking purposes (Li et al., 2024).

In recent years, a significant challenge has arisen concerning the provision of adequate water resources to meet the growing demands of an expanding population (Mishra et al., 2021). Millions of individuals worldwide are experiencing an escalating crisis due to the lack of access to safe water. This deficiency leads to detrimental health outcomes, loss of livelihoods, and unnecessary suffering, particularly among impoverished populations (Hanjra and Qureshi, 2010). Different wastewater treatment systems have been put into place to tackle these critical challenges and ensure the availability of clean drinking water. These systems include centralized wastewater treatment system (CWTS), decentralized wastewater treatment system (DEWATS), and on-site wastewater treatment system (OWTS).

3.2.1 Centralized Wastewater Treatment Systems

Centralized wastewater treatment systems (CWTS) are vital for addressing sanitation challenges in densely populated regions. These systems basically collect wastewater from different sources and direct it to a centralized treatment plant (Saadatinavaz et al., 2024). CWTS carries and processes sewage using a variety of treatment technologies, including gravity- or pressure-based sewer networks, to minimize negative environmental effects and impacts while enhancing public health outcomes (Brzusek et al., 2022). Environmental factors, such as rainfall, can significantly influence the efficacy of CWTS by affecting influent flow rates and wastewater composition. These changes directly impact effluent quality and the treatment processes' overall performance (Rashid and Liu, 2020). Although the implementation of CWTS is deemed essential for effective sanitation, their adoption is often contingent upon the community's willingness to pay. It is important to note that these systems can present several disadvantages, including high energy consumption and potentially significant costs (Jiao et al., 2022; González-Viar et al., 2016). Furthermore, CWTS may include advanced water reuse strategies, thereby improving sustainability by facilitating the recovery and reuse of treated water for a range of urban applications (Chen et al., 2017).

3.2.2 Decentralized Wastewater Treatment Systems

These systems are specifically engineered to manage wastewater and are very helpful in both urban and rural areas. They facilitate the treatment of various wastewater streams, including rainwater, blackwater, and greywater, which contribute to reducing environmental impacts, decreasing infrastructure costs, and improving local water quality (Chen et al., 2024; Giwa et al., 2024). To further optimize operational efficiency and resource recovery, DEWATS can integrate cutting-edge technologies, including anaerobic digestion, built wetlands, and multistep onsite treatment procedures (Avraham et al., 2023). However, obstacles like minimal knowledge, regulatory barriers, and financial limitations prevent their broader implementation (Giwa et al., 2024). Despite these challenges, DEWATS is gaining recognition for promoting sustainable water management and improving sanitation, particularly in developing nations (Ibrahim et al., 2024; Li et al., 2023).

3.2.3 On-site Wastewater System

These systems are specifically designed to manage waste materials at the designated disposal location for short durations. They are typically implemented in areas where insufficient housing density renders centralized wastewater treatment economically unfeasible (Yates, 2011). The effectiveness of these systems can vary greatly depending on their design and operational conditions. Commonly consisting of septic tanks and soil treatment units, they can efficiently minimize chemical and microbiological contamination (Wang et al., 2021; Dubber and Gill, 2014). The establishment of artificial wetlands that replicate the natural processes of natural wetlands and employing a combination of physical, chemical, and biological methods can facilitate the removal of pollutants from wastewater (Machineni, 2020). The integration of on-site wastewater system (OWTS) into urban design, viewed through a nature-based solutions perspective, can enhance soil health and water management while addressing ecological and infrastructure challenges (Daudin et al., 2022). However, significant environmental concerns are associated with OWTS, especially in socioeconomically disadvantaged areas where aging systems may contribute to heightened environmental degradation and health hazards (Capps et al., 2020). Therefore, enhancing public health protection and reducing ecological damage requires improving OWTS design and administration (Envall et al., 2023).

3.3 Regulatory Authority of India for Construction of Treatment Facility

In India, the construction of treatment facilities is governed by several independent regulatory agencies (IRAs) that are responsible for ensuring compliance with environmental and health standards. These agencies play a vital role in promoting autonomous operations and safeguarding public safety throughout such facilities' construction and operational phases.

Key Regulatory Agencies:

- The Central Pollution Control Board (CPCB) is tasked with monitoring and regulating environmental pollution levels. The CPCB establishes standards for effluent discharge and is responsible for overseeing the implementation of water treatment regulations.
- State Pollution Control Boards (SPCBs) function at the state level, implementing local regulations and ensuring that treatment facilities adhere to the national standards established by the CPCB. These entities are crucial in maintaining environmental compliance and protecting public health within their jurisdictions.
- The Ministry of Environment, Forest and Climate Change (MoEFCC) is responsible for formulating policies and regulations pertaining to environmental protection, including establishing and regulating treatment facilities.

Compliance and Standards:

- Environmental Impact Assessment (EIA) is amended for large-scale projects. These assessments are conducted to evaluate the potential environmental impacts prior to the construction activities.
- National Water Policy provides a comprehensive framework for effective and sustainable management of water resources, significantly influencing the design and operation of water treatment facilities.

Regulatory frameworks play a vital role in ensuring the safety and efficacy of treatment facilities. However, bureaucratic delays and insufficient enforcement mechanisms may impede their effectiveness (Datta et al., 2018).

3.3.1 Central Pollution Control Board

The CPCB is a statutory organization functioning under the MoEFCC, responsible for monitoring and controlling pollution across India. It plays a crucial role in establishing standards for effluent discharge and ensuring compliance among treatment facilities, thus serving as an essential entity for regulatory oversight in construction projects related to waste management. The CPCB formulates comprehensive environmental standards pertaining to air, water, and soil quality, thereby guiding industries and municipalities in compliance efforts to protect public health. Additionally, the CPCB is responsible for pollution monitoring through an extensive network of monitoring stations, providing essential data for evaluating environmental health and formulating effective policies aimed at combating pollution. In addition, the CPCB also implements key pollution control legislation, such as the Water (Prevention and Control of Pollution) Act and the Air (Prevention and Control of Pollution) Act, which are foundational for regulating pollution and safeguarding and ensuring environmental protection. Furthermore, the CPCB conducts the National Water Quality Monitoring Programme, which is pivotal in identifying pollution sources and ensuring safe water to communities across India. Through these initiatives, the CPCB significantly contributes to preserving and improving environmental quality in India (Siddiqui et al., 2022).

3.3.2 State Pollution Control Board

In India, SPCBs serve an essential role as regulatory agencies tasked with monitoring and mitigating pollution within their specific state jurisdictions. These boards operate within a legal framework established by key environmental legislation, notably the Water (Prevention and Control of Pollution) Act and the Air (Prevention and Control of Pollution) Act. These Acts grant SPCBs the authority to comply with established pollution standards, ensuring that industries and other entities adhere to permissible levels of pollutants released into the environment (Vallero, 2024). SPCBs employ various pollution control technologies to effectively manage environmental quality, including waste treatment processes and emission control systems. Additionally, they are integral to the EIA process, evaluating the potential environmental consequences of proposed projects to ensure compliance with regulations and promote sustainable development. A comprehensive understanding of the structure and functions of SPCBs is vital for grasping the complexities of environmental governance in India. These boards are pivotal in implementing policies to reduce pollution and safeguard public health (Rosencranz and Upadhyay, 2011).

3.3.3 Ministry of Environment, Forest, and Climate Change

The MoEFCC serves as the principal government authority in India responsible for environmental protection and sustainable development. This ministry is tasked with formulating policies and regulations that directly influence the construction of treatment facilities, ensuring that environmental considerations are seamlessly integrated into development initiatives. The MoEFCC plays a pivotal role in formulating and executing policies focused on environmental protection and sustainable development. This includes addressing critical issues such as pollution, biodiversity loss, and climate change through a comprehensive policy framework established by providing key documents, including the National Conservation Strategy and the National Forest Policy. Moreover, the MoEFCC is also charged with the responsibility of forest conservation, which is imperative for maintaining biodiversity and combating the effects of climate change, as forests serve as significant carbon sinks. Furthermore, the ministry also develops strategies for climate change mitigation, exemplified by the National Action Plan on Climate Change (NAPCC), which aims to reduce greenhouse gas emissions and enhance resilience to climate-related challenges. Another essential aspect of the MoEFCC's mandate is biodiversity conservation, which involves implementing programs designed to protect endangered species and their habitats, thereby ensuring ecological balance and sustainable resource management. Additionally, the ministry regulates the EIA process to evaluate the potential environmental effects associated with proposed projects, thereby ensuring compliance with legal standards and preventing environmental degradation. Through these multifaceted initiatives, the MoEFCC effectively addresses the dynamic environmental challenges faced by India (Chungyalpa, 2020).

3.3.4 Environmental Impact Assessment

EIA is a systematic process employed to evaluate a proposed project's potential environmental effects before its approval. For treatment facilities, the completion of an EIA is often mandated by regulatory requirements, ensuring that the environmental impacts are appropriately considered and mitigated. A significant challenge in integrating the EIA into an existing environmental management system (EMS) lies in achieving adequate control while allowing necessary flexibility to address site-specific project requirements. Environmental management plans (EMPs) represent one method of managing the environmental impacts associated with construction projects. These plans encompass project-specific protocols that interface the EIA and subsequent project planning and development phases (Marshall, 2002). Various key factors, including rational planning theory, technology assessment, risk assessment, and the environmental movement's policy objectives, have influenced EIA's evolution. Two critical aspects must be considered to thoroughly comprehend EIA and its relationship with more comprehensive, policy-relevant evaluations. First, the primary purpose of the EIA is to influence private sector behavior, even though it was initiated by U.S. legislation intended to change federal policy-making goals and processes that significantly affect the environment. Second, the interpretation of the term "environment" varied among some of the drafters of the National Environmental Policy Act (Caldwell, 1988).

3.3.5 Technological Standards

Upgrading wastewater treatment infrastructure, whether through new construction or the renovation of existing facilities, necessitates the integration of cutting-edge technologies to achieve greater efficiency and effectiveness in pollutant removal. This includes incorporating advanced filtration systems such as multimedia filters, which utilize layers of different materials (such as sand, gravel, and anthracite coal) to capture a broader spectrum of particulate matter. Furthermore, the inclusion of activated carbon filters is imperative for the adsorption of dissolved organic compounds, heavy metals, and other contaminants that are not effectively removed by conventional treatment methods. Apart from that, DEWATS and OWTS represent contemporary advancements in treatment technologies. These innovative technologies are essential for achieving elevated levels of pollutant removal, thereby improving water quality and ensuring compliance with increasingly stringent environmental regulations. By adopting and implementing these upgrades, wastewater treatment facilities can significantly enhance their operational performance and contribute to more effective overall water resource management (Cookey et al., 2016).

DEWATS is gaining increasing traction as an effective and preferred strategy for managing wastewater, particularly within the diverse contexts throughout Indian settings. This growing acceptance stems from several key advantages, primarily their cost-effectiveness compared to large, centralized treatment plants. DEWATS are often more economical to install and operate, making them particularly suitable, especially in areas with dispersed populations or those with limited existing infrastructure. Furthermore, the modular and adaptable nature of DEWATS enables customization for a wide range of Indian contexts, including rural villages, smaller towns, individual buildings, and industrial sites. These systems can be specifically tailored and designed to accommodate specific local conditions, such as varying wastewater volumes, land availability, and budgetary constraints.

Such adaptability facilitates the development of more efficient and sustainable wastewater management solutions that are better aligned with the specific needs of different communities and environments across India (Tortajada et al., 2024).

3.3.6 Compliance and Monitoring

India possesses a comprehensive legal and regulatory framework dedicated to environmental protection. This framework is anchored by key legislation, including the Environment (Protection) Act of 1986, the Water (Prevention and Control of Pollution) Act of 1974, and the Air (Prevention and Control of Pollution) Act of 1981. These legislative acts empower regulatory authorities to establish standards, issue permits, and undertake enforcement actions to address violations effectively.

- **Challenges in Compliance:** Despite the robust framework, ensuring compliance across all regions presents a notable challenge. This issue is particularly significant for small and medium-sized enterprises (SMEs) and in rural areas, where the availability of resources and the level of awareness may be limited.
- **Decentralized Systems in Rural Areas:** Rural areas commonly rely on decentralized treatment systems like individual septic tanks or community-based treatment facilities. The monitoring and enforcement of environmental regulations in these areas present significant challenges due to the extensive number of dispersed facilities and the limited resources available to local regulatory bodies.
- **Need for On-going Improvements:** Continuous improvement in regulatory practices is paramount. This includes strengthening monitoring mechanisms, promoting self-monitoring within industries, and enhancing the competencies of regulatory bodies at both the state and local levels. Therefore, continuous enhancement in these areas is vital for ensuring effective oversight and compliance with environmental regulations for the benefit of rural as well as urban societies and industries.
- **Technology Adoption:** The adaptation of advanced technological solutions can significantly improve regulatory oversight. This includes implementing online monitoring systems, utilizing remote sensing technologies, and employing data analytics to monitor and assess pollution levels and compliance status.

In the Indian context, the effective implementation of environmental regulations requires a multi-pronged approach that addresses the specific challenges encountered by various regions and industries. This strategy encompasses strengthening regulatory capacity, promoting awareness and compliance among industrial entities, and leveraging technology for more effective monitoring and enforcement processes.

3.4 Regulatory Authority in India for Quality and Maintenance Evaluation Treatment Facility

Government organizations, known as regulatory bodies, are in charge of enforcing laws and regulations across a range of industries. In India, a framework of key organizations is responsible for regulating and ensuring treatment facilities' operational integrity and quality. The MoEFCC is the federal government ministry overseeing hazardous waste treatment facilities. The MoEFCC establishes comprehensive guidelines for managing, storing, treating, and disposing hazardous waste. These organizations are responsible for issuing permits, monitoring effluents, ensuring compliance with environmental regulations, and evaluating treated water quality to ensure adherence to the standards outlined in the Environment (Protection) Act of 1986.

The Water (Prevention and Control of Pollution) Act of 1974 represents a foundational piece of environmental legislation in India. This act established Pollution Control Boards at both the federal and state levels. This government agency is authorized to enforce the law for the prevention of environmental pollution. Furthermore, it conferred upon these boards the authority to enforce regulatory measures, acting upon the recommendations generated by the National Committee on Environmental Planning and Coordination (NCEPC) (Chand, 2018).

3.4.1 Parameters of Water Quality

The degree of water contamination can be determined by looking at the nation's water quality. Water bodies are frequently affected by various pollutants, such as trace metals, pesticides, inorganic ions, and other micropollutants, which are present in water bodies. Establishing water quality criteria is essential, as these criteria are based on the various components characterizing the water quality. Guidelines for water quality may establish minimum permissible concentrations or identify a maximum allowable threshold for a specific concentration of any substance in a particular medium that is hazardous to exceeding. Such guidelines are typically formulated for raw water and its proportionate usage in drinking, domestic, irrigating, aquatic ecosystems, or as a habitat for biological organisms (Schweitzer and Noblet, 2018).

Wastewater recycling is equally essential for sustainable water utilization, as is the conservation of water resources. Presently, untreated wastewater is discharged into water bodies in approximately 76% of instances. A significant contributing factor to this issue is the inadequate availability of effluent treatment plants (ETPs) within various industrial sector segments. The significant capital expenditures necessary for establishing ETPs predominantly account for their scarcity and limited presence. Additionally, small-scale enterprises often find it financially unfeasible to establish their own recycling facilities. Consequently, these small-scale industrial units are choosing common effluent treatment plants (CETPs) because of the critical nature of wastewater recycling (Asiwal et al., 2016). In light of these challenges, it is imperative that all large-scale industries be mandated to set up ETPs. Additionally, the treatment of domestic sewage remains significantly insufficient, with only 19% of the total sewage generated undergoing treatment, despite the responsibility resting with municipal authorities. The CPCB is advocating for the adaptation of decentralized STPs to alleviate the financial burdens associated with establishing a comprehensive sewage network. For example, all new housing estates in Chennai are required to establish their own STPs. Moreover, it is advisable that freshwater should be supplied at a subsidized rate, with the development of tariff structures that establish a higher cost for freshwater compared to sewage-treated water. To effectively encourage the usage of recycled water, it is essential that these initiatives be supported by comprehensive awareness campaigns (Capodaglio et al., 2017).

3.4.2 Surveillance of Water Quality

Regular monitoring of water resources is essential to maintaining high water quality. Addressing key questions about specific water issues, their locations, severity, and causes is crucial for effective solutions. One of the significant challenges India faces in fighting water pollution is the limited number of monitoring stations and the infrequent nature of assessments. Currently, the CPCB monitors water quality at 2500 sites through a structured three-tiered approach that includes the Yamuna Action Plan (YAP), the Monitoring of Indian National Aquatic Resources System (MINARS), and the Global Environmental Monitoring System (GEMS). However, in terms of frequency, only 32% of the stations can evaluate water quality biannually, with the remaining 68% doing so yearly. Therefore, it is imperative that both the number and frequency of monitoring stations be increased. Moreover, monitoring organizations have often overlooked isolated settlements and remote communities due to the complicated and time-consuming process of collecting samples and shipping them to far-off labs because of communication and transportation issues. Therefore, it is necessary to establish monitoring stations in remote places and underserved areas, ensuring they have well-equipped laboratories, sufficient staff, and all the necessary infrastructure (Behmel et al., 2016). Strengthening water monitoring initiatives is crucial to ensuring cleaner water for marginalized communities.

3.4.3 Decentralized Water Regulatory Authority

The decentralization of water regulatory agencies and reorganization of governance structures present a wonderful opportunity to enhance wastewater management. Local organizations often have valuable insights and are more knowledgeable about water-related issues; by collaborating with them, we can develop impactful water-management plans. Involving communities and local institutions, such as panchayats, in the development and execution of water-supply programs is key to fostering effective actions. By prioritizing extensive training and capacity building, we can empower local communities to efficiently monitor, manage, and maintain water systems across India. Engaging the rural community at every level of community-based water-quality management is very important, as they have to monitor water sources and know how to take basic corrective action (Ismaeel, 2021). In present times, it has become imperative to conserve water and promote its judicious use. It is high time that treated wastewater reuse should be made mandatory. With proper treatment, this resource has the potential to be reused in a wide range of non-potable applications, including agriculture, industries, and even domestic use. By embracing treated wastewater, we can help conserve freshwater by replacing its utilization. To support effective wastewater management and utilization, we must cultivate an enabling environment at the national level. Establishing specific regulations and guidelines for safe handling, transportation, and disposal of wastewater in the country will set us up for success and environmental sustainability. By implementing the right policies, programs, laws, and regulations, we can make a difference in managing our precious water resources for future generations. It's high time to work together toward a sustainable and water-smart future.

The advancement of infrastructure and the allocation of funds should be encouraged and prioritized. It is imperative that local communities, non-governmental organizations, local and national governments, and international organizations are actively engaged as equal stakeholders in this endeavor. Since wastewater treatment is predominantly a local challenge, it is crucial to decentralize decision-making authority to the local authorities and implement a bottom-up approach. Since there is no one-stop answer or universal solution, tailoring wastewater treatment and reuse strategies to individual contexts is essential. Before recommending any solutions, it is necessary to consider the local conditions, the nature and volume of wastewater generated, and other pertinent factors (Singh et al., 2019).

The MoEFCC, along with the Ministry of Housing and Urban Affairs (MoHUA), and the recently established Ministry of Jal Shakti, collectively oversee pollution control initiatives in India. The CPCB and the MoEFCC, as the nodal agency, are in charge of establishing laws, regulations, and associated standards pertaining to pollution control. Effective wastewater management is essential, as it constitutes a critical aspect of societal infrastructure, and water is a valuable resource. This management necessitates the involvement and collaboration of diverse professions and perspectives. Nevertheless, it has been observed that other crucial sectors, including agriculture and public health, do not specifically contribute to these management efforts.

While the MoHUA does not directly oversee public health, the significance of public health and the increasing patterns of resource reuse are critical and warrant attention. The recent establishment of the Ministry of Jal Shakti signifies India's commitment to transitioning toward integrated water and wastewater management (Schellenberg et al., 2020).

India is home to approximately 1.38 billion individuals, representing 17.7% of the world's population, despite it possessing only 4% of the world's water resources. The country's population exceeds more than three times that of the combined population of the 17 other nations facing extreme water stress (Hofste et al., 2019). As reported by the Aqueduct's Water Risk Atlas, India ranks thirteenth worldwide in terms of water stress, categorizing it among the "very highly water-stressed countries" (Hofste et al., 2019; Qadir et al. 2020). The management of wastewater and sanitation in densely populated regions is an increasingly complex challenge. In numerous developing nations, only a small fraction of large urban areas is equipped with centralized sewage and wastewater treatment facilities, which renders on-site sanitation measures inadequate. There is an urgent need for complementary and intermediate solutions. In high-priority neighborhoods where the local government does not offer a comprehensive sanitation service, DEWATS is community-managed and utilizes anaerobic processes, offering a viable option for achieving significant improvements in sanitation in a relatively short timeframe. DEWATS typically incorporates several technological modules. Sedimentation ponds, settlers, septic tanks, or biodigesters make up the primary treatment; anaerobic baffled reactors, anaerobic filters, or anaerobic and facultative pond systems make up the secondary treatment; horizontal gravel filters make up the secondary aerobic/facultative treatment; and aerobic polishing ponds make up the post-treatment (Varma et al., 2022).

3.5 Conclusion

To achieve a sustainable future for sanitation, a comprehensive approach that encompasses the entire sanitation chain is essential. This effort begins with fostering greater public awareness and transforming societal attitudes from a passive "flush and forget" approach toward active and responsible engagement with sanitation practices. This entails embracing the "reduce, reuse, recycle" principle, which, in practical terms, translates to promoting the use of water-efficient toilets and implementing optimized sewerage systems featuring smaller, localized networks and separate collection systems for various waste streams. Decentralized, on-site, nature-based, and community-managed sanitation systems offer promising solutions to India's pressing urban sustainability challenges. However, the successful implementation of these innovative systems is contingent upon establishing a supportive environment at all levels of governance and within society as a whole.

Review Questions

1. What are the main infrastructure requirements for wastewater management systems?
2. What is the role of decentralized wastewater treatment systems (DEWATS) in wastewater management?
3. How do regulatory frameworks impact wastewater treatment infrastructure in India?
4. What are the key regulatory bodies involved in wastewater management in India?
5. What is the significance of Environmental Impact Assessments (EIAs) in the context of wastewater treatment facilities?
6. What are the challenges do State Pollution Control Boards (SPCBs) face while implementing wastewater management regulations?
7. In what ways can advanced technologies enhance wastewater treatment methods?
8. How does community involvement play a vital role in the establishment of decentralized sewage treatment systems?
9. What are the main challenges in developing wastewater treatment infrastructure in India?
10. Which governing bodies regulate wastewater treatment facilities in India?

References

Akula, L.K., Gaddam, V.B., Damaraju, M. et al. (2021). Domestic wastewater treatment in a coupled sequential batch reactor-electrochemical reactor process. *Water Environment Research*, **93**(6), 953–967. https://doi.org/10.1002/wer.1488

Alcaraz, C. and Zeadally, S. (2015). Critical infrastructure protection: requirements and challenges for the 21st century. *International Journal of Critical Infrastructure Protection*, **8**, 53–66.

Al-Taai, S.H.H. (2023). Wastewater: a study of its sources, methods of treatment, and impact on environmental pollution. *AIP Conference Proceedings*, **2475**, 050003. https://doi.org/10.1063/5.0102441

Arrington, D.A., Harris, R.J., Layman, C.A. et al. (2023). Leveraging green infrastructure for efficient treatment of reclaimed water. *Science of the Total Environment*, **872**, 162232. https://doi.org/10.2139/ssrn.4238461

Arroyo, C.A. and Vega, P.V. (2015). Legal regime of wastewater in Peru. 191–207. https://dialnet.unirioja.es/servlet/articulo?codigo=5518641

Asiwal, R.S., Sar, S.K., Singh, S. et al. (2016). Wastewater treatment by effluent treatment plants. *SSRG International Journal of Civil Engineering (SSRG–IJCE)*, **3**(12), 29–35.

Avraham, A., Nwaobi, A.F., Sela-Adler, M. et al. (2023). Towards a generic solution to onsite wastewater treatment. *Journal of Water Process Engineering*, **56**, 104517. https://doi.org/10.1016/j.jwpe.2023.104517

Behmel, S., Damour, M., Ludwig, R. et al. (2016). Water quality monitoring strategies—a review and future perspectives. *Science of the Total Environment*, **571**, 1312–1329.

Brzusek, A., Widomski, M.K., and Musz-Pomorska, A. (2022). Socio-economic aspects of centralized wastewater system for rural settlement under conditions of Eastern Poland. *Water*, **14**(10), 1667. https://doi.org/10.3390/w14101667

Caldwell, L.K. (1988). Environmental impact analysis (EIA): origins, evolution, and future directions. *Impact Assessment*, **6**(3–4), 75–83.

Capodaglio, A.G., Callegari, A., Cecconet, D. et al. (2017). Sustainability of decentralized wastewater treatment technologies. *Water Practice and Technology*, **12**(2), 463–477.

Capps, K.A. (2019). Wastewater infrastructure and the ecology and management of freshwater systems. *Acta Limnologica Brasiliensia*, **31**, e104. https://doi.org/10.1590/S2179-975X3719

Capps, K.A., Bateman McDonald, J.M., Gaur, N. et al. (2020). Assessing the socio-environmental risk of onsite wastewater treatment systems to inform management decisions. *Environmental Science & Technology*, **54**(23), 14843–14853. https://doi.org/10.1021/ACS.EST.0C03909

Chand, P. (2018). Environmental protection and regulations in India: role of the central pollution control board. *Indian Journal of Public Administration*, **64**(4), 645–663.

Chen, Y., Wong, K., Rudy, E. et al. (2024). Decentralized water infrastructure and peri-urban water security: policy challenges and opportunities in rainwater harvesting initiatives in Mexico City. *Deleted Journal*, **2**(3), 372–399. https://doi.org/10.3138/jccpe-2023-0015

Chen, Z., Wu, Q., Wu, G. et al. (2017). Centralized water reuse system with multiple applications in urban areas: lessons from China's experience. *Resources Conservation and Recycling*, **117**, 125–136. https://doi.org/10.1016/j.resconrec.2016.11.008

Chertes., K.L., Tupitsyna., O.V., Pystin., V.N. et al. (2022). Wastewater sediment treatment facilities in a large region as a multi-component system. *Vestnik MGSU*, **17**(2), 205–221. https://doi.org/10.22227/1997-0935.2022.2.205-221

Chungyalpa, W. (2020). Examination and analysis of the central pollution control board and the state pollution control board-Indian administrative arm for environmental protection. *Indian Journal of Sustainable Development*, **6**(1), 11.

Cookey, P.E., Koottatep, T., van der Steen, P. et al. (2016). Public health risk assessment tool: strategy to improve public policy framework for onsite wastewater treatment systems (OWTS). *Journal of Water, Sanitation and Hygiene for Development*, **6**(1), 74–88.

Cullis, J.D., Rossouw, N., Du Toit, G. et al. (2018). Economic risks due to declining water quality in the Breede River catchment. *Water SA*, **44**(3), 464–473. https://doi.org/10.4314/wsa.v44i3.14

Datta, P., Mohi, G., and Chander, J. (2018). Biomedical waste management in India: critical appraisal. *Journal of laboratory physicians*, **10**(01), 006–014.

Daudin, K., Weber, C., Fouché, O. et al. (2022). Soil-water adaptive management process: the case of on-site wastewater treatment systems in peri-urban areas in France. *Environmental Challenges*, **7**, 100506. https://doi.org/10.1016/j.envc.2022.100506

Dubber, D., and Gill, L. (2014). Application of on-site wastewater treatment in Ireland and perspectives on its sustainability. *Sustainability*, **6**(3), 1623–1642. https://doi.org/10.3390/su6031623

Envall, I., Fagerlund, F., Johansson Westholm, L. et al. (2023). Existing evidence related to soil retention of phosphorus from on-site wastewater treatment systems in boreal and temperate climate zones: a systematic map. *Environmental Evidence*, **12**(1), 1–29. https://doi.org/10.1186/s13750-023-00300-7

EPA, U.S.A. (2017). *2015 water and wastewater sector specific plan*. Retrieved December 16, 2018, from https://www.dhs.gov/sites/default/files/publications/nippssp-water-2015-508.pdf

Fadel, H.E., Merzouki, M., and Benlemlih, M. (2024). Treatment of industrial wastewater. In: *Advances and Challenges in Hazardous Waste Management*. IntechOpen. https://doi.org/10.5772/intechopen.1003924

Falinski, M.M., Turley, R.S., Kidd, J. et al. (2020). Doing nano-enabled water treatment right: sustainability considerations from design and research through development and implementation. *Environmental Science Nano*, **7**(11), 3255–3278. https://doi.org/10.1039/d0en00584c

Furlong, C., Dobbie, M., Morison, P. et al. (2019). Infrastructure and Urban planning context for achieving the visions of integrated urban water management and water sensitive Urban Design: the case of Melbourne. In: *Approaches to Water Sensitive*

Urban Design: Potential, Design, Ecological Health, Urban Greening, Economics, Policies, and Community Perceptions, 329–350. Elsevier. https://doi.org/10.1016/B978-0-12-812843-5.00016-2

Gadirova, E., Hajimahmud, V.A., Abuzarova, V.A. et al. (2024). Biological treatment of wastewater: a review. In: *Revolutionizing Automated Waste Treatment Systems: IoT and Bioelectronics*, 114–125. IGI Global Scientific Publishing. https://doi.org/10.4018/979-8-3693-6016-3.ch008

Gallagher, J. and Gill, L.W. (2021). The life cycle environmental performance of on-site or decentralised wastewater treatment systems for domestic homes. *Water*, **13**(18), 2542. https://doi.org/10.3390/w13182542

Giwa, H.N., Giwa, H.O., Alepu, S.O. et al. (2024). Decentralization of household wastewater streams: challenges and perspectives. *Public Economics*, 01–06. https://doi.org/10.70107/collectjenvironsci-art0011

González-Viar, M., Díez-Montero, R., Molinos-Senante, M. et al. (2016). Cost-effectiveness analysis of sewer mining versus centralized wastewater treatment: case study of the Arga river basin, Spain. *Urban Water Journal*, **13**(3), 321–330. https://doi.org/10.1080/1573062x.2014.991742

Hanjra, M.A. and Qureshi, M.E. (2010). Global water crisis and future food security in an era of climate change. *Food Policy*, **35**(5), 365–377. https://doi.org/https://doi.org/10.1016/j.foodpol.2010.05.006

Harris-Lovett, S., Lienert, J., and Sedlak, D.L. (2018). Towards a new paradigm of Urban water infrastructure: identifying goals and strategies to support multi-benefit municipal wastewater treatment. *Water*, **10**(9), 1127. https://doi.org/10.3390/w10091127

Hofste, R.W., Kuzma, S., Walker, S. et al. (2019). Aqueduct 3.0: updated decision-relevant global water risk indicators. *World Resources Institute*, 784–786.

Hukka, J.J. and Katko, T.S. (2015). Resilient asset management and governance for deteriorating water services infrastructure. *Procedia Economics and Finance*, **21**, 112–119. https://doi.org/https://doi.org/10.1016/S2212-5671(15)00157-4

Ibrahim, F.M., Hassan, B.M.E., Elmanssury, A.E. et al. (2024). Integrated, decentralized wastewater management use to improve the environmental health of Khartoum locality Sudan. *Journal of Health and Medical Sciences*, **7**(2), 97–108. https://doi.org/10.31014/aior.1994.07.02.321

Ismaeel, T.A. (2021). Designing a decentralized sewage treatment plant. *Iraqi Journal of Industrial Research*, **8**(1), 16–28.

Jacobsen, L.B. (2018). Keeping watch over our vital water infrastructure. *Journal-American Water Works Association*, **110**(8), 54–56.

Jiao, J., Yang, Z., Zhang, Y. et al. (2022). Are farmers willing to pay for centralized mode provision of rural domestic sewage treatment? A large-scale assessment in North China. *Frontiers in Environmental Science*, **10**, 861871. https://doi.org/10.3389/fenvs.2022.861871

Karamouz, M., Movahhed, M., and Elyasi, A.H. (2022). Financial allocation and network recovery for interdependent wastewater treatment infrastructure: development of resilience metrics. *Sustainable and Resilient Infrastructure*, **8**(sup1), 262–288. https://doi.org/10.1080/23789689.2022.2148447

Li, X., Zhang, X., Zhao, M. et al. (2024). Application of decentralized wastewater treatment technology in rural domestic wastewater treatment. *Sustainability*, **16**(19), 8635–8635. https://doi.org/10.3390/su16198635

Li, Z., Zheng, L., Koottatep, T. et al. (2023). Editorial: decentralized wastewater treatment technologies. *Frontiers in Environmental Science*, **11**, 1199552. https://doi.org/10.3389/fenvs.2023.1199552

Machineni, L. (2020). Review on biological wastewater treatment and resources recovery: attached and suspended growth systems. *Water Science and Technology*, **80**(11), 2013–2026. https://doi.org/10.2166/wst.2020.034

Marshall, R. (2002). Developing environmental management systems to deliver mitigation and protect the EIA process during follow up. *Impact Assessment and Project Appraisal*, **20**(4), 286–292.

Mishra, B.K., Kumar, P., Saraswat, C. et al. (2021). Water security in a changing environment: concept, challenges and solutions. *Water*, **13**(4), 490. https://doi.org/10.3390/w13040490

Monteith, H. and Sterne, L. (2001). Wastewater treatment technologies required for current and future regulatory requirements. *Canadian Journal of Civil Engineering*, **28**(S1), 81–91. https://doi.org/10.1139/l00-079

Muhammad, M.H., Idris, A.L., Fan, X. et al.(2020). Beyond risk: bacterial biofilms and their regulating approaches. *Frontiers in Microbiology*, **11**, 928. Frontiers Media S.A. https://doi.org/10.3389/fmicb.2020.00928

Neth, M., Mattsson, A., I'Ons, D. et al. (2022). A collaborative planning process to develop future scenarios for wastewater systems. *Journal of Environmental Management*, **316**, 115202. https://doi.org/10.1016/j.jenvman.2022.115202

Prescott, M.F., Dobbie, M.F., and Ramirez-Lovering, D. (2021). Green infrastructure for sanitation in settlements in the global south: a narrative review of socio-technical systems. *Sustainability*, **13**(4), 2071. https://doi.org/10.3390/SU13042071

Qadir, M., Drechsel, P., Cisneros, B.J. et al. (2020). Global and regional potential of wastewater as a water, nutrient and energy source. In *Natural resources forum* (Vol. 44, No. 1, pp. 40–51). Oxford, UK: Blackwell Publishing Ltd.

Rashid, S.S. and Liu, Y. (2020). Assessing environmental impacts of large centralized wastewater treatment plants with combined or separate sewer systems in dry/wet seasons by using LCA. *Environmental Science and Pollution Research*, **27**(13), 15674–15690. https://doi.org/10.1007/s11356-020-08038-2

Rosencranz, A. and Upadhyay, V. (2011). Some suggestions and recommendations towards a model State Pollution Control Board (SPCB) in India. *Environmental Law and Practice Review*, **1**, 106.

Saadatinavaz, F., Alomari, M.A., Ali, M. et al. (2024). Striking a balance: decentralized and centralized wastewater treatment systems for advancing sustainable development goal 6. *Advanced Energy and Sustainability Research*, **5**(10), 2400097. https://doi.org/10.1002/aesr.202400097

Sabliy, L. (2023). Development and implementation of industrial wastewater treatment technologies. *Problems of Water Supply Sewerage and Hydraulic*, **45**, 62–70. https://doi.org/10.32347/2524-0021.2023.45.62-70

Schellenberg, T., Subramanian, V., Ganeshan, G. et al. (2020). Wastewater discharge standards in the evolving context of urban sustainability–the case of India. *Frontiers in Environmental Science*, **8**, 30.

Schweitzer, L. and Noblet, J. (2018). Water contamination and pollution. In: *Green Chemistry*, 261–290. Elsevier.

Siddiqui, A., Chauhan, P., Halder, S. et al. (2022). Effect of COVID-19-induced lockdown on NO2 pollution using TROPOMI and ground-based CPCB observations in Delhi NCR, India. *Environmental Monitoring and Assessment*, **194**(10), 714.

Singh, S., and Farswan, S. (2023). Pivotal role of wastewater in era of circular economy. *Geodiversity and Impact on Environment*, **23**(2), 1–13.

Singh, P., Anand, A., Srivastava, P.K. et al. (2021). Delineation of groundwater potential zone and site suitability of rainwater harvesting structures using remote sensing and in situ geophysical measurements. In *Advances in Remote Sensing for Natural Resource Monitoring* (eds P.C. Pandey and L.K. Sharma). John Wiley & Sons. https://doi.org/10.1002/9781119616016.ch11

Singh, S., Kumar, V., Romero, R. et al. (2019). Applications of nanoparticles in wastewater treatment. In *Nanobiotechnology in bioformulations*, 395–418. Cham: Springer International Publishing.

Smail, L. (2024). Water and waste regulation. In: *Bloomsbury Publishing Plc eBooks*. London: Bloomsbury Professional. https://doi.org/10.5040/9781526519023

Srivastava, R.R. and Singh, P.K. (2022). Reuse-focused selection of appropriate technologies for municipal wastewater treatment: a multi-criteria approach. *International Journal of Environmental Science and Technology*, **19**(12), 12505–12522. https://doi.org/10.1007/s13762-021-03803-3

Tortajada, C., Hartley, K., Ong, C. et al. (2024). A review of alternative water supply systems in ASEAN. In: *Oxford Research Encyclopedia of Environmental Science*. Oxford university press.

Vallero, D.A. (2024). *Fundamentals of Water Pollution: Quantifying Pollutant Formation, Transport, Transformation, Fate and Risks*. Elsevier.

Varma, V.G., Jha, S., Raju, L.H.K. et al. (2022). A review on decentralized wastewater treatment systems in India. *Chemosphere*, **300**, 134462.

Wang, H., Wanakule, N., Asefa, T. et al. (2022). A risk-based framework to evaluate infrastructure investment options for a water supply system. *Journal of Environmental Engineering*, **148**(11), 05022008. https://doi.org/10.1061/(asce)ee.1943-7870.0002069

Wang, M., Zhu, J., and Mao, X. (2021). Removal of pathogens in onsite wastewater treatment systems: a review of design considerations and influencing factors. *Water*, **13**(9), 1190. https://doi.org/10.3390/w13091190

Wang, X., Daigger, G., Lee, D.J. et al. (2018). Evolving wastewater infrastructure paradigm to enhance harmony with nature. *Science Advances*, **4**(8), eaaq0210. https://doi.org/10.1126/sciadv.aaq0210

Werkneh, A.A. and Gebru, S.B. (2022). Development of ecological sanitation approaches for integrated recovery of biogas, nutrients and clean water from domestic wastewater. *Resources Environment and Sustainability*, **11**, 100095. https://doi.org/10.1016/j.resenv.2022.100095

Wysowska, E. and Kicińska, A. (2022). A model-based approach to risk evaluation and the assessment of protection provided by water intake and treatment systems. *Journal of Ecological Engineering*, **23**(7), 284–298. https://doi.org/10.12911/22998993/150272

Yates, M.V. (2011). On-site wastewater treatment. *Encyclopedia of Environmental Health*, **1–5**(4), 256–263. https://doi.org/10.1016/B978-0-444-52272-6.00046-5

Younas, F., Mustafa, A., Farooqi, Z.U.R. et al. (2021). Current and emerging adsorbent technologies for wastewater treatment: trends, limitations, and environmental implications. *Water*, **13**(2), 215. https://doi.org/10.3390/w13020215

Zainudeen, M.Y. (2007). Infrastructure for sustainable development and management of water resources. *Engineer: Journal of the Institution of Engineers, Sri Lanka*, **40**(2), 41. https://doi.org/10.4038/engineer.v40i2.7138

4

Sewage Composition and Treatment: Economic Burden and Future Implications

Soumyajit Chandra, Soumya Pandit, and Abdurrasheed Abubakar Yunus*

Department of Life Sciences, Sharda School of Basic Sciences and Research, Sharda University, Greater Noida, Uttar Pradesh, India

*Corresponding author: sounip@gmail.com

4.1 Introduction

A large percentage of untreated sewage contains pathogens, which include bacteria, viruses, and parasites, such as cholera, typhoid, and hepatitis. Proper treatment of sewage will ensure low health hazards to people. Disease prevention is fundamental to the functionality of sewage management in creating a smooth, healthy society where people get to enjoy living and improved quality of life as a result of good sewerage. Sewage management is equally important for environmental reasons due to its effects on cleansing water bodies, removing all harmful pollutants from their systems that would harm water life else (Smarzewska and Morawska, 2021). Untreated sewage has the potential to be an eco-disruptor and lead to problems like eutrophication and dead zones in aquatic ecosystems. Sewage treatment reduces the presence of materials like organic matter, heavy metals, nutrients including phosphorus and nitrogen, and pathogens. With these effects, protection toward the ecosystems and establishment toward environmental sustainability is realized. The natural resources and preservation of both human and wild populations are protected.

In this line, financially, the spent money on sewage treatment infrastructure would avoid a lot of medical expenditure, environmental restoration, and financial losses brought about by polluted sources of water (Starkl et al., 2022). Moreover, new sewage

Wastewater to Resource Recovery: Applying the Circular Economy Toward Sustainable Development, First Edition. Edited by Arvind Kumar Mungray et al. Published 2026 by John Wiley & Sons, Inc.

treatment technologies will enable the recovery of precious resources such as fertilizers meant for agriculture and biogas for energy generation, thereby supporting the circular economy. Effective sewage treatment results in reduced operating costs while ensuring the satisfaction of environmental requirements, which is good for businesses dependent on clean water for their operations. This further contributes to the growth of industries and therefore economic stability. There are also several desirable social impacts of sewage treatment. Clean water will be produced; offensive smells will be eradicated, and quality will be improved in the living environments through proper management of the same, which raises people's quality of living. Sewage treatment helps in making society more equitable just by removing such inequalities. Moreover, the reduction of greenhouse gas emissions is helpful in reducing the impacts of climate change. Advanced sewage treatment technology may capture methane and use it as a source of energy, hence reducing the carbon footprint as a whole. Well-designed sewage treatment systems also make metropolitan areas resilient to the impacts of climate change, including increased flooding and waterborne diseases as a result of climatic phenomena. This adaptability requires developing resilient and sustainable communities under the effects of climate change.

4.2 Definition and Composition of Sewage

4.2.1 Definition of Sewage

Wastewater, commonly referred to as sewage, is a complex mixture of varied waste products and water coming from stormwater runoff and residential, commercial, and industrial sources (Figure 4.1). It consists of all the waterborne wastes

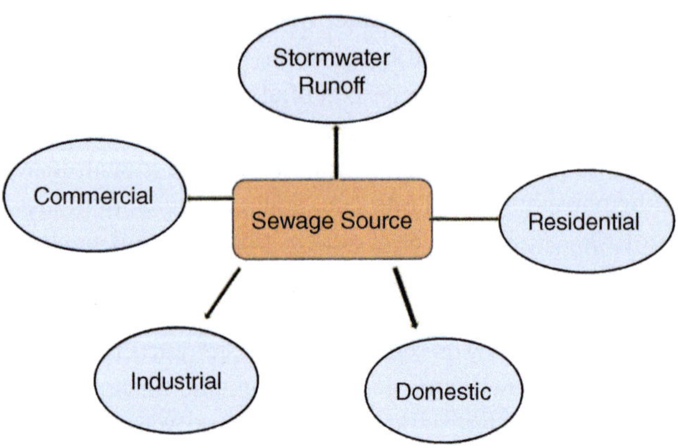

Figure 4.1 Source of sewage.

discharged from sinks, bathtubs, showers, and other domestic and commercial sources. The composition of sewage can be very different from location to location, but broadly, it consists of suspended particles, nutrients, pathogens, organic and inorganic elements, and chemicals.

About 99.9% of domestic sewage is made up of water, with the remaining 0.1% being made up of suspended and dissolved materials (Al-Mawla et al., 2023). These materials include food scraps, soaps, detergents, personal hygiene items, and human waste (urine and feces). Conversely, depending on the kind of industrial activity, industrial sewage may contain a wider range of contaminants, including heavy metals, chemicals, oils, and solvents (Khan et al., 2024). Treatment of sewage is a very complex process, necessitated by the requirement of tailor-making it for different characteristics that come from the wastewater because it presents differently in its composition. This water is known as stormwater runoff and is another component of sewage because it is caused by rain or snowmelt that flows over parking lots, roadways, and other surfaces; carries pollutants like oil, fertilizers, pesticides, and litter; and often enters the sewage system, thereby increasing the volume and complexity of wastewater to be treated (Muttil et al., 2023). Stormwater intrusion into sewers could create large variation in sewage volume and quality, which is a new challenge for the treatment plant.

In most cases, sewage is characterized by several key considerations: Biochemical Oxygen Demand (BOD): It measures the amount of oxygen required by the bacteria to break down organic matter in water; hence, it is an index of the quantity of organic pollution in water. Chemical Oxygen Demand (COD): It measures the total quantity of substances present in the water that are capable of oxidation. Total Suspended Solids (TSS) include silt, decaying plant and animal detritus, and other wastes suspended in water. Pathogens include microorganisms capable of causing diseases, such as bacteria, viruses, and parasites. Nutrients like nitrogen and phosphorus are potential contributors to eutrophication in water bodies in excess concentrations.

4.2.2 Composition of Sewage

Sewage is generally termed as wastewater and is a complex mixture of materials sourced from stormwater, industrial, and residential sources. According to the source, the composition of sewage can be quite different, but generally, it contains a mix of nutrients, pathogens, suspended sediments, chemicals, and organic and inorganic components.

Organic Matter: Generally, organic matter in sewage comprises food waste, soaps, detergents, and other biodegradable materials. Quantification of this organic stuff is done through parameters such as COD and BOD. BOD measures the amount of oxygen required by bacteria in decomposing organic materials. The higher the BOD, the higher organic pollution. COD gives an estimate of the general quality of the water by determining the net amount of chemicals present in water that are susceptible to oxidation.

Inorganic Matter: Sewage contains a range of metals, minerals, and salts that are considered inorganic compounds. Stormwater runoff, industrial operations, and domestic cleaning products can all be the source of these materials. Heavy metals like lead, mercury, and cadmium are common inorganic pollutants that, even at low quantities, may seriously endanger human health and the environment (Raj and Das, 2023).

Nutrients: Essential elements found in sewage include phosphate and nitrogen, which are leftovers from fertilizers, detergents, and human waste. Although these nutrients are necessary for plant development, an overabundance of them in water bodies can cause eutrophication, which can result in hazardous algal blooms, oxygen depletion, and the eventual death of aquatic life (Geletu, 2023).

Pathogens: Pathogens, which include bacteria, viruses, and parasites, are microorganisms that cause disease and are found in sewage. *Salmonella, Escherichia coli*, and many virus types that may cause hepatitis and gastroenteritis are common pathogens discovered in sewage. Since these bacteria provide serious health concerns to the public, efficient treatment procedures are required to guarantee the safety of the water.

Chemical and Toxic Substances: Additionally, sewage may contain a variety of hazardous materials and chemicals, particularly from industrial sources. These include personal hygiene items, insecticides, solvents, medicines, and other potentially harmful substances. Numerous of these chemicals may build up in the food chain and are persistent in the environment, endangering the long-term health of both humans and the ecosystem.

Suspended Solids: The particles that make up TSS in sewage include silt, decomposing plant and animal matter, industrial waste, and other detritus. Excessive TSS levels can clog rivers and treatment systems and lower water quality by obstructing light penetration, which is necessary for aquatic vegetation.

Variation by Source: Sewage's composition might differ greatly depending on where it comes from. While industrial sewage may include larger quantities of heavy metals, chemicals, and inorganic elements, domestic sewage often has a higher proportion of organic matter and nutrients. Particularly from metropolitan areas, stormwater runoff introduces extra contaminants, including oil, pesticides, and trash (Spahr et al., 2020).

4.3 Current State of Sewage Treatment

Although sewage treatment is an essential component of contemporary environmental and public health protection, its application differs greatly worldwide. There are significant disparities in sewage treatment between industrialized and developing nations, which stem from the intricate interaction of technological, economic, and regulatory issues. Primary, secondary, and tertiary treatment are the usual phases of sewage treatment. Large solids are physically removed during primary treatment by means of screening and sedimentation. In order to decrease BOD and break down organic materials, secondary treatment employs biological processes. By eliminating pathogens, nutrients, and other lingering impurities, tertiary treatment further purifies the water. This process frequently makes use of cutting-edge technologies, including chemical treatment, filtration, and disinfection.

4.3.1 Overview of Sewage Treatment Methods

In order to create a safe effluent that can be reused or released into the environment, wastewater must first undergo a crucial procedure called sewage treatment. Usually, there are several steps in the procedure, each intended to target a particular kind of pollution. These phases encompass primary, secondary, and tertiary care, in addition to advanced treatment technologies (Figure 4.2).

Primary Treatment: The initial step in treating sewage is called primary treatment, and it focuses on physically removing suspended debris and big particles. Several processes are involved at this stage: using grates and screens,

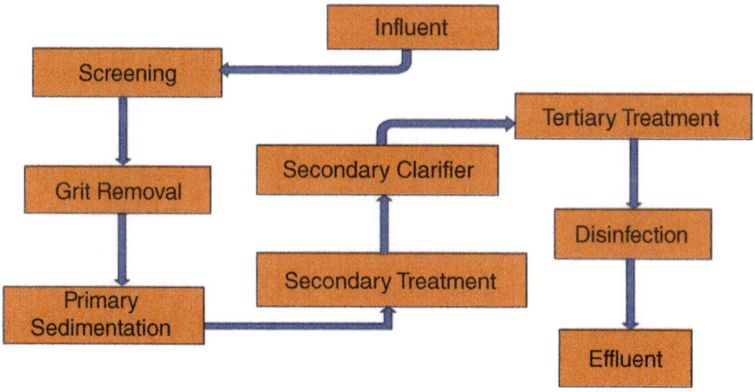

Figure 4.2 Mechanism of sewage treatment.

screening eliminates big items like sticks, rags, and other waste. Sewage can be allowed to settle in huge tanks due to sedimentation, where lighter materials like oils and grease float to the top and are skimmed off, while heavier contents sink to the bottom as sludge. About 50–60% of the suspended particles are eliminated during this procedure, and the BOD is decreased by 30–35% (Md Anawar and Chowdhury, 2020).

Secondary Treatment: The purpose of secondary treatment is to eliminate organic debris that has dissolved and become colloidal and evaded first treatment. The degradation of organic contaminants occurs at this stage through biological mechanisms. The process of activated sludge involves aerating sewage in large tanks in order to encourage the development of aerobic bacteria that break down organic waste. Secondary clarifiers are used to settle out the resultant sludge. Sewage is sprayed over a bed of stones or other medium that has been coated in microbial film using trickling filters, allowing microorganisms to decompose organic contaminants. Forced air is used in biological aerated filters (BAFs) to increase microbial activity, whereas sunshine, algae, and bacteria are naturally present in stabilization ponds and lagoons.

Tertiary Treatment: The last phase of sewage treatment, known as tertiary treatment, aims to polish the effluent by getting rid of any last bits of nutrients, pathogens, and inorganic compounds. To eliminate any residual suspended particulates, the effluent is filtered via materials like carbon or sand. By utilizing chemical disinfectants, ultraviolet (UV) light, or ozone, disinfection gets rid of germs and makes treated water safe to reuse or discharge (Zewde et al., 2020). Nutrient removal systems combine chemical precipitation and biological nutrient removal (BNR) to target nitrogen and phosphorus, which can induce eutrophication in receiving water bodies.

Advanced Treatment Technologies: Higher degrees of purification are attained by addressing certain contaminants with advanced treatment technology. Membrane bioreactors (MBRs) produce high-quality effluent with a less environmental footprint and more efficiency by combining biological treatment with membrane filtration. Semipermeable membranes are used in reverse osmosis (RO) to extract dissolved salts, metals, and other impurities. Chemical oxidants like hydrogen peroxide and ozone are used in advanced oxidation processes (AOPs) to break down complex organic contaminants that are hard for biological systems to break down. By up-taking, degrading, or stabilizing pollutants, plants and natural ecosystems are used in phytoremediation (Khan et al., 2023).

4.3.2 Global Sewage Treatment Statistics

Infrastructure for treatment of wastewater in many of the developed economies is considerably advanced and evolved. For most of these city centers, comprehensive tertiary wastewater treatments are available wherein all three stages of wastewater treatment processes are present. Most effectively, the legal regimes of a country like the United States, Germany, or Japan regarding such issues stringently lay stringent wastewater standards, thus resulting in maximum quality of protecting both environments and humans too. For example, Singapore is known for its high-tech sewage treatment plants that include massive-scale water recycling programs and MBR technology. Similarly, Sweden's approach incorporates resource recovery with state-of-the-art treatment methods that turn sewage into fertilizer and biogas. These countries are an example of how infrastructural and technological development can lead to efficient and sustainable sewage management (Table 4.1).

Conversely, most developing countries still experience raw sewage due to a meager budget, poor structural setup, and weak application of policies. These nations have huge percentages of untapped raw sewage that usually finds its

Table 4.1 Global sewage statistics.

Location	Improved sanitation accessibility (%)	Coverage of wastewater treatment (%)	Remarks
Europe	95–100	80–90	Strict environmental regulations; good adherence to wastewater treatment EU requirements.
America	100	85–95	High investment in infrastructure; extensive application of cutting-edge therapeutic technology.
South Asia	30–60	20–40	Low coverage in rural areas;
Africa	25–40	10–20	Limited access to sanitation; many areas just receive minimal care or none at all.
Global Average	~74	~56	Reflects overall progress; there are notable differences between developed and developing regions.

way into all water bodies, causing further pollution and posing a higher risk to human health in the community. For instance, sewage treatment mechanisms have not kept pace with fast urbanization and growth of population in India and Nigeria, which has led to widespread water pollution and even public health crises. International cooperation or aid is often applied for campaigns to improve sewage treatment of poor countries. Not even with these campaigns, development seem slow and the chasm between developed and underdeveloped regions is significantly considerable. About 60% of wastewater in the world is estimated to get into the environment without appropriate treatment. Approximately 70% of wastewater in high-income nations is treated, compared to only 8% in low-income nations. This discrepancy emphasizes how urgently wastewater management needs innovation and investment, especially in developing nations. Adoption of decentralized systems, which provide adaptable and affordable options for places without substantial infrastructure, is one of the latest developments in sewage treatment. Technologies that have the potential to offer efficient treatment in environments with limited resources, such as artificial wetlands, anaerobic digesters, and inexpensive filtering systems, are being investigated.

4.3.3 Case Studies

Examples of successful sewage treatment programs:

4.3.3.1 Singapore

Singapore is well known for its creative and effective sewage treatment systems, which have helped the city-state get past its small size and lack of water supplies. Key components of the sewage treatment system in Singapore include:

NEWater: High-grade reclaimed water that satisfies drinking water requirements is created from treated sewage and is known as NEWater in Singapore (Yeniçeri and Yeniçeri, 2023). RO, UV disinfection, and microfiltration are some of the sophisticated treatment techniques used by the NEWater program to create clean water from sewage.

Deep Tunnel Sewerage System (DTSS): The sewage from all around the island is collected and treated by Singapore's DTSS, an extensive network of deep tunnels and centralized wastewater treatment facilities (Kua et al., 2022). The DTSS reduces the need for energy-intensive pumping by using gravity-driven conveyance to move sewage to centralized treatment facilities. The DTSS reduces environmental impact and maximizes resource efficiency by centralizing treatment and optimizing land usage.

Sustainable Resource Management: Sustainability and resource recovery are on the top in the sewage treatment efforts of Singapore. Sewage sludge is digested anaerobically to produce biogas, which is further used in the generation of renewable energy besides producing NEWater. Dried sludge is incinerated, after which the ash is transformed into building materials. It therefore goes on to optimize resources with lesser waste.

4.3.3.2 Sweden

Sweden is known for their resource recovery and ecologically safe sewage treatment plants. Sweden actually emphasizes the recovery of resources and energy efficiency, thus also keeping up the quality of water. Some of the crucial ingredients to this effective Sweden's sewage treatment programs include:

Integrated Treatment System: The sewage treatment plants in Sweden apply integrated treatment systems that successfully remove contaminants from wastewater by combining chemical, biological, and physical processes. These facilities meet stringent environmental requirements and reach high treatment efficiency using advanced technologies such as nutrient removal, activated sludge, and biofiltration.

Nutrient Recycling: Recycling nutrients (especially nitrogen and phosphorus) from sewage is highly valued in Sweden. Phosphorus is extracted from sewage sludge through treatment, and this valuable fertilizer is then used to agriculture to minimize environmental contamination and lessen dependency on depleting phosphorus stocks. Furthermore, cutting-edge nutrient removal technologies are used to reduce nutrient releases into water bodies, reducing eutrophication and protecting aquatic environments.

Decentralized Treatment: Sweden uses decentralized treatment methods in rural and isolated locations in addition to centralized sewage treatment plants (Wittmanová et al., 2023). These systems, which include built wetlands and small-scale wastewater treatment facilities, offer affordable options for managing wastewater in places with little infrastructure or low population density. Decentralized treatment encourages local water reuse and recycling while lowering the environmental impact of sewage treatment.

4.3.3.3 India

India's fast urbanization, population increase, and poor infrastructure make sewage management extremely difficult (Schellenberg et al., 2020).

Ganga River Pollution: One of India's holiest rivers, the Ganga, is also one of the most contaminated (Khan et al., 2021). Significant urban areas beside the river, including Patna, Varanasi, and Kanpur, release copious volumes of raw sewage into the waterway. Millions of people who depend on the river for irrigation, drinking water, and bathing are at serious danger of health problems as a result of the elevated levels of pollutants, heavy metals, and infections (Lin et al., 2022).

Many health problems are the result of improper sewage treatment. Sewage-related illnesses such as diarrhea, cholera, and hepatitis are common in places with inadequate sanitation (Battersby, 2023).

Economic Consequences: India bears the heavy financial cost of inadequate sewage treatment (Reymond et al., 2020). Both the financial burden of treating waterborne infections on healthcare systems and the negative effects of sickness on labor force participation are significant. Furthermore, the tourism sector loses because dirty waterways discourage tourists, which has an impact on local economies that depend on tourism (Bhattarai and Conway, 2021).

4.3.3.4 Nigeria

Nigeria also has challenges with poor sewage treatment, which are exacerbated by the country's fast urbanization, limited infrastructure, and tax regulatory compliance. These difficulties are felt in many facets of Nigerian life, including:

Lagos Lagoon Pollution: One of the biggest cities in Nigeria, Lagos, releases a considerable amount of untreated sewage into the Lagos Lagoon (Olabode and Comte, 2024). As a result, there is a serious pollution problem with elevated levels of chemical and organic pollutants. The water quality has declined due to pollution, endangering aquatic life and upsetting the local fishing businesses, which provide a living for many communities (Kolawole and Iyiola, 2023).

Public Health Crisis: In Nigeria, there is no sewage treatment. This aggravates acute public health problems. Typhoid, cholera, and gastroenteritis outbreaks are common during the rainy season, especially during the flooding, which can overflow the sewage. There is a risk of poor health due to the scarcity of safe water and adequate sanitation facilities in densely populated urban centers and informal settlements.

Environmental Degradation: Sewage, if not treated and released in rivers and coastal water bodies, causes severe environmental devastation. Habitats are destroyed, while biodiversity is lost, as a result of disturbing the equilibrium of aquatic ecosystems. For the environment, this brings about an impact on human communities that rely on it for food and economic gain besides the environment (Reid et al., 2019).

Economic Costs: Poor sewage treatment in Nigeria, therefore, has a paramount economic implication. The bottom line is the high expense of healthcare associated with the treatment of diseases stemming from polluted water and productivity lost due to illness. Additionally, sectors within Nigeria's economy that focus on agriculture and fisheries experience loss due to natural resource exhaustion.

4.4 Economic Effects of Inadequate Wastewater Treatment

The world economy suffers majorly due to poor treatment of sewage. There is an impact on communities, governments, and companies. The impact extends beyond short-term effects on the environment and public health; they also have some influence on tourism, healthcare expenditure, economic productivity, and overall quality of living. It is important to study these financial implications so well to argue for better legal and regulatory frameworks on the water treatment of sewage.

4.4.1 Effects on Industries

This waterborne disease also costs the industries, as it renders workers incapable of performing their duties, thus absorbing a lot of manpower in absenteeism and decreased productivity. This is particularly destructive in developing countries where labor-based industries make up a considerable percentage of the GDP. Lost productive hours not only undermine economic efficiency, but also trickle down to affect household income, thereby worsening poverty and the general standard of living.

4.4.2 Effects on Communities

The most significant financial cost of untreated sewage is the high medical costs associated with infections acquired from water. The prevalence of diseases such as cholera, dysentery, and hepatitis is attributed to contaminated water sources, and these diseases are therefore very costly to treat. These diseases are more prevalent in countries with ineffective wastewater treatment facilities and hence create a greater need for medical care and medication. Public health structures tend to be severely cash-strapped, which diverts resources away from other critical services and infrastructure development.

Another industry that is badly impacted by insufficient sewage treatment is tourism (Khan et al., 2020). Beaches, rivers, and lakes that are dirty discourage travelers, which lowers tourism-related income (Pásková et al., 2024). Communities around rivers and coasts that depend mostly on tourism may see economic downturns if travelers choose greener, safer travel options. In certain areas, the bad press brought on by dirty settings can have a permanent impact on tourism, necessitating large expenditures for marketing and rehabilitation initiatives to make the areas appealing again.

4.4.3 Government and Municipal Expenditures

Large sums of money must be spent by governments and municipalities on maintaining and modernizing the infrastructure that insufficient sewage treatment has made necessary. Regular maintenance, quick fixes to stop disease outbreaks, and the requirement for backup water supply systems are expensive and take money away from other important development initiatives. Furthermore, increasing treatment capacity and investing in cutting-edge sewage treatment technology involve large financial outlays, which may be quite burdensome, particularly for developing countries with tight budgets (Aldalati, 2023).

Inadequate sewage treatment has a financial cost that frequently makes social and economic inequity worse. Poorer populations are usually the most affected by inadequate sanitation and sewage treatment because they cannot afford clean water and adequate medical care. Such inequality perpetuates the vicious cycle of poverty and disease, which hinders the progression of society and the economy. Hence, the investment in infrastructure for sewage treatment, aside from public health and environmental concerns, must be made to ensure social equity and economic stability.

4.5 Technologies and Innovations for Improving Sewage Treatment

Technological innovation in the field of sewage treatment will enhance effectiveness, sustainability, and efficiency in water management. Some of the emerging technologies aimed at breaking down the sophisticated nature of sewage; environmental degradation will be controlled, and money wastage brought about by inadequate treatment will be reduced. Some of the emerging new breakthroughs and latest technologies shaping the sewerage future are shown as follows:

4.5.1 Advanced Treatment Technologies

Membrane Bioreactors: In the case of MBRs, it produces high-quality effluent in a lower footprint by combining membrane filtration with biological treatment processes. A membrane is used for separation in MBR systems: it is either micro- or ultrafiltration, but there is a reactor with biomass that breaks the organic contaminants. Higher treatment efficiency, increased capacity to handle larger pollutant loads, and decreased sludge generation are just a few benefits of this technology (Ferrentino et al., 2023).

AOPs: Highly reactive radicals, including hydroxyl radicals, are produced during advanced oxidation procedures in order to break down complex organic contaminants that are impervious to traditional treatment techniques. Fenton's reagent, hydrogen peroxide under UV light, and ozone-based oxidation are examples of common AOPs (Savun-hekimoğlu, 2021). These procedures are quite good in degrading personal hygiene items, medications, and other micropollutants. AOPs can be added to current treatment systems to improve impurity removal and guarantee the creation of high-grade effluent.

Phytoremediation: Utilizing plants, phytoremediation eliminates, breaks down, or stabilizes pollutants found in wastewater. Through their roots, certain plants can absorb organic contaminants, heavy metals, and nutrients from water (Bhat et al., 2022). For the treatment of wastewater, this method is both environmentally friendly and visually beautiful, especially in places with low pollution loads. Green areas may be created in urban settings, and overall efficiency can be increased by integrating phytoremediation with other treatment techniques (Mocek-Płóciniak et al., 2023).

4.5.2 Innovations in Sewage Management

Wastewater treatment methods are changing as a result of innovations in sewage management, becoming more effective, long-lasting, and profitable. These developments tackle issues brought on by urbanization, population growth, and environmental concerns. Some of the most significant advancements in sewage control are listed here.

Decentralized Treatment Systems: An adaptable and scalable substitute for conventional centralized sewage treatment facilities is provided by decentralized treatment systems. These systems, which are intended to treat wastewater locally, at its source, minimize the need for large sewage networks and the associated transportation expenses. They are especially helpful in emerging countries, small towns, and rural locations where centralized infrastructure is either too expensive or unfeasible. Package treatment plants are one of the innovations in this field; they are small, modular systems that can be swiftly assembled and tailored to match particular local needs (Brar et al., 2022). Another example is constructed wetlands, which cleanse wastewater through natural processes including plants, soil, and microbes. In addition to offering other environmental benefits, including habitat formation and landscape enhancement, these wetlands are reasonably priced and need little upkeep.

Use of Artificial Intelligence and IoT in Sewage Management: The operation, monitoring, and process optimization of treatment facilities are being revolutionized by the incorporation of smart technology and digital solutions in sewage management (Pandey et al., 2022). These developments lower expenses, increase efficiency, and strengthen regulatory compliance. In order to facilitate proactive maintenance and early issue discovery, real-time monitoring makes use of modern sensors and Internet of Things (IoT) devices to provide data on water quality parameters, flow rates, and system performance (El-Shafeiy et al., 2023). Artificial intelligence (AI)-driven automated control systems optimize treatment procedures based on real-time data by modifying chemical doses, aeration rates, and other parameters to increase productivity and lower operating costs. These systems employ machine learning algorithms. By predicting equipment problems before they happen, predictive maintenance solutions reduce downtime, increase equipment longevity, and save maintenance costs.

4.5.3 Reducing Environmental Impact

Ecosystem preservation and sustainable development depend on sewage treatment's negative environmental effects being minimized. The reduction of pollution, preservation of resources, and enhancement of ecological health are contingent upon technological innovations and creative management strategies (Rehman et al., 2020). We can greatly reduce the environmental impact of sewage management by putting in place energy-efficient procedures, nutrient recovery methods, nature-based solutions, green infrastructure, and pollution control measures.

Nutrient Recovery and Recycling: The recovery and repurposing of vital nutrients such as nitrogen and phosphorus from sewage lower the demand for manufactured fertilizers and guard against nutrient contamination in water bodies. By recovering phosphorus and ammonia from sewage, a process known as struvite precipitation creates a slow-releasing fertilizer that may be applied to crops, completing the nutrient loop and lowering the danger of eutrophication in aquatic environments. Ammonia stripping removes ammonia from wastewater for reuse or conversion into nitrogen gas by nitrification–denitrification, hence lowering nitrogen burdens in effluents (Chen et al., 2021). Nitrification–denitrification procedures recover nitrogen from sewage.

Green Infrastructure and Stormwater Management: The environmental effects of urban runoff and sewage overflows can be greatly decreased by including green infrastructure into sewage control. Rainwater is absorbed by green walls and roofs, which also increase urban green areas, improve air quality, and reduce runoff and the strain on sewage systems (Nazif et al., 2021). Rainwater can percolate through permeable pavements, lowering runoff and enhancing groundwater recharge. This helps control stormwater runoff, lessen floods, and alleviate the strain on sewage treatment facilities. In order to enhance water quality and lessen sewage overflows, bioswales and rain gardens collect and treat stormwater runoff using soil and plant (Deksissa, 2023). This process slows down runoff, filters pollutants, and increases groundwater recharge.

Pollution Prevention and Source Control: Reducing the environmental impact of wastewater treatment may be achieved by effectively keeping contaminants out of the sewage system. The load of hazardous contaminants may be greatly reduced by enforcing companies to treat their effluents prior to releasing them into municipal sewage systems. This keeps out heavy metals, toxic compounds, and other pollutants from getting to treatment facilities. Campaigns for education and public awareness are also very important. The entry of hazardous materials into wastewater can be decreased by educating the public on correct waste disposal practices and the effects of common home chemicals on sewage systems (Ahmed et al., 2021). Pollution prevention requires actions that encourage the adoption of environmentally friendly goods and safe disposal techniques for hazardous and prescription items.

4.6 Future Projections and Potential Solutions

Future sewage treatment and management will be influenced by population increase, urbanization, and the pressing need to reduce environmental consequences. Forecasts suggest that there will be a sustained increase in the need for effective and environmentally friendly wastewater treatment methods (Khan et al., 2022). Adopting cutting-edge technology, integrated management strategies, and aggressive policy actions will be necessary to address these issues. Below are some significant forecasts for the future as well as possible remedies to lessen the financial strain and environmental effects of sewage treatment.

4.6.1 Future Projections

Predicted Trends in Sewage Generation and Treatment Needs: Global trends including population expansion, urbanization, climate change, and technological improvements will have a big impact on sewage generation and treatment in the future. The amount of sewage produced will rise along with the world's population, especially in metropolitan areas. The current sewage treatment system will be under more stress as a result of this surge, requiring renovations and extensions to accommodate increased capacities. At the same time, conventional sewage treatment methods will face difficulties due to the introduction of new pollutants such as industrial chemicals, medicines, personal care items, and microplastics (Hamidian et al., 2021). Therefore, in order to properly handle these new toxins, treatment methods in the future will need to change. Sewage treatment is predicted to become more effective and efficient due to the rapid improvements in treatment technology. Advanced oxidation techniques, MBRs, and electrochemical treatments are examples of innovations that will proliferate and provide enhanced water quality along with increased degrees of contamination removal. Furthermore, the recovery of valuable resources from sewage will receive more attention in the future. Treatment facilities will be built to recover nutrients (nitrogen and phosphorus), energy (biogas), and water for reuse, which will lessen the environmental impact of sewage treatment and have positive economic effects (Rufí-Salís et al., 2022).

Impact of Population Growth and Urbanization: The current sewage infrastructure will continue to be strained by rapid urbanization. The need for effective and efficient sewage treatment technologies will grow as more people migrate to cities. Many metropolitan areas, especially in developing nations, currently struggle with insufficient sewage systems. This will necessitate large expenditures for the modernization and expansion of infrastructure. Urbanization and population expansion will raise the pollutant loads in sewage, requiring more sophisticated treatment methods to comply with environmental laws and safeguard public health. Decentralized sewage treatment systems, which treat wastewater locally and eliminate the need for large sewer networks while offering more flexibility in controlling sewage in densely populated metropolitan areas, will proliferate in response to these issues. The problems with sewage management that now exist will get worse due to climate change. Sewage systems can be overwhelmed by extreme weather events like flooding and heavy rain, which can contaminate water bodies and cause overflows. Additionally, biological processes in treatment facilities can be impacted by rising temperatures. In order to overcome these obstacles, future sewage management plans will need to take climate resilience into account (Dickin et al., 2020).

4.6.2 Potential Solutions

4.6.2.1 Policy and Regulatory Frameworks for Improved Sewage Management

Ensuring compliance with environmental standards and enhancing sewage management need efficient policies and regulatory frameworks. In sewage treatment, governments and regulatory agencies are essential in establishing guidelines, upholding laws, and encouraging best practices. Important policy initiatives consist of:

Strict Water Quality Requirements: To safeguard the environment and public health, severe water quality requirements for treated effluent must be established and strictly enforced.

Pollution Prevention and Regulation: Putting laws into place to stop contaminants from entering sewage systems, such as industrial pretreatment programs that control the discharge of industrial wastewater.

Resource Recovery Incentives: Encouraging the adoption of circular economy techniques by providing incentives and subsidies for the recovery of valuable resources from sewage, such as water, energy, and nutrients.

Investment Incentives: Financing investments in sewage infrastructure upgrades, technology adoption, and research and development through grants, tax breaks, and other forms of financial assistance.

Investment in Infrastructure and Technology: To solve the issues brought on by population increase, urbanization, and environmental deterioration, as well as to modernize sewage treatment facilities, significant investments in infrastructure and technology are needed. Important areas for investments consist of:

Infrastructure Upgrades: Putting money into sewage treatment plant and sewer network development, renovation, and modernization to boost capacity, boost efficiency, and lessen environmental effect.

Technology Adoption: To improve treatment efficiency and resource recovery, consider investing in cutting-edge sewage treatment technology, including MBRs, sophisticated oxidation processes, and decentralized treatment systems.

Climate-resilient Designs: To maintain the dependability and continuity of sewage treatment operations, investments should be made in climate-resilient infrastructure designs that can resist extreme weather events like storms and floods.

Research and Development: Putting money into research and development to create innovative, more economical, ecologically friendly, and efficient sewage treatment technologies, procedures, and materials.

4.6.2.2 Public–Private Partnerships and Community Involvement

The community's engagement and public–private partnerships (PPPs) are crucial in tackling the intricate problems associated with sewage management. Effective sewage management solutions may be developed and implemented by collaborating with local communities, governments, non-profit organizations, and the commercial sector (Dagilienė et al., 2021). This is possible because these partnerships can pool resources, skills, and local knowledge. Important tactics consist of:

Joint Financing Operation: Establishing PPPs to pool resources and skills from the private sector to jointly fund, design, develop, manage, and maintain sewage treatment facilities while guaranteeing public accountability and supervision.

Community Engagement: Planning, decision-making, and implementation of sewage management should involve local people to guarantee that the solutions are sustainable from an environmental, social, and cultural perspective.

Capacity Building: Enhancing the ability of local people, stakeholders, and sewage management experts to successfully engage in sewage management projects through training, education, and technical support.

Public Awareness and Campaign: Launching public awareness programs to inform people about the value of resource recovery and circular economy techniques, as well as the significance of efficient sewage management, water conservation, and pollution control.

4.7 Conclusion and Recommendations

Proper sewage management would ensure sustainable growth, conserve the environment, and retain public health. This chapter has discussed the composition of sewage, the present scenario of sewage treatment in the world, the monetary expense of poor sewage treatment, advanced technology for its enhancement, and future outlook about sewage management.

4.7.1 Conclusion

The problems associated with wastewater treatment are becoming more complicated with urbanization, population growth, and the introduction of new pollutants. Not only does poor sewage treatment pose a threat to the environment and public health, but it also costs businesses and communities a lot of money. However, there are ways of dealing with these issues: changing policies, investing in technology and infrastructure, and involving the community.

4.7.2 Recommendations

Policy Reforms: In order to ensure the environmental compliance of wastewater treatment, governments must establish tough pollution control legislations and standards related to water quality. Reforms must also facilitate incentives on resource recovery and circular economy concepts to trigger sustainable management of the system.

Investment in Infrastructure and Technology: Modernizing the infrastructure for sewage treatment and using cutting-edge technologies need large expenditures. Funding for R&D projects and infrastructure projects should be mobilized through cooperation between governments, corporations, and international organizations.

Public–Private Partnership: It is necessary to create PPPs to fund, plan, construct, and run sewage treatment plants. Developing socially and ecologically acceptable solutions requires active collaboration from stakeholders and the community.

Capacity Building and Awareness: The implementation of capacity development programs is necessary to educate the public about the significance of appropriate sewage management, water conservation, and pollution control, as well as to train experts in sewage management.

Review Questions

1. What are the major constituents of sewage? What is the impact of these constituents on treatment?
2. Why is sewage treatment described as an economic burden? Describe the costs associated.
3. What will be the long-term risks to public health and the environment if sewage is not treated?
4. What should be expected in the future relating to sewage treatment technology, such that it reduces the economic and environmental impacts?
5. What is the role of decentralized systems for addressing future challenges in sewage management?
6. How is investment in sewage treatment infrastructure crucial for sustainable urban development?
7. To what extent could the recovery of resources such as biogas and fertilizers from sewage lower economic burdens?
8. How do membrane bioreactors influence the efficiency of sewage treatment?
9. What are dead zones, and to what extent is sewage pollution the cause of such dead zones?
10. To what extent might sewage treatment assist in creating a more resilient, climate-adapted community?

References

Ahmed, J., Thakur, A., and Goyal, A. (2021). Industrial wastewater and its toxic effects. In: *Biological Treatment of Industrial Wastewater* (ed. M.P. Shah), 1–14. The Royal Society of Chemistry. https://doi.org/10.1039/9781839165399-00001

Aldalati, O. (2023). *Investigating the existing state of the municipal water and sewage infrastructure (VA) in Sweden and identifying privatization & partnership opportunities*. https://www.gotriple.eu/documents/ftuppsalauniv%3Aoai%3ADiVA.org%3Auu-499954

Al-Mawla, Y., Jaeel, A.J., Fadous, N.J. et al. (2023). Characteristics of raw domestic sewage for Al Kut city. In: *AIP Conference Proceedings*. AIP Publishing. https://doi.org/10.1063/5.0140880

Battersby, S. (2023). *Sewerage and Sewage as an Environmental Health Issue*. Taylor & Francis.

Bhat, S.A., Bashir, O., Ul Haq, S.A. et al. (2022). Phytoremediation of heavy metals in soil and water: an eco-friendly, sustainable and multidisciplinary approach. *Chemosphere*, **303**, 134788. https://doi.org/10.1016/j.chemosphere.2022.134788

Bhattarai, K. and Conway, D. (2021). Impacts of economic growth, transportation, and tourism on the contemporary environment. In: *Contemporary Environmental Problems in Nepal: Geographic Perspectives* (eds. K. Bhattarai and D. Conway), 563–662. Cham: Springer International Publishing. https://doi.org/10.1007/978-3-030-50168-6_7

Brar, S.K., Kumar, P., and Cuprys, A. (2022). *Modular Treatment Approach for Drinking Water and Wastewater*. Elsevier.

Chen, T.L., Chen, L.-H., Lin, Y.J. et al. (2021). Advanced ammonia nitrogen removal and recovery technology using electrokinetic and stripping process towards a sustainable nitrogen cycle: a review. *Journal of Cleaner Production*, **309**, 127369.

Dagilienė, L., Varaniūtė, V., and Bruneckienė, J. (2021). Local governments' perspective on implementing the circular economy: a framework for future solutions. *Journal of Cleaner Production* **310**, 127340. https://doi.org/10.1016/j.jclepro.2021.127340

Deksissa, T. (2023). Urban Stormwater Runoff for potable use: potential and challenges. In: *Alternative Water Sources for Producing Potable Water: Advances in Research & Technology* (eds. T. Younos, J. Lee, and T.E. Parece), 71–96. Cham: Springer Nature Switzerland. https://doi.org/10.1007/698_2023_1020

Dickin, S., Bayoumi, M., Giné, R. et al. (2020). Sustainable sanitation and gaps in global climate policy and financing. *NPJ Clean Water*, **3**, 24.

El-Shafeiy, E., Alsabaan, M., Ibrahem, M.I. et al. (2023). Real-time anomaly detection for water quality sensor monitoring based on multivariate deep learning technique. *Sensors*, **23**, 8613. https://doi.org/10.3390/s23208613

Ferrentino, R., Langone, M., Fiori, L. et al. (2023). Full-scale sewage sludge reduction technologies: a review with a focus on energy consumption. *Water*, **15**, 615. https://doi.org/10.3390/w15040615

Geletu, T.T. (2023). Lake eutrophication: control of phytoplankton overgrowth and invasive aquatic weeds. *Lakes & Reservoirs: Research & Management*, **28**, e12425. https://doi.org/10.1111/lre.12425

Hamidian, A.H., Ozumchelouei, E.J., Feizi, F. et al. (2021). A review on the characteristics of microplastics in wastewater treatment plants: a source for toxic chemicals. *Journal of Cleaner Production*, **295**, 126480. https://doi.org/10.1016/j.jclepro.2021.126480

Khan, A.S., Anavkar, A., Ali, A. et al. (2021). A review on current status of riverine pollution in India. *Biosciences Biotechnology Research Asia*, **18**, 9–22. https://doi.org/10.13005/bbra/2893

Khan, M.J., Wibowo, A., Karim, Z. et al. (2024). Wastewater treatment using membrane bioreactor technologies: removal of phenolic contaminants from oil and coal refineries and pharmaceutical industries. *Polymers*, **16**, 443. https://doi.org/10.3390/polym16030443

Khan, N., Hassan, A.U., Fahad, S. et al. (2020). Factors affecting tourism industry and its impacts on global economy of the world. *Available at SSRN 3559353*. https://doi.org/10.2139/ssrn.3559353

Khan, S., Masoodi, T.H., Pala, N.A. et al. (2023). Phytoremediation prospects for restoration of contamination in the natural ecosystems. *Water*, **15**, 1498. https://doi.org/10.3390/w15081498

Khan, S.A.R., Ponce, P., Yu, Z. et al. (2022). Environmental technology and wastewater treatment: strategies to achieve environmental sustainability. *Chemosphere*, **286**, 131532.

Kolawole, A.S. and Iyiola, A.O. (2023). Environmental pollution: threats, impact on biodiversity, and protection strategies. In: *Sustainable Utilization and Conservation of Africa's Biological Resources and Environment* (eds. S.C. Izah and M.C. Ogwu), 377–409. Singapore: Springer Nature. https://doi.org/10.1007/978-981-19-6974-4_14

Kua, H.W., He, X., Tian, H. et al. (2022). Life cycle climate change mitigation through next-generation urban waste recovery systems in high-density Asian cities: a Singapore case study. *Resources, Conservation and Recycling*, **181**, 106265.

Lin, L., Yang, H., and Xu, X. (2022). Effects of water pollution on human health and disease heterogeneity: a review. *Frontiers in Environmental Science*, **10**, 880246. https://doi.org/10.3389/fenvs.2022.880246

Md Anawar, H. and Chowdhury, R. (2020). Remediation of polluted river water by biological, chemical, ecological and engineering processes. *Sustainability*, **12**, 7017. https://doi.org/10.3390/su12177017

Mocek-Płóciniak, A., Mencel, J., Zakrzewski, W. et al. (2023). Phytoremediation as an effective remedy for removing trace elements from ecosystems. *Plants*, **12**, 1653. https://doi.org/10.3390/plants12081653

Muttil, N., Nasrin, T., and Sharma, A.K. (2023). Impacts of extreme rainfalls on sewer overflows and WSUD-based mitigation strategies: a review. *Water*, **15**, 429. https://doi.org/10.3390/w15030429

Nazif, S., Razavi, S.G., Soleimani, P. et al. (2021). Rainwater and green roofs. In: *Handbook of Water Harvesting and Conservation* (ed. S. Eslamian), 355–372. Wiley. https://doi.org/10.1002/9781119478911.ch23

Olabode, O.F. and Comte, J.C. (2024). Water scarcity in the fast-growing megacity of Lagos, Nigeria and opportunities for managed aquifer recharge. *WIREs Water*, **11**, e1733. https://doi.org/10.1002/wat2.1733

Pandey, S., Twala, B., Singh, R. et al. (2022). Wastewater treatment with technical intervention inclination towards smart cities. *Sustainability*, **14**, 11563. https://doi.org/10.3390/su141811563

Pásková, M., Štekerová, K., Zanker, M. et al. (2024). Water pollution generated by tourism: review of system dynamics models. *Heliyon*, **10**, e23824. https://doi.org/10.1016/j.heliyon.2023.e23824

Raj, K. and Das, A.P. (2023). Lead pollution: impact on environment and human health and approach for a sustainable solution. *Environmental Chemistry and Ecotoxicology*, **5**, 79–85. https://doi.org/10.1016/j.enceco.2023.02.001

Rehman, S.U., Bhatti, A., Kraus, S. et al. (2020). The role of environmental management control systems for ecological sustainability and sustainable performance. *Management Decision*, **59**, 2217–2237. https://doi.org/10.1108/MD-06-2020-0800

Reid, A.J., Carlson, A.K., Creed, I.F. et al. (2019). Emerging threats and persistent conservation challenges for freshwater biodiversity. *Biological Reviews*, **94**, 849–873. https://doi.org/10.1111/brv.12480

Reymond, P., Chandragiri, R., and Ulrich, L. (2020). Governance arrangements for the scaling up of small-scale wastewater treatment and reuse systems – lessons from India. *Frontiers in Environmental Science*, **8**, 72. https://doi.org/10.3389/fenvs.2020.00072

Rufí-Salís, M., Petit-Boix, A., Leipold, S. et al. (2022). Increasing resource circularity in wastewater treatment: environmental implications of technological upgrades. *Science of the Total Environment*, **838**, 156422.

Savun-hekimoğlu, B. (2021). Comparison of advanced oxidation processes (sonication, fenton reaction and ozone-based processes) for water decolorization. *International Journal of Environment and Geoinformatics*, **8**, 166–171. https://doi.org/10.30897/ijegeo.797764

Schellenberg, T., Subramanian, V., Ganeshan, G. et al. (2020). Wastewater discharge standards in the evolving context of urban sustainability–the case of India. *Frontiers in Environmental Science*, **8**, 30. https://doi.org/10.3389/fenvs.2020.00030

Smarzewska, S. and Morawska, K. (2021). Wastewater treatment technologies. In: *Handbook of Advanced Approaches Towards Pollution Prevention and Control* (eds. R.O.A. Rahman and C.M. Hussain), 3–32. Elsevier. https://doi.org/10.1016/B978-0-12-822121-1.00001-1

Spahr, S., Teixidó, M., Sedlak, D.L. et al. (2020). Hydrophilic trace organic contaminants in urban stormwater: occurrence, toxicological relevance, and the need to enhance green stormwater infrastructure. *Environmental Science: Water Research & Technology*, **6**, 15–44. https://doi.org/10.1039/C9EW00674E

Starkl, M., Brunner, N., Das, S. et al. (2022). Sustainability assessment for wastewater treatment systems in developing countries. *Water*, **14**, 241. https://doi.org/10.3390/w14020241

Wittmanová, R., Hrudka, J., Raczková, A. et al. (2023). Small domestic wastewater treatment plants in rural areas in the conditions of the Slovak republic. *International Multidisciplinary Scientific GeoConference: SGEM*, **23**, 141–147.

Yeniçeri, M. and Yeniçeri, K.T. (2023). Recycling wastewater with membrane technology and the case of Singapore. *Bitlis Eren Üniversitesi Fen Bilimleri Dergisi*, **12**, 478–484. https://doi.org/10.17798/bitlisfen.1242232

Zewde, A.A., Li, Z., Zhang, L. et al. (2020). Utilisation of appropriately treated wastewater for some further beneficial purposes: a review of the disinfection method of treated wastewater using UV radiation technology. *Reviews on Environmental Health*, **35**, 139–146. https://doi.org/10.1515/reveh-2019-0066

5

Conventional Physical and Chemical Treatment Technologies and Economic Implications

Mohamed Sameh El-harty[1]*, Ahmed Salah Elkomy*[1]*, and Nabila Shehata*[2]*

[1]*Materials Science and Nanotechnology Department, Faculty of Postgraduate Studies for Advanced Sciences, Beni-Suef University, Beni-Suef, Egypt*
[2]*Environmental Science and Industrial Development Department, Faculty of Postgraduate Studies for Advanced Sciences (PSAS), Beni-Suef University, Beni-Suef, Egypt*

*Corresponding author: nabila.shehata@psas.bsu.edu.eg

5.1 Introduction

Spatial variations result from the components of wastewater discharged into water bodies reflecting the community's civilization and residents' metabolisms. The management procedures and techniques are constant and unaltered in spite of these constituent differences. Untreated wastewater can release a lot of foul-smelling gases, reduce dissolved oxygen, which can kill fish, cause eutrophication from the fast growth of aquatic plants and algae blooms, and contaminate water bodies with a variety of harmful substances and disease-causing microorganisms that are found in the human digestive system. The number of fecal coliforms in streams or other water bodies increases the amount of domestic sewage because wastewater contains large levels of bacteria, especially coliforms, which come from the fecal waste of humans and other warm-blooded animals. When pathogens including salmonella, hepatitis, *Escherichia coli*, and fecal coliforms are screened for or decreased before disposal, WWT is deemed effective.

Wastewater to Resource Recovery: Applying the Circular Economy Toward Sustainable Development, First Edition. Edited by Arvind Kumar Mungray et al. Published 2026 by John Wiley & Sons, Inc.

Physical, chemical, and biological processes—which frequently appear in different combinations within treatment systems—are used to remove contaminants from wastewater. Different techniques have been used in the past to treat wastewater, and new techniques have been developed as a result of technological breakthroughs. The goal of treatment techniques is to get rid of nutrients like nitrogen and phosphate, disease-causing pathogenic microbes, suspended particulates, and biodegradable organics (BOD). Each treatment approach's standard operating procedures must be understood. Whereas the chemical treatment approach is predicated on chemical reactions, the physical treatment method depends on physical forces such as screening, mixing, flocculation, sedimentation, and filtration. In contrast, biological treatment methods use biological activity. Poor sanitation practices, such as incorrect wastewater disposal, have a negative impact on urban populations and require significant state expenditure to preserve public health. Although good WWT technologies exist to address these challenges, some are frequently costly and do not provide direct economic rewards, discouraging many local governments from deploying them. Subsequently, searching for cost-effective, simple WWT process is mandatory (Quansah et al., 2018).

Several factors influence treatment method selection, including cost considerations, waste kind and quantity, effluent heterogeneity, and effluent standards. The removal and recovery of hazardous or valuable metals from industrial wastewater using biological substances, such as living and nonliving microbes, has become more and more popular because of its increased efficacy, accessibility, and inexpensive raw material prices. Bacteria, algae, fungi, and yeasts may effectively degrade organic substances in municipal wastewater and absorb heavy metals from industrial wastewater (Rajasulochana and Preethy, 2016). The choice of procedure is influenced by several factors, such as pollutant load and flow rate, sewage structure, cost of treatment, and the presence of other contaminants in the wastewater. Treatment systems that have high installation and operational costs, require long processing periods, generate limited output, or emit harmful by-products are frequently less desirable for industrial applications. Therefore, it is essential to determine alternate treatment techniques capable of totally dissolving or eliminating pollutants (Ranjit et al., 2021). However, WWT procedures vary into three primary categories: physical, chemical, and biological. Based on the wastewater's properties, each category offers unique benefits. To improve the removal of pollutants, physical and chemical treatment techniques are frequently combined with biological treatment (Sravan et al., 2024).

The main aim of this chapter is to report the current physical and chemical processes for WWT along with the associated challenges. Moreover, the main factors affecting the economics of these sewage treatment processes and the effect of economic analysis on sewage treatment strategy have also been discussed.

5.2 Conventional Physical Methods Used for Sewage Treatment

Physical treatment primarily aims to eliminate big settleable organic waste and smaller inorganic matter, resulting in a 60–65% reduction in suspended solids and a 30–35% reduction in BOD. This physical process mainly involves sedimentation, flocculation, and coagulation (Shekho and Hassan, 2024). Physical pollution removal strategies are based on mass transfer principles. These methods are favored due to their ease of use, adaptability, effectiveness, and capacity to recycle contaminants. Physical approaches also have the benefit of requiring fewer chemicals. Since physical treatments don't rely on living things, they are frequently thought to be more dependable than other approaches. Because adsorption is so inexpensive and effective, researchers have been using it more and more as a physical technique in recent years. The main physical processes used in wastewater are listed in the following section.

5.2.1 Screening

In WWT, screening is an essential process that removes floating materials such as tree branches, logs, dead animals, pieces of wood, paper, clothing, diapers, paper towels, and plastic and glass bottles. A screener is a device with evenly sized apertures used to remove floating debris from wastewater; it is the initial stage of the therapeutic procedure. The substances that the screens retain are known as screening. These materials have the ability to block and harm pipelines, pumps, valves, and other equipment if they are not removed. Holes in plates, wire networks, gratings, rods of steel, and parallel bars are examples of screening devices. Depending on their size, the apertures are usually round or rectangular. Screeners are divided into coarse, medium, and fine screens according to the size of their holes.

5.2.2 Sedimentation

One common physical treatment method is sedimentation; it separates suspended materials from wastewater using gravitation. It is the most widely employed unit operation in WWT, with the main objective being the production of cleared effluent of WTPs (Shekho and Hassan, 2024).

Long, narrow tanks called grit chambers are designed to reduce water flow so particles like coffee grounds, eggshells, and sand can settle out. Using the sedimentation principle, which is based on gravity, these chambers efficiently remove heavy inorganic contaminants like sand and ash. Grit must be removed because it can seriously deteriorate pumps and other equipment, particularly in urban areas with integrated sewer systems that transport significant volumes of gravel, sand,

and silt, which are washed off streets after storms. Grit chambers are typically positioned before pumps or other machinery used for grinding, besides coarse bar racks would come before them if they are mechanically cleaned.

The efficiency of sedimentation tanks varies based on design and operational parameters. This method effectively removes settleable solids and heavy particles, is relatively simple to operate, and requires low energy. Despite its advantages, sedimentation has limitations in removing fine suspended solids and colloidal particles. It also necessitates a large area for settling tanks and may require chemical pretreatment to enhance removal efficiency (Sravan et al., 2024). In the sedimentation process, solid particles in sewage settle due to gravitational forces. The majority of these solid particles, which float in flowing sewage, are organic molecules. The sewage flow is stopped and held in a tank called a sedimentation tank to let these particles settle. Particle behavior allows for the classification of the settling process into four types: compression, zone or inhibited settling, flocculant settling, and discrete settling.

i. Discrete Settling: This process uses constant particles, also known as small or uniform particles, that remain the same in terms of dimensions, form, and weight. The use of a grit chamber in sewage is one instance.
ii. Flocculant Settling: Particles of this sort vary in mass, area, or form, losing their original characteristics. Bioflocs settling in sedimentation tanks is one example.
iii. Zone or Hindered Settling: In this case, flocculation is followed by particle settlement as a period or block, maintaining a fixed position relative to one another. This process results in the thickening of sludge, with particle concentration increasing from top to bottom. An example is the secondary sedimentation tank used in the activated sludge process.
iv. Compression: Sludge thickens at the bottom of this process as a result of particles in the lower layers settling because of the additional weight of the particles above them. Sludge thickening in a WWT facility serves as one illustration.

Sedimentation tanks can be classified by: (i) form (square and round), (ii) sewage flow (longitudinal, horizontal, radial, and spiral), (iii) purpose and location (primary, secondary, sewage tanks, coagulation-cum-sedimentation tanks, grit chambers, and Imhoff tanks), and (iv) method (batch and continuous flow) (Ranjit et al., 2021).

5.2.3 Membrane-based Technology

The technology of membrane separation is a method of treatment and physical process involving selectively allowing components of the sewage to pass through by generating a pressure difference across the membrane. As sewage flows through the separation membrane, certain substances are selectively permeable while others are retained, achieving separation and purification. Commonly used membrane separation technologies in sewage treatment include microfiltration (MF), nanofiltration (NF), ultrafiltration (UF), reverse osmosis (RO), and electrodialysis (Edris, 2024). Recent developments have concentrated on increasing the effectiveness of contaminant removal, even though membrane-based separation is a frequently utilized technology for pollutant removal. Frequent improvements, as well as the addition or removal of particular constituents, are examples of these improvements. The pore size of membrane technology can vary; for example, UF has even smaller pores than MF, which has pores that range from 0.001 to 0.1 μm. To get rid of particles smaller than 1 nm, RO uses a semipermeable membrane, whereas NF has pores that range from 1 to 10 nm. The kind of membrane employed and the kind of contaminants present can affect how well a particular filtration technology works. On a smaller scale, contaminants of emerging concern (CEC) can also be successfully eliminated using traditional granular filtration techniques. Water passes through granular filter beds in this system, trapping and separating contaminants. Therefore, separating and extracting CEC from wastewater is facilitated by passing it through a membrane or granules (Shekho and Hassan, 2024). The main applications for this method are desalinating brackish water and treating seawater. Depending on the material, the driving force, the detachment mechanism, and the minimal size of separation attained, there are several types of membranes.

5.2.4 Filtration

Filtration is highly effective for removing fine suspended solids and microorganisms from wastewater. It can be integrated with other treatment methods to enhance the overall performance. Filtration is adaptable to various water qualities and can be customized to meet specific requirements. However, it necessitates frequent maintenance and periodic replacement of filter media, resulting in elevated operational and maintenance costs. If not properly maintained, filtration systems may become clogged, leading to reduced flow rates. These physical treatment methods are essential for eliminating suspended solids and particles from wastewater (Friman, 2024). MF is a physical separation technique that uses membranes with porosities ranging from 0.1 to 10 μm to remove microorganisms and suspended particles from water streams. It is frequently used to treat municipal and industrial wastewater in addition to potable water. The use of UF in providing drinkable water occurs via purging raw water of particles and macromolecules. To facilitate the process of separation, filters usually function at pressures between 1 and 10 bar and have membranes with pore diameters between 100 and 10 nm. NF is another effective technique for treating sewage. This membrane-based method operates under pressure and allows relatively pure water to pass through while selectively blocking harmful substances such as organic materials, nutrients, turbid microbes, ions of inorganic metals, and other substances that deplete oxygen agents. These membranes are particularly effective for removing heavy metals in WWT. Carbon derivatives are commonly used as adsorbents in this process. Adsorbing membranes are essential due to the presence of functional adsorption groups that capture heavy metals. This technique is considered the optimal solution for water filtration, with polyamide membranes exhibiting reduced rejection of salt and different molecular weight-based cut-offs, achieving a rejection rate of over 92%. Recently, NF technology has enabled pollution removal from wastewater, demineralization of the dairy sector, pulp-bleaching effluents, and textile effluents, and separation of medicines from fermentation broths. Surface water treatment, for example, uses NF to treat wastewater with low amounts of total dissolved solids. The dairy sector uses this technology a lot to partially demineralize its products. This technique mostly eliminates organic by-products and softens water. For particle separation, nano-filters work at pressures between 3 and 20 bar and have pores that range in size from 1 to 2 nm. A partially permeable membrane is used in RO to filter out undesirable ions, salts, and compounds from water. It is crucial for desalinating seawater and is frequently used to treat freshwater. The pore diameters of the membranes used in this procedure are smaller than 1 nm. RO usually occurs at high pressures, up to 80 bar, because of the poor porosity of these membranes (Ranjit et al., 2021).

5.2.5 Flotation

Flotation assists in separating solid and liquid particles by adding air bubbles to the wastewater to produce a skimable layer of floating particles (Sravan et al., 2024). Flotation is another effective method that excels at removing small and light particles from wastewater. It features a compact system design with a smaller footprint than sedimentation, making it particularly suitable for wastewater with high oil and grease content. However, flotation requires the addition of chemicals to optimize performance, which can increase capital and operational costs. Additionally, there may be challenges associated with managing and disposing of flotation sludge. A gravity separation method called flotation is mostly used to separate minerals like sulfides. This method is essential for the economies of developed countries and is increasingly used to recover metal ions from industrial wastewater, addressing various waste streams containing heavy metals. Ion and dissolved-air flotation are the two most important types of flotation separation that have been documented. By adding bubbles of gas to the wastewater, dissolved-air flotation raises metal ions to its surface in a way that makes the sludge readily cleaned. This method efficiently removes heavy metals from wastewater, including Cu, Cd, Fe, Ni, Mo, Mn, and As. On the other hand, ion flotation is used to remove Cr^{3+}, Pb^{2+}, and Cu^{2+} from wastewater when different surfactants are present. This method achieves a 95% removal efficiency for these metal ions. Flotation has drawbacks, such as a lower removal rate and inefficiency when handling large amounts of metals, in spite of its ease of use, specificity, and high elimination efficiencies. Furthermore, excessive toxicity might block the process if biosurfactants are not utilized.

5.2.6 Adsorption

A physical phenomenon known as adsorption occurs when soluble molecules stick to solid surfaces and are therefore eliminated. It is a well-known physical therapy technique that makes use of adsorbents' remarkable surface area. Prior to use, this surface area must be activated by eliminating impurities from the adsorbents. Consequently, activated carbon is commonly employed to enhance the efficiency of the adsorption process. Many different types of hazardous, inorganic, and organic contaminants can be successfully eliminated with this method. The size, amount, temperature, molecular weight, and other chemical characteristics of the pollutant all affect the adsorption capacity. When it comes to the selective removal of CEC, activated carbons outperform alternative adsorbents such as carbon nanotubes and charcoal. To increase the effectiveness of CEC removal, the adsorption-based treatment technique can be readily combined with other reliable WWT techniques. Microplastics and CEC recovered from different systems are currently being employed as adsorbents to guarantee the environmentally friendly nature of the adsorption process. Investigating the adsorption potential of substitute materials is therefore essential for efficient CEC removal. In order to eliminate organic dyes and pollutants containing heavy metals from water and wastewater, nanosorbents have been used extensively. Because of the tiny dimensions and huge ratio of area of surface to volume, nanoparticles are particularly effective for certain contaminants due to their biocompatibility and adaptability. Metal oxides at the nanoscale demonstrate higher adsorption levels compared to bulk sizes due to metal-ligand precipitation. The pH level of wastewater significantly affects the removal efficiency of metal pollutants through adsorption; higher pH levels improve the effectiveness of heavy metal ion removal. Increasing the wastewater pH enhances the attractive interactions between metallic ions that are highly charged and the surface of nanosorbents. As the quantity of negatively charged adsorbent sites increase, the adsorption capacity rises as well. Nanosorbents can be classified into various categories based on their use and surface properties, such as (i) carbon nanomaterials, (ii) silicon nanomaterials, (iii) metallic nanomaterials, and (iv) nanoparticles used as adsorbents (Rathod et al., 2024).

5.3 Conventional Chemical Methods Used for Sewage Treatment

Chemical treatment involves adding chemicals to change the physical and chemical characteristics of pollutants, including their pH, charge, solubility, and reactivity, in order to make their removal easier using substances including lime, alum, iron salts, UV radiation, chlorine, polymers, and ozone, and strategies include adsorption, ion exchange, precipitation, coagulation/flocculation, neutralization, and disinfection. Organic materials, pathogens, poisonous compounds, and nutrients are among the colloidal or dissolved contaminants that can be effectively removed using these techniques. Chemical treatment, however, can be costly and complicated, producing secondary wastes that need to be disposed of or treated further (Sravan et al., 2024). Both organic and inorganic particles, as well as total dissolved solids and total suspended solids, are successfully eliminated from secondary effluent through chemical treatment. Additionally, it can get rid of some minerals, viruses, and organic and inorganic components. There are five categories for tertiary therapies.

5.3.1 Chemical Precipitation

A chemical treatment method called chemical precipitation efficiently eliminates phosphorus and dissolved heavy metals, as well as other inorganic contaminants from wastewater. It can be combined with other treatment methods to achieve the desired effluent quality. Chemical precipitation is relatively straightforward to operate and utilizes readily available chemicals. Nonetheless, precise control of pH, dosage, and mixing conditions is necessary for optimal results. This process generates significant amounts of sludge that must be properly managed, and additional treatment may be required for residual chemical removal (Friman, 2024). However, it is widely employed in many different industries and is regarded as one of the most well-established and effective methods. In order to promote precipitation, the process transforms dissolved metal ions into solid particles. A coagulant can precipitate metal ions by changing the electro-oxidation potential, pH, or co-precipitation and then remove the deposits that occur. Due to its affordability, ease of use, and pH-adjustable nature, hydroxide precipitation is frequently used. To do this, hydroxide is added to the wastewater that has been agitated, which causes insoluble metal hydroxide precipitates to develop. For example, calcium hydroxide (lime) and metal ions can react to form calcium ions and metal hydroxide precipitates. The technique of precipitation has a number of disadvantages despite its high effectiveness, which is usually between 95% and 99% for the majority of heavy metals in wastewater. Notable issues include its dependence on high pH levels and the challenges associated with removing large volumes of sludge from the water (Rajasulochana and Preethy, 2016).

5.3.2 Ion Exchange

One technique for treating water is ion exchange which removes undesirable ionic contaminants by exchanging them with less objectionable ionic substances. This technology is particularly effective for removing inorganic ions that have dissolved. It is frequently used in WWT facilities to soften water and remove heavy metals, nitrogen, and total solids that are dissolved. Ion exchange resins have the ability to transport soluble anions or cations into the sludge by exchanging them with electrolyte solutions. The exchange of ions is aided by these resins' particular metal absorption capabilities. In this procedure, an electrolyte is used to suspend the ion exchange resin. Ions cling to the resin beads as water passes across the resin bed, releasing solutions that are loosely contained into the water. As soon as the resin beds are saturated and lose their effectiveness, they must be recharged. This involves flushing using a salt brine solution to dissolve the resin, which permits ion exchange between the salt brine and the cleansed water (Ranjit et al., 2021). Metals that contaminate wastewater are frequently precipitated using sulfides. This reversible chemical reaction can convert unwanted metallic ions into safer kinds, making the exchange of ions treatment an effective substitute for already mentioned methods. This process effectively removes heavy metallic ions from wastewater, achieving an efficiency of approximately 95–99% for most heavy metals. However, the precipitation method has notable drawbacks, including high pH dependence and challenges related to managing the large volumes of sludge generated during treatment. The ion exchange process involves bonding to a stationary solid particle, allowing for the removal of various sewage containing heavy metal ions such as Pb^{2+}, Hg^{2+}, Ni^{2+}, Cu^{2+}, Cd^{2+}, Zn^{2+}, V^{5+}, V^{4+}, Cr^{4+}, and Cr^{3+}. When the cations of the sewage contaminant are swapped out for the cation of an ion exchange mechanism, an ion exchange occurs. The reversible chemical reaction enables the conversion of undesirable metallic ions into fewer harmful forms, enhancing the attractiveness of ion exchange treatment as an alternative method. Various resins have been evaluated for their effectiveness in removing cations, including the Amberlite IRC86 chelating exchange resin and the Diaion CR11 weak cationic resin. Both resins have shown approximately 90% efficiency in eliminating trivalent chromium from aqueous solutions (Ayach et al., 2024).

5.3.3 Advanced Oxidation

Numerous catalytic applications have been demonstrated using chemical oxidation techniques. It is commonly acknowledged that the Advanced Oxidation Process (AOP) is a crucial WWT technique. The AOPs include a number of WWT methods that are all aimed at producing oxidizing species, such as hydroxyl radicals (•OH). Numerous processes, such as photo-electrochemical oxidation, electrochemical oxidation, ozonation, and UV-assisted Fenton's oxidation, can cause oxidation. Catalysts and pH are essential variables in oxidation processes employed for treating industrial wastewater effluent. Oxidation via electrochemistry is a potential method for treating wastewater. Electrochemical AOPs are distinguished by their high efficiency and ease of equipment operation. Nevertheless, issues like decreased oxidation efficiency and growing energy prices continue to exist. Enhancing the stability and catalytic activity of electrode materials has been the focus of recent research. Using a pulsed current supply, pulse electrolysis is a useful technique for increasing energy efficiency (Shekho and Hassan, 2024). Because AOPs can decompose contaminants and reach high removal efficiency, oxidation via photo-electrochemistry is being used more and more. Among these, electrochemical techniques based on hydroxyl (•OH) are increasingly crucial for limiting pollution levels below permissible limits. By irradiating a semiconductor anode with sunlight, photo-electrocatalysis makes it easier to remove dyes. Hydroxyl radicals generated through water oxidation play a crucial role in oxidizing the dyes, leading to the breakdown of aromatic rings and degradation of azo bonds. High dye removal rates can be achieved at elevated currents and low initial concentrations, while significant reductions in dye concentrations can occur at low currents and low concentrations (Shekho and Hassan, 2024). UV, visible, and infrared radiation can all affect how quickly a chemical reaction proceeds. A substance that absorbs light and promotes the chemical transformation of products is a photocatalyst. This process typically involves five key steps: (i) reactions occurring on the surface; (ii) adherence of reaction products to the surface; (iii) transformation of the surface; (iv) removal of products from the surface of the catalyst; and (v) the products' evaporation. Materials for semiconductors with nanostructures are considered the most effective photocatalysts due to their high availability of surface electrons and holes produced by photosynthesis. Both organic and inorganic contaminants, such as bacteria and heavy metal ions, can be removed by photocatalysis. Heterogeneous photocatalysis, an advanced oxidation process, utilizes the transformation of photon energy to chemical energy (Ayach et al., 2024). AOPs can effectively break down textile effluent that contains organic components or colors and might produce free radicals with a high oxidation potential, including hydroxyl radicals. Fenton's oxidation is acknowledged as a practical and cutting-edge method of treating water (Shekho and Hassan, 2024).

5.3.4 Ozonation

An extensive range of chemical and inorganic substances can react with ozone, a powerful oxidizing agent. Its significance in WWT has grown over the past decade, primarily due to the lack of harmful by-products in reactions involving ozone. Ozonation is a complex oxidation process that introduces ozone, greatly improving the biodegradability of wastewater. Numerous studies have shown that ozonation effectively removes personal care and pharmaceutical products, with many substances being completely eliminated. However, ozone has a short half-life, and concentrations exceeding approximately 23% are considered hazardous. Recent research indicates that ozone has low utilization efficiency; while it can oxidize and mineralize organic contaminants, it is less effective at generating unsafe by-products. The technology of catalytic ozonation has emerged as a potential WWT approach (Shekho and Hassan, 2024).

5.3.5 Electrochemical Separation

Electrochemical techniques have been attracting a lot of focus as a dependable method for treating industrial wastewater. Compared to different systems, electrochemical methods offer many benefits, such as consistent performance, the ability to accommodate changes in influent speed and composition, and the ability to operate under surrounding pressure and temperature by precipitating them as hydroxides in pH-neutral or slightly acidic electrolytes. To improve the removal capabilities for different contaminants, the electrode material selection is essential. Additionally, the amount of ions produced by the electric current has a direct impact on the treated effluent's quality. Therefore, any element influencing the process's charge will also have an effect on its overall efficiency. By that method, numerous metal ions, including Hg^{2+}, Cd^{2+}, Pb^{2+}, and Cu^{2+}, have been effectively and efficiently eliminated. However, addressing energy consumption is essential for making this feasible for use in industrial settings (Ayach et al., 2024). Direct reactions involving oxidation and reduction are used in electrochemical processes to produce reactive chemical species or compounds that can eliminate contaminants from wastewater. There are two main types of electrochemical techniques: electrodialysis and electrocoagulation.

In electrodialysis, salt ions are transported from wastewater when an electric potential difference is applied across ion-exchange membranes. Usually, this procedure takes place inside an electrodialysis cell, which is composed of two sections: a brine portion and a diluent part, which are divided by anionic exchange membranes. The two electrodes are separated by the cationic membrane. The anionic membrane prevents further migration of negatively charged ions toward the anode due to the electric potential difference, whereas the cationic membrane restricts the movement of positively charged ions toward the cathode. An electric current flows across the cathode and anode as a result of this ion movement. Consequently, the brine compartment's ion concentration rises while it decreases in the dilute compartment.

Electrocoagulation (EC) involves giving water an electrical charge, which changes the particle surface charge and promotes the agglomeration of suspended materials. This cutting-edge, cost-effective, and multipurpose treatment method efficiently eliminates germs, emulsified oils, suspended particles, heavy metals, grease, latex, and total petroleum hydrocarbons, among other impurities from water (Ranjit et al., 2021). The EC process employs physical and chemical methods to treat organic matter in sewage, including filtration, coagulation, and electrochemical treatment. Coagulation and filtration can target wastewater with low organic matter content and high suspended solids, serving as pretreatment for high chemical oxygen demand (COD) wastewater. In EC, iron or aluminum is typically used as the anode, and a polymerization reaction occurs under alkaline conditions, generating hydroxide precipitation. During this process, the effluent quality is further enhanced through net trapping. The EC-electrolysis hybrid technology was used to treat domestic sewage. In this approach, EC served as a pretreatment method, while electrolysis acted as an advanced treatment technique to further enhance effluent quality. Under optimal conditions such as pH and electrolysis duration, COD removal rates can reach up to 93% (Edris, 2024). This approach is extremely effective at eliminating colloidal particles, turbidity, and dissolved organic matter. It also improves the elimination of specific contaminants like heavy metals and increases the effectiveness of downstream treatment operations. However, for best effectiveness, coagulants must be carefully selected and dosed. The sludge produced during coagulation-flocculation requires additional treatment and disposal, therefore this process may result in increased chemical usage and expenses (Friman, 2024).

5.4 Physicochemical Processes Methods Used for Sewage Treatment

Some of the treatment processes, which are mainly physical, can involve chemical reactions like chemical adsorption. Moreover, there are some treatment processes that combine physical and chemical mechanisms for the removal of pollutants from wastewater.

5.4.1 Coagulation/Flocculation

The process by which particles in a liquid group together to form bigger clumps that finally settle as a result of gravity is known as coagulation/flocculation. By neutralizing charges, the coagulation/flocculation technique produces gelatinous agglomerates that are used to clean municipal and industrial wastewaters. This technique develops connections between suspended particles and is quick, easy, and economical. Coagulants that link with pollutants and dispersed colors include aluminum sulfate ($Al_2(SO_4)_3{\cdot}18H_2O$), ferric sulfate ($Fe_2(SO_4)_3{\cdot}7H_2O$), lime ($Ca(OH)_2$), and ferric chloride ($FeCl_3{\cdot}7H_2O$). The use of high molecular weight polymers and proton amine groups, as well as some techniques like sorption, electrostatic forces, and bridging, aids in the removal of impurities and the coagulation of colors, lowering chemical concentrations in wastewater (Shekho and Hassan, 2024).

Adding chemical agents to charged particles causes them to become unstable, a process known as coagulation. Chemical coagulants are occasionally required to improve sedimentation since simple sedimentation may not always be enough to remove colloidal waste from sewage. Coagulation-aided sedimentation or chemical precipitation are terms used to describe this process. About 60–80% of the suspended matter with this technique may be effectively removed. Coagulants are positively charged substances that combine with negatively charged colloidal particles to generate insoluble, gelatinous precipitates. Chemical agents employed in coagulation include lime, calcium carbonate, sodium aluminate, iron salts, and alum (aluminum sulfate). Additionally, coagulant-aids aid in the coagulation process, increasing coagulants' efficacy and lowering the volume of sludge generated. Activated silica and weighing agents such as silica, powdered limestone, and polyelectrolytes are a few examples. While flocculation concentrates on aggregating unstable particles, coagulation destabilizes colloids by reducing their repulsive forces. On the other hand, flocculation involves the combination of particles to develop larger aggregates. Materials like polyacrylamide, polyferric sulfate, polyaluminum chloride, and different polymer flocculants can be used to do this. Coagulation and flocculation techniques have proven to be simple and successful in eliminating various contaminants from wastewater. One of the greatest flocculants is polyethylene glycol, according to earlier research. With removal efficiencies ranging from 95% to 99%, this approach effectively removes As, Se, Cr, Sb, Ag, Cu, Pb, Ni, and other heavy metals. However, flocculation has a number of disadvantages, such as the risks to health posed by inorganic coagulants, the production of large amounts of sludge, the limited selectivity for particular metals, the ineffectiveness of addressing new pollutants, the possibility of sewage color increases, the decreased efficacy of natural coagulants, and difficulties in scaling up the process.

5.5 Limitations of Conventional Methods for Sewage Treatment

Traditional methods of treating wastewater face significant limitations that compromise their ecological sustainability and effectiveness. Because they require significant financial commitments for both initial setup and continuing operations, these conventional systems frequently present economic issues in places with little financial resources, particularly developing nations; such financial burdens may be very burdensome (Fetouh et al., 2021; Öberg et al., 2020). Furthermore, traditional methods are time-consuming and rigid, which makes them unsuitable for rapidly expanding metropolitan regions that need flexible and quick fixes. Additionally, the ecological and resource recovery requirements may not be sufficiently addressed by conventional WWT techniques, which could result in inefficiencies in nutrient and energy recovery, which are considered the two crucial aspects of sustainable sanitation practices (Werkneh and Gebru, 2023). Last but not least, a move toward more inventive and nature-based treatment methods may be necessary because the dependence on automated procedures may restrict this conventional system's suitability in various geographic situations (Gedda et al., 2021). The following sections outline the key limitations associated with conventional sewage treatment methods.

i. Inefficiency in Removal of Micropollutants: Traditional sewage treatment techniques may not adequately eliminate various types of pollutants, especially CEC like pharmaceuticals, personal care products, microplastics, PFAS, and intricate organic compounds. These substances can persist through treatment processes and end up in the environment, posing concerns to both public health and water quality. This issue is implied in the debates about the difficulty of treating wastewater and the variable properties of effluents that require specific treatments (Corpus et al., 2024).
ii. Technological and Economic Hurdles: Although traditional approaches might be more economical in the short run, their incapacity to handle newly discovered contaminants may result in increased long-term expenses for environmental cleanup and public health effects (Corpus et al., 2024).
iii. Need for Multidisciplinary Approaches: The challenges posed by CEC necessitate a multidisciplinary approach that conventional methods may not support. There is a growing recognition of the need for collaboration among researchers, policymakers, and industry stakeholders to develop innovative solutions that go beyond traditional treatment methods (Corpus et al., 2024).
iv. High Energy Consumption: Conventional sewage treatment facilities consume substantial energy, notably in aeration systems utilized in methods such as activated sludge, leading to increased operational expenses and environmental repercussions (Wang, 2024; Gedda et al., 2021).
v. Lack of Skilled Manpower: Skilled personnel are typically essential for the performance and maintenance of conventional treatment systems. A scarcity of trained workers can impede the optimal performance of these systems (Gedda et al., 2021).

vi. Odor Issues: Traditional treatment methods that incorporate bacterial growth systems possess the capacity to produce unpleasant odors, which can inconvenience nearby communities (Gedda et al., 2021).

vii. Sludge Disposal Issues: Traditional sewage treatment methods frequently produce substantial volumes of sludge, necessitating supplementary treatment and disposal measures. Ensuring the appropriate disposal of this sludge is crucial to avert environmental pollution, yet managing it can be both costly and intricate (Gedda et al., 2021; Crini and Lichtfouse, 2019).

viii. Large Space Requirement: Conventional methods often necessitate a significant physical footprint, especially in biological treatment procedures such as activated sludge systems. This spatial demand can pose a challenge in urban settings where land availability is limited. The substantial space requirements can impede the adoption of these techniques in densely populated areas (Wang, 2024; Crini and Lichtfouse, 2019).

ix. High Chemical Consumption: Conventional treatment procedures heavily utilize chemicals for coagulation, disinfection, and pH control. This heavy reliance gives rise to worries regarding chemical remnants in the treated water and the possibility of harmful side products, which could compromise the safety of the treated effluent (Crini and Lichtfouse, 2019; Wang, 2024).

x. High Operational Costs: Traditional treatment methods like coagulation/flocculation and biological treatment can incur substantial operational and maintenance expenses due to energy consumption, chemical usage, limitations in handling wastewater volumes surpassing the fixed design capacity of treatment plants, and the requirement for skilled personnel to oversee the systems. These factors can render such methods economically less feasible, particularly for smaller treatment facilities (Gedda et al., 2021; Crini and Lichtfouse, 2019).

xi. Environmental Impact: Certain traditional methods may result in negative environmental consequences, like the emission of greenhouse gases or other pollutants during treatment operations. These impacts can counteract the sustainability objectives of wastewater management and give rise to regulatory hurdles (Crini and Lichtfouse, 2019).

xii. Poor Adaptability: Traditional methods frequently encounter difficulties in accommodating changes in both water quality and quantity. This inflexibility can impede their efficiency in treating sewage with varying characteristics. When wastewater properties fluctuate, the treatment systems may struggle to adapt accordingly, resulting in inefficiencies and potential breaches of discharge regulations (Wang, 2024; Crini and Lichtfouse, 2019).

5.6 Factors Affecting Economics of Sewage Treatment Method

The economic viability and overall financial implications associated with the various methodologies employed in sewage treatment are profoundly shaped by a multitude of critical elements, including but not limited to the operational expenditures incurred during the treatment processes and the specific methodologies utilized for treating wastewater, as well as an array of additional factors that collectively exert a significant influence on the overall life cycle cost (LCC) associated with WWT systems. The following is an elaboration of the primary factors that are essential to determining the economics related to the diverse sewage treatment methods:

i. Operational Costs: The operational cost structure, which comprises maintenance, chemical, energy, and sludge disposal expenses, is one of the main determinants. Energy costs are frequently shown to be a significant contributor in case studies, making up to 36% of overall operating costs, while personal expenses might contribute up to 48% of the average cost. Operational expenses of wastewater treatment plants (WWTPs) might differ greatly. For example, operating costs can range from 0.1158 €/m^3 at the minimum to 0.7491 €/m^3 at the most, indicating a variability of up to 6.5 times. The weighted average cost is 0.2200 €/m^3, which reflects the volume of wastewater treated. Different treatment types incur different costs. Primary treatment averages 0.12 €/m^3, secondary treatment 0.26 €/m^3, and tertiary treatment 0.32 €/m^3, showing that the complexity of the treatment process directly affects costs (Molinos-Senante et al., 2010; Ilyas et al., 2021).

ii. Initial Construction Cost: The expensive initial expenditure for WWTPs is a significant hurdle, especially in poor nations. The technical choices made on the technology to meet environmental criteria have a direct impact on this expense. Also, setting extremely high standards frequently results in higher expenses, which may prevent the construction of essential treatment facilities. A significant amount of the overall LCC is made up of the initial building costs. For example, it has been found that the initial building costs account for approximately 22.2% of the overall LCC, whereas replacement costs are often lower, at approximately 4.6%. Ultimately, the technology selection is also important; waste stabilization ponds (WSP) have been found to be the most economical system, followed by several types of activated sludge systems (Ilyas et al., 2021).

iii. Geographical Location: The economics of sewage treatment are also impacted by the geographic location and the particular regulatory framework. Significant variations in the total cost-effectiveness of various treatment techniques may result from regional variations in labor, material, and energy costs. Furthermore, it is becoming more and more crucial to incorporate economic and environmental evaluations into decision-making procedures (Ilyas et al., 2021; Junior et al., 2012).

iv. Quality of Sewage Water: The quality of sewage water has a considerable impact on operating costs. Better effluents might need less rigorous treatment, which would save money. Additionally, efficient pollutant load management might result in comparative advantages when implementing WWTPs. This results in unique sewage treatment methods. Therefore, in order to find the most effective treatment approaches, the economic evaluations must take local factors into account (Molinos-Senante et al., 2010; Junior et al., 2012).

v. Treatment Technology: Different WWT technologies have varying economic implications. For instance, advanced technologies may have higher initial costs but could lead to lower operational costs over time. The literature indicates that technology at the community level is frequently more financially viable than centralized systems (Molinos-Senante et al., 2010).

vi. Environmental Benefits: It is crucial to quantify the environmental benefits that are frequently disregarded due to their absence of market value. By determining shadow prices for pollutants eliminated during treatment processes, one can attain greater comprehension of the financial feasibility of WWTPs. This assessment of environmental externalities is imperative for substantiating the financial investments associated with WWT (Crini and Lichtfouse, 2019; Junior et al., 2012).

vii. Public Policy and Market Mechanisms: Market mechanisms ought to be seamlessly integrated into public policies concerning WWT. This necessitates a thorough examination of costs and benefits by all relevant stakeholders, including users, planners, and financial agents, to guarantee that financial resources are allocated prudently and effectively (Molinos-Senante et al., 2010; Ilyas et al., 2021).

viii. Social Impacts: In the context of analyzing and evaluating the economic ramifications associated with various sewage treatment methodologies, it is imperative to comprehensively account for the multifaceted social implications that arise, particularly emphasizing the critical aspects of public health and overall quality of life within affected communities. The inadequacies and inefficiencies inherent in wastewater management systems can precipitate a plethora of grave health complications, especially among populations that are already vulnerable due to socioeconomic factors, which consequently exerts a detrimental influence on both economic productivity and the equitable distribution of social resources (Molinos-Senante et al., 2010; Junior et al., 2012).

ix. Cost-benefit Analysis: How a cost-benefit analysis (CBA) is carried out serves as a robust methodological framework that facilitates a thorough and systematic assessment of the economic viability associated with WWT initiatives and undertakings.

Through careful comparison of the expected ecological gains against the calculated fiscal implications and through a comprehensive examination of diverse potential situations, it has been established that a phased and gradual method of introducing treatment standards may offer greater social and environmental returns for every financial unit invested, particularly in the early stages of this process. This insight underscores the notion that adopting a more adaptable and flexible approach to regulatory standards can yield significant economic advantages, thereby empowering stakeholders to make well-informed and strategic decisions regarding the implementation and operational management of WWTPs (Molinos-Senante et al., 2010; Lokhande and Kalbar, 2024).

5.7 Effect of Economic Analysis on Sewage Treatment Strategy

Economic analysis assumes an indispensable position in the formulation and refinement of sewage treatment strategies by establishing a comprehensive framework that facilitates the thorough evaluation of the multifaceted expenses and advantages related to a plethora of treatment options, particularly in developing nations where there exists a pronounced scarcity of resources and financial constraints are prevalent. This rigorous analytical process is of paramount importance for decision-makers, as it empowers them to discern the most economically efficient solutions that not only satisfy regulatory mandates but also consider a broad spectrum of environmental and social ramifications that may arise from the application of certain therapeutic methods. Furthermore, the systematic review and assessment of life cycle cost analysis (LCCA) in the WWT setting underscores the critical significance of economic evaluations in steering and informing these strategic approaches (Ćetković et al., 2022; Lopez-Serrano et al., 2024).

One of the most significant ramifications of economic analysis is its profound capacity to guide and inform the selection of various treatment technologies that may be employed in sewage processing. It is imperative to recognize that different treatment technologies manifest distinctly varied capital investments and operational expenditures, which can exert a considerable influence on the overall LCC associated with sewage treatment systems. In addition, economic analysis serves to facilitate the rigorous comparison of alternative sewage treatment strategies, thereby providing a structured approach to evaluating differing methodologies. Using refined methods like multi-criteria decision analysis (MCDA), individuals responsible for making choices can examine a range of treatment options founded on an extensive set of criteria, which cover economic viability, environmental responsibility, and social equity aspects. This integrative methodology fosters a more holistic and nuanced evaluation of the trade-offs and compromises inherent in diverse strategic options, ultimately culminating in the identification of solutions that are more economically sustainable and viable. Moreover, this integration significantly enhances the comprehension of the economic ramifications of various environmental impacts. Consequently, economic analysis not only plays a pivotal role in pinpointing cost-effective technologies but also promotes the adoption of operational practices that are in alignment with overarching sustainability objectives (Ilyas et al., 2021; Meneses et al., 2016; Ćetković et al., 2022; López-Morales and Rodríguez-Tapia, 2019). The salient effects of economic analysis on sewage treatment strategies are summarized as follows:

i. Informed Decision-making: The process of economic analysis, which is particularly exemplified through the comprehensive framework of CBA, enables various stakeholders to meticulously assess and evaluate the intricate financial implications and ramifications associated with the diverse sewage treatment methods available for implementation. Through a thorough analysis of the estimated costs, which cover both primary and secondary spending, against the predicted ecological gains, which could involve advancements in public well-being and environmental sustainability, leaders are given the chance to determine the strategy that is both economically feasible and supportive of larger societal aims and environmental care (Molinos-Senante et al., 2010; Ilyas et al., 2021).
ii. Valuation of Environmental Benefits: This analysis underscores the critical significance of accurately quantifying environmental benefits, which are frequently overlooked in traditional economic assessments due to their inherent lack of market value and the challenges associated with their measurement. Through rigorous economic analysis, these benefits can be ascribed a monetary value, thereby providing a robust justification for the financial outlays and expenditures that are associated with WWT and facilitating a more comprehensive understanding of the true costs and benefits of such environmental interventions (Crini and Lichtfouse, 2019; Molinos-Senante et al., 2010; Lokhande and Kalbar, 2024).
iii. Optimization of Treatment Processes: Economic analysis is essential to the identification and elucidation of the most cost-effective treatment processes available for wastewater management. To give an example, an exploration of the assorted treatment categories, including primary, secondary, and tertiary therapies, along with their corresponding expenses, brings forth key observations about their economic practicality and performance. By gaining a thorough understanding of these financial implications, stakeholders are empowered to optimize treatment strategies that are tailored to meet the specific needs and financial constraints of their respective communities, ultimately enhancing the overall effectiveness of wastewater management initiatives (Molinos-Senante et al., 2010; Junior et al., 2012).

iv. Revenue Generation from Reclaimed Water: The potential financial returns derived from the sale of reclaimed water are essential to shaping the overall economic viability and sustainability of WWTPs, as this aspect significantly influences their operational and capital investment decisions. A comprehensive analysis demonstrates that when the reclaimed water is effectively marketed and sold, there is a notable increase in the average net profit margins, thereby rendering the entire treatment process more economically appealing and advantageous for stakeholders involved. This positive financial outlook serves to stimulate the adoption of innovative strategies that prioritize the integration of water reuse practices, ultimately contributing to enhanced environmental sustainability and resource conservation (Molinos-Senante et al., 2010).

v. Identification of Economic Viability: It is evident that a substantial number of WWTPs may struggle to maintain economic viability if they do not take into account the myriad environmental benefits as well as the potential revenue streams generated from the sale of reclaimed water. Through rigorous economic analysis, stakeholders can effectively identify which WWT facilities possess the attributes necessary for financial sustainability, while simultaneously pinpointing those plants that may necessitate additional investigation or targeted improvements to bolster their economic performance (Molinos-Senante et al., 2010; Lokhande and Kalbar, 2024).

vi. Policy Development and Implementation: The process of economic analysis plays an integral role in the development of effective policy frameworks by furnishing policymakers with evidence-based recommendations that are essential for the formulation of robust wastewater management strategies. This systematic assessment points to the crucial necessity of executing efficient and sustainable practices in water resource management, thus safeguarding that both existing and anticipated water needs are addressed without harming environmental quality (Molinos-Senante et al., 2010; Lopez-Serrano et al., 2024).

vii. Investment Decisions: The substantial financial burdens associated with the initial capital investments necessary for WWTP construction and operation frequently act as a significant deterrent to their implementation and deployment within various regions. Through the utilization of comprehensive economic analysis, stakeholders are able to discern and identify the technologies and strategic approaches that exhibit the highest levels of financial viability, thereby facilitating a more judicious allocation of the limited financial resources that are often available. By prioritizing the development and implementation of cost-effective solutions, governmental entities possess the ability to significantly enhance the coverage and efficacy of WWT services without jeopardizing the integrity of their fiscal budgets (Crini and Lichtfouse, 2019; Junior et al., 2012).

viii. Local Adaptation: The process of economic analysis meticulously considers the distinctive local specificities, which include the unique characteristics of various water bodies, as well as the diverse regional needs that can vary widely from one locale to another. This localized analytical approach serves to ensure that the strategies employed for sewage treatment are meticulously tailored to align with the particular economic and environmental contexts of each area, ultimately leading to outcomes that are not only more effective in their implementation but also exhibit a greater degree of sustainability over time (Junior et al., 2012).

ix. Public Policy Formulation: The deliberate integration of economic analysis within the framework of public policy formulation can significantly enhance the decision-making processes that pertain to the management of wastewater systems and resources. By comprehensively analyzing the extensive range of costs and benefits tied to the multiple treatment options at hand, policymakers are positioned to develop and implement strategies that are not only founded on solid economic principles but also showcase a commitment to social equity and environmental sustainability, thereby fostering a balanced approach to resource management (Junior et al., 2012).

x. Long-term Planning: The application of economic analysis plays a pivotal role in supporting long-term planning initiatives by providing projections of the future costs and benefits that are likely to be associated with the implementation of different sewage treatment strategies over extended periods. Such foresight enables decision-makers to undertake the gradual and systematic implementation of treatment systems, which can contribute to cumulative environmental improvements over time, as clearly indicated in the findings articulated within the referenced paper (Junior et al., 2012; Lokhande and Kalbar, 2024).

In conclusion, economic analysis plays a pivotal role in shaping sewage treatment methodologies by facilitating informed decision-making, directing the selection of appropriate technologies, optimizing treatment processes, enabling comparative evaluations among various alternatives, assessing environmental benefits, fostering the integration of economic and ecological considerations, enhancing cost-efficiency, managing pollution loads, adapting to localized conditions, informing public policy directives, and bolstering long-term strategic planning. This holistic framework is indispensable for formulating effective and sustainable sewage treatment solutions that simultaneously address both financial and ecological imperatives.

5.8 Conclusion

The conventional physical (screening, sedimentation, filtration, flotation, adsorption, and membrane-based technologies) and chemical (chemical precipitation, ion exchange, advanced oxidation, ozonation, and electrochemical separation) technologies for WWT provide effective means for managing contaminants from wastewater. Optimizing factors like pH, dosage, pollutant initial conditions, and mixing conditions is essential to achieve the desired treatment outcomes and minimize related challenges. Several factors influence treatment method selection, including waste nature and concentration, effluent heterogeneity, effluent standards, and cost concerns. Physical treatment primarily aims to remove small-sized inorganic matter and large-sized settle able organic matter, achieving a reduction of 60–65% of suspended solids and 30–35% of BOD. Chemical treatment involves adding chemicals or agents to facilitate the removal of pollutants by altering their physical and chemical properties, such as pH, charge, solubility, and reactivity. Strategies include chemical oxidation, coagulation, flocculation, precipitation, ion exchange, neutralization, and disinfection, using agents like alum, lime, iron salts, polymers, chlorine, ozone, UV light, and various catalysts. Unfortunately, conventional WWT methods face significant limitations that compromise their ecological sustainability and effectiveness. Because traditional sewage treatment techniques may not adequately eliminate various types of pollutants, especially CEC like pharmaceuticals, personal care products, micro plastics, PFAS, and intricate organic compounds. These substances can persist through treatment processes and end up in the environment, posing concerns to both water quality and public health, consume substantial energy, produce substantial volumes of sludge, large space requirement, high operational costs, and high chemical consumption. This heavy reliance gives rise to worries regarding chemical remnants in the treated water and the possibility of harmful side products, which may result in negative environmental consequences, like the emission of greenhouse gases or other pollutants during treatment operations, lack of skilled manpower, and odor issues. Economic analysis assumes an indispensable position in the formulation and refinement of sewage treatment strategies by establishing a comprehensive framework that facilitates the thorough evaluation of the multifaceted costs and benefits associated with a plethora of treatment options.

Review Questions

1. Which one of the waste treatment plants (WTPs) is more economic?
2. Which one of the WTPs is more eco-friendly?
3. Which one of the WTPs is more easily to be integrated with other processes?
4. What are the economic benefits of wastewater treatment?

5 Are there any negative impacts of CBA on wastewater treatment plants (WWTPs)?
6 What are the major factors impacting investment decisions?
7 What are the economic and environmental benefits of WWTPs?
8 What is the mechanism of physical treatment methods and their efficiencies and limitations?
9 What is the mechanism of advanced oxidation processes and their efficiencies and limitations?
10 What are the different types of membranes and accordingly their pore size utilized in primary treatment processes?

Acknowledgments

And when Jesus brought clear proofs, he said, “I have come to you with wisdom and to make clear to you some of that over which you differ, so fear Allah and obey me.”

References

Ayach, J., El Malti, W., Duma, L. et al. (2024). Comparing conventional and advanced approaches for heavy metal removal in wastewater treatment: an in-depth review emphasizing filter-based strategies. *Polymers (Basel)*, **16**(14), 1959.

Ćetković, J., Knežević, M., Lakić, S. et al. (2022). Financial and economic investment evaluation of wastewater treatment plant. *Water (Basel)*, **14**(1), 122.

Corpus, R.M.B., Bayani, M.S., Aguilar, J.L. et al. (2024). Emerging pollutants in waste water: challenges and advancements in treatment technology. In: *IOP Conference Series: Earth and Environmental Science* (Vol. 1372, No. 1, p. 012037). IOP Publishing.

Crini, G. and Lichtfouse, E. (2019). Advantages and disadvantages of techniques used for wastewater treatment. *Environmental Chemistry Letters*, **17**, 145–155.

Edris, A. (2024). The evolving trend of mobile integrated domestic sewage treatment equipment. *Frontiers in Environmental Research*, **2**(2), 15–24. ISSN: 2960-026X. https://doi.org/10.61784/ferv2n212

Fetouh, M., Rustum, R., and Adeloye, A.J. (2021). A comparative study of conventional extended aeration and modern membrane bioreactor methods of sewage treatment from the environmental and financial perspectives. *GEOMATE Journal*, **20**(78), 50–56.

Friman, H. (2024). Harnessing the potential of renewable energy sources to reduce fossil energy consumption in the wastewater treatment process. *International Journal of Environmental and Ecological Engineering*, **18**(5), 138–142.

Gedda, G., Balakrishnan, K., Devi, R.U. et al. (2021). Introduction to conventional wastewater treatment technologies: limitations and recent advances. *Materials Research Foundations*, **91**, 1–36.

Ilyas, M., Kassa, F.M., and Darun, M.R. (2021). Life cycle cost analysis of wastewater treatment: a systematic review of literature. *Journal of Cleaner Production*, **310**, 127549.

Junior, A.M.J., Imbroisi, D., and Nogueira, J.M. (2012). Economics of wastewater treatment: cost-effectiveness, social gains and environmental standards. *Environmental Economics*, **3**(3), 17–32.

Lokhande, S. and Kalbar, P.P. (2024). Economic and environmental benefits of natural treatment systems for sewage treatment: a life cycle perspective. *Water Research*, **257**, 121710.

López-Morales, C.A. and Rodríguez-Tapia, L. (2019). On the economic analysis of wastewater treatment and reuse for designing strategies for water sustainability: lessons from the Mexico Valley Basin. *Resources Conservation and Recycling*, **140**, 1–12.

Lopez-Serrano, M.J., Lakho, F.H., Van Hulle, S.W.H. et al. (2024). Economic implications and scope of wastewater management through life cycle cost assessment. In: *Waste Management and Life Cycle Assessment for Sustainable Business Practice*, 65–85. IGI Global.

Meneses, M., Concepción, H., and Vilanova, R. (2016). Joint environmental and economical analysis of wastewater treatment plants control strategies: a benchmark scenario analysis. *Sustainability*, **8**(4), 360.

Molinos-Senante, M., Hernández-Sancho, F., and Sala-Garrido, R. (2010). Economic feasibility study for wastewater treatment: a cost–benefit analysis. *Science of the Total Environment*, **408**(20), 4396–4402.

Öberg, G., Metson, G.S., Kuwayama, Y. et al. (2020). Conventional sewer systems are too time-consuming, costly and inflexible to meet the challenges of the 21st century. *Sustainability*, **12**(16), 6518.

Quansah, A., Ntaryamira, T., and Rwemera, J. (2018). Sludge wastewater management by conventional treatment process: case study-Bujumbura municipal sewage. *International Journal of Sciences*, **4**, 52–65.

Rajasulochana, P. and Preethy, V. (2016). Comparison on efficiency of various techniques in treatment of waste and sewage water–a comprehensive review. *Resource-Efficient Technologies*, **2**(4), 175–184.

Ranjit, P., Jhansi, V., and Reddy, K.V. (2021). Conventional wastewater treatment processes. In: *Advances in the Domain of Environmental Biotechnology: Microbiological Developments in Industries, Wastewater Treatment and Agriculture*, 455–479. Springer.

Rathod, S., Preetam, S., Pandey, C. et al. (2024). Exploring synthesis and applications of green nanoparticles and the role of nanotechnology in wastewater treatment. *Biotechnology Reports*, **41**(5), e00830.

Shekho, M.S. and Hassan, N.E. (2024). A review on techniques for the cleaning of wastewater. *GSC Advanced Research and Reviews*, **18**(1), 118–128.

Sravan, J.S., Matsakas, L., and Sarkar, O. (2024). Advances in biological wastewater treatment processes: focus on low-carbon energy and resource recovery in biorefinery context. *Bioengineering*, **11**(3), 281.

Wang, Q. (2024). Comparison of the typical traditional and advanced sewage treatment technologies. *Highlights in Science, Engineering and Technology*, **83**, 1–7.

Werkneh, A.A. and Gebru, S.B. (2023). Development of ecological sanitation approaches for integrated recovery of biogas, nutrients and clean water from domestic wastewater. *Resources, Environment and Sustainability*, **11**, 100095.

6

Conventional Biological Treatment Technologies and Economic Implications

*Sachin Chunchu[1], Soubhagya Nayak[2], Nayan Babulal Gadling[2], Sudipa Bhadra[2], Soujanya Jonnalagadda[1], Venkateswara Rao Polisetty[1], and Surajbhan Sevda[2]**

[1]*Department of Civil Engineering, National Institute of Technology Warangal, Warangal, Telangana, India*
[2]*Department of Biotechnology, National Institute of Technology Warangal, Warangal, Telangana, India*

*Corresponding author: sevdasuraj@gmail.com; sevdasuraj@nitw.ac.in

6.1 Introduction

The global demand for efficient and sustainable wastewater treatment systems is increasing due to the rising population and industrial activity. Inadequate treatment of wastewater before its discharge into natural water bodies poses significant risks to both human health and the environment, as it often contains a range of organic pollutants, nutrients, and pathogens. Biological wastewater treatment, which relies on the natural metabolic activities of microorganisms to break down contaminants, has emerged as one of the most widely adopted approaches to wastewater management due to its cost-effectiveness and environmental sustainability (Kronenberg et al., 2017; Sevda et al., 2018; Kabutey et al., 2019; Molina et al., 2020).

Conventional biological wastewater treatment methods such as the activated sludge process, trickling filters, rotating biological contactors (RBCs), Aerated Lagoons, and Membrane Bioreactors (MBRs) are widely employed in both municipal and industrial wastewater treatment facilities. These technologies have proven effective in reducing total suspended solids, chemical oxygen demand (COD), biochemical oxygen demand (BOD), as well as nutrients like phosphorus

Wastewater to Resource Recovery: Applying the Circular Economy Toward Sustainable Development, First Edition. Edited by Arvind Kumar Mungray et al. Published 2026 by John Wiley & Sons, Inc.

and nitrogen. However, the effectiveness, cost-efficiency, and suitability of each technology can vary depending on factors such as plant size, influent characteristics, and local regulatory requirements (Park and Yoo, 2009; Zhang et al., 2020; Bhadra et al., 2024).

In addition to technical considerations, the economic implications of wastewater treatment technologies play a crucial role in decision-making for both municipalities and industries. The costs related to the construction, operation, and maintenance of treatment plants can vary significantly, directly influencing the selection of technology and overall system design. Moreover, balancing environmental protection objectives with financial limitations remains a persistent challenge—particularly in developing regions where funding for large-scale infrastructure projects may be scarce (Feng et al., 2010; Pradhan et al., 2017; Rao et al., 2023; Sohn et al., 2023; Yilmaz et al., 2024).

This chapter provides a comprehensive detail of the most widely used biological wastewater treatment technologies, focusing on their operational principles, performance, and economic implications. It will also examine the cost-benefit considerations involved in the selection and implementation of these systems, along with the challenges and limitations that may affect their overall efficiency and financial viability. Finally, the chapter will explore emerging trends and innovations that have the potential to improve both the technical effectiveness and economic sustainability of biological wastewater treatment processes (Gao et al., 2024).

6.2 Overview of Wastewater Treatment Needs

The rapid expansion of cities continues to drive the growing demand for water across industrial, agricultural, and domestic sectors. In this context, reclaimed wastewater has emerged as a promising unconventional resource to help address the current challenges of water scarcity (Sleutels et al., 2012; Singh et al., 2024). In response to a decade of growing environmental awareness, authorities around the world are increasingly prioritizing the treatment and prevention of environmental contamination and degradation. However, rapid industrial and economic development has intensified water pollution, making it a critical global concern. Wastewater from domestic and agricultural activities is often discharged directly into the soil, while increased human activity continues to contribute to the pollution of rivers and lakes (Behera et al., 2021).

In response to stringent effluent discharge regulations and increasing pollution in receiving water bodies, a range of treatment methods have been developed and implemented. Nitrogen and phosphorus are among the primary nutrients present in contaminated water, and their removal is essential for improving water quality. Water pollution affects the health of billions of people in both developed and developing countries, contributing to the spread of waterborne diseases such as cholera, cryptosporidiosis, typhoid, diarrhea, and ascariasis caused by exposure to contaminated water (White et al., 1996; Hemdan et al., 2021; Rusyn, 2021).

With the rapid advancement of technology, wastewater treatment methods are continually evolving. A variety of approaches—including physical, chemical, biological, and hybrid processes—are employed to remove pollutants effectively. Wastewater treatment plants (WWTPs) utilize these methods to treat wastewater through various biological processes, ultimately producing water suitable for its intended use. The data generated from these processes serves as a foundation for developing customized hybrid treatment solutions that are both energy-efficient and environmentally sustainable (Arun et al., 2024).

6.3 Importance of Biological Wastewater Treatment

The primary purpose of WWTPs is to mitigate the impact of wastewater generated in urban areas before it is discharged into receiving water bodies. While these facilities offer significant environmental benefits, they also pose certain negative impacts throughout their lifecycle—including during construction, operation, and decommissioning (Thevenon et al., 2011; Pathy et al., 2021; Yilmaz et al., 2024). Life cycle assessment is the most widely used methodology for evaluating the global environmental impacts of WWTPs. It encompasses the full range of impacts associated with a WWTP's entire lifecycle—from construction and operation to decommissioning or recycling—although not all stages are consistently analyzed in detail. In our view, the construction phase has received relatively limited attention in previous studies. Therefore, the chapter focuses specifically on that phase, examining both civil works and equipment-related impacts (Chen et al., 2015; Ho, 2022; Afonso et al., 2024).

According to some studies, the construction phase contributes to less than 5% of the total environmental impacts associated with WWTPs (Kusumawardani et al., 2021; Shen et al., 2021). Moreover, creating comprehensive inventories requires significant effort and time, which encourages some practitioners and researchers to streamline the process. One potential solution is to estimate the volume of reinforced concrete, which can then be used as a basis for estimating the quantities of other construction materials. To eliminate environmental contaminants (ECs) and other impurities, as well as to dispose of the broken-down materials, biological wastewater treatment relies on biological processes and organisms (Grandcl′ement et al., 2017). Biomineralization is the process by which bacteria, fungi, and algae break down ECs into smaller molecules, ultimately converting them into water and carbon dioxide. Through microbial metabolism, organic materials and nutrients such as phosphorus and nitrogen are degraded and assimilated. These biological processes are commonly employed to treat wastewater containing organic pollutants (Christensen et al., 2009).

Biological processes can be categorized as either anaerobic or aerobic. Aerobic techniques include methods such as MBRs, sequential batch reactors, and activated sludge systems. In contrast, anaerobic methods encompass technologies like anaerobic sludge reactors and anaerobic film reactors (Zhang and Angelidaki, 2013; Kang et al., 2014; Krzeminski et al., 2017; Mannina et al., 2019). Both household and municipal WWTPs extensively utilize biological methods for wastewater treatment. According to Ahmed, anaerobic treatment effectively produces CO_2, methane, and biomass, while aerobic treatment results in CO_2, water, and biomass as the end products (Ahn and Logan, 2010). The anaerobic and aerobic processes can be succinctly compared. Additionally, biological wastewater treatment methods are further categorized into conventional and non-conventional treatment methods.

6.4 Conventional Biological Treatment Technologies

Conventional biological treatment (CBT) technologies are widely used in wastewater treatment due to their proven effectiveness, cost-efficiency, and simplicity. These methods typically rely on the natural metabolic processes of microorganisms to break down contaminants in wastewater. It uses techniques such as microorganism-based treatment, bioremediation, nitrification–denitrification, aerobic/anaerobic treatment, biofilters, and biofilm reactors. These technologies are suitable for a wide range of wastewater treatment applications, both in municipal and industrial settings, and are designed to reduce pollutants such as organic matter, nitrogen, and phosphorus.

6.4.1 Aerobic Treatment

6.4.1.1 Nitrification–Denitrification

While nitrates and nitrites are converted into nitrogen gas during the denitrification process, ammonium is converted into nitrite during nitrification. Denitrification occurs in an anoxic environment, whereas nitrification takes place in the presence of oxygen. The MBR technique is commonly used in conjunction with denitrification to remove ECs. High efficiency was observed in the removal of endocrine-disrupting chemicals (EDCs), including pentachlorophenol (PCPs), bisphenol A, 4-tert-octylphenol, 17-beta-estradiol, and estrone (E1), with removal rates ranging from 82% to 100% through denitrification (Kumar et al., 2019; Baby and Ahammed, 2022; Prathiba et al., 2022).

However, studies have shown that the denitrification process achieved an 88–89% elimination rate for PCP and triclosan. The removal efficiency of PCPs decreased during nitrification, while the removal of EDCs improved. Additionally, EDCs exhibited the highest removal efficiency during denitrification, followed by pesticides and PCPs, with pharmaceutical compounds showing the lowest removal efficiency (Maiti et al., 2019; Yuan et al., 2023). Denitrification outperformed nitrification in terms of EC elimination. This highlights how a biologically based non-aerated filter can be integrated into a nitrification–denitrification system to effectively treat nitrogen-heavy wastewater. The removal efficiencies of NH_4^+, COD, and total nitrogen were improved to 88.62 ± 0.81%, 76.12 ± 0.57%, and 50.48 ± 1.02%, respectively, when nitrogen-limited autotrophic microbes were examined (Li et al., 2023).

6.4.1.2 Activated Sludge Process

In the wastewater aeration tank, biomass is generated as microorganisms grow in the presence of oxygen. The tank treats wastewater from sewage and other industrial sources by using air, biological flocs, bacteria, and protozoa. These organisms decompose organic waste into carbon dioxide, water, and other chemical by-products (Kumar et al., 2019). This technique could effectively remove over 10 different types of surfactants.

The activated sludge process is demonstrated by supplying oxygen to wastewater fed into a biological reactor, which stimulates the growth of microorganisms. These microorganisms help reduce contamination in the aeration tank (Kumar et al., 2019; Sonawane et al., 2022). As a result, activated sludge is produced and transferred to the secondary settling tank for clarification. In this tank, the sludge is separated, with most of it being returned to the biological reactor and the remainder removed. According to the activated sludge process, both organic and inorganic waste are extracted from wastewater using aerobic organisms (Siwek et al., 2023). Activated sludge treatment is a commonly used method for treating industrial and municipal wastewater. The effectiveness of the activated sludge process is primarily influenced by the metabolism and interactions of the microbial population, particularly the bacteria essential to the treatment process. Recent studies have explored the potential applications of activated sludge in biological treatment (Maw et al., 2024).

6.4.1.3 Biofilter

Biofilters are biological reactors capable of breaking down pollutants through the action of biological organisms (Wang et al., 2018; Khanongnuch et al., 2022). Biofilters can be further categorized based on the types of organisms present in the reactor. One common type of biofilter is the biological trickling filter, which consists of a three-phase system made up of biofilm carriers. Wastewater is introduced into the bioreactor via a distribution system, where it trickles down over the biofilm surface. In the third phase, air flows either upward or downward through the system. Trickling filters can be used alone or in combination with other treatment methods (Ahn et al., 2005).

This method can also significantly reduce salinity, indicating that activated sludge is much more effective at eliminating organic pollutants from wastewater compared to trickling filter beds, which are more focused on removing ionic compounds. The removal of nitrogen and toxicity from municipal wastewater was successfully achieved using a sulfur-based denitrification biofilter combined with micro-electrolysis. The results showed a decrease in toxicity by 8–17%. Mei and colleagues developed the membrane-aerated biofilter by combining membrane aeration with biofilter technology to treat wastewater containing p-nitroaniline (Hidayaturrahman and Lee, 2019). At the pilot scale, the system achieved complete (100%) removal of p-nitroaniline at an influent sulfur content of 120 mg/L, while removal rates for o-nitroaniline ranged from 56% to 86%. (Ardakani and Gholikandi, 2020).

6.4.1.4 Rotating Biological Contactors and Its Advantages

It is one of the effective secondary treatments that utilizes microbial communities to treat wastewater from households and businesses. This traditional aerobic method achieves BOD removal, nitrification, and denitrification. It is a biofilm process commonly integrated into modern industrial and municipal wastewater treatment facilities (Hassard et al., 2015; Han et al., 2019). In an RBC, the contactor's shaft axis has a number of discs. (Hassard et al., 2015). The contactor receives a specific volume of water, causing the disc to become partially submerged in the effluent. In this position, the disc interacts with the microbial film present in the wastewater. As the disc rotates, the portion that was previously submerged is exposed to atmospheric oxygen, stimulating the metabolism of the microorganisms. This action allows the organic matter in the wastewater to serve as a nutrient, promoting the growth and breakdown of the microbial community (Hassard et al., 2015; Han et al., 2019).

A layer of microorganisms, known as sludge, is created and maintained through regular operation. The functionality of a four-stage laboratory cross-flow scale system is also influenced by this microbial layer (Singh et al., 2024). The study's findings demonstrated that, for all anaerobic baffled reactor (ABR) alternatives employed in this investigation, trapping dissolved methane as gaseous methane and converting it into heat and energy can enhance the benefits for ecosystem quality, human health, and resource management. Additionally, the performance of the RBC and ABR systems was compared under various environmental conditions (Hassard et al., 2015; Han et al., 2019). The impact of climate change is reduced by 87% when comparing the ABR system to the simulated

ABR-RBC system. This reduction occurs because the RBC oxidizes 80% of the dissolved methane (Md Khudzari et al., 2019). Less biosolid production results in a smaller negative impact, as the trickling filter system experiences damage from biosolid disposal that is approximately five times greater than the damage caused by the ABR system, in addition to the issues related to dissolved methane (Van Groenestijn et al., 2002; Winfield et al., 2012).

6.4.2 Anaerobic Treatment

Anaerobic wastewater treatment is a biological process that occurs in the absence of oxygen, where microorganisms break down organic pollutants in wastewater to produce biogas, primarily methane and carbon dioxide. This method is often used for treating high-strength wastewater, such as that from industrial processes, agricultural runoff, and municipal waste. Unlike aerobic treatment, which relies on oxygen for microbial metabolism, anaerobic treatment is more energy-efficient because it requires no external oxygen supply and generates renewable energy in the form of methane.

6.4.2.1 Membrane Bioreactor

It is a hybrid wastewater treatment method that combines membrane technology with biological reactors. This method is used to treat wastewater containing pollutants that are not effectively addressed by constructed wetlands or activated sludge systems. Suspended solids are collected and settled in sedimentation tanks. The wastewater is then directed into a biologically activated sludge tank, which incorporates an integrated membrane system and fine bubble aerators to enhance the treatment process (Ardakani and Gholikandi, 2020).

MBR employs ozonation in combination with adsorption and biodegradation to remove ECs. The effectiveness of MBR in removing contaminants, ranging from analgesics and antibiotics to anti-inflammatory drugs, stimulants, and other pharmaceutical compounds, gradually decreases as the complexity of the pollutants increases (Kitade et al., 2013; Mutamim et al., 2013; Samsami et al., 2020).

The MBR approach demonstrates that the overall efficiency is highest for EDCs, with efficiency declining for beta blockers, PCPs, pharmaceuticals, and pesticides. Due to their appealing benefits—such as energy savings, biogas recovery, and reduced sludge production—anaerobic biological technologies have garnered significant attention in the wastewater treatment sector (Kumar et al., 2023).

6.4.2.2 Microbial Fuel Cell

Microbial fuel cell (MFC) is a relatively new technique that operates as a bioelectrochemical system, overcoming many of the constraints associated with traditional anaerobic wastewater treatment methods (Sevda et al., 2016; Bhadra et al., 2024).

Biodegradable organic substrates are anaerobically oxidized, with microbes in the anodic compartment generating protons and electrons as they break down organic materials (Schievano et al., 2018). The fundamental principle behind MFC functioning is that microorganisms utilize the organic substrate in the anode zone as a source (donor) of carbon and electrons. Electrons are transferred to non-biological cathodes, where they are accepted by oxygen in the cathode. MFCs are bioelectrochemical systems designed to generate electricity sustainably. Electrochemically active bacteria play a key role in this process by producing electricity through their metabolic activities. Biomass oxidation is achieved via biological reactions within the MFC. Depending on the type of catalysts used, MFCs can be classified into categories based on organisms or enzymes. By employing living microorganisms as electrode catalysts, MFCs effectively generate power (Mohan and Das, 2009; Srikanth et al., 2010; Teng et al., 2010). The method, which can utilize both organic and inorganic materials, has gained significant attention as one of the most promising and environmentally beneficial energy sources. By directly converting waste into clean power or high-value energy, it helps address energy-related challenges posed by traditional wastewater systems, such as excessive sludge formation (Li et al., 2023).

6.4.2.3 Composting

Traditional wastewater treatment typically involves several physical, chemical, and biological processes to remove contaminants from wastewater before it is either released into the environment or reused (Cucina et al., 2021; Baldera-Moreno et al., 2022; Cazaudehore et al., 2022). Primary treatment is the first step in wastewater treatment, focusing on the physical separation of solids from liquids. Methods such as sedimentation and filtration are used to remove large particles and debris. The goal is to reduce the wastewater's organic matter and total suspended solids (Vuorinen and Saharinen, 1997; Ding et al., 2021). In secondary treatment, organic matter and nutrients are further reduced through biological processes. Microorganisms, such as bacteria, break down organic pollutants. Techniques like trickling filters, biofilm reactors, and activated sludge systems are commonly used to achieve this. As a result, the BOD and COD are significantly reduced (Gao et al., 2010; Lim et al., 2016). The goal of tertiary treatment is to eliminate any remaining pollutants, including pathogens and nutrients like nitrogen and phosphorus. Methods such as filtration, chemical treatment, and UV or chlorination-based disinfection are used to achieve this. After tertiary treatment, the wastewater can be safely released into the environment or reused. Although not a direct component of traditional wastewater treatment, composting plays a crucial role in managing organic waste within the system (Afonso et al., 2024).

6.5 Economic Implications of Conventional Biological Treatment

The implementation of conventional biological wastewater treatment methods, such as activated sludge, trickling filters, and biological aerated filters, comes with both capital and operational costs. These costs vary depending on the scale of the treatment plant, the technology employed, and the specific wastewater characteristics being treated.

6.5.1 Capital Costs

In CBT, capital costs refer to the expenses incurred during the construction and setup of the treatment facility, including infrastructure, equipment, and land acquisition (Suarez et al., 2010; Bhuyar et al., 2018; Mohammadi et al., 2020). A few examples of capital costs in CBT include land acquisition, which involves purchasing or leasing land for the treatment facility. Construction costs refer to expenses related to building the treatment plant, including foundations, buildings, and other structures. Equipment costs involve the purchase of items such as pumps, blowers, aeration systems, clarifiers, and sludge handling systems. Installation costs cover the price of installing the infrastructure and equipment, including instrumentation, electrical wiring, and piping. Permitting and regulatory fees are the costs for obtaining the necessary permits and ensuring compliance with regulations. Design and engineering fees include costs for consulting, designing the treatment plant, and creating construction plans. Finally, contingency costs refer to additional funds set aside to cover unforeseen construction-related expenses or cost overruns. Several factors can influence the capital costs of CBT systems. Operational scale is one such factor. While larger facilities typically have higher initial costs, they benefit from economies of scale. As capacity increases, the cost per unit of treated wastewater tends to decrease, making larger facilities more cost-effective in the long run (Badmus et al., 2018). The initial investment for different treatment technologies varies. For example, activated sludge systems typically require more infrastructure compared to simpler techniques like constructed wetlands (Yilmaz et al., 2024).

6.5.2 Operational Costs

CBT plant operating costs include recurring expenditures such as energy consumption, which is significant due to the aeration, pumping, and other operations required. For instance, aeration can account for as much as 60% of the energy used in activated sludge systems. Labor costs are another major expense, covering the pay and benefits for employees such as operators, maintenance workers, and administrative staff. This also includes the costs of retaining and training qualified personnel. Maintenance costs are incurred for the upkeep of equipment such as blowers, pumps, and clarifiers, which require both routine and emergency maintenance. This includes the price of consumables and replacement parts. Chemical expenses cover the costs of coagulants, flocculants, and disinfectants used in the treatment process. Sludge handling and disposal also represent a significant expense, as it involves the transportation and disposal of the sludge produced during treatment. Additionally, there are costs associated with monitoring water quality to ensure compliance with regulations, which include reporting and laboratory analysis. Other miscellaneous costs, such as utilities (electricity and water), insurance, and administrative overheads, also contribute to operating expenses. For example, continuous operation of blowers and aerators in a CBT plant using activated sludge may lead to high electricity costs, and the labor costs for operators working in shifts to monitor the process can be substantial. Operational costs can vary significantly depending on the size of the plant (Ibáñez et al., 2024).

6.5.3 Revenue Generation

There are various ways that CBT can generate revenue. One key method is energy production. Biogas produced during anaerobic digestion can be captured and used to create sellable heat or electricity. This not only reduces the energy costs associated with the treatment process but also provides a potential revenue stream. In addition to energy, other revenue-generating opportunities can include the recycling of treated wastewater for industrial, agricultural, or landscaping purposes, as well as the sale of by-products such as biosolids, which can be used as fertilizers. These avenues help offset the costs of operation and make CBT a more financially viable option (Escapa et al., 2016).

Resource recovery is another significant revenue-generating opportunity in CBT. Phosphorus, nitrogen, and other nutrients can be recovered from wastewater and processed into fertilizers. This not only helps reduce environmental pollution by preventing these nutrients from being released into natural water bodies but also creates a valuable product that can be sold to agricultural industries. The recovery of these nutrients from wastewater offers both environmental and economic benefits, contributing to the sustainability of wastewater treatment operations and creating new revenue streams for facilities (Li et al., 2023).

Reusing treated wastewater for purposes such as irrigation, industrial operations, or even as drinking water can generate revenue through water sales. By providing an alternative water source, facilities can contribute to water conservation while creating a revenue stream. After being treated, sludge can be converted into biosolids and sold as fertilizers or soil conditioners. These by-products can be used in agriculture or landscaping, offering an additional income source for WWTPs. Wastewater treatment facilities can participate in carbon markets, selling the carbon credits they earn by reducing greenhouse gas emissions. Techniques like biogas recovery and energy-efficient operations can contribute to lowering the carbon footprint, enabling the sale of credits that provide additional financial returns (Jiménez-Benítez et al., 2023). By incorporating these revenue-generating methods—nutrient recovery, water reuse, biosolid sales, and carbon credits—wastewater treatment facilities can offset operational costs and improve their financial sustainability.

6.5.4 Externalities and Social Costs

By treating wastewater before it's released into natural water bodies, CBT significantly reduces pollution. This improvement in water quality supports healthier ecosystems and lowers the costs associated with environmental cleanup efforts, contributing to long-term environmental sustainability. By minimizing water pollution, CBT helps reduce the incidence of waterborne diseases, leading to better public health outcomes. This, in turn, can result in reduced healthcare costs for communities and governments, making CBT not only environmentally beneficial but also economically advantageous in terms of public health (Ibáñez et al., 2024). Cleaner water and improved environmental conditions can have positive social impacts, such as increasing property values and creating more recreational opportunities. Communities benefit from enhanced aesthetics and outdoor spaces, which contribute to overall quality of life and can stimulate local economies.

6.5.5 Assessment of Economic Value

Cost savings resulting from the prevention of environmental damage can be a valuable way to assess the economic benefits of wastewater treatment. By improving water quality through wastewater treatment, the demand for additional water treatment processes in the future is reduced, leading to long-term cost savings for municipalities and industries. Proper wastewater treatment helps prevent pollution that can cause fish kills. Fewer fish kills mean less financial loss for the fishing industry, tourism, and local economies dependent on healthy aquatic ecosystems. Effective wastewater treatment can preserve ecosystems and support biodiversity, contributing to the health of wildlife and the broader environment. This can have direct economic benefits through sustainable tourism, agriculture, and ecosystem services (Pergola et al., 2018; Pandit et al., 2021; Carruthers and Lee, 2022).

Improved public health as a result of reduced waterborne diseases directly translates into greater productivity, reduced healthcare expenses, and longer life expectancies. For example, a reduction in waterborne illnesses means fewer medical treatments and less strain on healthcare systems. A study could estimate the savings in healthcare costs and the value of healthier individuals contributing more to the economy, thereby justifying the financial viability of CBT systems. This involves comparing the costs of building, operating, and maintaining CBT facilities with the economic value of external benefits such as improved public health, better water quality, and environmental protection. For instance, a CBT facility might lower healthcare expenses by $1 million annually by reducing waterborne diseases. Additionally, enhanced water quality could boost local property values by $500 000 each year. The net economic impact can be determined by subtracting the operational and maintenance costs of the plant from these benefits (Ibáñez et al., 2024). By applying this method, communities and policymakers can evaluate the broader economic impacts of CBT technologies, demonstrating not only the direct benefits but also the long-term advantages of investing in wastewater treatment infrastructure.

6.6 Challenges and Future Directions

Biological treatment technologies, despite their advantages, face several issues when it comes to broad implementation. One major concern is the high costs associated with these systems. Both the initial setup and ongoing operational costs can be prohibitive, making them less accessible for certain municipalities or industries. Efficiency is another challenge, as traditional biological methods are often ineffective at removing certain contaminants, especially more complex or persistent pollutants. These systems can also consume a substantial amount of energy, adding to both operational costs and environmental impact. Excess sludge production is another common issue with biological treatment systems. The accumulation of sludge requires additional handling, disposal, or treatment, which can be both costly and environmentally problematic. Furthermore, emerging pollutants, such as pharmaceuticals and personal care products, present a unique challenge. These compounds are difficult for traditional systems to remove, making it harder to meet the latest effluent quality standards. These factors combine to present obstacles in the widespread adoption of biological treatment technologies, requiring further research and development to overcome the limitations (Jia et al., 2014; Tervahauta et al., 2014).

Combining biological treatment with physical, chemical, or other biological techniques can significantly enhance the efficiency and effectiveness of wastewater treatment. For instance, integrating advanced oxidation processes (AOPs) with biological methods can improve the removal of stubborn contaminants that traditional biological systems may struggle with. AOPs, which generate highly reactive species like hydroxyl radicals, can break down organic pollutants, making them more amenable to biodegradation by microorganisms. This hybrid approach offers a more comprehensive solution, optimizing contaminant removal across a wider range of pollutants while potentially reducing the overall energy and maintenance costs associated with traditional methods. The synergy between biological and other treatment techniques provides a promising path for more effective and sustainable wastewater management (Samsami et al., 2020). Membrane technology is another promising innovation that enhances wastewater treatment efficiency using advanced membranes. These membranes can filter out contaminants more effectively while also reducing energy consumption, which makes the process more cost-effective and energy-efficient. MFCs also hold great potential, as they treat wastewater while simultaneously generating bioenergy, providing a dual benefit of waste management and energy production. Genetically engineered microorganisms offer another avenue for improving biological treatment processes, as these modified organisms can accelerate the breakdown of specific pollutants, making the treatment process more efficient. Nanotechnology, with its ability to enhance biological processes through the use of nanomaterials, can further improve the efficacy and efficiency of wastewater treatment by offering enhanced adsorption, filtration, and contaminant removal properties. Looking ahead, the focus is shifting toward sustainable practices, with an emphasis on circular approaches to waste management. These practices promote the reuse and recycling of resources, turning wastewater treatment systems into sustainable, closed-loop systems that minimize waste and maximize resource recovery. This shift aligns with global trends toward more eco-friendly, resource-efficient solutions that aim to reduce environmental impact while optimizing treatment performance (Shemfe et al., 2018; Bhardwaj et al., 2021; Yilmaz et al., 2024).

6.7 Conclusion

CBT technologies, such as aerobic and anaerobic processes, are effective at decomposing organic pollutants in wastewater. These methods use naturally occurring microorganisms, making them a reliable choice for companies and municipalities managing wastewater of moderate organic strength. In terms of economic considerations, these technologies have significant financial implications. This chapter suggests that the choice of technology can be influenced by the wide variations in construction, operation, and maintenance costs. It emphasizes that a cost-benefit analysis must be conducted to ensure the financial viability of treatment systems. It also discusses opportunities for revenue generation. For example, anaerobic digestion can produce biogas, which can be captured and used for energy production. Additionally, treated wastewater can be recycled for industrial and agricultural use, creating additional revenue streams. In terms of environmental benefits, these technologies improve water quality by treating wastewater before it is discharged into natural water bodies, thus reducing environmental pollution and minimizing the costs of cleanup.

Review Questions

1. What are the main biological mechanisms underlying conventional wastewater treatment technologies such as activated sludge and trickling filters?
2. How do operational parameters like hydraulic retention time (HRT) and sludge retention time (SRT) affect treatment performance in conventional biological systems?
3. What are the major energy demands in conventional biological wastewater treatment systems, and how can these be optimized for cost savings?
4. How does sludge production and disposal affect the overall cost and environmental footprint of conventional biological treatment (CBT)?
5. How does nutrient removal (N and P) efficiency vary across different CBT systems?
6. What are the economic trade-offs between capital investment and operational cost in centralized versus decentralized biological treatment systems?
7. What is the difference between aerobic and anaerobic systems for wastewater treatment?
8. What is the difference between the mechanism of energy generation between anaerobic systems (like up-flow anaerobic sludge blanket) and bioelectrochemical systems (microbial fuel cells)?
9. What are the limitations of conventional wastewater treatment systems toward energy generation?
10. What are the potentials of bioelectrochemical systems and their limitations, and accordingly how advanced technologies can overcome these issues?

References

Afonso, V. Borges, R., Rodrigues, B. et al. (2024). Are native microalgae consortia able to remove microplastics from wastewater effluents? *Environmental Pollution*, **349**, 123931. https://doi.org/10.1016/j.envpol.2024.123931

Ahn, Y. and Logan, B.E. (2010). Bioresource Technology Effectiveness of domestic wastewater treatment using microbial fuel cells at ambient and mesophilic temperatures. *Bioresour Technol*, **101**, 469–475. https://doi.org/10.1016/j.biortech.2009.07.039

Ahn, Y., Park, E.-J., Oh, Y.-K. et al. (2005). Biofilm microbial community of a thermophilic trickling biofilter used for continuous biohydrogen production. *FEMS Microbiology Letters*, **249**, 31–38. https://doi.org/10.1016/j.femsle.2005.05.050

Ardakani, M.N. and Gholikandi, G.B. (2020). Microbial fuel cells (MFCs) in integration with anaerobic treatment processes (AnTPs) and membrane bioreactors (MBRs) for simultaneous efficient wastewater/sludge treatment and energy recovery – a state-of-the-art review. *Biomass and Bioenergy*, **141**, 105726. https://doi.org/10.1016/j.biombioe.2020.105726

Arun, J., SundarRajan, P., Pavithra, K.G. et al. (2024). New insights into microbial electrolysis cells (MEC) and microbial fuel cells (MFC) for simultaneous wastewater treatment and green fuel (hydrogen) generation. *Fuel*, **355**, 129530. https://doi.org/10.1016/j.fuel.2023.129530

Baby, M.G. and Ahammed, M.M. (2022). Nutrient removal and recovery from wastewater by microbial fuel cell-based systems – a review. *Water Science and Technology*, **86**(1), 29–55. https://doi.org/10.2166/wst.2022.196

Badmus, K.O, Tijani, J.O., Massima, E. et al. (2018). Treatment of persistent organic pollutants in wastewater using hydrodynamic cavitation in synergy with advanced oxidation process. *Environmental Science and Pollution Research*, **25**(8), 7299–7314. https://doi.org/10.1007/s11356-017-1171-z

Baldera-Moreno, Y., Pino, V., Farres, A. et al. (2022). Biotechnological aspects and mathematical modeling of the biodegradation of plastics under controlled conditions. *Polymers*, **14**(3), 1–22. https://doi.org/10.3390/polym14030375

Behera, B. Unpaprom, Y., Ramaraj, R. et al. (2021). Integrated biomolecular and bioprocess engineering strategies for enhancing the lipid yield from microalgae. *Renewable and Sustainable Energy Reviews*, **148**, 111270. https://doi.org/10.1016/j.rser.2021.111270

Bhadra, S., Nayak, S., and Sevda, S. (2024). Simultaneous organic wastewater treatment and bioelectricity production in a dual chamber microbial fuel cell with Scenedesmus obliquus biocathode. *Energy Conversion and Management*, **316**, 118849. https://doi.org/10.1016/j.enconman.2024.118849

Bhardwaj, R., Sharma, T., Nguyen, D.D. et al. (2021). Integrated catalytic insights into methanol production: sustainable framework for CO2 conversion. *Journal of Environmental Management*, **289**, 112468. https://doi.org/10.1016/j.jenvman.2021.112468

Bhuyar, P., Muniyasamy, S., and Govindan, N. (2018). Green revolution to protect environment – an identification of potential micro algae for the biodegradation of plastic waste in Malaysia. *World Congress on BIOPOLYMERS AND BIOPLASTICS & RECYCLING Expert Opin Environ Biol 2018*, **7**, 1. https://doi.org/10.4172/2325-9655-C5-033

Carruthers, D.N. and Lee, T.S. (2022). Translating advances in microbial bioproduction to sustainable biotechnology. *Frontiers in Bioengineering and Biotechnology*, **10**, 1–22. https://doi.org/10.3389/fbioe.2022.968437

Cazaudehore, G., Guyoneaud, R., Evon, P. et al. (2022). Can anaerobic digestion be a suitable end-of-life scenario for biodegradable plastics? A critical review of the current situation, hurdles, and challenges. *Biotechnology Advances*, **56**, 107916. https://doi.org/10.1016/j.biotechadv.2022.107916

Chen, X., Vollertsen, J., Nielsen, J.L. et al. (2015). Degradation of PPCPs in activated sludge from different WWTPs in Denmark. *Ecotoxicology*, **24**(10), 2073–2080. https://doi.org/10.1007/s10646-015-1548-z

Christensen, A.M., Markussen, B., Baun, A., Halling-Sørensen, B. (2009). Probabilistic environmental risk characterization of pharmaceuticals in sewage treatment plant discharges. *Chemosphere*, **77**, 351–358. https://doi.org/10.1016/j.chemosphere.2009.07.018

Cucina, M., de Nisi, P., Tambone, F. et al. (2021). The role of waste management in reducing bioplastics' leakage into the environment: a review. *Bioresource Technology*, **337**, 125459. https://doi.org/10.1016/j.biortech.2021.125459

Ding, Y., Zhao, J., Liu, J.-W. et al. (2021). A review of China's municipal solid waste (MSW) and comparison with international regions: management and technologies in treatment and resource utilization. *Journal of Cleaner Production*, **293**, 126144. https://doi.org/10.1016/j.jclepro.2021.126144

Escapa, A., Mateos, R., Martínez, E.J. et al. (2016). Microbial electrolysis cells: an emerging technology for wastewater treatment and energy recovery. From laboratory to pilot plant and beyond. *Renewable and Sustainable Energy Reviews*, **55**, 942–956. https://doi.org/10.1016/j.rser.2015.11.029

Feng, C.-H., Li, F.-B., Mai, H.-J. et al. (2010). Bio-electro-Fenton process driven by microbial fuel cell for wastewater treatment. *Environmental Science & Technology*, **44**(5), 1875–1880. https://doi.org/10.1021/es9032925

Gao, M., Li, B., Yu, A. et al. (2010). The effect of aeration rate on forced-aeration composting of chicken manure and sawdust. *Bioresource Technology*, **101**(6), 1899–1903. https://doi.org/10.1016/j.biortech.2009.10.027

Gao, R., Gao, S.-H., Li, J. et al. (2024). Emerging technologies for the control of biological contaminants in water treatment: a critical review. *Engineering*, **48**, 185–204. https://doi.org/10.1016/j.eng.2024.08.022

Grandcl′ement, C., Seyssiecq, I., Piram A. et al. (2017). From the conventional biological wastewater treatment to hybrid processes, the evaluation of organic micropollutant removal: a review. *Water Research*, **111**, 297–317.

Han, Y., Ma, J., Xiao, B. et al. (2019). New integrated self-refluxing rotating biological contactor for rural sewage treatment. *Journal of Cleaner Production*, **217**, 324–334. https://doi.org/10.1016/j.jclepro.2019.01.276

Hassard, F., Biddle, J., Cartmell, E. et al. (2015). Rotating biological contactors for wastewater treatment–a review. *Process Safety and Environmental Protection*, **94**, 285–306. https://doi.org/10.1016/j.psep.2014.07.003

Hemdan, B.A., Gamila, E.E.-T., Goswami, P. et al. (2021). The role of biofilm in the development and dissemination of ubiquitous pathogens in drinking water distribution systems: an overview of surveillance, outbreaks, and prevention. *World Journal of Microbiology and Biotechnology*, **37**(2), 1–18. https://doi.org/10.1007/s11274-021-03008-3

Hidayaturrahman, H. and Lee, T.G. (2019). A study on characteristics of microplastic in wastewater of South Korea: identification, quantification, and fate of microplastics during treatment process. *Marine Pollution Bulletin*, **146**, 696–702. https://doi.org/10.1016/j.marpolbul.2019.06.071

Ho, D.T.K. (2022). Abundance of microplastics in wastewater treatment sludge. *Journal of Human, Earth, and Future*, **3**(1), 138–146. https://doi.org/10.28991/HEF-2022-03-01-010

Ibáñez, J., Pérez-de-Mora, A., Santiago-Herrera, M. et al. (2024). Environmental and socio-economic evaluation of a groundwater bioremediation technology using social Cost-Benefit Analysis: application to an in-situ metal(loid) precipitation case study. *Science of the Total Environment*, **954**, 176720. https://doi.org/10.1016/j.scitotenv.2024.176720

Jia, Q., Xiang, H., Wang, H. et al. (2014). Production of polyhydroxyalkanoates (PHA) by bacterial consortium from excess sludge fermentation liquid at laboratory and pilot scales. *Bioresource Technology*, **171**(1), 159–167. https://doi.org/10.1016/j.biortech.2014.08.059

Jiménez-Benítez, A., Ruiz-Martínez, A., Ferrer, J. et al. (2023). Life cycle costing of AnMBR technology for urban wastewater treatment: a case study based on a demo-scale AnMBR system. *Journal of Environmental Chemical Engineering*, **11**(3), 110267. https://doi.org/10.1016/j.jece.2023.110267

Kabutey, F.T., Zhao, Q., Wei, L. et al. (2019). An overview of plant microbial fuel cells (PMFCs): configurations and applications. *Renewable and Sustainable Energy Reviews*. https://doi.org/10.1016/j.rser.2019.05.016

Kang, C.S., Eaktasang, N., Kwon, D.-Y. et al. (2014). Enhanced current production by desulfovibrio desulfuricans biofilm in a mediator-less microbial fuel cell. *Bioresource Technology*, 165, 27–30. https://doi.org/10.1016/j.biortech.2014.03.148

Khanongnuch, R., Abubackar, H.N., Keskin, T. et al. (2022). Bioprocesses for resource recovery from waste gases: current trends and industrial applications. *Renewable and Sustainable Energy Reviews*, **156**, 111926. https://doi.org/10.1016/j.rser.2021.111926

Kitade, T., Wu, B., Chong, T.H. et al. (2013). Fouling reduction in MBR-RO processes: the effect of MBR F/M ratio. *Desalination and Water Treatment*, **51**(25–27), 4829–4838. https://doi.org/10.1080/19443994.2013.776243

Kronenberg, M., Trably, E., Bernet, N. et al. (2017). Biodegradation of polycyclic aromatic hydrocarbons: using microbial bioelectrochemical systems to overcome an impasse. *Environmental Pollution*, **231**, 509–523. https://doi.org/10.1016/j.envpol.2017.08.048

Krzeminski, P., Leverette, L., Malamis, S. et al. (2017). Membrane bioreactors – A review on recent developments in energy reduction, fouling control, novel configurations, LCA and market prospects. *Journal of Membrane Science*, **527**, 207–227. https://doi.org/10.1016/j.memsci.2016.12.010

Kumar, S., Priyadarshini, M., Ahmad, A. et al. (2023). Advanced biological and non-biological technologies for carbon sequestration, wastewater treatment, and concurrent valuable recovery: a review. *Journal of CO2 Utilization*, **68**, 102372. https://doi.org/10.1016/j.jcou.2022.102372

Kumar, S.S., Kumar, V., Malyan, S.K. et al. (2019). Microbial fuel cells (MFCs) for bioelectrochemical treatment of different wastewater streams. *Fuel*, **254**, 115526. https://doi.org/10.1016/j.fuel.2019.05.109

Kusumawardani, Y., Subekti, S., Astuti, W. et al. (2021). Portable wastewater treatment plant using banana stem filter media in small scale motor vehicle washing services. *IOP Conference Series: Earth and Environmental Science*, **746**(1), 012039. IOP Publishing. https://doi.org/10.1088/1755-1315/746/1/012039

Li, L., Liao, Q., Liu, C. et al. (2023). Enhanced biological wastewater treatment supplemented with anaerobic fermentation liquid of primary sludge. *Journal of Environmental Management*, **347**, 119086. https://doi.org/10.1016/j.jenvman.2023.119086

Lim, S.L., Lee, L.H., and Wu, T.Y. (2016). Sustainability of using composting and vermicomposting technologies for organic solid waste biotransformation: recent overview, greenhouse gases emissions and economic analysis. *Journal of Cleaner Production*, **111**, 262–278. https://doi.org/10.1016/j.jclepro.2015.08.083

Maiti, D., Ansari, I., Rather, M.A. et al. (2019). Comprehensive review on wastewater discharged from the coal-related industries – characteristics and treatment strategies. *Water Science and Technology*, **79**(11), 2023–2035. https://doi.org/10.2166/wst.2019.195

Mannina, G., Presti, D., Montiel-Jarillo, G. et al. (2019). Bioplastic recovery from wastewater: a new protocol for polyhydroxyalkanoates (PHA) extraction from mixed microbial cultures. *Bioresource Technology*, **282**, 361–369. https://doi.org/10.1016/j.biortech.2019.03.037

Maw, M.M., Boontanon, N., Aung, H.K.Z.Z. et al. (2024). Microplastics in wastewater and sludge from centralized and decentralized wastewater treatment plants: effects of treatment systems and microplastic characteristics. *Chemosphere*, **361**, 142536. https://doi.org/10.1016/j.chemosphere.2024.142536

Md Khudzari, J., Gariépy, Y., Kurian, J. et al. (2019). Effects of biochar anodes in rice plant microbial fuel cells on the production of bioelectricity, biomass, and methane. *Biochemical Engineering Journal*, **141**, 190–199. https://doi.org/10.1016/j.bej.2018.10.012

Mohammadi, L., Rahdar, A., Bazrafshan, E. et al. (2020). Petroleum hydrocarbon removal from wastewaters: a review. *Processes*, **8**(4), 447. https://doi.org/10.3390/pr8040447

Mohan, Y. and Das, D. (2009). Effect of ionic strength, cation exchanger and inoculum age on the performance of microbial fuel cells. *International Journal of Hydrogen Energy*, **34**(17), 7542–7546. https://doi.org/10.1016/j.ijhydene.2009.05.101

Molina, M.C., Bautista, L.F., Catalá, M. et al. (2020). From laboratory tests to the ecoremedial system: the importance of microorganisms in the recovery of PPCPs-disturbed ecosystems. *Applied Sciences (Switzerland)*, **10**(10), 3391. https://doi.org/10.3390/APP10103391

Mutamim, N.S.A., Noor, Z.Z, Hassan, M.A.A. et al. (2013). Membrane bioreactor: applications and limitations in treating high strength industrial wastewater. *Chemical Engineering Journal*, **225**, 109–119. https://doi.org/10.1016/j.cej.2013.02.131

Pandit, S., Savla, N., Sonawane, J.M. et al. (2021). Agricultural waste and wastewater as feedstock for bioelectricity generation using microbial fuel cells: recent advances. *Fermentation*, **7**(3), 169. https://doi.org/10.3390/fermentation7030169

Park, J.Y. and Yoo, Y.J. (2009). Biological nitrate removal in industrial wastewater treatment: which electron donor we can choose. *Applied Microbiology and Biotechnology*, **82**(3), 415–429. https://doi.org/10.1007/s00253-008-1799-1

Pathy, A., Ray, J., and Paramasivan, B. (2021). Challenges and opportunities of nutrient recovery from human urine using biochar for fertilizer applications. *Journal of Cleaner Production*, **304**, 127019. https://doi.org/10.1016/j.jclepro.2021.127019

Pergola, M., Persiani, A., Palese, A.M. et al. (2018). Composting: the way for a sustainable agriculture. *Applied Soil Ecology*, **123**, 744–750. https://doi.org/10.1016/j.apsoil.2017.10.016

Pradhan, D., Sukla, L.B., Sawyer, M. et al. (2017). Recent bioreduction of hexavalent chromium in wastewater treatment: a review. *Journal of Industrial and Engineering Chemistry*, **55**, 1–20. https://doi.org/10.1016/j.jiec.2017.06.040

Prathiba, S., Kumar, P.S., and Vo, D.V.N. (2022). Recent advancements in microbial fuel cells: a review on its electron transfer mechanisms, microbial community, types of substrates and design for bio-electrochemical treatment. *Chemosphere*, **286**(P3), 131856. https://doi.org/10.1016/j.chemosphere.2021.131856

Rao, A., Varshney, S., Bhadra, S. et al. (2023). Use of biofilm bacteria to enhance overall microbial fuel cell performance. In: *Understanding Microbial Biofilms*, 699–712. Elsevier.

Rusyn, I. (2021). Role of microbial community and plant species in performance of plant microbial fuel cells. *Renewable and Sustainable Energy Reviews*, **152**, 111697. https://doi.org/10.1016/j.rser.2021.111697

Samsami, S., Mohamadizaniani, M., Sarrafzadeh, M.-H. et al. (2020). Recent advances in the treatment of dye-containing wastewater from textile industries: overview and perspectives. *Process Safety and Environmental Protection*, **143**, 138–163. https://doi.org/10.1016/j.psep.2020.05.034

Schievano, A., Colombo, A., Cossettini, A. et al. (2018). Single-chamber microbial fuel cells as on-line shock-sensors for volatile fatty acids in anaerobic digesters. *Waste Management*, **71**, 785–791. https://doi.org/10.1016/j.wasman.2017.06.012

Sevda, S., Dominguez-Benetton, X., Graichen, F.H.M. et al. (2016). Shift to continuous operation of an air-cathode microbial fuel cell long-running in fed-batch mode boosts power generation. *International Journal of Green Energy*, **13**(1), 71–79. https://doi.org/10.1080/15435075.2014.909363

Sevda, S., Sharma, S., Joshi, C. et al. (2018). Biofilm formation and electron transfer in bioelectrochemical systems. *Environmental Technology Reviews*, **7**, 220–234. Taylor and Francis Ltd. https://doi.org/10.1080/21622515.2018.1486889

Shemfe, M., Gadkari, S., Yu, E. et al. (2018). Life cycle, techno-economic and dynamic simulation assessment of bioelectrochemical systems: a case of formic acid synthesis. *Bioresource Technology*, **255**, 39–49. https://doi.org/10.1016/j.biortech.2018.01.071

Shen, M., Hu, T., Huang, W. et al. (2021). Removal of microplastics from wastewater with aluminosilicate filter media and their surfactant-modified products: performance, mechanism and utilization. *Chemical Engineering Journal*, **421**(P1), 129918. https://doi.org/10.1016/j.cej.2021.129918

Singh, D., Mishra, V, Kushwaha, J. et al. (2024). Strategies for biological treatment of waste water: a critical review. *Journal of Cleaner Production*, **454**, 142266. https://doi.org/10.1016/j.jclepro.2024.142266

Siwek, M., Edgecock, T., Chmielewski, A.G. et al. (2023). The potential of electron beams for the removal of microplastics from wastewater and sewage sludge. *Environmental Challenges*, **13**, 100760. https://doi.org/10.1016/j.envc.2023.100760

Sleutels, T.H.J.A., Heijne, A.T., Buisman, C.J.N. et al. (2012). Bioelectrochemical systems: an outlook for practical applications. *ChemSusChem*, **5**(6), 1012–1019. https://doi.org/10.1002/cssc.201100732

Sohn, W., Jiang, J., Phuntsho, S. et al. (2023). Nutrients in a circular economy: role of urine separation and treatment. *Desalination*, **560**, 116663. https://doi.org/10.1016/j.desal.2023.116663

Sonawane, J.M., Mahadevan, R., Pandey, A. et al. (2022). Recent progress in microbial fuel cells using substrates from diverse sources. *Heliyon*, **8**(12), e12353. https://doi.org/10.1016/j.heliyon.2022.e12353

Srikanth, S., Pvani, T., Sarma, P.N. et al. (2010). Synergistic interaction of biocatalyst with bio-anode as a function of electrode materials. *International Journal of Hydrogen Energy*, **36**(3), 2271–2280. https://doi.org/10.1016/j.ijhydene.2010.11.031

Suarez, S., Omil, F., and Lema, J.M. (2010). Fate and removal of pharmaceuticals and personal care products (PPCPs) in a conventional activated sludge treatment process. *WIT Transactions on Ecology and the Environment*, **135**, 255–265. https://doi.org/10.2495/WP100221

Teng, S., Tong, Z., Li, W. et al. (2010). Electricity generation from mixed volatile fatty acids using microbial fuel cells. *Applied Microbiology and Biotechnology*, **87**, 2365–2372. https://doi.org/10.1007/s00253-010-2746-5

Tervahauta, T., Bryant, I.M., Leal, L.H. et al. (2014). Improved energy recovery by anaerobic grey water sludge treatment with black water. *Water (Switzerland)*, **6**(8), 2436–2448. https://doi.org/10.3390/w6082436

Thevenon, F., Graham, N.D., Chiaradia, M. et al. (2011). Local to regional scale industrial heavy metal pollution recorded in sediments of large freshwater lakes in central Europe (lakes Geneva and Lucerne) over the last centuries. *Science of the Total Environment*, **412–413**, 239–247. https://doi.org/10.1016/j.scitotenv.2011.09.025

van Groenestijn, J.W., Hazewinkel, J.H.O., Nienoord, M. et al. (2002). Energy aspects of biological hydrogen production in high rate bioreactors operated in the thermophilic temperature range. *International Journal of Hydrogen Energy*, **27**, 1141–1147. https://doi.org/10.1016/S0360-3199(02)00096-4

Vuorinen, A.H. and Saharinen, M.H. (1997). Evolution of microbiological and chemical parameters during manure and straw co-composting in a drum composting system. *Agriculture, Ecosystems and Environment*, **66**(1), 19–29. https://doi.org/10.1016/S0167-8809(97)00069-8

Wang, Z., He, S., Huang, J. et al. (2018). Comparison of heterotrophic and autotrophic denitrification processes for nitrate removal from phosphorus-limited surface water. *Environmental Pollution*, **238**, 562–572. https://doi.org/10.1016/j.envpol.2018.03.080

White, D.C., Suttont, S.D., and Ringelberg, D.B. (1996). The genus Sphingomonas: physiology and ecology. *Current Opinion in Biotechnology*, **7**(3), 301–306. https://doi.org/10.1016/S0958-1669(96)80034-6

Winfield, J., Ieropoulos, I., and Greenman, J. (2012). Investigating a cascade of seven hydraulically connected microbial fuel cells. *Bioresource Technology*, **110**, 245–250. https://doi.org/10.1016/j.biortech.2012.01.095

Yilmaz, M., Guven, H., Ozgun, H. et al. (2024). The application of life cycle assessment (LCA) to anaerobic technologies for the treatment of municipal wastewater: a review. *Process Safety and Environmental Protection*, **182**, 357–370. https://doi.org/10.1016/j.psep.2023.11.078

Yuan, Y., Leng, C., Zhou, Y. et al. (2023). Impact of separate concentrations of polyethylene microplastics on the ability of pollutants removal during the operation of constructed wetland-microbial fuel cell. *Journal of Environmental Management*, **341**, 118107. https://doi.org/10.1016/j.jenvman.2023.118107

Zhang, Y. and Angelidaki, I. (2013). A new method for in situ nitrate removal from groundwater using submerged microbial desalination-denitrification cell (SMDDC). *Water research*, **47**(5), 1827–1836. https://doi.org/10.1016/j.watres.2013.01.005

Zhang, Y.T., Wei, W., Sun, J. et al. (2020). Long-term effects of polyvinyl chloride microplastics on anaerobic granular sludge for recovering methane from wastewater. *Environmental Science and Technology*, **54**(15), 9662–9671. https://doi.org/10.1021/acs.est.0c02433

7

Composition of Industrial Wastewater and Its Conventional Treatment Facility

*Heba A. Younes and Nabila Shehata**

Environmental Science and Industrial Development Department, Faculty of Postgraduate Studies for Advanced Sciences, Beni-Suef University, Beni-Suef, Egypt

*Corresponding author: nabila.shehata@psas.bsu.edu.eg

7.1 Introduction

Clean water is an important resource for the survival of humanity on Earth. The rapid expansion of the industrial revolution and urbanization has produced massive amounts of wastewater worldwide containing significant levels of hazardous chemicals and heavy metals. Several industries rely substantially on water to carry out their procedures (Singh et al., 2023). Textile manufacturing involves significant use of water for dyeing, printing, and finishing operations. For example, more than 2500 L of water can be consumed to manufacture just one cotton T-shirt (Singh et al., 2019). The generation of hydroelectric and geothermal power also requires water. Cleaning, cooling, and material purification are examples of water-intensive processes used in solar panel manufacturing. Numerous amounts of water are needed in the soft drink industries for the washing of equipment, the processing of ingredients, and fermentation. Water is required for the processing of agricultural products such as fruits, vegetables, and cereals. Washing, rinsing, and heating are common steps in the processing of goods like meat, dairy, and fruits, all of which add significantly to water use (Sodhi et al., 2020). Chemical processes like refining petrochemicals as well as producing drugs need a lot of water to cool down, remove solvents, and carry out chemical reactions. The extraction of ore and processing of minerals are two mining processes that need

Wastewater to Resource Recovery: Applying the Circular Economy Toward Sustainable Development, First Edition. Edited by Arvind Kumar Mungray et al. Published 2026 by John Wiley & Sons, Inc.

water to keep the dust down, separate the materials, and treat them (Norgate and Haque, 2010). Pulping and bleaching are two steps in the paper-making process that require a lot of water. The paper industry is recognized as a significant user of water (Man et al., 2018). Water is used extensively in the mining and preparation of coal to wash and separate impurities from it. The oil and gas industry utilizes hydraulic fracturing, or "fracking," which includes large amounts of water mixed with chemicals, to recover hydrocarbons from subsurface deposits.

In the same way, water is necessary for cooling in thermal power facilities, including those that operate on gas, nuclear, and coal (Fthenakis and Kim, 2010). Using the best method available to address water treatment concerns satisfies the purifying goal. Commercially available techniques for handling the purification process include membrane technology, physical, chemical, physiochemical, and hybrid approaches. However, these procedures have certain drawbacks or defects and should be carefully assessed before being put into use. The method that has been widely employed in the commercial sector to modify the turbidity, pH, total dissolved solids (TDS), and total suspended solids (TSS) of water through chemical dosing is the most efficient. Ozonation, UV radiation, coagulation/flocculation, and chlorination are a few more processes. All of these technologies do, however, come with maintenance, recycling, and cost concerns. The biological technique uses microorganisms' metabolic activity to break down and transform contaminants (such as toxins) into biomass and associated gases like CH_4, CO_2, N_2, CO, and SO_2 and is the most widely used technology for large-scale water and wastewater treatments (Hashmi et al., 2023).

Due to poor wastewater treatment facilities, untreated wastewater is frequently released into aquatic ecosystems, resulting in environmental contamination and endangering both aquatic life and human health. Therefore, the global management of industrial wastewater has become a significant issue. Multiple studies and research have been carried out to remediate wastewater utilizing various technologies (physical, chemical, and biological processes) (Mustafa and Asmatulu, 2020). This chapter highlights some of the pollutants found in different industrial wastewater and their adverse effects, along with knowledge of how they affect the choice of treatment strategy. Also, different traditional industrial wastewater treatment methods have been applied and their disadvantages.

7.2 Composition of Industrial Wastewater

Industrial wastewater pollutants differ from industry to industry based on the chemicals used or produced and the technology applied. Different industries contribute to water pollution by contaminants, including inorganic, organic, and biological, at varying concentrations. Some of these pollutants that pose much environmental risk are explained in the following sections.

7.2.1 Heavy Metals

Metallic elements with a density over five times higher than water are classified as heavy metals (Duffus, 2002). Contamination of the environment by heavy metals can originate from either naturally occurring substances or human activities. Heavy metals, in contrast to other pollutants, are resistant to degradation and destruction, facilitating their accumulation throughout the food chain. These metals can be found in an elemental state (i.e. ore deposits) or a chemical compound. The mining process is the most significant contributor to heavy metal contamination. Also, improperly treated wastewater contributes to ecosystem pollution by heavy metals. Metal manufacturing processes (e.g. electroplating, refining), the textile, paint, and mining sectors are the primary heavy metal sources, including nickel, lead, and arsenic, which are frequently detected in industrial wastewater.

Nevertheless, a subclass of heavy metals known as toxic heavy metals has detrimental effects that can result in bio-intoxication and poisoning. At deficient concentrations, some of them are harmful. Osteoporosis, cardiac dysfunction, proteinuria, and elevated blood pressure have been associated with low-dose prolonged exposure to cadmium (Nogawa et al., 2004). This metal is frequently linked to advanced technology and power-generating parts, including pigments, solar cells, batteries, and alloys. Because of its corrosion resistance, it is also utilized in the automotive industry. It is also present in phosphate fertilizers because, like many heavy metals, it hinders the growth of organisms (Niño-Savala et al., 2019). Lead is not an excellent electrical conductor. It is utilized in lead-acid batteries, metallic alloys, construction, ammunition, fuses, pigments, X-ray and radiation protection, and cable sheaths. Drinkable water tainted with lead can have some adverse effects, including anemia, damage to the kidneys and brain, infertility, miscarriage, and behavioral abnormalities like hypersensitivity, impulsivity, and aggression. Although oral exposure is the most frequent, the same kinds of effects can also be obtained through vaginal, anal, cutaneous, and respiratory routes. The long-term toxicity of this metal is well recognized, which can readily cause harm to the nervous system at extremely low doses. Acute poisoning manifests muscle weakness, discomfort, constipation, diarrhea, loss of weight, blood in the urine, and other symptoms (Aziz et al., 2023). Lead concentrations in industrial wastewater range from 200 to 500 mg/L. Its levels must be drastically decreased to reach 0.05–0.10 mg/L before discharge, per EPA and WHO guidelines (Dongre, 2020). Several techniques for treating industrial wastewater, including coagulation, ion exchange, precipitation, adsorption, and reverse osmosis (RO), have been implemented to reduce lead pollution (Khaled et al., 2019; Qasem et al., 2021). Mercury is the only hazardous element present in liquid form at room temperature. Industrial waterways containing mercury are typically related to paints, fertilizers, and dental care. Acute exposure usually causes fever and vomiting. However, extended exposure leads to symptoms such as tremors, pink sickness, and renal issues (Budnik and Casteleyn, 2019).

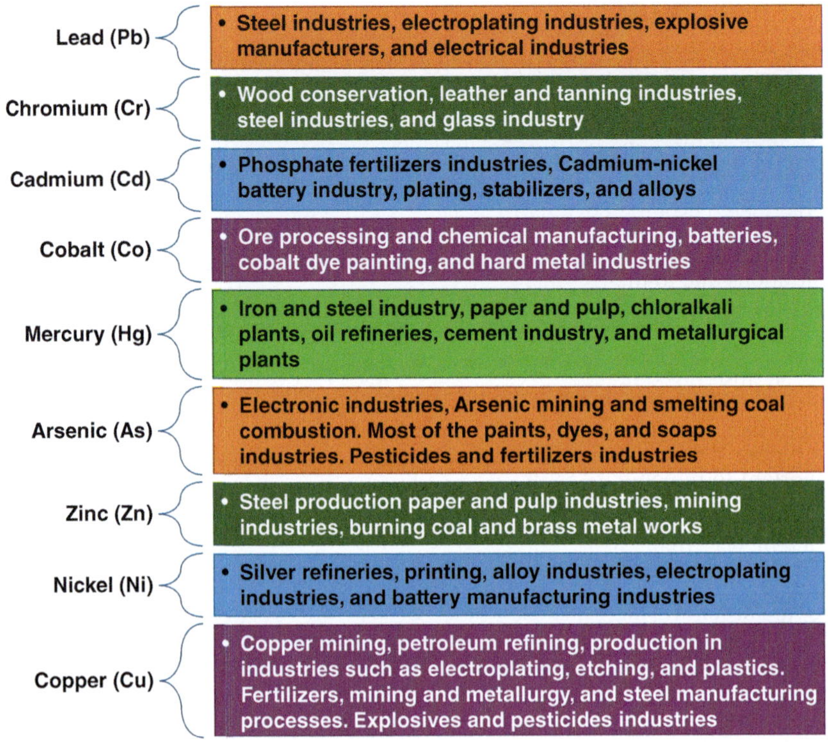

Figure 7.1 Major industries that contribute to water pollution.

Arsenic, a toxic metalloid, is globally distributed due to natural and human-caused processes. It is possible to find arsenic in food, tobacco goods, drinking water, and industrial processes. Wastewaters generated during the production of pesticides, glasses, and pharmaceuticals may contain the metalloid; however, arsenic compounds are also encountered in the microelectronics and optical industries. Additionally, gallium arsenide and arsine exhibit high electron mobility and are utilized in light-emitting diodes and superconductor applications (Sevak and Pushkar, 2024). Figure 7.1 includes the main heavy metal ions discharged from the different industries (Kumar et al., 2023).

7.2.2 Dyes

Dyes are soluble organic compounds that are difficult to extract from solutions using traditional techniques because they contain hydrophilic groups like auxochromes. In the environment, they inhibit photosynthesis, block solar radiation, lower dissolved oxygen levels, and endanger aquatic life in water bodies because of the chromophoric groups. Also, their detrimental consequences include toxicity, carcinogenicity, mutagenicity, persistence, and inhibition of plant growth. Furthermore, they raise industrial wastewater's chemical and biological oxygen demands, increasing treatment costs (Mota et al., 2023). Dyes may be produced from both natural and synthetic substances. Petroleum is the primary source of synthetic colors, containing several hazardous substances. Dyes are primarily used in the food, cosmetic, textile, paper, and tannery industries (Islam et al., 2023). Textile dyes frequently contain azo compounds, which are hazardous at low concentrations due to their solid and complex nitrogen–nitrogen bond structures. In the food sector, colors are categorized as certified or non-certified. As previously mentioned, synthetic dyes like azo and indigoid must undergo a certification process because of their increased potential for toxicity. Conversely, natural colors derived from plants, insects, or minerals might not require a license, but their use still requires adherence to purity standards and concentration limitations. It is important to remember that different countries have different licensing requirements. Dyes can be used with pharmaceuticals for the same objective but in a different way than they are typically applied. Because dyes can interact with drug-resistant organisms and produce severe toxicity for bacteria but not humans, they are becoming increasingly popular in therapy.

7.2.3 Oil and Grease

The petroleum industry is the primary source of pollutants related to oil and grease (O&G) (Abd El-Gawad, 2014). Lubricating grease is made of petroleum oils and additives and is used to lubricate heavy machinery and equipment in different industries. On the other hand, heat, friction, and combustion processes can transfer metals, making them more poisonous. Wastewater from various businesses, including manufacturing crude oil, refineries, petrochemicals, and car washes, contributes to O&G pollutants. Due to population growth and industrial demand, O&G is becoming more prevalent in wastewater streams. Three types of oil are found in wastewater: an emulsified oil mixture, a free oil phase, and an oil–water mixture. Oil droplets must be bigger than 150 microns, less than 20 microns in emulsified mixtures, or between 20 and 150 microns in oil–water mixtures. These pollutants can lead to imbalances in the food chain, skin irritation, allergic responses, and liver damage. They also harm wastewater treatment facilities, resulting in treatment failure (Nowak et al., 2019). Accumulation of wastewater impurities known as O&G forms a top layer in bodies of water, impacting the ecology and perhaps leading to cancerous mutations. These contaminants result in high chemical oxygen demand (COD) levels, such as dyed wastewater. The oily top layer compromises aquatic life and dissolved oxygen levels, obstructing oxygen transport between the atmosphere and the water. In addition to being combustible and containing hazardous polycyclic aromatic hydrocarbons (PAHs), metals raise water temperature by absorbing solar energy.

7.2.4 Organic Compounds

Many pollutants are present in industrial wastewater, but organic pollutants are the most harmful. Phenols, azo dyes, pesticides, and polychlorinated biphenyls (PCBs) are examples of organic pollutants. Some of these substances are resistant to biological breakdown. The development of contemporary technologies, such as high-performance liquid chromatography, has made it possible to identify chemicals in various environmental matrices at lower concentrations.

Organic contaminants of emerging concern (CECs) are not routinely monitored in the environment and may originate from human activity or natural processes. They are caused mainly by the discharge of either household or industrial wastewater, and standard treatment methods cannot break down these chemicals. They can bioaccumulate and endanger ecosystems and human health at low concentrations (ng/L or μg/L). The CECs include pharmaceutical and personal care goods, endocrine-disrupting chemicals (EDCs), pesticides, persistent organic pollutants (POPs), and industrial chemicals (Feng et al., 2023). This classification covers many chemicals, and as new compounds and their effects are identified over time, the list of CECs is updated continuously. CECs might have home, industrial, agricultural, or commercial sources like any other pollution. Their long-distance migration and persistence allow them to be found in improbable and distant locations.

POPs are among the most concerning organic pollutants. They are carbon compounds that do not change over time when left in the environment. They can build up in adipose tissues of living things and are poisonous to humans and animals alike, causing nervous system alterations, various cancers, and troubles with reproduction (Yadav et al., 2021). Pesticides, alkylphenols, polyaromatics, organic oxygen compounds, and steroids are all classified as EDCs.

7.3 Traditional Industrial Wastewater Treatment Methods

Preliminary, primary, secondary, and tertiary treatment are the four main steps in wastewater treatment. Floating debris, such as paper, glass bottles, sticks, plastics, wood, and heavy inorganic substances, will be eliminated during the preliminary treatment. Screeners, grit chambers, and skimming tank devices have been used in preliminary treatment (Verbyla, 2017). The screener is a tool that is designed to remove floating effluent materials by providing a consistent-size opening.

They are divided into three categories: fine, medium, and coarse screens, depending on the opening size (Crini and Lichtfouse, 2018). The thin, long grit chambers are tanks designed to limit the flow of water to settle solids like sand. They remove heavy inorganic substances, like sand and ash. The foundation of this method is the idea of sedimentation brought on by gravitational pull. Finally, the function of the skimming tank is to remove different types of oily and greasy contaminants that are poured into the sewage system from residential and industrial sources. A skimming tank has walls made of baffles that split it. There are three interconnected sections within it. As a result of the compressed air being forced out of the tank floor, the greasy and oily components of the waste solidify and coagulate into rising air bubbles. The material that has been removed is pushed into the stilling compartment, a side chamber, where it is either mechanically or manually removed (Gedda et al., 2021).

7.3.1 Primary Treatment

Coagulation and sedimentation are physical processes used in primary treatment that eliminate suspended particulates (60–65%) and biochemical oxygen demand (BOD) (30–35%). Firstly, the sedimentation process uses gravity to remove suspended solids from water. It occurs in two stages: primary and secondary sedimentation. Most solid particles in moving waste are composed of suspended organic compounds. A sedimentation tank stops the flow and allows the solid particles to settle to the bottom. Particles settle into four types based on their tendency: discrete, flocculant, zone or hindered, and compression. Secondly, coagulation involves adding chemical agents to destabilize charged particles. This method is known as coagulation or chemical precipitation. Sedimentation alone cannot effectively remove colloidal trash from wastewater. Chemical coagulants may be required in some cases to aid sedimentation. This strategy allows suspended particles (60–80%) to be readily eliminated. Two types of chemical agents are employed in coagulation. The first one is the coagulants like sodium silicate and aluminum sulfate that form precipitates through electrostatic attraction between them (positive charge) and negatively charged colloidal particles. The second substance is coagulant aids that improve coagulant efficiency and result in less sludge formation such as activated silica, powdered limestone, and polyelectrolytes.

7.3.2 Secondary Treatment

Biological wastewater treatment, or secondary treatment, is appropriate for industrial wastewater with a BOD/COD ratio of more than 0.6. It reduces suspended particles and organic pollutant concentration. Microorganisms decompose organic waste, including bacteria, algae, fungi, and protozoa. To promote balanced microbial development in biological treatments, the ratio of BOD: N: P for aerobic treatment must be 100:5:1, while for anaerobic treatment, it is 100:2.5:0.5. Microorganisms convert unstable organic materials into solid inorganic forms. Wastewater secondary treatment procedures can be characterized as aerobic or anaerobic (Pachwarya et al., 2021).

7.3.2.1 Aerobic Method

Aerobic wastewater treatment degrades organic pollutants as well as nitrogen and phosphorus compounds by oxygen in either suspended or attached growth treatment. The treatment of aerobic suspended growth can be accomplished through a variety of techniques, including batch reactors, activated sludge, and aerated lagoons. Activated sludge is the most widely utilized current biological treatment technique. Mechanical aerators are used to aerate wastewater containing organic materials with bacteria in an aeration tank. Microorganisms metabolize suspended particulate particles when ideal aerobic conditions develop (Ahansazan et al., 2014). Microbes produce new bacterial cells from organic matter; the rest are converted to water and carbon dioxide. Sludge, or flocs, are formed when newly generated bacterial cells clump together.

After settling, activated sludge is moved to the aeration tank, where it can be recycled. Maintaining a steady oxygen supply is crucial for the successful activation of sludge. It can be performed using revolving paddles or mechanical aeration. Protozoa growth is a performance indicator for activated sludge. Eliminating waste sludge allows for its usage as fertilizer in agriculture. Two types of activated sludge processes exist: conventional and modified. The traditional activated sludge technique requires a settling tank and a sludge removal line. After initial treatment, wastewater is released to the tank's head, and oxygen is evenly distributed throughout. Several adjustments were made during the modified activated sludge method to increase aeration efficiency. Aeration methods include step, tapered, long-term, and high-rate aeration with full mixing. Several variables affect the activated sludge process, including the reactor type, the aeration tank's air supply, the food-to-microbes ratio, the nutrients' availability, and the rate of sludge recirculation. The presence of acidic conditions, low temperatures, extended sludge retention in a secondary settling tank, and industrial effluent discharge promote the proliferation of filamentous fungus.

Aerated lagoons are ponds that have been artificially aerated. Surface aerators are introduced in these aerated lagoons. These aerators can reduce the foul odors caused by organic compounds. Temperature-dependent continuous nitrification is possible in aerated lagoons. Compared to activated sludge, this process's microbial treatment has a larger surface area and is more responsive to temperature impacts. Aerated lagoons are classified as either facultative anaerobic or aerobic (Panchangam et al., 2024). Facultative aerated lagoons are bodies of water that allow aerobic and anaerobic microorganisms to grow. The lower section is typically oxygen-free in these lagoons, while the upper layers contain oxygen. Lagoons that can function with low oxygen and aeration are called partially mixed aerated lagoons, where some solids settle. On the other hand, aerobic aerated lagoons have powerful aeration that maintains them completely oxygenated without any solid particles settling (Fonade et al., 2000). Subsequent treatment is conducted following the lagoons to reduce effluent solids.

Sequencing batch reactors are modified activated sludge systems where the aeration and sedimentation operations are done sequentially in one tank. The treatment steps in this method begin with opening the valve to fill the tank and then mechanically mixing in the absence of air (anoxic). The second step is to add air to the mixed liquor through mechanical pumps. Afterward, suspended solids are settled before effluent release. In the idle stage, extra sludge can be removed.

Aerobic attached growth treatment is a technique that eliminates organic matter and promotes nitrification from wastewater. It comprises trickling filters, rotating biological contactors (RBCs), roughing filters, and packed bed reactors. Trickling filters, also known as sprinkler filters, are typically utilized for secondary home and industrial wastewater treatment. A typical trickling filter unit uses porous media like pebbles, coke, gravel, and peat moss. Wastewater spray causes microbial slime to form on the medium bed. This slime contains bacteria such as *Pseudomonas*, fungi, and yeasts. The unit maintains optimal aerobic conditions through splashing and diffusion. The microorganisms consume organic materials as the waste passes over the freshly developed microbial slime layer. Microbes on the film's upper layer cause oxidation. As a result, the biofilm grows in width over time and settles near the tank's bottom.

Roughing filters are mainly employed to diminish the presence of suspended solid organic materials in wastewater. They can reduce harmful microorganisms, iron, and manganese in wastewater (Affam and Ezechi, 2019). If there is a high amount of organic material, the slimy substance produced by microorganisms will constantly be released. According to the flow direction, roughing filters are classified as downflow with sand media and upflow with gravel media filters.

RBCs is a system for treating wastewater using spinning discs or media to promote organic matter-degrading microorganism growth (Waqas et al., 2023). This system is crucial for treating processing industrial pollutants, such as those produced by the vegetable sector.

Packed (fluidized) bed reactors contribute to nitrification as well as BOD removal. This technology utilizes a reactor loaded with a medium in which aeration begins in the reactor's lower region. When wastewater enters this reactor, microorganisms develop biofilms on the media. Once formed, biofilm consumes the floating organic waste in wastewater as a food source.

7.3.2.2 Anaerobic Method

A biological process that occurs without oxygen is called an anaerobic method. This method breaks down organic materials into by-products in four stages (Hydrolysis, Acidogenesis, Acetogenesis, and Methanogenesis) (Mara and Horan, 2003). Hydrolysis uses enzymes to decompose large molecules, including polysaccharides, proteins, and lipids, into amino acids and fatty acids. These large molecules act as substances that microorganisms can use for their growth and as an energy source. During acidogenesis, the low-molecular-weight molecules transform into acidic metabolites such as lactate and butyrate. Methanogenesis is the concluding phase in which gases like carbon dioxide and methane are generated from intermediary compounds produced during acidogenesis. A consortium is a collection of anaerobic organisms that collaborate to decompose waste.

It is additionally categorized into two subgroups: anaerobic suspended growth and anaerobic contact digesters. The anaerobic suspended growth treatment is a process that occurs without oxygen and involves using microorganisms suspended in a liquid medium. The anaerobic contact digester method primarily treats industrial waste with high BOD. This procedure is conducted in a series of sequentially arranged reactors. The sludge is piped and digested in these reactors under anaerobic conditions. Following this, a clarifier separates two layers, with the resulting supernatant being discharged while the sludge settles and is prepared for recycling (Álvarez et al., 2008). An attached growth system is a method that is used in the anaerobic fixed-film process. This approach involves the attachment of microbial organisms that are responsible for waste remediation to a medium in a reactor. The fundamental part of anaerobic attached-growth treatment is filtration. The expanded bed process is a fluidized bed operation method that involves the extension of the bed to accommodate a greater volume of particles.

7.3.2.3 Pond Treatment Methods

Humans construct specialized biologically mediated ponds to facilitate the reduction of organic waste and the elimination of pathogens from wastewater. This is accomplished through the collaborative action of algae and bacteria. These ponds, commonly known as stabilization ponds, are typically expansive and shallow basins. They are often utilized to treat municipal water and industrial effluents. A section of the stabilization pond is used to treat the influent, while the other section is used to release the treated effluent. After treatment, purified water can be utilized in irrigation or discharged into different water streams. The treatment by the pond system may occur in one pond or multiple ponds, where each pond has its own function. These ponds can be classified as aerobic ponds, anaerobic ponds, and facultative ponds (Butler et al., 2017). They are typically 0.5–1.5 feet deep and allow light to travel across the full liquid depth.

Normally, oxygen can be generated by cultivating algae in a pond. In addition to this, it is also sustained through constant atmospheric dispersion facilitated by pumps. The bacteria and algae in these ponds demonstrate a symbiotic relationship. The pond's oxygenic conditions are maintained by bacteria, which decompose organic materials to produce carbon dioxide and nutrients that algae feed on. The effluent purification of aerobic ponds is significantly impacted by the presence of rotifers and protozoa, light, temperature, and nutrients. Anaerobic ponds are kept in circumstances without oxygen and are primarily used for treating effluents containing concentrated organic materials. Heat conservation is achievable because of their significant depth, reaching up to 30 feet, around 9 m. Given ideal circumstances, approximately 75% of BOD can be eliminated with anaerobic treatment. The anaerobic ponds are influenced by various parameters like duration of hydraulic retention, pH, temperature, sulfide, hazardous chemicals, and the level of mixing. Facultative ponds are primarily used for the remediation of industrial and municipal effluent. These ponds utilize distinct categories of microorganisms (aerobic, anaerobic, and facultative).

7.3.3 Tertiary Treatment

Through the use of tertiary treatment, secondary effluent will be treated from contaminants such as organic and inorganic components of nutrients and eradicate harmful microorganisms. Different techniques have been applied in tertiary treatment, such as ion exchange, membranes, and advanced oxidation processes (AOPs). The technique by which ions move from a solid to a liquid or gas is called ion exchange. It is a water treatment technology that implies mitigating or reducing the level of an undesired ionic pollutant in water by substituting it with another ionic substance. Ion exchange is a highly suitable technology for successfully removing dissolved inorganic ions. The capacity of resins to assimilate metals is unique, and they have the potential to effectively perform ion exchange. In this process, the ion exchange resin is uniformly disseminated in an electrolyte solution. Water flows through the resin bed, attracting ions to the resin particles, which results in the discharge of the loosely held solution into the water. Over time, beds get overloaded and lose their efficacy, requiring recharging. A salt brine solution is used to cleanse the resin during recharge, which results in a movement of ions between the saline solution and the flushed-out water (Huang et al., 2020).

Membrane filtration techniques employ a semipermeable membrane to separate wastewater, successfully removing dissolved particles with a size range of 0.0001–1 μm. This membrane serves as a selective barrier under intense pressure (1 and 10 bar), permitting specific components to pass while retaining others in the liquid. It primarily treats seawater and brackish water desalination (Van der Bruggen and Vandecasteele, 2002). Membrane filtrations can be categorized according to the membrane material, motive power, and the minimum separation size obtained. Microfiltration (MF) employs a membrane with pores of between 0.1 and 10 μm in size to isolate bacteria and suspended particles from flowing water, while ultrafiltration (UF) separates bigger molecules and particles from liquids. On the other hand, nanofiltration (NF) uses a membrane (1–2 nm) to isolate ions and tiny molecules, so it can be applied to surface water remediation and other applications where the concentration of TDS is low to partially demineralize products such as in the dairy industry. This procedure primarily facilitates the reduction of water hardness and eliminates the organic by-products. RO is a method of filtration that employs the application of pressure as high as 80 bars to a membrane having a pore size smaller than 1 nm to extract particles and solutes from a liquid. Although its main use is in treating freshwater, it is an essential component in the desalination of saltwater.

Electrochemical techniques involve direct oxidation and reduction reactions to eliminate contaminants from wastewater, such as electrodialysis and electrocoagulation. This oxidation and reduction reaction can be achieved by either employing chemicals that physically destroy the contaminants or generating reactive chemical species (Drogui et al., 2008). The electrodialysis method entails the migration of salt ions from the effluent by the application of an electric potential difference through ion exchange membranes. An electrodialysis cell is a typical setup for this, with an anion exchange membrane separating the brine and diluent portions, and the cationic membrane situated in the middle of the two electrodes within the system. The electric potential gradient causes anions to flow toward the anode while cations go toward the cathode. Their movements are hindered by anionic and cationic membranes, respectively. As a result of electrodialysis, the concentration of ions in the saline section rises while those in the dilute section fall (Valero et al., 2011). Electrocoagulation is a method where an electrical charge is applied to water to alter particles' surface charge. This causes suspended materials to come together and form bigger particles (Khandegar and Saroha, 2013). This technology is innovative, cost-effective, and has many applications.

The AOP chemically converts complex organic wastewater substances that resist biological degradation into simpler final products. The process entails producing and utilizing oxygen and hydroxide free radicals ($HO^{\cdot}$) (Khandegar and Saroha, 2013). The three most commonly used advanced oxidation techniques are ozone/UV, ozone/hydrogen peroxide, and hydrogen peroxide/UV.

7.4 Effect of Varied Composition on the Choice of Treatment Strategy

Applying the treatment process for wastewater depends on different factors such as pollutant types, concentration, wastewater quantity, and required effluent quality. As presented in Table 7.1, Preliminary treatment is beneficial for removing

Table 7.1 Different treatment methods for eliminating different contaminants.

Treatment type	Treatment process	Pollutants removals	References
Preliminary	Screen	Particles Solid materials	Topare et al. (2011)
Primary	Screen Grit chamber Skimming tank	Particles Oil Grease	Verbyla (2017)
Secondary	Activated sludge Oxidation bond Trickling filter	Biodegradable organic pollutants	Sonune and Ghate (2004)
Tertiary	Membranes Advanced oxidation process Electrodialysis and coagulation	Organic Inorganic Pathogens Suspended particles	Zagklis (2022)

particles found in wastewater. Therefore, this elimination enhances the efficiency and upkeep of succeeding treatment units. The primary treatment effectively removes settleable inorganic and organic pollutants that account for up to 50% of BOD and more than 50% of O&G material. Furthermore, secondary treatment effectively eliminates 95% of residual organic matter and suspended particles. Different advanced processes can be used in tertiary treatment to remove organic and inorganic pollutants and disinfection, where the AOP and membranes are more suitable and applied for organic pollutant degradation. On the other hand, an adsorption process using different materials can be employed to get rid of heavy metals, anions, and organic pollutants.

7.5 Limitations of Conventional Methods

Currently, decentralized units of treating industrial wastewater are encountering numerous challenges related to increased chemical usage, sludge disposal, energy consumption, and space requirements. In addition, they have significant operational challenges, such as the inability to handle wastewater volumes beyond the fixed design capacity and a shortage of qualified workers (Crini and Lichtfouse, 2019). The biological and physical approaches are undesirable and result in the production of more waste. On the other hand, the physical separation method also not eliminate all organic molecules; instead, it only converts them to a different

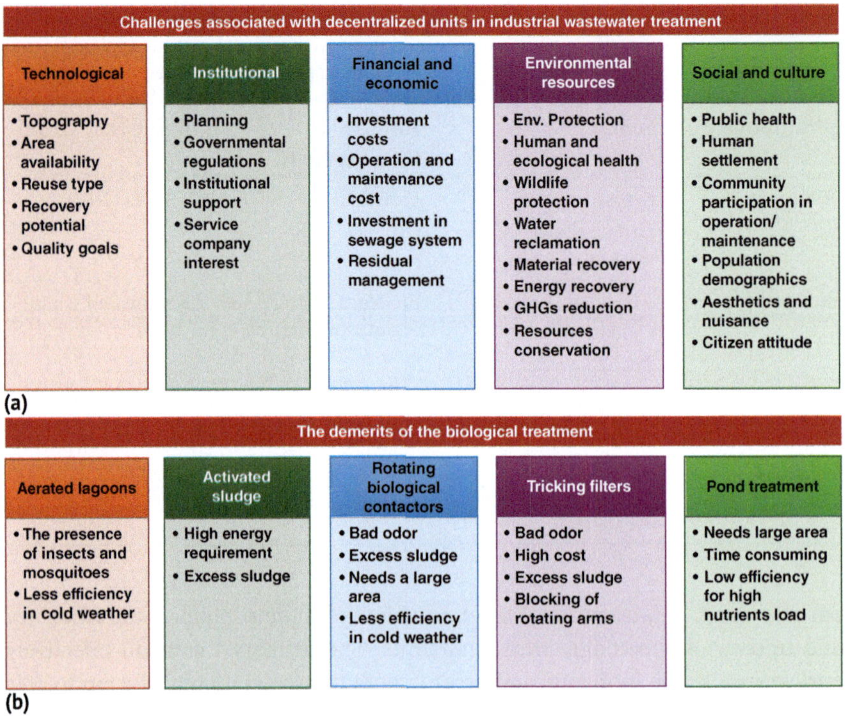

Figure 7.2 The challenges associated with the current units used for industrial wastewater treatment plants (a) and the demerits of the biological treatment (b).

phase. Additionally, its effectiveness is limited when dealing with a high volume of effluent. As shown in Figure 7.2, the decentralized units still have numerous challenges (Panchangam et al., 2024).

7.6 Modified Methods for Treatment of Industrial Wastewater; A Case Study

Rapid industrialization worldwide generates wastewater with numerous pollutants, causing severe water contamination. The textile industry produces textiles through various processes that use a lot of water and produce a lot of waste. Textile wastewater is made from multiple operations like washing, bleaching, mercerizing, desizing, and dyeing. Pollution is mainly caused by bleaching, dyeing, and mercerizing. Sizing agents, primarily natural starches, modified celluloses, and synthetic materials, are used in synthetic yarns, increasing organic pollution and TSS. Reactive dyes, such as pyrimidine, dichloro-fluoropyrimidine, and sulfone, are rapidly growing in the dyeing process due to the demand for vibrant-colored cloth. Synthetic anionic and cationic dyes, like rhodamine B and methylene blue, are resistant to degradation and can cause genetic alterations. Exposure to these harmful substances can lead to neurological diseases, thyroid gland dysfunction, and potentially fatal outcomes (Körbahti et al., 2007). The dyeing phase in wet processing involves hundreds of colors and chemicals, requiring 36–176 L/kg of water (Yaqub et al., 2023). Desizing alone accounts for 50% of the total wastewater (Ghaly et al., 2013).

Consequently, it is critical to implement suitable treatment methods. Wastewater treatment and reuse are crucial for converting colored wastewater from textile and dye effluents into minerals. Critical factors for effective dye wastewater treatment include the process's effectiveness, efficiency, recycling, and appropriate cost and environmental impact (Pang and Abdullah, 2013). Methods like coagulation, anaerobic processes, and membrane separation have some effectiveness but some problems hinder the complete removal (Shang et al., 2006). Various approaches, such as adsorption (Qin et al., 2024), oxidation/reduction (Boczkaj and Fernandes, 2017; Li et al., 2024), biotechnology (Bharathi et al., 2022), and membrane filtration (Nascimbén Santos et al., 2020), have been developed to treat wastewater. By integrating new wastewater technologies with conventional methods, it is possible to achieve efficient wastewater purification as well as enhance the recycling of treated effluent. Membrane-based bioreactor (MBR) technology integrates activated sludge with membrane filtration, a very efficient approach to treating industrial wastewater. It has a high removal rate, low hydraulic retention period, low sludge yield, adequate energy, and resource recovery (Deschamps et al., 2021; An et al., 2023). It has been applied to real textile effluent in pilot-scale reactors to eliminate all traces of color and persistent chemicals (Carvalho et al., 2020). Despite MBR's benefits, a significant obstacle to their widespread use is membrane fouling caused by solutes, colloids, and particles. Because anaerobic and aerobic oxygen environments result in distinct microbial communities and metabolic/foulant behavior, different MBR types exhibited varying fouling potential. Several techniques are used to control membrane fouling and attain energy savings. Three main categories define the present approaches for reducing membrane fouling: physical, chemical, and biological. Integrating the AOP with membrane filtration can effectively mineralize organic pollutants and improve fouling control. AOP by UV/O_3 and UV/H_2O_2 has shown to be quite successful in treating textile wastewater and lowering membrane fouling (Mehmood et al., 2023).

Moreover, to replace the expensive decolorization agents, a hybrid electrochemical/Fenton system was proposed. The hybrid system's primary benefits are its low cost, high efficiency, and lack of sludge formation. To lower energy consumption and increase the electrochemical oxidability index and pollutant removal efficiency, a circulatory system was used. Taking reactor performance and operating costs into account, this system is deemed suitable as a post-treatment technology to support the textile WWTP's anaerobic-aerobic unit.

7.7 Summary

Wastewater purification is critical to guarantee a safe environment and protect public health. There exist multiple traditional techniques for the treatment of wastewater. The choice of a particular treatment is primarily based on the characteristics of wastewater, whether it comes from a municipality or an industrial source. Occasionally, it is also determined by the financial consequences and level of technical proficiency required for operation and upkeep. Contaminants of rising concern need to be addressed, taking into account a more modern perspective. They can cause other, as-yet-unknown effects and disturb the endocrine system, as demonstrated by their detrimental impacts on humans at low doses. These pollutants result from human activity and are found in the environment for a long time. Consequently, the list of these substances is ever-expanding and requires extensive research. The effects and complexity of the toxins that the industrial sector produces for both the environment and people must be acknowledged. As a result, appropriate treatment approaches need to be considered and created. Hazardous effects from these behaviors will no longer be a threat if this concept is widely accepted; instead, they can be effectively regulated and rendered harmless. It has been challenging to identify a single treatment approach that can effectively eliminate all contaminants due to the complexity and diversity of pollutants. Therefore, adequate wastewater treatment methods must be used to remove multicomponent contaminants from water resources. As a result, different treatment technologies were developed to guarantee the elimination of all pollutants. However, industrial wastewater treatment can be improved by combining new technologies with traditional techniques and reducing operation costs. Future studies should focus on maximizing the hybrid systems for industrial wastewater treatment plants along with minimizing their cost.

Review Questions

1. What are the impacts of industrial wastewater composition on wastewater treatment efficiency?
2. What are the main micropollutants in industrial wastewater?
3. What are the criteria for selecting the appropriate wastewater treatment method that ensures high treatment efficiency?

4 Suggest methods by which sludge can be reused?
5 Discuss how some materials/energy can be recycled from water during water treatment process?
6 List which technologies can be integrated to increase the overall treatment performance?
7 Does the integration of renewable energies improve the sustainability of wastewater treatment?
8 How to conserve energy in industrial wastewater treatment plants?
9 Can the wastewater treatment begin inside the industrial plant via choosing the green raw materials and green industrial production technologies?
10 What are the main challenges in industrial wastewater treatment?

References

Abd El-Gawad, H.S. (2014). Oil and grease removal from industrial wastewater using new utility approach. *Advances in Environmental Chemistry*, **2014**, 1–6. https://doi.org/10.1155/2014/916878

Affam, A.C. and Ezechi, E.H. (2019). Application of graded limestone as roughing filter media for the treatment of leachate. *Handbook of Research on Resource Management for Pollution and Waste Treatment*, **4**(9), 176–219. https://doi.org/10.4018/978-1-7998-0369-0.ch009

Ahansazan, B., Afrashteh, H., Ahansazan, N. et al. (2014). Activated sludge process overview. *International Journal of Environmental Science and Development*, **5**(1), 81–85. https://doi.org/10.7763/IJESD.2014.V5.455

Álvarez, J.A., Ruíz, I., and Soto, M. (2008). Anaerobic digesters as a pretreatment for constructed wetlands. *Ecological Engineering*, **33**(1), 54–67. https://doi.org/10.1016/j.ecoleng.2008.02.001

An, Z., Zhu, J., Zhang, M. et al. (2023). Anaerobic membrane bioreactor for the treatment of high-strength waste/wastewater: a critical review and update. *Chemical Engineering Journal*, **470**, 144322. https://doi.org/10.1016/j.cej.2023.144322

Bharathi, D., Nandagopal, J.G.T., Ranjithkumar, R. et al. (2022). Microbial approaches for sustainable remediation of dye-contaminated wastewater: a review. *Archives of Microbiology*, **204**(3), 1–11. https://doi.org/10.1007/s00203-022-02767-3

Boczkaj, G. and Fernandes, A. (2017). Wastewater treatment by means of advanced oxidation processes at basic pH conditions: a review. *Chemical Engineering Journal*, **320**, 608–633. https://doi.org/10.1016/j.cej.2017.03.084

Budnik, L.T. and Casteleyn, L. (2019). Mercury pollution in modern times and its socio-medical consequences. *Science of the Total Environment*, **654**, 720–734. https://doi.org/10.1016/j.scitotenv.2018.10.408

Butler, E., Hung, Y.-T., Al Ahmed, M.S. et al. (2017). Oxidation pond for municipal wastewater treatment. *Applied Water Science*, **7**(1), 31–51. https://doi.org/10.1007/s13201-015-0285-z

Carvalho, J.R.S., Amaral, F.M., Florencio, L. et al. (2020). Microaerated UASB reactor treating textile wastewater: the core microbiome and removal of azo dye Direct Black 22. *Chemosphere*, **242**, 125157. https://doi.org/10.1016/j.chemosphere.2019.125157

Crini, G. and Lichtfouse, E. (2018). *Green Adsorbents for Pollutant Removal: Fundamentals and Design*, 1–22. *Springer Nature.*

Crini, G. and Lichtfouse, E. (2019). Advantages and disadvantages of techniques used for wastewater treatment. *Environmental Chemistry Letters*, **17**(1), 145–155. https://doi.org/10.1007/s10311-018-0785-9

Deschamps, L., Merlet, D., Lemaire, J. et al. (2021). Excellent performance of anaerobic membrane bioreactor in treatment of distillery wastewater at pilot scale. *Journal of Water Process Engineering*, **41**, 102061. https://doi.org/10.1016/j.jwpe.2021.102061

Dongre, R.S. (2020). Lead: toxicological profile, pollution aspects and remedial solutions. In *Lead Chemistry*, 1–22. https://doi.org/10.5772/intechopen.93095

Drogui, P., Blais, J.-F., and Mercier, G. (2008). Review of electrochemical technologies for environmental applications. *Recent Patents on Engineering*, **1**(3), 257–272. https://doi.org/10.2174/187221207782411629

Duffus, J.H. (2002). "heavy metals" – a meaningless term? (IUPAC technical report). *Pure and Applied Chemistry*, **74**(5), 793–807. https://doi.org/10.1351/pac200274050793

Feng, W., Deng, Y., Yang, F. et al. (2023). Systematic review of contaminants of emerging concern (CECs): distribution, risks, and implications for water quality and health. *Water (Switzerland)*, **15**(22), 3922. https://doi.org/10.3390/w15223922

Fonade, C., Rols, J.L., Goma, G. et al. (2000). Improvement of industrial wastewater treatment by aerated lagoon: case studies. *Water Science and Technology*, **42**(5–6), 193–200. https://doi.org/10.2166/wst.2000.0514

Fthenakis, V. and Kim, H.C. (2010). Life-cycle uses of water in U.S. electricity generation. *Renewable and Sustainable Energy Reviews*, **14**(7), 2039–2048. https://doi.org/10.1016/j.rser.2010.03.008

Gedda, G., Balakrishnan, K., Devi, R.U. et al. (2021). Introduction to conventional wastewater treatment technologies: limitations and recent advances. *Materials Research Foundation*, **91**, 1–36. https://doi.org/10.21741/9781644901144-1

Ghaly, A.E., Ananthashankar, R., Alhattab, M.V.V.R. et al. (2013). Production, characterization and treatment of textile effluents: a critical review. *Journal of Chemical Engineering & Process Technology*, **5**(1), 1–18. https://doi.org/10.4172/2157-7048.1000182

Aziz, K.H.H., Mustafa, F.S., Omer, K.M. et al. (2023). Heavy metal pollution in the aquatic environment: efficient and low-cost removal approaches to eliminate their toxicity: a review. *RSC Advances*, **13**(26), 17595–17610. https://doi.org/10.1039/d3ra00723e

Hashmi, Z., Bilad, M.R., Fahrurrozi et al. (2023). Recent progress in microalgae-based technologies for industrial wastewater treatment. *Fermentation*, **9**(3), 1–25. https://doi.org/10.3390/fermentation9030311

Huang, X., Guida, S., Jefferson, B. et al. (2020). Economic evaluation of ion-exchange processes for nutrient removal and recovery from municipal wastewater. *NPJ Clean Water*, **3**(1), 7. https://doi.org/10.1038/s41545-020-0054-x

Islam, T., Sarwar, Z., Rahman, M.M. et al. (2023). Impact of textile dyes on health and ecosystem: a review of structure, causes, and potential solutions. *Environmental Science and Pollution Research*, **30**, 9207–9242. Springer Berlin Heidelberg. https://doi.org/10.1007/s11356-022-24398-3

Khaled, B., Bouguerra, W., Hamroouni, B. et al. (2019). Investigation of electrocoagulation reactor design parameters effect on the removal of cadmium from synthetic and phosphate industrial wastewater. *Arabian Journal of Chemistry*, **12**(8), 1848–1859. https://doi.org/10.1016/j.arabjc.2014.12.012

Khandegar, V. and Saroha, A.K. (2013). Electrocoagulation for the treatment of textile industry effluent – a review. *Journal of Environmental Management*, **128**, 949–963. https://doi.org/10.1016/j.jenvman.2013.06.043

Körbahti, B.K., Aktaş, N., and Tanyolaç, A. (2007). Optimization of electrochemical treatment of industrial paint wastewater with response surface methodology. *Journal of Hazardous Materials*, **148**(1–2), 83–90. https://doi.org/10.1016/j.jhazmat.2007.02.005

Kumar, A., Thakur, A., and Panesar, P.S. (2023). A review on the industrial wastewater with the efficient treatment techniques. *Chemical Papers*, **77**, 4131–4163. Versita. https://doi.org/10.1007/s11696-023-02779-3

Li, X., Liu, H., Zhang, Y. et al. (2024). A review of metallurgical slags as catalysts in advanced oxidation processes for removal of refractory organic pollutants in wastewater. *Journal of Environmental Management*, **352**, 120051. https://doi.org/10.1016/j.jenvman.2024.120051

Man, Y., Han, Y., Wang, Y. et al. (2018). Woods to goods: water consumption analysis for papermaking industry in China. *Journal of Cleaner Production*, **195**, 1377–1388. https://doi.org/10.1016/j.jclepro.2018.06.026

Mara, D. and Horan, N. (2003). *Handbook of Water and Wastewater Microbiology, Handbook of Water and Wastewater Microbiology*. Elsevier. https://doi.org/10.1016/B978-0-12-470100-7.X5000-6

Mehmood, C.T., Qiu, H., Chen, L. et al. (2023). Ceramic membrane reactor integrated with UV/O3/Catalyst Beads for treating real textile wastewater: enhanced effluent quality, fouling control and molecular transformations of DOM. *Journal of Environmental Chemical Engineering*, **11**(5), 110832. https://doi.org/10.1016/j.jece.2023.110832

Mota, I.G.C., Neves, R.A.M.D., Nascimento, S.S.D.C. et al. (2023). Artificial dyes: health risks and the need for revision of international regulations. *Food Reviews International*, **39**(3), 1578–1593. https://doi.org/10.1080/87559129.2021.1934694

Mustafa, R. and Asmatulu, E. (2020). Preparation of activated carbon using fruit, paper and clothing wastes for wastewater treatment. *Journal of Water Process Engineering*, **35**, 101239. https://doi.org/10.1016/j.jwpe.2020.101239

Nascimbén Santos, É., László, Z., Hodúr, C. et al. (2020). Photocatalytic membrane filtration and its advantages over conventional approaches in the treatment of oily wastewater: a review. *Asia-Pacific Journal of Chemical Engineering*, **15**(5), 1–29. https://doi.org/10.1002/apj.2533

Niño-Savala, A.G., Zhuang, Z., Ma, X. et al. (2019). Cadmium pollution from phosphate fertilizers in arable soils and crops: an overview. *Frontiers of Agricultural Science and Engineering*, **6**(4), 419–430. https://doi.org/10.15302/J-FASE-2019273

Nogawa, K., Kobayashi, E., Okubo, Y. et al. (2004). Environmental cadmium exposure, adverse effects and preventive measures in Japan. *BioMetals*, **17**(5), 581–587. https://doi.org/10.1023/B:BIOM.0000045742.81440.9c

Norgate, T. and Haque, N. (2010). Energy and greenhouse gas impacts of mining and mineral processing operations. *Journal of Cleaner Production*, **18**(3), 266–274. https://doi.org/10.1016/j.jclepro.2009.09.020

Nowak, P., Kucharska, K., and Kamiński, M. (2019). Ecological and health effects of lubricant oils emitted into the environment. *International Journal of Environmental Research and Public Health*, **16**(16), 1–13. https://doi.org/10.3390/ijerph16163002

Pachwarya, R.B., Mina, U., Ramanathan, A.L. et al. (2021). Advanced and traditional methods of wastewater treatment. *Geodiversity & Impact on Environment*, **24**(3), 11–12. https://www.researchgate.net/publication/350516003

Panchangam, S.C., Sarma, L.S., Vijayarengan, P. et al. (2024). Common decentralized treatment and disposal methods. In: *Decentralized Sanitation and Water Treatment*, 53–71. CRC Press.

Pang, Y.L. and Abdullah, A.Z. (2013). Current status of textile industry wastewater management and research progress in Malaysia: a review. *Clean – Soil, Air, Water*, **41**(8), 751–764. https://doi.org/10.1002/clen.201000318

Qasem, N.A.A., Mohammed, R.H., and Lawal, D.U. (2021). Removal of heavy metal ions from wastewater: a comprehensive and critical review. *npj Clean Water*, **4**(1). 1–15. https://doi.org/10.1038/s41545-021-00127-0

Qin, W., Dong, Y., Jiang, H. et al. (2024). A new approach of simultaneous adsorption and regeneration of activated carbon to address the bottlenecks of pharmaceutical wastewater treatment. *Water Research*, **252**, 121180. https://doi.org/10.1016/j.watres.2024.121180

Sevak, P. and Pushkar, B. (2024). Arsenic pollution cycle, toxicity and sustainable remediation technologies: a comprehensive review and bibliometric analysis. *Journal of Environmental Management*, **349**, 119504. https://doi.org/10.1016/j.jenvman.2023.119504

Shang, N.C., Chen, Y.-H, Yang, Y.-P. et al. (2006). Ozonation of dyes and textile wastewater in a rotating packed bed. *Journal of Environmental Science and Health – Part A Toxic/Hazardous Substances and Environmental Engineering*, **41**(10), 2299–2310. https://doi.org/10.1080/10934520600873043

Singh, B.J., Chakraborty, A., and Sehgal, R. (2023). A systematic review of industrial wastewater management: evaluating challenges and enablers. *Journal of Environmental Management*, **348**, 119230. https://doi.org/10.1016/j.jenvman.2023.119230

Singh, R.P., Singh, P.K., Gupta, R. et al. (2019). Treatment and recycling of wastewater from textile industry. *Advances in Biological Treatment of Industrial Waste Water and Their Recycling for a Sustainable Future*, 225–266. Springer. https://doi.org/10.1007/978-981-13-1468-1_8

Sodhi, H.S., Singh, D., and Singh, B.J. (2020). A conceptual examination of lean, six sigma and lean six sigma models for managing waste in manufacturing SMEs. *World Journal of Science, Technology and Sustainable Development*, **17**(1), 20–32. https://doi.org/10.1108/WJSTSD-10-2019-0073

Sonune, A. and Ghate, R. (2004). Developments in wastewater treatment methods. *Desalination*, **167**(1–3), 55–63. https://doi.org/10.1016/j.desal.2004.06.113

Verbyla, M., Sperling, M.V., and, Maiga, Y. (2017). Part four, management of risk from excreta and wastewater. In *Waste Stabilization Ponds* (eds. J.B. Rose and B. Jiménez-Cisneros). Water and Sanitation for the 21st Century: Health and Microbiological Aspects of Excreta and Wastewater Management (Global Water Pathogen Project). https://doi.org/10.14321/waterpathogens.65

Topare, N.S., Attar, S.J., and Manfe, M.M. (2011). Sewage/wastewater technologies: a review. *Scientific Reviews and Chemical Communications*, **1**(1), 18–24. www.sadgurupublications.com

Valero, F., Barcelo, A., and Arbos, R. (2011). Electrodialysis technology – theory and applications. *Desalination, Trends and Technologies*, **28**, 3–20. https://doi.org/10.5772/14297

Van der Bruggen, B. and Vandecasteele, C. (2002). Distillation vs. membrane filtration: overview of process evolutions in seawater desalination. *Desalination*, **143**(3), 207–218. https://doi.org/10.1016/S0011-9164(02)00259-X

Waqas, S., Harun, N.Y., Sambudi, N.S. et al. (2023). A review of rotating biological contactors for wastewater treatment. *Water (Switzerland)*, **15**(10), 1–19. https://doi.org/10.3390/w15101913

Yadav, D., Rangabhashiyam, S., Verma, P. et al. (2021). Environmental and health impacts of contaminants of emerging concerns: recent treatment challenges and approaches. *Chemosphere*, **272**, 129492. https://doi.org/10.1016/j.chemosphere.2020.129492

Yaqub, M., Celebi, M.D., Dilaver, M. et al. (2023). Treating textile wastewater to achieve zero liquid discharge: a comprehensive techno-economic analysis. *Water, Air, and Soil Pollution*, **234**(10), 1–26. https://doi.org/10.1007/s11270-023-06646-5

Zagklis D.P. and Bampos, G. (2022). Tertiary wastewater treatment technologies: technical, economic, and life cycle aspects. *Processes*, **10**(11), 2304. https://doi.org/10.3390/pr10112304

8

Legal Framework and Industrial Insight for Treatment of Waste

Saji Simon[1], *Bibin K. Suresh*[1], *Anusree P. C.*[1], *Anantha Singh T. S.*[1], *George K. Varghese*[1]*, *and Nidheesh P. V.*[2]

[1]*Department of Civil Engineering, National Institute of Technology Calicut, Kozhikode, Kerala, India*
[2]*National Environmental Engineering Research Institute, Nagpur, Maharashtra, India*

*Corresponding author: gkv@nitc.ac.in

8.1 Introduction

Industrial waste is a by-product of various industrial processes, and its characteristics, environmental impacts, and resource potentials vary widely depending on the type of industry, the processes involved, and the treatment methods used. Understanding these aspects is crucial for developing effective treatment, recycling, and resource recovery strategies. Discharging toxic pollutants from the industries into the water bodies harms aquatic life, reducing biodiversity and aquatic productivity. Excess nutrients (nitrogen and phosphorus) from agricultural, food processing, or chemical industries can cause eutrophication, a process where water bodies receive excess nutrients, leading to excessive growth of algae, oxygen depletion, and eventual harm to aquatic organisms. High levels of organic waste (measured as BOD or/and COD) can deplete oxygen in water, leading to hypoxic or anoxic conditions, which are detrimental to aquatic life. The increase in suspended solids can increase turbidity in water bodies, reducing light penetration, affecting aquatic plants, and harming organisms that rely on clear water for feeding (Bilotta and Brazier, 2008). Toxic metals and persistent organic pollutants (POPs) from industrial wastewater can be absorbed by soil, leading to soil contamination and adversely affecting the plant growth. The runoff from mining operations can introduce toxic metals like mercury and lead into

Wastewater to Resource Recovery: Applying the Circular Economy Toward Sustainable Development, First Edition. Edited by Arvind Kumar Mungray et al.

the surrounding soil and groundwater (Dehkordi et al., 2024). These chemicals can accumulate in the food chain, affecting fish and other aquatic organisms, eventually leading to human consumption and consequent health risks (Adnan et al., 2024).

Considering the impacts of industrial discharges on the environment, it is very much evident that these need to be regulated. Regulations to manage industrial waste not only protect human health and environment, but also promote sustainability. They help prevent pollution, minimize the risk of accidents, and reduce harmful exposure to toxins. Regulations also encourage resource recycling, waste reduction, and the adoption of cleaner technologies, resulting in more sustainable industrial practices (United States Environmental Protection Agency, 2000). Regulations ensure industries are held accountable for their environmental impact, fostering public trust and compliance with international agreements. Overall, regulations are crucial for balancing industrial growth with long-term ecological and social responsibility. In this context, this chapter discusses the legal instruments relevant to industrial waste management. Although detailed discussions are limited to Indian regulations, considering the international audience, regulations from different parts of the world are also discussed. Aligning with the theme of the book, discussion is made on regulations that promote recovery of resources from industrial waste.

8.2 Industrial Activity and Waste Generation

Industries across various sectors generate different types of waste based on their operations, materials used, and products manufactured. A brief discussion of the sectors and the waste produced is given as follows.

Manufacturing industries generate solid waste like scrap metal, wood, plastic, rubber, and textiles, as well as chemical waste including solvents, paints, and lubricants. Hazardous wastes such as heavy metals, acids, and alkalis are also common based on the type of products they manufacture (Ogunwumi and Salami, 2023). The *construction sector* produces hazardous wastewater, including paints, solvents, adhesives, and asbestos (US-EPA, 2024). Acid mine drainage is an important waste from the *mining sector*. It is formed through a chemical reaction between surface water/shallow subsurface water and rocks that contain sulfur-bearing minerals, resulting in sulfuric acid. Specific types of mining, such as gold mining, produce hazardous residues like mercury and cyanide (*Abandoned Mine Drainage*, 2023). *Power industries* produce sludge from water treatment in hydroelectric and nuclear plants, and thermal power plants produce wet scrubber wash water that contains particulate matter and spent water. Produced water generated in oil and *petroleum industries* is highly hazardous (Senapati, 2011). *Agriculture and food processing* create chemical waste from pesticides, fertilizers, and preservatives used for food processing. The waste from slaughterhouses contributes to methane emissions, while the fruit and vegetable processing produces peels and pulp containing wastewater (Riseh et al., 2024). The *chemical industry* generates chemical solvents, by-products, residues, and hazardous waste, including pesticides, dyes, refractory compounds, etc. (Nagarajan et al., 2022).

Pharmaceuticals and healthcare facilities generate chemical waste, expired drugs, blood tissues, floor wash water with high disinfectant content, etc. The drug manufacturing produces waste from production processes, including solvents, materials discarded during manufacturing, and expired drugs (Shinde et al., 2020). *Textile manufacturing* produces wastewater contaminated with highly toxic and complex dyes, chemicals, detergents, etc. The textile manufacturing is one of the highly polluting industries (Chequer et al., 2013).

It is clear that each industry has specific waste streams that result from its operations, and the types of waste vary greatly in terms of volume, composition, and potential environmental impact. Proper management and recycling of these wastes are essential to minimize their impact on the environment and human health. In many cases, industries are increasingly adopting sustainable practices such as waste reduction, resource recovery, and cleaner production technologies to mitigate the negative effects of their waste generation.

8.2.1 Recovery of Resources from Industrial Waste

Despite the environmental challenges it poses, industrial waste also holds significant resource potential. Several valuable resources can be recovered from the waste, supporting sustainability, reducing costs, and promoting circular economy practices (Hajam et al., 2023).

a) Water Recovery

The most abundant resource to recover from wastewater is perhaps the water itself. Advanced treatment technologies like reverse osmosis, ultrafiltration, and other membrane processes (Aguiar et al., 2016) can treat industrial wastewater for reuse in the same facility or in some other facility. Water recovery is common in industries like textiles, food and beverage, thermal power plants, and chemical manufacturing, where large amounts of water are used and can be treated and reused for cooling, cleaning, or even as process water (Wang et al., 2024).

b) Energy Recovery

Anaerobic digestion of organic-rich industrial wastewater (e.g. from food processing or agricultural industries) or sludge can produce biogas (mainly methane). Biogas is a renewable energy source that can be used for heating, electricity generation, or fueling industrial processes. In power plants and some chemical industries, heat recovery systems capture and reuse the excess thermal energy from the wastewater for the other industrial processes, improving overall energy efficiency (Zandi Lak et al., 2024).

c) Nutrient Recovery

Wastewater, particularly from agricultural runoff, food processing, and sewage treatment plants, contains high concentrations of phosphorus. Through processes like struvite precipitation (Vasa and Chacko, 2021), phosphorus can be recovered and used as a slow-release fertilizer. This reduces the environmental impact of excess phosphorus while mitigating the need for mined phosphate (Cardoso et al., 2019). Techniques like ammonia stripping or nitrification/denitrification can recover nitrogen from wastewater, which can also be used as a fertilizer (Rahimi et al., 2020).

d) Heavy Metals Recovery

Spent petroleum catalysts, medical waste, battery waste, metal finishing industry waste, and fly ash contain recoverable amounts of metals. E-waste is another repository of valuable metals. Technologies like electrowinning, electrochemical recovery, and ion exchange allow the extraction of these metals, reducing the environmental impact of metal contamination while providing valuable resources for reuse (Krishnan et al., 2021).

e) Chemicals recovery

Some industrial wastewater contains valuable organic chemicals, solvents, or acids that can be recovered for reuse. Techniques like solvent extraction, distillation, or adsorption are commonly used in industries like pharmaceuticals, petrochemicals, and paints and coatings to recover valuable chemicals. Biosurfactants produced during microbial treatment of wastewater can also be recovered for use in detergents, cosmetics, or cleaning products (Dutta et al., 2021).

f) Carbon Recovery

Biochar is a carbon-rich material that can be synthesized from industrial sludge and used in a variety of applications, including sewage treatment (Kulkarni et al., 2024). Carbon dioxide recovery is particularly relevant in industries like brewing or fermentation, where CO_2 is a by-product. Capturing this CO_2 and utilizing it for applications, such as carbonation in beverages, can create a sustainable and profitable material loop.

The state-of-the-art technologies are available for the recovery of nutrients and minerals from the industrial wastewater. The ammonia stripping process can recover 100% of the ammonia from dairy wastewater (Beckinghausen et al., 2020), fluidized bed reactor based process can recover 97% of phosphorus from municipal wastewater (Liu et al., 2021), precipitation can recover 90% of the potassium from industrial wastewater (Arslanoğlu and Tümen, 2021), bio-electrodialysis can recover 59% of nitrogen from industrial wastewater (Beckinghausen et al., 2020), electroplating process can recover up to 96.5% of the chromium from tannery wastewater (da Silva et al., 2020), etc. The implementation of the desirable process for the recovery of resources from wastewater depends on the prevailing regulations and its execution by the concerned authorities.

Instead of operating in a sustainable manner, industries may prioritize profit maximization over environmental concerns, for example, by not implementing a cleaner technology because of the higher cost or by disposing of waste in a cheaper, environmentally harmful way. These practices can increase their profits in the short term, as they avoid the additional costs associated with pollution control, waste management, or emission reduction. However, the harmful effects of pollution are borne by the public. Pollutants in air, water, and soil lead to health problems, environmental degradation, and even economic impacts, such as lower property values and increased healthcare costs. Since these costs fall on individuals, communities, and governments, they are known as "externalities"—costs that are not accounted for in the industry's balance sheet. This creates an imbalance where the industry enjoys profits from unsustainable practices while the public is left to shoulder the environmental and economic consequences. This has resulted in governments regulating industries, requiring them to internalize some of these costs through pollution controls, fines, or incentives for resource recovery and cleaner production methods.

8.3 Existing Regulations and Guidelines on Industrial Waste Management

Regulations on industrial waste management play a critical role in promoting sustainable practices, reducing environmental pollution, and ensuring the desirable treatment and disposal of waste generated by industries worldwide. Along with explicit statutory provisions, there are provisions that are implicit in many national statutes and international conventions/protocols that can be invoked to ensure proper management of industrial waste.

8.3.1 International Regulations and Guidelines

International regulations are essential to control pollution because environmental issues like water pollution, air pollution, climate change, and ocean pollution cross borders, impacting countries globally. Without these regulations, nations might reduce environmental standards to stay competitive, leading to a "race to the bottom" in which pollution escalates worldwide. Global standards create a fairer economic playing field, prevent competitive harm to nations with stricter laws, and protect shared ecosystems and biodiversity. The international regulations and guidelines on waste management are designed to set global standards, to promote innovations in sustainable technologies by collaborative research, and to encourage compliance, especially in an era of growing industrial activities and interconnected global markets. Table 8.1 lists some of the key international frameworks and conventions relevant to industrial waste management:

Table 8.1 International frameworks and conventions related to industrial waste management.

Framework/Convention	Key provisions	Relevance to industrial waste management
Basel Convention on the Control of Transboundary Movements of Hazardous Wastes and Their Disposal (1989) Adopted in 1989 and entered into force in 1992, it is one of the most important international treaties regulating the movement of hazardous waste between nations.	• *Transboundary Movements:* The Basel Convention requires countries to control the movement of hazardous waste across borders. Waste can only be moved from one country to another if both the exporting and importing countries give consent, and the waste is destined for environmentally sound disposal or recycling. • *Environmentally Sound Disposal:* The treaty mandates that hazardous waste be treated and disposed of in an environmentally safe manner, promoting methods like incineration in high-temperature facilities or secure landfilling. • *Monitoring and Reporting:* Countries must establish systems to monitor and report the movement and disposal of hazardous waste, ensuring compliance with the treaty's provisions.	The Basel Convention has helped curb the illegal trafficking of hazardous waste from the developed nations to developing countries (EPA, 2024). It also encourages industries to adopt cleaner production technologies and better waste management practices (Alter, 2000).
Rotterdam Convention on the Prior Informed Consent Procedure for Certain Hazardous Chemicals and Pesticides in International Trade (1998) The Rotterdam Convention seeks to promote shared responsibility and cooperative efforts among countries in the international trade of hazardous chemicals and pesticides.	• *Prior Informed Consent (PIC)*: Before a hazardous chemical or pesticide can be exported, the importing country must agree to receiving it after being informed about its potential hazards. • *Notification and Consent Process*: Exporting countries must notify importing countries when they intend to ship hazardous chemicals or pesticides, ensuring the latter are aware of the risks. • *Risk Reduction:* The convention aims to reduce the risks associated with industrial chemicals and pesticides by promoting safer alternatives and encouraging the safe management of chemicals throughout their lifecycle.	The Rotterdam Convention helps control the movement of chemicals that might end up as industrial waste, ensuring that countries are prepared for safe management and disposal of hazardous materials. It also emphasizes the need for industries to consider environmental and health risks in their chemical and waste management practices (International Institute for Sustainable Development, 2024).

Stockholm Convention on Persistent Organic Pollutants (POPs) (2001) The Stockholm Convention is a global treaty aimed at eliminating or restricting the production and use of certain persistent organic pollutants (POPs), which are highly toxic chemicals that persist in the environment, accumulate in living organisms, and pose serious health risks to humans and wildlife.	• *Elimination and Reduction:* The treaty calls for the phase-out of POPs and the reduction of their emissions into the environment. It specifically targets chemicals like dioxins, furans, and certain pesticides, which are often by-products of industrial processes. • *Waste Management:* The Stockholm Convention mandates that industries manage POPs-containing waste, including residues from the incineration of POPs, through safe and environmentally sound disposal techniques. • *Monitoring and Reporting:* Countries are required to report on the production, use, and disposal of POPs and to establish measures for the safe management of POPs-containing waste.	The Stockholm Convention plays a key role in encouraging industries to eliminate or replace POPs in manufacturing processes. It also fosters the development of waste treatment technologies that safely neutralize these chemicals to prevent environmental contamination (US EPA, 2024).
The Kyoto Protocol (1997) and Paris Agreement (2015)—Climate Change Frameworks While not directly focused on industrial waste management, both the Kyoto Protocol and the Paris Agreement on climate change have significant implications for industrial waste management, particularly in reducing greenhouse gas emissions and to limit global warming to below 2 °C above pre-industrial levels and promoting sustainable industrial practices.	• *Greenhouse Gas Emissions:* Industrial processes, including waste incineration and landfill operations, are sources of methane and carbon dioxide. International agreements like the Kyoto Protocol and the Paris Agreement encourage industries to reduce these emissions, often through better waste management practices. • *Sustainable Practices:* The global focus on sustainability in waste management aligns with goals set by the Paris Agreement, which advocates for the reduction of carbon footprints across all sectors, including waste treatment and disposal.	• These treaties indirectly affect industrial waste management by pushing industries to reduce carbon emissions from their operations, including from waste disposal practices (Nyirenda, 2023). • They promote technologies such as waste-to-energy (WTE) that can help reduce the environmental impact of waste while producing energy.

Apart from those listed in Table 8.1, there are a few other global agreements, standards, etc. that promote proper waste management. The Global Environment and Health Management Standards (ISO 14001 and ISO 45001), which talks about the environmental management and health and safety of workers, The World Trade Organization (WTO) (WTO, 2024) Agreements, which encourages the countries to follow international agreement (like the Basel Convention) to avoid hindrance in the global trade, The United Nations Environment Programme (UNEP) and United Nations Industrial Development Organization (UNIDO), which works on the global environment and industrial development issues including waste management, are examples (UNIDO, 2017).

8.3.2 Indian Regulations and Guidelines

India has a comprehensive regulatory framework for managing industrial waste. They aim at reducing the adverse environmental and health impacts of industrial activities. Specifically, they address waste reduction, proper treatment and disposal of waste, recycling, and resource recovery. The regulations in India are governed by a mix of environmental laws, policies, and guidelines enforced by various national and state/provincial authorities. Table 8.2 provides an overview of Indian regulations governing industrial waste management.

8.3.3 Regulations Promoting Resource Recovery

Regulations on resource recovery from industrial waste aim to ensure that valuable materials and resources are retrieved from waste streams, reducing the need for virgin raw materials, conserving natural resources, and minimizing the environmental impact of waste disposal. These regulations often focus on promoting recycling, reusing materials, and supporting the development of a circular economy. Various countries have specific frameworks for managing industrial waste recovery, with an emphasis on reducing waste, recovering materials, and promoting sustainable practices. The regulatory frameworks from different developed nations like Japan's "Waste Management and Public Cleansing Law (1970)," United States' "Resource Conservation and Recovery Act (RCRA), 1976"(US EPA, 2024), United Kingdom's "The Environmental Protection Act (1990)," South Korea's, "Act on the Promotion of Saving and Recycling of Resources (2007)," European Union's "Waste Framework Directive (WFD) (2008/98/EC)," and Australia's "National Waste Policy—Action Plan (2018)," mandate the industries to practice the waste hierarchy, resource recovery (Stanchev et al., 2017), recycling process, hazardous waste recovery, zero-waste, and waste-to-energy (WTE) practices (Department of Climate Change, Energy, the Environment and Water, 2022).

Table 8.2 Indian regulations and their relevance to industrial waste management.

The Water (Prevention and Control of Pollution) Act, 1974	The **Water Act** governs the treatment and disposal of industrial effluents and wastewaters. It is designed to ensure that industries do not pollute water bodies through the discharge of untreated or inadequately treated wastewater (Water (Prevention and Control of Pollution) Act, 1974, 1974).

- The Central Pollution Control Board (CPCB) and State Pollution Control Boards (SPCBs) that regulate pollution from industries were established under the Water Act.
- The CPCB set strict standards for effluents, emissions, and handling of hazardous waste. These standards vary from industry to industry, and industries must ensure that their effluent meets the prescribed limits.
- Industries that discharge liquid waste are required to install effluent treatment plants (ETPs) to treat wastewater before disposal. In some cases, industries are required to recycle or reuse treated water.
- SPCBs enforce environmental standards at the state level and are responsible for ensuring that industries comply with environmental regulations.

The Air (Prevention and Control of Pollution) Act, 1981	The **Air Act** regulates the emissions of air pollutants, including those resulting from industrial activities.

- The CPCB sets emission standards for various industrial activities, requiring industries to install air pollution control equipment like scrubbers, electrostatic precipitators, and bag filters to limit emissions.
- Industries are required to monitor their air emissions and submit compliance reports. Those that exceed permissible emission limits face penalties and/or may be asked to shut down until corrective actions are taken (CPCB, 2016).

The Environment Protection Act (EPA), 1986	This is perhaps the most important and comprehensive environmental legislation in India.

- Section 8.3 of the EPA empowers the central government to set standards and issue directions for the handling, treatment, and disposal of industrial waste.
- The Act has provisions to set standards for air and water quality, hazardous waste disposal, and guidelines on the management of industrial effluents and emissions.

National Green Tribunal (NGT) 2010	The **National Green Tribunal (NGT)** was established in 2010 to provide a specialized forum for environmental justice. It has technical members along with judicial members to ensure effective redress in matters involving scientific and technical issues.

- The NGT has the authority to hear cases concerning violations of the Acts listed above in the table. It can also order compensation to the victims of pollution by the industries. It can impose fines, order the closure of non-compliant industries, or mandate corrective actions.
- The NGT has become a crucial player in holding industries accountable for violations of laws related to waste management and pollution.

(Continued)

Table 8.2 (Continued)

The Hazardous Waste (Management, Handling, and Transboundary Movement) Rules, 2016	The **Hazardous Waste Management Rules** regulate the generation, storage, treatment, and disposal of hazardous waste in India. These rules are crucial for managing the waste produced by industries such as chemical manufacturing, petroleum, textiles, and electronics that produce hazardous by-products.
The rules define hazardous waste as any waste that contains dangerous substances that pose a risk to human health or the environment. According to the Rule: • Industries that generate hazardous waste must seek permission from the SPCB. • Industries are encouraged to minimize hazardous waste generation by adopting cleaner production technologies and recycling waste whenever possible. • These rules mandate that hazardous waste must be disposed of in an environmentally sound manner, either through incineration, landfilling in designated hazardous waste landfills, or using other treatment technologies. • The rules comply with the provisions of the Basel Convention, effectively controlling the international transportation of hazardous waste and ensuring adequate management and disposal when exported from India.	
The Solid Waste Management Rules, 2016	The **Solid Waste Management Rules (SWM)**, 2016, provide guidelines for the management of municipal solid waste but are also relevant to industrial waste when it overlaps with municipal waste streams.
• The rules require waste generators, including industries, to segregate waste at the point of generation. This applies to biodegradable, non-biodegradable, and hazardous waste. • Industries generating large amounts of solid waste are required to adopt systems to process their waste in a way that reduces the environmental impact. This includes composting, recycling, and using waste as a resource (e.g. converting it to energy). • The rules emphasize proper waste treatment and disposal to avoid contamination of the environment. Industries must follow specific guidelines when disposing of solid waste in landfills, ensuring that the landfill design prevents leachate and gas leakage.	
The Bio-Medical Waste Management Rules, 2016	While primarily aimed at healthcare facilities, the **Bio-Medical Waste Management Rules, 2016,** also apply to industries that produce bio-medical waste, such as pharmaceutical companies, medical device manufacturers, and research labs.
• Bio-medical waste must be segregated into categories like infectious, non-infectious, hazardous, and sharps and stored in color-coded containers. • The rules require that bio-medical waste be treated through methods like autoclaving, incineration, or deep burial, depending on the nature of the waste. • Industries involved in the generation of bio-medical waste must obtain authorization from the appropriate pollution control authorities to handle and dispose of waste in accordance with prescribed guidelines (CPCB, 2016).	

Extended Producer Responsibility (EPR) Under the Plastic Waste Management Rules, 2016

The **Plastic Waste Management Rules, 2016,** include provisions for **Extended Producer Responsibility (EPR)**, which is a key regulatory framework for industrial sectors involved in the production and use of plastic products.

- The Rules make the industries accountable for the goods during their entire lifespan, including post-consumer waste.
- The Rules require industries to collect used plastic products, encourage recycling, and set up collection centers for waste plastic.
- Producers are mandated to clearly label plastic products to facilitate recycling and submit annual reports to the relevant authorities on their compliance with the EPR guidelines.

The "Extended Producer Responsibility" scheme was first introduced by Japan in 1970s itself, and the "Waste Management and Public Cleansing Law" of Japan acts as a guide for the other nations, which was subsequently adopted by most of the nations including South Korea in 2007, European Union in 2008, India in 2016, etc. The policy of circular economy was adapted by Japan as early as 1970, which mandated most of the industries in Japan to implement closed-loop recycling system. The other developed and developing nations started implementing the closed-loop recycling system (circular economy) only in the early 2000s. The developed and developing nations are now moving forward with a commitment toward reducing the environmental burden caused by pollution. Accordingly, the European Union targets to recycle 65% of the municipal waste by 2025, and Australia aims to recycle 80% of plastic, paper, and other packaging waste by 2025.

8.4 Impacts of Regulations

The impact of industrial waste management regulations on industries, the environment, and society has been significant, resulting in waste reduction and resource recovery to improved economic performance and public health. Effective regulations are also supporting transition to a more circular economy. The following are some of the major impacts of industrial waste management regulations:

Reduction in Pollution

Where industrial waste management regulations are implemented strictly, there is a significant reduction in the release of harmful pollutants into the air, water, and soil. By enforcing proper waste treatment, recycling, and disposal, regulations limit the contamination of ecosystems and improve the overall quality of the environment. For example, studies have shown that environmental regulations have been effective in industrial pollution control in China, although the impact has not been uniform throughout the country (Xiong and Wang, 2020; Yang et al., 2021); Similarly, stringent regulations and its implementation have resulted in a decreasing trend of SO_2 concentrations in the air in India (Kuttippurath et al., 2022). Direct regulation is found to be an effective approach for multi-pollutant control in an industrial context, and the efficiency of such regulation may be underestimated if the analysis does not consider the full range of pollutants and operational effects (Testa et al., 2012).

Reduced Landfill Use

Regulations that mandate recycling and resource recovery reduce the volume of industrial waste sent to landfills. By pushing industries to adapt recycling technologies and WTE systems, these regulations help divert large amounts of waste from disposal, reducing the landfill use. A sector that could significantly reduce the amount of landfilled waste in response to regulatory control is the construction industry. For instance, implementation of the landfill diversion policies under the EU WFD, 2008, has resulted in the United Kingdom in an increase in Construction and Demolition (C&D) waste recovery to nearly 90% in 2017 from 69% in 2005, which is above the target (70%) set by WFD (Au et al., 2018). In Norway, the implementation of the Directive resulted in the recovery of 74% of the total amount of C&D waste generated (Calvo et al., 2014).

Encouragement of Cleaner Production Technologies

By imposing standards and incentivizing industries to adopt cleaner production technologies, waste regulations lead to reduced generation of hazardous waste. An example is the Restriction of Hazardous Substances (RoHS) Directive of the European Union (2002/95/EC) or similar regulations existing in other countries that regulate the use of hazardous materials in electronic and electrical equipment (EEE). The Directive restricts the use of Lead, Mercury, Cadmium, Hexavalent chromium, Polybrominated biphenyls, Polybrominated diphenyl ether, Bis(2-ethylhexyl) phthalate, Butyl benzyl phthalate, Dibutyl phthalate, and Diisobutyl phthalate. The regulation of these toxic substances would mean lower health risk, especially for those involved in the end-of-life handling of these products. Although some "non-toxic" alternatives to these materials have shown higher potential environmental impact in upstream life cycle assessment (LCA) (Anders, 2020), considering the end-of-life stages in the EEE life cycle, environmental advantages can be there for the restriction.

Cost Savings Through Resource Efficiency

Regulations encouraging resource recovery and recycling help industries recover valuable materials (e.g. metals, plastics, chemicals) from waste. This reduces the need for raw material extraction and lowers the costs associated with purchasing new materials. Zero discharge regulations for industrial units have been shown to result in significant reduction in water and energy consumption (Panagopoulos and Haralambous, 2020) in some cases as high as 95% reduction in water consumption and 80% reduction in energy consumption (Lee and Okos, 2011).

Promotion of Innovation in Waste Management Technologies

Regulations incentivize innovation in waste management technologies, such as advanced recycling, WTE, and waste minimization technologies. This can lead to the development of new industries and jobs in the green technology sector, fostering economic growth (Jiang et al., 2020). The circular economy initiatives have driven investment in new recycling processes, the development of biodegradable materials, and innovative waste treatment technologies. The new technologies promise economic and environmental benefits while fostering sustainable consumption and collaborative innovation (Sánchez-García et al., 2024).

Reduction in Legal Liabilities and Penalties

Compliance with local, national, and international regulations reduces the legal and financial risks for businesses, encouraging them to take a more proactive approach to waste management. Adhering to industrial waste regulations helps companies avoid fines, penalties, and legal actions associated with improper waste disposal and could serve as a good defense against claims from the public. The Central Pollution Control Board of India has a comprehensive guideline on penalties to be paid by industries that fail to comply with the hazardous management regulations resulting in environmental damage (CPCB, 2016).

On the other hand, there are incentives for getting involved in sustainable waste management practices. In India, for example, businesses that are concerned with the collecting, processing, or treating biodegradable waste enjoy, subject to some conditions, 100% exemption from paying income tax under section 80JJA of the Income Tax Act, 1961. South Korea's EPR regulations and WTE policies offer tax incentives, grants, and subsidies to industries that comply with waste management regulations and adopt innovative recycling and recovery solutions. Such incentives lower the cost of compliance for businesses, encouraging them to invest in sustainable practices.

In short, the impact of industrial waste management regulations is diverse, offering significant benefits to the environment, economy, public health, and industry practices. By enforcing standards for waste disposal, recycling, and resource recovery, it is possible to drive sustainability and create new economic opportunities. However, to draw the full benefits, it is essential to address challenges related to infrastructure, enforcement, and equity, particularly in developing countries where industrial waste management systems may still be evolving.

8.5 Challenges to the Implementation of Regulations

The implementation of industrial waste regulations, while essential for sustainable development, faces a range of challenges that can hinder their effectiveness. These challenges are diverse, involving technological, financial, regulatory, and socio-political factors. The following are the key challenges to the implementation of industrial waste management regulations:

Lack of Adequate Infrastructure

Many regions, especially in developing countries, lack the necessary infrastructure to collect, treat, recycle, and safely dispose of industrial waste. The absence of efficient recycling plants, landfills with proper containment systems, and WTE technologies can significantly hinder compliance with regulations. Industrial waste, particularly hazardous waste, often requires specialized handling and transport to appropriate treatment or disposal facilities. Inadequate transportation systems or the lack of access to such facilities in rural or remote areas can lead to improper disposal or illegal dumping of waste (Kumari and Raghubanshi, 2023; Maalouf and Agamuthu, 2023).

Industries in some regions may not have access to or be able to afford advanced waste recovery and recycling technologies that are required by regulations. This can make it difficult for companies to meet resource recovery or recycling targets. Many industrial operations, especially in older industries, rely on outdated equipment or practices that are not aligned with modern waste management standards. Retrofitting these systems with newer technologies can be costly and logistically difficult, resulting in poor performance (Singh et al., 2023).

Financial Constraints

Many industries, especially small and medium-sized enterprises (SMEs), face significant upfront costs associated with implementing waste management measures. These may include investments in waste treatment facilities, recycling equipment, and compliance systems for tracking and reporting waste management activities. Even after initial investment, industries must bear ongoing operational costs, such as paying for proper waste disposal at designated sites, hiring waste management staff, and maintaining compliance with environmental standards. When it comes to hazardous waste, the costs associated with safe handling and treatment can be prohibitively expensive for small businesses (Peng et al., 2023).

In some regions, governments may not provide adequate financial incentives or subsidies to support the transition to sustainable waste management practices. Without these incentives, businesses may be reluctant to make the necessary investments in recycling or waste recovery technologies (Yadav et al., 2024). In many developing countries, there is a lack of cooperation between the public and private sectors in waste management. Public–private partnerships (PPPs) could play a vital role in bridging the financial gap by investing in waste infrastructure, but their absence limits the availability of resources for implementing regulatory measures effectively (Taron et al., 2023; Mmereki, 2018).

Weak Enforcement and Monitoring

One of the major reasons for the low effectiveness of industrial waste management regulations is inadequate enforcement. In some countries, particularly developing ones, regulatory mechanisms often lack the capacity and resources needed for effective implementation. This leads to non-compliance, illegal dumping, and ineffective waste management practices (Sakamoto et al., 2019). In certain regions, corruption can undermine enforcement efforts. Industries may bribe local authorities to overlook violations, such as improper waste disposal or failure to comply with waste treatment standards. This reduces the effectiveness of regulations and incentivizes non-compliance (D'Amato et al., 2018).

Monitoring and tracking industrial waste generation, disposal, and recovery rates are essential for ensuring compliance with regulations. However, many countries lack robust waste management data systems, making it difficult for authorities to monitor progress or identify violations (Jonášová et al., 2024; Iqbal et al., 2015). Without accurate data on waste flows, it is challenging to measure the effectiveness of policies or to ensure that waste is being handled properly. Regulatory agencies may be understaffed or underfunded, which limits their ability to carry out inspections, conduct environmental audits, or ensure proper enforcement of waste management laws (Agunwamba, 1998). This results in delayed actions, gaps in enforcement, and insufficient consequences for noncompliance.

Legal and Regulatory Gaps

In some countries, industrial waste management regulations are fragmented, with different laws covering different aspects of waste management (e.g. hazardous waste, plastic waste, e-waste). This creates confusion for businesses and regulatory bodies, making compliance and enforcement difficult. Some industrial waste management regulations are overly complex or difficult for industries to understand, particularly for small businesses. The complexity of compliance procedures may lead to delays, mistakes, or complete avoidance of compliance.

Even when waste management regulations are in place, the penalties for non-compliance may be weak or insufficient to act as a deterrent. Without the threat of significant fines or legal action, industries may not feel compelled to comply with environmental standards. Some regulations lack clear, enforceable guidelines on best practices for waste disposal, recycling, or resource recovery. This ambiguity can make it difficult for industries to comply with regulations and for regulatory bodies to enforce them.

Socio-cultural and Political Challenges

In many areas, there is local resistance to waste management projects such as landfills, waste treatment plants, or recycling facilities, often referred to as the NIMBY (not in my backyard) syndrome. Communities may oppose the establishment of waste management infrastructure near their homes, citing concerns about pollution, health risks, or declining property values (Kikuchi and Gerardo, 2009). In some regions, there may be a cultural tendency to undervalue waste management or recycling. Industries and the general public may not view waste reduction or resource recovery as a priority, leading to reluctance in adopting sustainable practices. A study by (Kirchherr et al., 2017) identified "culture" as the most significant constraint to circular economy practices. Business and policy experts who took part in their study highlighted the preference of the consumers for new products. This discourages investments on enhancing goods' durability. Another factor is the lower cost associated with an unsustainable option that lures the public toward opting it, for example, oil-based plastics in place of bio-based plastics.

The political landscape can have a significant impact on the implementation of waste management regulations. Changes in government or shifting priorities may lead to inconsistency in policy implementation or a lack of long-term commitment to waste management reforms (Costa and Ferreira Dias, 2020). Powerful industrial lobbies may influence government policies, leading to weaker regulations or slower implementation of waste management laws. In some cases, industries may lobby for leniency in regulations to minimize compliance costs, undermining the intent of the legislation.

Global and Cross-border Challenges

Industrial waste, particularly hazardous materials, is often transported across borders. In many cases, this waste is shipped to the developing countries with lax environmental standards, where inadequate resources lead to its improper disposal. Despite international agreements like the Basel Convention, illegal dumping and transboundary waste movement of hazardous waste continue to pose a significant challenge to waste management efforts (Ahmed, 2019; Khan, 2016). Moreover, multinational corporations may move production to countries with lower environmental (Nidheesh et al., 2022) standards, increasing the likelihood that wastes from their operations are not properly managed. Waste management regulations may be circumvented or poorly enforced in such regions, exacerbating global environmental pollution.

The challenges to the implementation of industrial waste management regulations are diverse but interconnected. Addressing these challenges requires coordinated efforts at the local, national, and international levels, including improved infrastructure, enhanced financial support, better enforcement, and cultural change brought about by increased awareness. Policymakers must ensure that regulations are clear and practicable, and adaptable to new industrial developments. Collaboration between governments, industries, and the civil society is important for the effective implementation of the regulations leading to a sustainable, resource-efficient future.

8.6 Summary and Conclusion

Industrial waste generation is a growing global challenge, with significant environmental and health impacts. While regulations at both the national and international levels play an essential role in mitigating these risks, their effectiveness is often hindered by issues like poor enforcement, resource constraints, and resistance from industries. Nonetheless, the impacts of well-enforced regulations are clear: they protect public health, preserve the environment, and drive innovations in waste reduction and resource recovery.

In spite of the progress made in regulating industrial waste, many challenges remain. Strengthening enforcement mechanisms, fostering international collaboration, and providing support to industries to help them comply with regulations are essential next steps. As industries continue to evolve, regulations must also adapt, encouraging more sustainable production processes and technologies that minimize waste generation. The future of industrial waste management will be decided by the effectiveness of the efforts to balance environmental protection, public health, and economic growth.

Review Questions

1. Discuss the different industrial activities that generate waste.
2. Why is resource recovery from industrial wastewater important?
3. Describe the major resources that can be recovered from industrial wastewater and their significance.
4. Explain the different resource recovery technologies that are employed for the recovery of resources from wastewater.
5. Explain the different international frameworks and conventions related to industrial waste management.
6. List the different environmental regulations and guidelines in India and explain their relevance to industrial waste management.
7. Explain the different regulations and guidelines implemented by different developed nations on resource recovery.

8 "The adoption of resource recovery regulations and guidelines has a positive impact on the pollution prevention and industrialist profit." Justify the above statement with proper evidence.
9 What are all the challenges faced by the developed and developing nations in implementing the regulations and guidelines on industrial waste management?
10 How regional and national politics impact the implementation of environmental regulation?

References

Abandoned Mine Drainage [WWW Document]. (2023). *Polluted runoff nonpoint source pollution.* https://www.epa.gov/nps/abandoned-mine-drainage

Adnan, M., Xiao, B., Ali, M.U. et al. (2024). Heavy metals pollution from smelting activities: a threat to soil and groundwater. *Ecotoxicology and Environmental Safety*, **274**, 116189. https://doi.org/10.1016/j.ecoenv.2024.116189

Aguiar, A.O., Andrade, L.H., Ricci, B.C. et al. (2016). Gold acid mine drainage treatment by membrane separation processes: an evaluation of the main operational conditions. *Separation and Purification Technology*, **170**, 360–369. https://doi.org/10.1016/j.seppur.2016.07.003

Agunwamba, J.C. (1998). Solid waste management in Nigeria: problems and issues. *Environmental Management*, **22**(6), 849–856. https://doi.org/10.1007/s002679900152

Ahmed, I. (2019). The Basel Convention on the control of transboundary movements of hazardous wastes and their disposal: a legal misfit in global ship recycling jurisprudence. *Washington International Law Journal*, **29**, 411.

Alter, H. (2000). Environmentally sound management of the recycling of hazardous wastes in the context of the Basel Convention. *Resources, Conservation and Recycling*, **29**(1–2), 111–129. https://doi.org/10.1016/S0921-3449(99)00061-0

Anders, A.S. (2020). Does the Restriction of Hazardous Substances (RoHS) directive help reduce environmental impacts? *International Journal of Green Technology*, **6**, 24–37.

Arslanoğlu, H. and Tümen, F. (2021). Potassium struvite (slow release fertilizer) and activated carbon production: resource recovery from vinasse and grape marc organic waste using thermal processing. *Process Safety and Environmental Protection*, **147**, 1077–1087. https://doi.org/10.1016/j.psep.2021.01.025

Au, L.S., Ahn, S., and Kim, T.W. (2018). System dynamic analysis of impacts of government charges on disposal of construction and demolition waste: a Hong Kong case study. *Sustainability*, **10**(4), 1077.

Beckinghausen, A., Odlare, M., Thorin, E. et al. (2020). From removal to recovery: an evaluation of nitrogen recovery techniques from wastewater. *Applied Energy*, **263**, 114616. https://doi.org/10.1016/j.apenergy.2020.114616

Bilotta, G.S. and Brazier, R.E. (2008). Understanding the influence of suspended solids on water quality and aquatic biota. *Water Research*, **42**(12), 2849–2861. https://doi.org/10.1016/j.watres.2008.03.018

Calvo, N., Varela-Candamio, L., and Novo-Corti, I. (2014). A dynamic model for construction and demolition (C&D) waste management in Spain: driving policies based on economic incentives and tax penalties. *Sustainability*, **6**(1), 416–435.

Cardoso, M., Scholz, M., and Antunes, M. (2019). Phosphorus recovery from municipal wastewater treatment: critical review of challenges and opportunities for developing countries. *Journal of Environmental Management*, **248**, 109268. https://doi.org/10.1016/j.jenvman.2019.109268

Chequer, F.M.D., de Oliveira, G.A.R., Ferraz, E.R.A. et al. (2013). Textile dyes: dyeing process and environmental impact. In: *Eco-Friendly Textile Dyeing and Finishing*. IntechOpen. https://doi.org/10.5772/53659

Costa, I.M. and Dias, F.M. (2020). Evolution on the solid urban waste management in Brazil: a portrait of the Northeast Region. *Energy Reports*, **6**, 878–884. https://doi.org/10.1016/j.egyr.2019.11.033

CPCB. (2016). Guidelines for management of healthcare waste BMW 2016. *Central Pollution Control Board*, 13–35, 25–56.

D'Amato, A., Mazzanti, M., Nicolli, F. et al. (2018). Illegal waste disposal: enforcement actions and decentralized environmental policy. *Socio-Economic Planning Sciences*, **64**, 56–65. https://doi.org/10.1016/j.seps.2017.12.006

da Silva, G.S., dos Santos, F.A., Roth, G. et al. (2020). Electroplating for chromium removal from tannery wastewater. *International Journal of Environmental Science and Technology*, **17**(2), 607–614. https://doi.org/10.1007/s13762-019-02494-1

Dehkordi, M.M., Pournuroz Nodeh, Z., Soleimani Dehkordi, K. et al. (2024). Soil, air, and water pollution from mining and industrial activities: sources of pollution, environmental impacts, and prevention and control methods. *Results in Engineering*, **23**, 102729. https://doi.org/10.1016/j.rineng.2024.102729

Department of Climate Change, Energy, the Environment and Water. (2022). Annual Climate Change Statement 2022: The first Annual Climate Change Statement to Parliament as required by the Climate Change Act 2022. Canberra: Australian Government.

Dutta, D., Arya, S., and Kumar, S. (2021). Industrial wastewater treatment: current trends, bottlenecks, and best practices. *Chemosphere*, **285**, 131245. https://doi.org/10.1016/j.chemosphere.2021.131245

EPA. (2024). *International Agreements on Transboundary Shipments of Hazardous Waste | US EPA*. US environment Protection agency.

Hajam, Y.A., Kumar, R., and Kumar, A. (2023). Environmental waste management strategies and vermi transformation for sustainable development. *Environmental Challenges*, **13**, 100747. https://doi.org/10.1016/j.envc.2023.100747

International Institute for Sustainable Development. (2024). *The Rotterdam Convention on the Prior Informed Consent Procedure for Certain Hazardous Chemicals and Pesticides in International Trade*. International Institute for Sustainable Development.

Iqbal, M., Breivik, K., Syed, J.H. et al. (2015). Emerging issue of e-waste in Pakistan: a review of status, research needs and data gaps. *Environmental Pollution*, **207**, 308–318. https://doi.org/10.1016/j.envpol.2015.09.002

Jiang, Z., Wang, Z., and Zeng, Y. (2020). Can voluntary environmental regulation promote corporate technological innovation? *Business Strategy and the Environment*, **29**(2), 390–406. https://doi.org/10.1002/bse.2372

Jonášová, S.K., Moldan, B., and Veselá, A. (2024). Bridging the gaps: recommendations for improving the status quo pertaining to textile waste data discrepancies in the Czech Republic. *Waste Forum*, 2024(3), 223–244.

Khan, S.A. (2016). E-products, E-waste and the basel convention: regulatory challenges and impossibilities of international environmental law. *Review of European, Comparative & International Environmental Law*, **25**(3), 248–262. https://doi.org/10.1111/reel.12163

Kikuchi, R. and Gerardo, R. (2009). More than a decade of conflict between hazardous waste management and public resistance: a case study of NIMBY syndrome in Souselas (Portugal). *Journal of Hazardous Materials*, **172**(2–3), 1681–1685. https://doi.org/10.1016/j.jhazmat.2009.07.062

Kirchherr, J., Reike, D., and Hekkert, M. (2017). Conceptualizing the circular economy: an analysis of 114 definitions. *Resources, Conservation and Recycling*, **127**, 221–232. https://doi.org/10.1016/j.resconrec.2017.09.005

Krishnan, S., Zulkapli, N.S., Kamyab, H. et al. (2021). Current technologies for recovery of metals from industrial wastes: an overview. *Environmental Technology & Innovation*, **22**, 101525.

Kulkarni, S.S., Mayilswamy, N., Sidharth, S. et al. (2024). Sustainable wastewater management via biochar derived from industrial sewage sludge. *Circular Economy and Sustainability*, **4**(1), 163–200.

Kumari, T. and Raghubanshi, A.S. (2023). Waste management practices in the developing nations: challenges and opportunities. In *Waste Management and Resource Recycling in the Developing World*, 773–797. Elsevier. https://doi.org/10.1016/B978-0-323-90463-6.00017-8

Kuttippurath, J., Patel, V.K., Pathak, M. et al. (2022). Improvements in SO2 pollution in India: role of technology and environmental regulations. *Environmental Science and Pollution Research*, **29**(52), 78637–78649. https://doi.org/10.1007/s11356-022-21319-2

Lee, W. and Okos, M.R. (2011). Sustainable food processing systems – Path to a zero discharge: reduction of water, waste and energy. *Procedia Food Science*, **1**, 1768–1777. https://doi.org/10.1016/j.profoo.2011.09.260

Liu, Z.G., Min, X B., Feng, F. et al. (2021). Development and simulation of a struvite crystallization fluidized bed reactor with enhanced external recirculation for phosphorous and ammonium recovery. *Science of the Total Environment*, **760**, 144311. https://doi.org/10.1016/j.scitotenv.2020.144311

Maalouf, A. and Agamuthu, P. (2023). Waste management evolution in the last five decades in developing countries – A review. *Waste Management and Research*, **41**(9), 1420–1434. https://doi.org/10.1177/0734242X231160099

Mmereki, D. (2018). Current status of waste management in Botswana: a mini-review. *Waste Management and Research*, **36**(7), 555–576. https://doi.org/10.1177/0734242x18772097

Nagarajan, K.S.K., Geor, G.C., Rajalakshmi, M.S. et al. (2022). A comprehensive review on comparison among effluent treatment methods and modern methods of treatment of industrial wastewater effluent from different sources. *Applied Water Science*, **12**(4), 70. https://doi.org/10.1007/s13201-022-01594-7

Nidheesh, P.V., Ganiyu, S.O., Martínez-Huitle, C.A. et al. (2022). Recent advances in electro-Fenton process and its emerging applications. *Critical Reviews in Environmental Science and Technology*. https://doi.org/10.1080/10643389.2022.2093074

Nyirenda, H. (2023). Fiddling at the conference of the parties? Peeping into the highs and lows of the post-Kyoto climate change conferences: a review on contexts, decisions and implementation highlights. *Environment, Development and Sustainability*, **27**, 8155–8185. /https://doi.org/10.1007/s10668-023-04277-8

Ogunwumi, O.T. and Salami, L. (2023). Perspective chapter: industrial waste landfills. In: *Solid Waste and Landfills Management – Recent Advances*. IntechOpen. https://doi.org/10.5772/intechopen.108787

Panagopoulos, A., and Haralambous, K. J. (2020). Minimal Liquid Discharge (MLD) and Zero Liquid Discharge (ZLD) strategies for wastewater management and resource recovery-Analysis, challenges and prospects. *Journal of Environmental Chemical Engineering*, **8**(5), 104418. https://doi.org/10.1016/j.jece.2020.104418

Peng, H., An, C., Ng, K.T.W. et al. (2023). Advancing cleaner municipal waste transport through carbon accounting in the cap-and-trade system. *Transportation Research Part D: Transport and Environment*, **114**, 103560. https://doi.org/10.1016/j.trd.2022.103560

Rahimi, S., Modin, O., and Mijakovic, I. (2020). Technologies for biological removal and recovery of nitrogen from wastewater. *Biotechnology Advances*, **43**, 107570. https://doi.org/10.1016/j.biotechadv.2020.107570

Riseh, R.S., Vazvani, M.G., Hassanisaadi, M. et al. (2024). Agricultural wastes: a practical and potential source for the isolation and preparation of cellulose and application in agriculture and different industries. *Industrial Crops and Products*, **208**, 117904. https://doi.org/10.1016/j.indcrop.2023.117904

Sakamoto, M., Ahmed, T., Begum, S. et al. (2019). Water pollution and the textile industry in Bangladesh: Flawed corporate practices or restrictive opportunities? *Sustainability (Switzerland)*, **11**(7). https://doi.org/10.3390/su11071951

Sánchez-García, E., Martínez-Falcó, J., Marco-Lajara, B., et al. (2024). Revolutionizing the circular economy through new technologies: a new era of sustainable progress. *Environmental Technology and Innovation*, **33**(September 2023). https://doi.org/10.1016/j.eti.2023.103509

Senapati, M.R. (2011). Fly ash from thermal power plants – waste management and overview. *Current Science*, **100**(12), 1791–1794.

Shinde, A.R., Patil, A.P., Patil, T.M. et al. (2020). Pharmaceutical waste management: significant for health of nature and future. *International Journal of Pharmaceutical Sciences Review and Research*, **65**(2), 68–74. https://doi.org/10.47583/ijpsrr.2020.v65i02.011

Singh, B. J., Chakraborty, A., and Sehgal, R. (2023). A systematic review of industrial wastewater management: Evaluating challenges and enablers. *Journal of Environmental Management*, **348**, 119230. https://doi.org/10.1016/j.jenvman.2023.119230

Stanchev, P., Katsou, E., Pons, S. et al. (2017). Municipal solid waste management and waste-to-energy in the context of a circular economy and energy recycling in Europe. **141**. https://doi.org/10.1016/j.energy.2017.11.128

Taron, A., Majumder, A., Bodach, S. at al. (2023). Public-private partnerships for the circular bio-economy in the global south: Lessons learned. In *Resource Recovery and Reuse* (Vol. 2023, Issue 22). https://doi.org/10.5337/2023.205

Testa, F., Styles, D., and Iraldo, F. (2012). Case study evidence that direct regulation remains the main driver of industrial pollution avoidance and may benefit operational efficiency. *Journal of Cleaner Production*, **21**(1), 1–10. https://doi.org/10.1016/j.jclepro.2011.09.002

UNIDO. (2017). *CIRCULAR ECONOMY*. https://www.unido.org/sites/default/files/2017-07/Circular_Economy_UNIDO_0.pdf

United States Environmental Protection Agency. (2000). Guide for industrial waste management. *EPA*, 1–478. https://www.epa.gov/sites/production/files/2016-03/documents/industrial-waste-guide.pdf

US-EPA. (2024). *Sustainable Management of Construction and Demolition Materials*. Sustainable Materials Management. https://www.epa.gov/smm/sustainable-management-construction-and-demolition-materials

US EPA. (2024). *Persistent Organic Pollutants: A Global Issue, A Global Response*. https://www.epa.gov/international-cooperation/persistent-organic-pollutants-global-issue-global-response US EPA

Vasa, T.N. and Chacko, S.P. (2021). Recovery of struvite from wastewaters as an eco-friendly fertilizer: review of the art and perspective for a sustainable agriculture practice in India. *Sustainable Energy Technologies and Assessments*, **48**, 101573. https://doi.org/10.1016/j.seta.2021.101573

Wang, H., Yang, J., Zhang, H. et al. (2024). Membrane-based technology in water and resources recovery from the perspective of water social circulation: a review. *Science of the Total Environment*, **908**, 168277. https://doi.org/10.1016/j.scitotenv.2023.168277

Water (Prevention and Control of Pollution) Act, 1974, Pub. L. No. EIF 23 March 1974 (1974). https://www.indiacode.nic.in/handle/123456789/1612?sam_handle=123456789/1362

WTO. (2024). *Principles of the trading system*. https://www.wto.org/english/thewto_e/whatis_e/tif_e/fact2_e.htm

Xiong, B. and Wang, R. (2020). Effect of environmental regulation on industrial solid waste pollution in China: From the perspective of formal environmental regulation and informal environmental regulation. *International Journal of Environmental Research and Public Health*, **17**(21), 1–17. https://doi.org/10.3390/ijerph17217798

Yadav, S., Kumar Mangla, S., Priyamvada, P. et al. (2024). An energy-efficient model for PPE waste management in a closed-loop supply chain. *Business Strategy and the Environment*, **33**(2), 1191–1207. https://doi.org/10.1002/bse.3541

Yang, M., Yan, X., and Li, Q. (2021). Impact of environmental regulations on the efficient control of industrial pollution in China. *Chinese Journal of Population Resources and Environment*, **19**(3), 230–236. https://doi.org/10.1016/j.cjpre.2021.12.025

Zandi Lak, S., Rahimpour, M.R., and Bonyadi, M. (2024). Liquid bio-fuels from algal materials. In: *Encyclopedia of Renewable Energy, Sustainability and the Environment*, 433–442. Elsevier. https://doi.org/10.1016/B978-0-323-93940-9.00142-0

9

Industrial Wastewater Treatment: Environmental and Economic Implications

Mona Ebaid[1], *Ghada Abdallah*[2], *Nabila Shehata*[2], *and Michael Kornaros*[3]*

[1]*Material Science and Nanotechnology Department, Faculty of Postgraduate Studies for Advanced Sciences, Beni-Suef University, Beni-Suef, Egypt*
[2]*Environmental Science and Industrial Development Department, Faculty of Postgraduate Studies for Advanced Sciences, Beni-Suef University, Beni-Suef, Egypt*
[3]*Laboratory of Biochemical Engineering & Environmental Technology (LBEET), Department of Chemical Engineering, University of Patras, Patras, Greece*

*Corresponding author: kornaros@chemeng.upatras.gr

9.1 Introduction

The current era of rapid economic and industrial development has led to the emergence of significant environmental concerns related to the treatment of industrial wastewater. Contamination of water sources by heavy metals such as mercury, lead, zinc, copper, cadmium, and nickel has been identified as a major public health risk with a range of toxicological implications. In response, a number of remediation strategies have been studied and developed with the goal of revitalizing contaminated wastewater without causing harm to the surrounding environment (Kumar et al., 2023). One strategy for reducing water waste, reducing pressure on natural water sources, and facilitating the transition to renewable energy is wastewater treatment. By maximizing energy recovery, reducing demand for natural resources, and supporting agriculture, water treatment promotes sustainable resource management. One of the most environmentally friendly ways to generate electricity, conserve water, and increase agricultural productivity is to optimize wastewater treatment (Silva, 2023).

Wastewater to Resource Recovery: Applying the Circular Economy Toward Sustainable Development, First Edition. Edited by Arvind Kumar Mungray et al. Published 2026 by John Wiley & Sons, Inc.

The lack of adequate sanitation increases the risk of infectious diseases, including typhoid and cholera, as well as gastrointestinal, skin, and kidney diseases. These diseases are responsible for the deaths of 297,000 children under the age of five each year, or 800 per day (Tariq and Mushtaq, 2023). Therefore, it is crucial to consider the methods and technologies associated with industrial wastewater treatment and the key role it plays in maintaining environmental balance and promoting a sustainable economy. Furthermore, the implementation of industrial wastewater treatment has led to significant economic benefits, including the emergence of the circular economy (CE). The CE represents the primary objective for wastewater utilities and is considered the optimal means to achieve resource, energy, and CO_2 neutrality (Faragò et al., 2021). The aim is to describe the economic consequences of industrial wastewater treatment and to identify the measures that can be taken to move toward a green economy. The majority of wastewater treatment plants (WWTPs) use older technologies, such as anaerobic sludge digestion, which require higher energy consumption. As a result, conventional technologies have only been able to recover a portion of the energy contained in wastewater. In contrast to these technologies, the next generation of WWTPs will achieve near-total recovery of energy, water, and nutrients, with a particular focus on nitrogen and phosphorus. These plants will also play a key role in managing emerging contaminants and recovering valuable products, including heavy and radioactive metals, hormones, fertilizers, and pharmaceutical chemicals (Kesari et al., 2021). Figure 9.1 represents a schematic representation of various sources of wastewaters, their treatment chains, and finally applications of treated water for irrigation, drinking, or other domestic uses. It seems to be a very costly affair.

In short, the treatment of industrial wastewater has significant environmental and economic impacts. It is therefore imperative that companies adopt the most cost-effective, environmentally benign, and efficient wastewater management techniques in order to ensure that their operations are conducted in an environmentally sustainable manner (Thanigaivel et al., 2022). Accordingly, the main objective of this chapter is to examine the environmental and economic impacts of wastewater treatment through a number of lenses, including environmental impact assessment, the economic benefits of water reuse and resource recovery, and the adoption of sustainable wastewater management strategies.

9.2 The Use of Water in Industrial Processes

Water's unique properties, including its ability to dissolve a wide range of substances and its high specific heat capacity, make it a valuable resource in many industrial applications. Its ability to absorb and dissipate heat makes it an effective

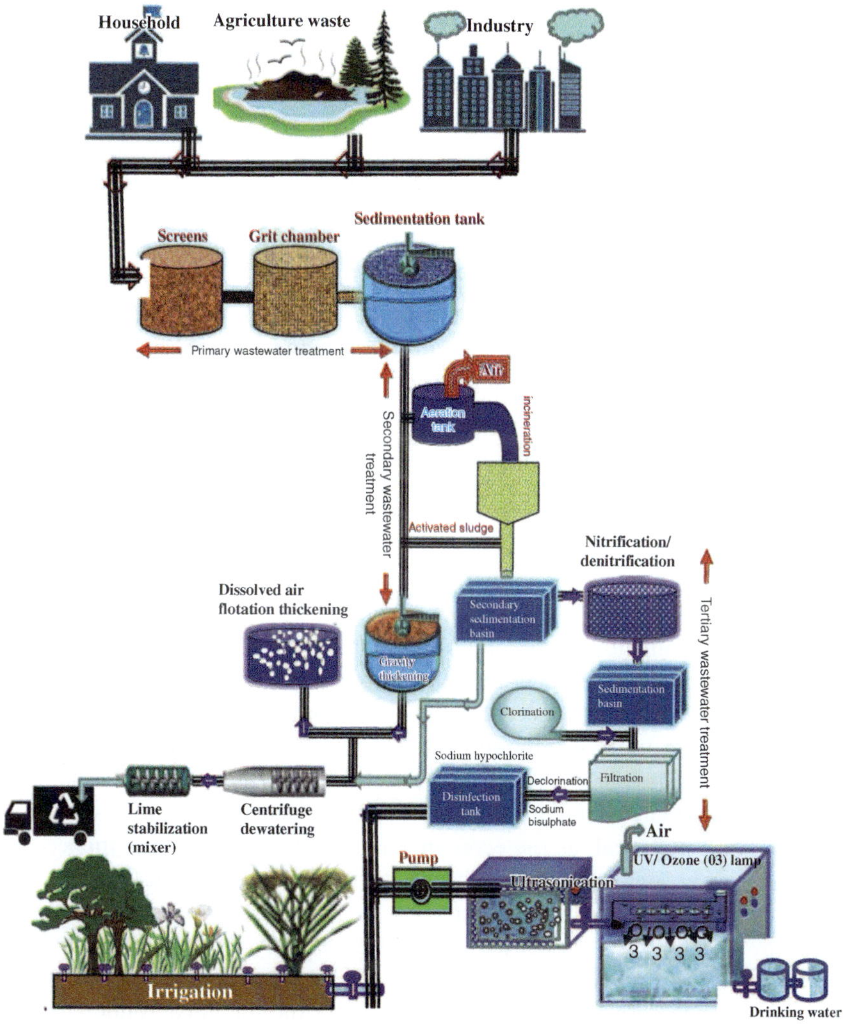

Figure 9.1 A schematic diagram of the different processes employed in wastewater management that yield higher quality effluent from different sources to a final destination, such as irrigation. *Source:* Kesari et al. (2021)/Springer Nature/CC BY 4.0.

coolant, particularly in industrial contexts where significant amounts of heat are generated. The quality of water used in industrial processes is of paramount importance (Yang et al., 2020). The pharmaceutical industry, which requires exceptionally high levels of purity, requires a much higher quality of water than the oil industry. In addition, the corrosive nature of saltwater requires the use of freshwater to protect machinery from damage. Industrial processes require vast quantities of water for a variety of purposes, including cleaning, processing, dilution, cooling, and transportation. The amount of water used per unit of product varies widely from industry to industry.

Water is a vital resource used by various sectors, including industry, agriculture, and the general public. The increasing global demand for water (Jain et al., 2021), coupled with the adverse effects of climate change (Isaac et al., 2022), has caused excessive water resource use in numerous areas. To overcome these obstacles, integrated industrial water treatment strategies, such as water recycling/reuse and the use of alternative water sources, are essential (Sharma et al., 2023). The future of integrated water treatment will continue to be shaped by the creation of novel ideas and technological, digital, and non-technological solutions, as well as by an emphasis on strategic priorities, especially in industrial settings (Becker et al., 2019).

Unfortunately, the industrial sector is frequently held responsible for increased water consumption and water pressure. Energy production is expanding to meet growing demand. By 2035, Asia's water demand for energy production is expected to exceed 230 billion m^3, an increase of 73 billion m^3 from 2010 (Weerasooriya et al., 2021). The mining industry faces significant challenges in ensuring environmental sustainability, particularly with respect to water use. It is estimated that approximately 19 billion tons of solid tailings will be generated globally by 2025, raising concerns about their management. Tailings dams are the largest water users in mining operations. Inadequate water management in tailings can destabilize these deposits, leading to potential dam failures and subsequent environmental disasters. Prudent management of watersheds in such areas is critical, as the economic and environmental consequences of increased water consumption are significant (Hamraoui et al., 2024).

In the seafood industry, the largest water use is for fish processing (42%), followed by washing (38%). Water used for on-site ancillary services such as toilets, washroom sinks, and canteen sinks is categorized separately. As shown in the healthier production assessment for fish processing, the amount of water required per ton of fish is 5–11 m^3, 15 $m^{3,}$ and 0.5 m^3 for filleting, canning, and fishmeal, respectively. Similarly, clean-in-place (CIP) and various technologies for the fish sector consume significant amounts of water (Murali et al., 2021). The problems of scaling and corrosion in industrial water systems are significant challenges because of their negative impact on plant performance, maintenance requirements, and production costs. In particular, scaling can reduce water flow rates in pipes and cause blockages in heating systems, including steam boilers, industrial equipment, and domestic appliances such as water heaters. This can ultimately affect the efficiency of these systems (Zaman et al., 2018). It is imperative to develop strategies that simultaneously protect freshwater resources and promote the use of unconventional water sources, especially wastewater and desalinated water (Kesari et al., 2021). Integrated water reuse systems are a promising solution because they enable closed-loop water reuse and overcome the limitations of traditional water management practices (Morseletto, 2020). These systems not only conserve water but also optimize its comprehensive use and improve its sustainability and resilience (Sharma et al., 2024). Therefore, enhancing and encouraging water efficiency is essential to achieving Sustainable Development Goal 6 (SDG 6) of ensuring availability and sustainable management of water and sanitation for all (Lozano and Borrego-Marín, 2024). In addition, the nature of the steelmaking industry necessitates the extensive use of water due to its high specific heat capacity, making it a particularly attractive medium for heat transfer in industrial operations. The importance of cooling cannot be overstated, especially in the context of steel production (Tong et al., 2018).

It is critical to identify strategies that can simultaneously reduce consumption and improve the security of water resources. One potential solution is to diversify water supply sources (Meese et al., 2021). Therefore, water treatment in industrial activities has the potential to significantly reduce freshwater consumption and wastewater discharge. For example, (i) filtration is the removal of contaminants based on their size. The removal of contaminants from wastewater allows the water to be reused for a variety of purposes; (ii) chemical processes, combined with physical or biological approaches, are used to reduce pollutants discharged into water and wastewater. A variety of chemical processes are used to convert contaminants into useful products or to remove contaminants for safe disposal (Saravanan et al., 2021). Given the pivotal role of the industrial sector in global water consumption and the increasing demand to reduce its environmental impact, a comprehensive plan for industrial water conservation can be achieved by combining the concept of water footprint (WF) with sustainability goals (Kumar and Thakur, 2024).

Water is an essential component for sustainable development, from health to social well-being and the economy (United Nations, 2023; Liu et al., 2020). A significant amount of research has been devoted to exploring the relationship between water use and industrial expansion, with a particular focus on the energy, iron and steel, and mining sectors. As the world's population continues to grow and industrialization accelerates, pressure on water resources increases, raising concerns about future water availability. Some studies have used modified versions of the Water Stress Index (WSI) to assess water scarcity. However, as urbanization and industrialization continue, the industrial sector's demand for water resources has increased. This is evidenced by findings reported in the China Water Resources Bulletin (Li et al., 2023a).

Industrial symbiosis (IS) is a key idea in the development of green industries. The exchange of materials as well as energy between industrial facilities is a fundamental aspect of IS, where waste from one facility is transformed into material inputs for another (Tiu and Cruz, 2017). The effective control of water exchange within an IS mechanism, coupled with the optimization of manufacturing techniques to reduce water use, can significantly improve the overall ecological viability of the industrial complex (Yu and Ding, 2021). Accurate calculation of end-use water consumption is critical for the adoption of water-saving technologies with efficient water systems (Mazzoni et al., 2023). A thorough knowledge of the factors influencing water use is required for the effective implementation of successful water resource management strategies (Grespan et al., 2022). Identifying water use patterns is critical for determining the effectiveness of potential efficiency measures and for analyzing the elements that influence water use (Mohan et al., 2021).

The UN World Water Development Report 2023 predicts an 80% increase in urban water demand by 2050, with the number of urban dwellers facing water scarcity expected to reach 1.7–2.4 billion. As populations grow and demand for water increases, the issue of water scarcity becomes more urgent. The issue stems from the recognition of the detrimental impact of water scarcity on environmentally sustainable growth (Liu et al., 2024). While recent efforts have focused on reducing water consumption in industrial processes, industrial buildings remain significant water consumers, particularly due to the high water demands of their large workforces. This represents a significant opportunity to optimize water use in urban areas. Consequently, industrial buildings are strategic targets for the implementation of water conservation measures (Matos et al., 2023).

9.3 Environmental Impact of Industrial Activity

For more than two decades, green chemistry has been a driving force in transforming chemical and product development, with a focus on minimizing toxicity and environmental impact (Anastas and Eghbali, 2010). While the safe-by-design approach aims to reduce the environmental impact of chemicals, it does not ensure that alternative chemicals are inherently less harmful to the environment (Maertens et al., 2021). This shift represents a broader, more sustainable approach to technology development in the field of green technology.

The adoption of clean technology is a critical step in transforming mining practices and reducing the industry's environmental impact. Mining plays a central role in modern society, providing raw materials essential to a wide range of industries, including construction, electronics, and medical equipment. However, this reliance on mining has significant environmental consequences (Li et al., 2023b). There are significant environmental costs associated with this development, including pollution. Toxic metals like lead, cadmium, zinc, and

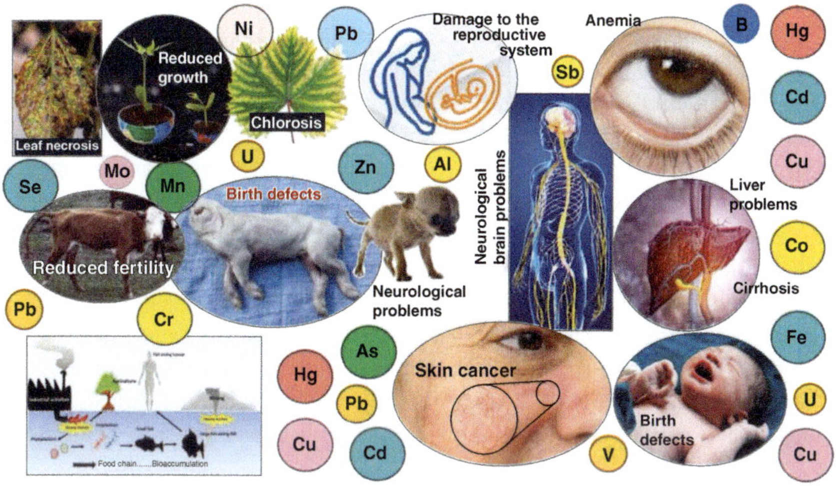

Figure 9.2 Environmental impact of mining activities, including the generation of pollution and associated consequences. *Source:* Dehkordi et al. (2024)/with the permission of Elsevier.

mercury are released into the air, water, and soil throughout the mining and mineral processing processes. As shown in Figure 9.2, these heavy metals seriously endanger biodiversity and humans.

Water flow may be impacted by heavy metal-contaminated soils, such as those containing cadmium, lead, zinc, and mercury, through a number of mechanisms. One such mechanism is direct leaching, whereby metal ions can leach directly into groundwater or surface water. This is particularly common in areas with unstable soils or where heavy metal application is common practice (Li et al., 2021). Another mechanism of contamination is surface runoff, which can transfer soil-borne toxins to water bodies such as lakes or rivers (Haque et al., 2021). In addition, saturation and deposition contribute to water contamination. Rainfall can mobilize soil contaminants and transport them to water bodies (Zhang et al., 2020). In addition, plants can take up heavy metals in the soil and release them into waterways, either directly or through biological accumulation (Ketterer et al., 2001).

Recently, there has been a notable trend in the manufacturing industry to replace conventional manufacturing (CM) processes with additive manufacturing (AM) (Kokare et al., 2023). While traditional methods such as machining and forging have been the dominant approach to component manufacturing, AM is emerging as a key driver of innovation (Jemghili et al., 2021). This trend is reflected in the scientific community, as evidenced by the increased global investment in AM research. The global AM market is expected to expand significantly, from a valuation of $11 billion in 2019 to an estimated $35 billion in 2024. This represents a projected compound annual growth rate of approximately 26% (Gonçalves et al., 2023). This technology has the ability to reduce the environmental impact of both aeronautical and industrial machinery by more than 60%, primarily by reducing material consumption. In the case of aeronautical components, the use phase also contributes to environmental and economic benefits through fuel and carbon footprints. Rising atmospheric CO_2 levels pose a significant threat to global climate change. As the effects of global warming intensify, the immediate desire to reduce the release of CO_2 has become paramount (Zhang et al., 2023).

A variety of strategies exist to protect human well-being and manage the risks associated with air and soil pollution from lead, mercury, arsenic, cadmium, and zinc. These strategies include a range of measures designed to control and reduce these contaminants (Barhoum et al., 2023):

- Pollution management: This method involves reducing the discharge of pollutants into the environment. The measures implemented to achieve this objective include the installation and maintenance of pollution control mechanisms in the workplace and mines, the installation of on-site or decentralized treatment plants, and the upgrading of technological advances.
- Assessment and sampling methods: This technique requires the monitoring of water, air, and soil quality in regions with documented pollution, the periodic collection and analysis of samples to determine pollutant concentrations, and the use of remote sensing techniques for comprehensive macroscale monitoring. It also involves the identification of susceptible populations, including local farmers, residents, and individuals (Yu et al., 2023).
- Public health interventions: This strategy includes public education and awareness campaigns about the dangers of wastewater contamination and the implementation of preventive measures. It also includes ensuring access to potable water through the distribution of water purification filters in polluted areas and the provision of medical and health services, including frequent assessment and treatment for those at risk (Wang et al., 2023).
- The research strategy: This includes three primary objectives: developing novel methods for remediating contaminated water and soil, investigating the long-term health effects of heavy metal contamination through epidemiological studies, and advancing advanced water and soil purification technologies (Soni et al., 2022).
- Promoting cooperation: It is of paramount importance to foster collaboration among industry, academia, government, and non-governmental organizations to facilitate knowledge sharing, exchange of expertise, and allocation of resources for research and remediation programs (Dehkordi et al., 2024).

9.4 Generation of Industrial Wastewater

A comprehensive solution to the emerging problem of water scarcity will require the incorporation of many water management techniques and the installation of cutting-edge infrastructure for wastewater treatment and reuse (Bauer et al., 2020). Water scarcity has persisted due to increases in freshwater use and wastewater discharge on a global scale. The current global volume of wastewater generation is estimated at 380 billion m^3 per year, with projections indicating an expected increase of 51% by 2050 (Qadir et al., 2020). As noted in a United Nations World Water Assessment Programme (WWAP) report (WWAP, 2017), the release of wastewater without treatment continues to be an international problem. Typically, WWTPs are designed to achieve a balance between cost-effectiveness and achieving optimal water quality standards (Crini and Lichtfouse, 2019). Numerous technologies are frequently used in wastewater treatment, each of which is designed to remove particular pollutants. This multi-stage process often includes preliminary, primary, secondary, and tertiary treatment stages. Figure 9.3 illustrates the overall process of wastewater treatment and the approaches used at each stage.

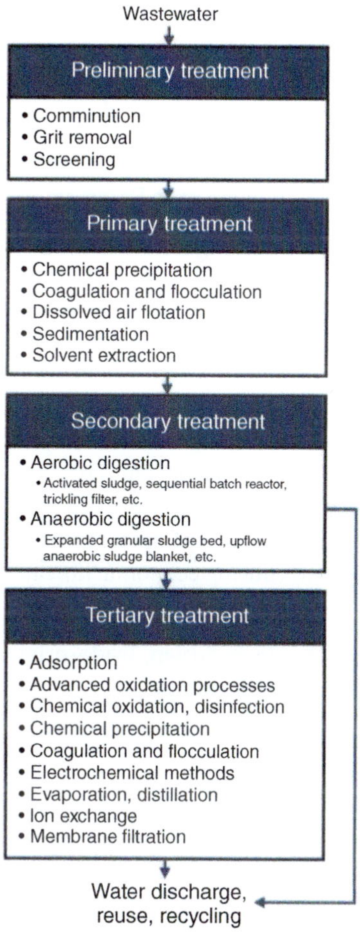

Figure 9.3 An outline of the primary processes utilized in a general wastewater treatment plant. *Source:* Kato and Kansha (2024)/Springer Nature/CC BY 4.0.

Despite the lack of comprehensive documentation of global industrial wastewater generation and discharge volumes, reports from the European Union (EU) indicate that manufacturing is the primary industrial source of wastewater in developed countries (WWAP, 2017). The main manufacturing sectors responsible for industrial wastewater pollution include pharmaceuticals, petrochemicals, textiles, paper, pulp, steel and ferrous metal production, and food processing (WWAP, 2017).

Industrial wastewater streams are often more hazardous than municipal wastewater. The composition and concentrations of pollutants vary considerably between different industrial sectors. As a result, industrial wastewater may not be adequately treated by municipal wastewater treatment facilities, which are primarily intended to remove biochemical oxygen demand (BOD). Industrial wastewater treatment necessitates a thorough examination of its characteristics and the application of tailored treatment processes. Due to the diverse nature of industrial wastewater, WWTPs are often designed on a case-by-case basis, taking into account factors such as the characteristics of the wastewater to be treated, regulatory requirements, land availability, feasibility of the technology to be used, and economic considerations (Kato and Kansha, 2024).

9.5 Economic Burden of Industrial Activity and Strategies

The global economic landscape has changed significantly over the past 40 years. Rapid economic expansion has been strongly associated with institutional reform, and current research has focused significantly on the impact of both formal and informal institutions on economic performance (Barg et al., 2021). At the same time, national governments have had to adapt their role in response to these evolving domestic and international dynamics (Tassinari, 2019). Developing countries face significant competitive pressures as many firms struggle to adapt to the demands of open markets, both in terms of exporting and competing with imports during economic liberalization (Buira, 2005).

9.5.1 Tech-driven, Trade-oriented Industries: A Focus on National Industrial Strategy

Industries that rely on global trade and technological advancement are ideal targets for a successful national industrial strategy. These industries often include sectors such as aerospace, computers, semiconductors, biopharmaceuticals, transportation equipment, advanced machinery, software, and instrumentation (Atkinson, 2020).

China's economic growth in recent decades has been driven by industrialization, which is seen as critical to achieving national goals. Industrialization can be understood in two ways: (i) narrowly: the rapid growth of the industrial sector leading to a decline in the share of agriculture in the economy, and (ii) broadly: a broader process of economic modernization involving the mechanization of all sectors of the economy.

Soviet economic thought emphasized the importance of heavy industry, especially engineering, as the foundation of socialist industrialization. Thus, industrialization involves not only the growth of industrial sectors, but also the transformation and modernization of the entire economic structure (Jiejiao and Qin, 2021). China recognizes that technological innovation is essential to becoming a global industrial leader. To sustain industrial growth and achieve high-income status, China has acquired critical technologies. Despite being the world's second-largest economy, China acknowledges its weak innovation capacity, which it views as a major vulnerability (Xi, 2019). The global dominance of neoliberal economic policies is largely attributed to the strategic influence of the United States. By promoting free market ideology and using its bargaining power in international trade agreements, the United States has shaped the global economic landscape to favor its economic interests. Furthermore, the United States has often used protectionist measures and government intervention to protect its domestic industries while advocating for deregulation and reduced government intervention in other countries (Tassinari, 2019).

Entrepreneurship, which encompasses technological, organizational, and commercial aspects, also serves as a driver of competition and industrial competitiveness (Laperche et al., 2011). The ongoing process of globalization has made regional interactions an important driver of economic growth (Wang et al., 2024). The relationship between market reforms and venture capital investment is complex and multidimensional. First, changes in commercialization can alter the distribution of capital by removing administrative barriers and financial repression, thereby facilitating venture capital investment in small and medium-sized enterprises. Second, these reforms can also affect the demand for venture capital by strengthening corporate governance and reducing soft budget constraints, which may encourage entrepreneurs to seek external financing. Finally, the effectiveness of venture capital in driving innovation and economic growth depends on a supportive regulatory environment and a strong entrepreneurial ecosystem (Ren et al., 2024).

To ensure the long-term viability of industries, strategic measures such as automation, sustainable energy practices, market diversification, resilient supply chains, and lean manufacturing are essential (Hosen et al., 2025). These approaches minimize waste and optimize processes, contributing to economic sustainability (Jabari et al., 2024). In addition, improving energy efficiency is paramount to reducing energy consumption and carbon footprints. This is especially important given the increasing energy demand and environmental concerns associated with industrial growth.

The energy sector has been significantly affected by international economic sanctions. These sanctions, combined with reduced gross domestic product (GDP), limited access to energy-efficient technologies, and capital outflows, create a complex challenge for improving energy efficiency in the country (Balali et al., 2024). International sanctions can have a negative impact on a country's economy by reducing productivity and hindering technological progress (Kelishomi and Nisticò, 2024). The implementation of sanctions can lead to a reduction in formal sector output and an expansion of the informal sector, which could potentially lead to an increase in energy consumption per unit of output. This can ultimately lead to reduced energy efficiency and lower overall economic growth (Demir and Tabrizy, 2022). The adoption of innovative technologies by manufacturing units can significantly reduce product prices and increase competitiveness. This provides an incentive for additional units to adopt energy-efficient innovations, thereby maximizing resource utilization and improving overall energy consumption (Zhang and Fu, 2022).

9.6 Reuse and Reduce Approach Applicable to Industry

9.6.1 Applying Water Reuse Strategies in Industry

The effectiveness of water reuse strategies has been evaluated and implemented in a variety of industrial contexts. Illustrative examples include: (i) The potential for water reuse in the tanning process as a means of conserving natural water resources has been investigated. This involves the use of methods and technologies that limit the impact of process effluents in leather manufacturing, while reducing the high chromium concentration in industrial tannery effluents. In addition to conserving potable water supplies through wastewater reuse, it is proposed to reuse the residual chromium in tannery floats that would otherwise be treated and disposed of in the environment. In fact, recycling and reuse techniques of tanning floats would undoubtedly contribute to increasing the use of chromium (de Aquim et al., 2019). (ii) Water reuse in the food sector: When selecting potential water sources for reuse, the water requirements and characteristics of a food plant at different process stages are evaluated. The reuse of water, either directly or indirectly, is of great importance as the economy is constantly improving and water is purified before being discharged into aquatic ecosystems. Volume, chemical properties and concentrations, acceptable treatment processes, process economics, and local discharge regulations and laws are the major factors influencing water reuse. A comprehensive water reuse strategy should consider all available options, including promoting the reuse of treated water and minimizing the waste of high-quality water (Shrivastava et al., 2022).

Water consumption in industry has increased as industrialization has progressed in developed countries. With today's advanced technology, there are various ways to reuse water. In addition, an enormous task in industrial management is the intelligent use and reuse of water, with the ultimate goal of using water that is scarce in the environment. The implementation of water recycling plans is an essential part of a sustainable water management strategy due to the limited availability of freshwater sources and the need to protect them. Wastewater reuse is one strategy for reducing water consumption. Water withdrawals from local resources can be reduced if industrialization reduces water consumption. Therefore, the most important elements are improving social interactions and water availability, increasing the efficiency of water input, reducing wastewater output and its pollutant load, lowering thermal energy prices, and reducing the process of primary costs. Last but not least, wastewater reuse, in addition to wastewater treatment, offers sufficient opportunities to reduce project costs (Ostad-Ali-Askari and Eslamian, 2021).

9.6.2 Reduction Strategies for Industrial Water Footprint

A major advance in the development of techniques, policies, and indicators for calculating freshwater use and assessing wastewater discharge is the concept of the WF. The WF uses indicators to calculate freshwater use and analyze wastewater discharge. Industry's knowledge of its role in pollution and water scarcity has increased, creating a need to address these issues. The Industrial Water Footprint (IWF) technique provides a graphical representation of freshwater consumption and wastewater discharge impacts at the product and environmental levels. This technique focuses on industrial production steps (Kumar et al., 2021).

Minimizing WF is a successful approach to improving water sustainability. This can be achieved by focusing on the products and industries that consume the most water and contribute the most to water pollution (Malla et al., 2024). The use of advanced water treatment technologies, water-efficient cooling systems, and water recycling strategies can help reduce demand on freshwater supplies using treated wastewater for non-drinking purposes such as industrial processes and irrigation. This can be achieved by promoting the use of treated wastewater for non-potable uses. Establishing regulations and infrastructure for water reuse facilitates more beneficial water management and conservation practices (Malla et al., 2024).

9.6.3 Environmental and Economic Advantages of Water Recycling for Industries

The Reduction, Reuse, and Recycling of wastewater have a significant economic and environmental impact:

I. Economic benefits:
   - Cost savings, where reduced reliance on external water sources reduces costs in addition to reduced disposal costs (fees associated with wastewater treatment and disposal are reduced) (Arena et al., 2020).
   - Increased efficiency, where resource optimization maximizes the use of available water sources while reducing overall consumption.
   - Efficient water use can save energy in cooling and heating water (Colella et al., 2021).
   - Wastewater treatment ensures the recovery of valuable products and increases the CE of the industrial plant.

II. Environmental benefits:
   - Reduced water use puts less strain on local and underground water bodies and aquifers.
   - Sustainable use ensures the long-term availability of water for biodiversity.
   - Wastewater treatment contributes to achieving sustainability, with a particular focus on clean water and sanitation (Goal 3), clean energy (Goal 7), climate action (Goal 13), life under water (Goal 14), and life on land (Goal 15).

9.7 Conclusion

Industrial wastewater treatment technologies are constantly evolving, with new technologies emerging and developing at a rapid pace. In order to develop and implement these technologies more effectively, it is essential to consider the key trends that are shaping the industry. This will allow for the creation of solutions that are not only more efficient but also have a tangible impact on both the economy and the environment. Focusing on treatment technologies, there is a need to further develop research in advanced and bioremediation technologies. Much more work needs to be done on energy recovery, developing automation and AI optimization to reduce costs. However, to remain competitive and ensure a prosperous future, countries need to adopt strategic industrial policies that foster innovation, technological progress, and sustainable growth. By prioritizing technology-driven, trade-oriented industries and creating an enabling environment for business, nations can navigate the complexities of the global economy and achieve long-term economic prosperity. The future of industrialization lies in sustainable practices and technological innovation. By integrating clean technology, green chemistry, and CE principles, industries can minimize their environmental impact and achieve long-term sustainability.

Review Questions

1. What are the primary sources of industrial wastewater, and how do they impact the environment?
2. Describe the key stages in the industrial wastewater treatment process and their importance.
3. What are the major contaminants found in industrial wastewater, and how are they typically removed?
4. Explain the differences between physical, chemical, and biological wastewater treatment methods.
5. What role do regulatory standards play in industrial wastewater treatment, and what are some key regulations?
6. What are the environmental and economic benefits of wastewater reuse and recycling in industries?
7. Discuss the significance of sludge management in wastewater treatment and the methods used for its disposal.
8. What are some advanced treatment technologies used in industrial wastewater treatment, and how do they improve efficiency?
9. How can industries minimize wastewater generation and improve sustainability in wastewater management?
10. What challenges do industries face in wastewater treatment, and what future advancements might help address them?

References

Anastas, P. and Eghbali, N. (2010). Green chemistry: principles and practice. *Chemical Society Reviews*, **39**, 301–312. https://doi.org/10.1039/B918763B

Arena, C., Genco, M., and Mazzola, M.R. (2020). Environmental benefits and economical sustainability of urban wastewater reuse for irrigation—a cost-benefit analysis of an existing reuse project in Puglia, Italy. *Water*, **12**, 2926.

Atkinson, R.D. (2020). *The Case for a National Industrial Strategy to Counter China's Technological Rise*. Information Technology and Innovation Foundation.

Balali, H., Farzanegan, M.R., Zamani, O. et al. (2024). Economic sanctions, energy consumption, and CO_2 emissions in Iran: a system dynamics model. *Policy Studies*, 1–30. https://doi.org/10.1080/01442872.2024.2358842

Barg, J.A., Drobetz, W., and Momtaz, P.P. (2021). Valuing start-up firms: a reverse-engineering approach for fair-value multiples from venture capital transactions. *Finance Research Letters*, **43**, 102008. https://doi.org/10.1016/j.frl.2021.102008

Barhoum, A., Hamimed, S., Slimi, H. et al. (2023). Modern designs of electrochemical sensor platforms for environmental analyses: principles, nanofabrication opportunities, and challenges. *Trends in Environmental Analytical Chemistry*, **38**, e00199. https://doi.org/10.1016/j.teac.2023.e00199

Bauer, S., Linke, H.J., and Wagner, M. (2020). Combining industrial and urban water-reuse concepts for increasing the water resources in water-scarce regions. *Water Environmental Research*, **92**, 1027–1041. https://doi.org/10.1002/wer.1298

Becker, D., Jungfer, C., and Track, T., (2019). Integrated industrial water management–challenges, solutions, and future priorities. *Chemie Ingenieur Technik*, **91**, 1367–1374.

Buira, A. (2005).*The IMF and the World Bank at Sixty*. Anthem Press.

Colella, M., Ripa, M., Cocozza, A. et al. (2021). Challenges and opportunities for more efficient water use and circular wastewater management. The case of Campania Region, Italy. *Journal of Environmental Management*, **297**, 113171. https://doi.org/10.1016/j.jenvman.2021.113171

Crini, G. and Lichtfouse, E. (2019). Advantages and disadvantages of techniques used for wastewater treatment. *Environmental Chemistry Letters*, **17**, 145–155. https://doi.org/10.1007/s10311-018-0785-9

de Aquim, P.M., Hansen, É., and Gutterres, M. (2019). Water reuse: an alternative to minimize the environmental impact on the leather industry. *Journal of Environmental Management*, **230**, 456–463. https://doi.org/10.1016/j.jenvman.2018.09.077

Demir, F. and Tabrizy, S.S. (2022). Gendered effects of sanctions on manufacturing employment: evidence from Iran. *Review of Development Economics*, **26**, 2040–2069.

Dehkordi, M.M., Pournuroz Nodeh, Z., Soleimani Dehkordi, K. et al. (2024). Soil, air, and water pollution from mining and industrial activities: sources of pollution, environmental impacts, and prevention and control methods. *Results in Engineering*, **23**, 102729. https://doi.org/10.1016/j.rineng.2024.102729

Faragò, M., Damgaard, A., Madsen, J.A. et al. (2021). From wastewater treatment to water resource recovery: environmental and economic impacts of full-scale implementation. *Water Research*, **204**, 117554.

Gonçalves, A., Ferreira, B., Leite, M. et al. (2023). Environmental and economic sustainability impacts of metal additive manufacturing: a study in the industrial machinery and aeronautical sectors. *Sustainable Production and Consumption*, **42**, 292–308. https://doi.org/10.1016/j.spc.2023.10.004

Grespan, A., Garcia, J., Brikalski, M.P. et al. (2022). Assessment of water consumption in households using statistical analysis and regression trees. *Sustainable Cities and Society*, **87**, 104186. https://doi.org/10.1016/j.scs.2022.104186

Hamraoui, L., Bergani, A., Ettoumi, M. et al. (2024). Towards a circular economy in the mining industry: possible solutions for water recovery through advanced mineral tailings dewatering. *Minerals*, **14**, 319.

Haque, E., Thorne, P.S., Nghiem, A.A. et al. (2021). Lead (Pb) concentrations and speciation in residential soils from an urban community impacted by multiple legacy sources. *Journal of Hazardous Materials*, **416**, 125886. https://doi.org/10.1016/j.jhazmat.2021.125886

Hosen, M.I., Tushar, S.R., Alam, M.F. et al. (2025). Strategies for economic sustainability: an empirical study on Muri, Mura, and Muda in the readymade garment sector. *Green Technologies and Sustainability*, **3**, 100115. https://doi.org/10.1016/j.grets.2024.100115

Isaac, R., Viaro, V., Fonseca, C. et al. (2022). Identification of key factors for Urban—industrial water reuse: a multi-criteria analysis case study. *Water*, **14**, 1314.

Jabari, L., Salem, A.A., Zamani, O. et al. (2024). Economic sanctions and energy efficiency: evidence from Iranian industrial sub-sectors. *Energy Economics*, **139**, 107920. https://doi.org/10.1016/j.eneco.2024.107920

Jain, R., Nigam, H., Mathur, M. et al. (2021). Towards green thermal power plants with blowdown water reuse and simultaneous biogenic nanostructures recovery from waste. *Resources, Conservation and Recycling*, **168**, 105283. https://doi.org/10.1016/j.resconrec.2020.105283

Jemghili, R., Ait Taleb, A., and Khalifa, M. (2021). A bibliometric indicators analysis of additive manufacturing research trends from 2010 to 2020. *Rapid Prototyping Journal*, **27**, 1432–1454.

Jiejiao, L. and Qin, S. (2021). A review and outlook of the centennial of China's industrialization: the CPC's industrialization strategy. *China Economist*, **16**, 2–31.

Kato, S. and Kansha, Y. (2024). Comprehensive review of industrial wastewater treatment techniques. *Environmental Science and Pollution Research*, **31**, 51064–51097. https://doi.org/10.1007/s11356-024-34584-0

Kelishomi, A.M. and Nisticò, R. (2024). Economic sanctions and informal employment. *Labour Economics*, **89**, 102581. https://doi.org/10.1016/j.labeco.2024.102581

Kesari, K.K., Soni, R., Jamal, Q.M.S. et al. (2021). Wastewater treatment and reuse: a review of its applications and health implications. *Water, Air, & Soil Pollution*, **232**, 1–28.

Ketterer, M.E., Lowry, J.H., Simon, J. et al. (2001). Lead isotopic and chalcophile element compositions in the environment near a zinc smelting–secondary zinc recovery facility, Palmerton, Pennsylvania, USA. *Applied Geochemistry*, **16**, 207–229. https://doi.org/10.1016/S0883-2927(00)00029-9

Kokare, S., Oliveira, J.P., and Godina, R. (2023). Life cycle assessment of additive manufacturing processes: a review. *Journal of Manufacturing Systems*, **68**, 536–559. https://doi.org/10.1016/j.jmsy.2023.05.007

Kumar, A. and Thakur, A. (2024). Industrial water conservation by water footprint and sustainable development goals. In: *Current Directions in Water Scarcity Research*, 87–117. Elsevier.

Kumar, A., Thakur, A., and Panesar, P.S. (2023). A review on the industrial wastewater with the efficient treatment techniques. *Chemical Papers*, **77**, 4131–4163.

Kumar, P.S., Prasanth, S.M., Harish, S. et al. (2021). Industrial water footprint: case study on textile industries. In *Water Footprint. Environmental Footprints and Eco-design of Products and Processes* (ed. S.S. Muthu), 35–60. Singapore: Springer. https://doi.org/10.1007/978-981-33-4377-1_2

Laperche, B., Lefebvre, G., and Langlet, D. (2011). Innovation strategies of industrial groups in the global crisis: rationalization and new paths. *Technological Forecasting and Social Change*, **78**, 1319–1331.

Li, M., Yang, J., Liang, S. et al. (2021). Ammonia chloride assisted air-chlorination recovery of tin from pyrometallurgical slag of spent lead-acid battery. *Resources, Conservation and Recycling*, **170**, 105611. https://doi.org/10.1016/j.resconrec.2021.105611

Li, Y., Zhang, Y., Yang, L. et al. (2023a). Decoupling analysis of China's mining industrial development and water usage: based on production-based and consumption-based perspectives. *Journal of Cleaner Production*, **385**, 135668. https://doi.org/10.1016/j.jclepro.2022.135668

Li, H., Yao, J., Sunahara, G. et al. (2023b). Quantifying ecological and human health risks of metal(loid)s pollution from non-ferrous metal mining and smelting activities in Southwest China. *Science of The Total Environment*, **873**, 162364. https://doi.org/10.1016/j.scitotenv.2023.162364

Liu, B., Zhang, L., Wang, W. et al. (2024). Development of a conceptual regional industrial water use efficiency model driven by economic development level. *Journal of Hydrology: Regional Studies*, **55**, 101926. https://doi.org/10.1016/j.ejrh.2024.101926

Liu, K., Yang, G., and Yang, D. (2020). Investigating industrial water-use efficiency in mainland China: an improved SBM-DEA model. *Journal of Environmental Management*, **270**, 110859. https://doi.org/10.1016/j.jenvman.2020.110859

Lozano, S. and Borrego-Marín, M.M. (2024). A metafrontier network DEA approach for water usage efficiency assessment in the light of SDG target 6.4. *Environmental Science & Policy*, **160**, 103857. https://doi.org/10.1016/j.envsci.2024.103857

Maertens, A., Golden, E., and Hartung, T. (2021). Avoiding regrettable substitutions: green toxicology for sustainable chemistry. *ACS Sustainable Chemistry & Engineering*, **9**, 7749–7758. https://doi.org/10.1021/acssuschemeng.0c09435

Malla, F.A., Tamana, M., Rahman, F. et al. (2024). Strategies for the reduction of water footprints. In: *Current Directions in Water Scarcity Research*, 187–201. Elsevier.

Matos, C., Nogueira, P., Magalhães, P. et al. (2023). Water consumption assessment and classification in industrial buildings – the case study of Continental Advanced Antenna, Vila Real, Portugal. *Science of The Total Environment*, **902**, 166445. https://doi.org/10.1016/j.scitotenv.2023.166445

Mazzoni, F., Alvisi, S., Franchini, M. et al. (2023). Exploiting high-resolution data to investigate the characteristics of water consumption at the end-use level: a dutch case study. *Water Resources and Industry*, **29**, 100198. https://doi.org/10.1016/j.wri.2022.100198

Meese, A.F., Kim, D.J., Wu, X. et al. (2021). Opportunities and challenges for industrial water treatment and reuse. *ACS ES&T Engineering*, **2**, 465–488.

Mohan, G., Chapagain, S.K., Fukushi, K. et al. (2021). An extended input–output framework for evaluating industrial sectors and provincial-level water consumption in Indonesia. *Water Resources and Industry*, **25**, 100141. https://doi.org/10.1016/j.wri.2021.100141

Morseletto, P. (2020). Targets for a circular economy. *Resources, Conservation and Recycling*, **153**, 104553. https://doi.org/10.1016/j.resconrec.2019.104553

Murali, S., Krishnan, V.S., Amulya, P.R. et al. (2021). Energy and water consumption pattern in seafood processing industries and its optimization methodologies. *Cleaner Engineering and Technology*, **4**, 100242.

Ostad-Ali-Askari, K. and Eslamian, S. (2021). Water reuse in industry: necessities and possibilities. *American Journal of Engineering and Applied Sciences*, **14**, 94–102. https://doi.org/10.3844/ajeassp.2021.94.102

Qadir, M., Drechsel, P., Jiménez Cisneros, B. et al. (2020). Global and regional potential of wastewater as a water, nutrient and energy source. In: *Natural Resources Forum*, 40–51. Wiley Online Library.

Ren, Y., Liu, X., and Zhu, Y. (2024). Incremental marketization reforms and venture capital strategy adjustments: based on industrial chain innovation development. *Finance Research Letters*, **70**, 106346. https://doi.org/10.1016/j.frl.2024.106346

Saravanan, A., Kumar, P.S., Jeevanantham, S. et al. (2021). Effective water/wastewater treatment methodologies for toxic pollutants removal: processes and applications towards sustainable development. *Chemosphere*, **280**, 130595.

Sharma, K., Nagabhooshanam, N., Dasi, S. et al. (2024). Design and implementation of closed-loop water reuse systems in Urban and industrial settings for maximizing resource recovery and minimizing waste. *Desalination and Water Treatment*, **320**, 100850. https://doi.org/10.1016/j.dwt.2024.100850

Sharma, M., Joshi, S., Prasad, M. et al. (2023). Overcoming barriers to circular economy implementation in the oil & gas industry: environmental and social implications. *Journal of Cleaner Production*, **391**, 136133. https://doi.org/10.1016/j.jclepro.2023.136133

Shrivastava, V., Ali, I., Marjub, M.M. et al. (2022). Wastewater in the food industry: treatment technologies and reuse potential. *Chemosphere*, **293**, 133553. https://doi.org/10.1016/j.chemosphere.2022.133553

Silva, J.A. (2023). Wastewater treatment and reuse for sustainable water resources management: a systematic literature review. *Sustainability*, **15**, 10940.

Soni, A., Das, P.K., Hashmi, A.W. et al. (2022). Challenges and opportunities of utilizing municipal solid waste as alternative building materials for sustainable development goals: a review. *Sustainable Chemistry and Pharmacy*, **27**, 100706. https://doi.org/10.1016/j.scp.2022.100706

Tariq, A. and Mushtaq, A. (2023). Untreated wastewater reasons and causes: a review of most affected areas and cities. *International Journal of Chemical and Biochemical Sciences*, **23**, 121–143.

Tassinari, M. (2019). Introduction. In: *BT – Capitalising Economic Power in the US: Industrial Strategy in the Neoliberal Era* (ed. M. Tassinari), 1–16. Cham: Springer International Publishing. https://doi.org/10.1007/978-3-319-76648-5_1

Thanigaivel, S., Priya, A.K., Gnanasekaran, L. et al. (2022). Sustainable applicability and environmental impact of wastewater treatment by emerging nanobiotechnological approach: future strategy for efficient removal of contaminants and water purification. *Sustainable Energy Technologies and Assessments*, **53**, 102484.

Tiu, B.T.C. and Cruz, D.E. (2017). An MILP model for optimizing water exchanges in eco-industrial parks considering water quality. *Resources, Conservation and Recycling*, **119**, 89–96. https://doi.org/10.1016/j.resconrec.2016.06.005

Tong, Y., Zhang, Q., Cai, J. et al. (2018). Water consumption and wastewater discharge in China's steel industry. *Ironmaking & Steelmaking*, **45**, 868–877.

United Nations (2023). Department of Economic and Social Affairs. SDGs Goals. https://sdgs.un.org/goals (accessed 25 August 2025).

Wang, H., Han, L., Li, T. et al. (2023). Temporal-spatial distributions of road silt loadings and fugitive road dust emissions in Beijing from 2019 to 2020. *Journal of Environmental Sciences*, **132**, 56–70. https://doi.org/10.1016/j.jes.2022.07.007

Wang, P., Li, H., Wang, L. et al. (2024). The impact of teleconnections of built-up land on regional carbon burden under the shared socio-economic pathways. *Science of the Total Environment*, **906**, 167589.

Weerasooriya, R.R., Liyanage, L.P.K., Rathnappriya, R.H.K. et al. (2021). Industrial water conservation by water footprint and sustainable development goals: a review. *Environment, Development and Sustainability*, **23**, 12661–12709.

WWAP (United Nations World Water Assessment Programme) (2017). *The United Nations World Water Development Report 2017. Wastewater: The Untapped Resource*. Paris, UNESCO.

Xi, J. (2019). *Xi Jinping: The Governance of China*. Foreign Languages Press.

Yang, L., Yang, Y., Lv, H. et al. (2020). Water usage for energy production and supply in China: decoupled from industrial growth? *Science of The Total Environment*, **719**, 137278. https://doi.org/10.1016/j.scitotenv.2020.137278

Yu, D. and Ding, T. (2021). Assessment on the flow and vulnerability of water footprint network of Beijing city, China. *Journal of Cleaner Production*, **293**, 126126. https://doi.org/10.1016/j.jclepro.2021.126126

Yu, H., Zahidi, I., and Liang, D. (2023). Spatiotemporal variation of vegetation cover in mining areas of Dexing City, China. *Environmental Research*, **225**, 115634. https://doi.org/10.1016/j.envres.2023.115634

Zaman, M., Shahid, S.A., and Heng, L. (2018). *Guideline for Salinity Assessment, Mitigation and Adaptation Using Nuclear and Related Techniques*. Springer Nature.

Zhang, R., and Fu, Y. (2022). Technological progress effects on energy efficiency from the perspective of technological innovation and technology introduction: an empirical study of Guangdong, China. *Energy Reports*, **8**, 425–437. https://doi.org/10.1016/j.egyr.2021.11.282

Zhang, Y., McDermott, S., Davis, B. et al. (2020). High incidence of brain and other nervous system cancer identified in two mining counties, 2001–2015. *Spatial and Spatio-temporal Epidemiology*, **32**, 100320. https://doi.org/10.1016/j.sste.2019.100320

Zhang, Y., Yu, L., Cui, K. et al. (2023). Carbon capture and storage technology by steel-making slags: recent progress and future challenges. *Chemical Engineering Journal*, **455**, 140552. https://doi.org/10.1016/j.cej.2022.140552

10

Case Study for Wastewater Treatment

Nabila Shehata[1]* *and Rania Osama*[2]

[1]*Environmental Science and Industrial Development Department, Faculty of Postgraduate Studies for Advanced Sciences (PSAS), Beni-Suef University, Beni-Suef, Egypt*
[2]*Department of Civil Engineering, Faculty of Engineering, Minia University, Minia, Egypt*

*Corresponding author: nabila.shehata@psas.bsu.ed.eg

10.1 Introduction

Water contamination is one of the most severe problems of the 21st century, causing the deaths of 780,000 people annually due to lack of good access to potable drinking water and poor conditions of sanitation. Water consumption has been growing at a rate of 1% annually since 1980 and will still continue to do so up to the year 2050. Residential, agricultural, and industrial uses are all rather demanding on water supplies. Treating wastewater could be the solution to the water crisis and ensure safe drinking water for humanity (Almukdad et al., 2021). Industrial wastewater entering the food chain and damaging the ecosystem highlights the importance of effective treatment for a sustainable environment.

Large milk production all over the world is attributed to the contribution of more wastewater due to this industry (Pathak et al., 2016). The EU-27 countries produce 1.925×10^8 m^3/y of dairy wastewater. The cheeses, acidified milk, drinking milk, and butterfat products form 45, 18, 19, and 13%, respectively of this wastewater. More than 70% of this wastewater is discharged from six countries: the Netherlands, Germany, Italy, Spain, Poland, and France (Stasinakis et al., 2022). All aspects of the dairy industry, including milk production products, their packaging, storage, effective marketing, and distribution, affect the environment (Kushwaha et al., 2010). Dairy industry wastewater poses challenges due to high organic content: carbohydrates, proteins, grease/oil, and fluctuations in temperature, pH, phosphate, and nitrate levels. Special techniques are necessary for effective contamination management (Ahmaruzzaman and Gupta, 2011).

There is increasing interest in reusing or recycling industrial wastewater to achieve the principles of circular economy and sustainable development. Dairy wastewater doesn't contain toxic or harmful compounds; instead, it contains lactose, proteins, minerals, as well as fats (Mukhopadhyay et al., 2003). To meet the effluent standards, different treatment methods have been applied on-site such as segregation of flows for recovery (e.g. recovery of salt from saline wastewater) and reuse (whey from wastewater), neutralization and equalization of flow, dissolved air flotation to remove fat, and aerobic or/and anaerobic activated sludge to manage the organic (Stasinakis et al., 2022). Moreover, adsorption, membrane-based technologies, advanced oxidation, and flocculation/coagulation have been implemented to manage the wastewater. However, the treatment of industrial wastewater impacts the economy of the associated plant. Moreover, sludge discharged from dairy plants has environmental and economic concerns. Hence, proposing new hybrid technologies for water, materials, and energy recovery from dairy wastewater is strongly proposed to minimize the overall cost of the dairy plant and achieve the principles of circular economy, economic growth, and sustainable development. So, it is important to consider the implementation of the 10Rs of sustainability in the planning and design phases of any dairy plant.

The main aims of this chapter are to highlight the structural information required for scaling up as well as the economics behind construction, operation, and maintenance of wastewater treatment plant (WWTP). The treatment of dairy wastewater was used as a case study owing to the challenges associated with its treatment.

10.2 Structural Information of the Wastewater Treatment Plant

WWTPs should be built to meet effluent and biosolids legislative requirements and reuse goals with appropriate simplicity and expense. The design must face the challenges associated with wastewater with different organic load and flow rates as well as continuous changes in the legislative laws. Effective design and planning of a WWTP lie in identifying the physicochemical properties of the raw wastewater and effluent requirement, proposing different alternative treatment processes, selecting the operating processes, and developing a detailed design of plant operation and maintenance (O&M).

Hence, the structural information must cover the plant planning, treatment route selection with justification, mass- and heat-balance calculations, and size and description of the different equipment including but not limited to settlers, reactors, and filters. Scientists, analysts, and engineers should use concepts from a variety of disciplines, including chemistry, engineering, microbiology, nanotechnology, architecture, economics, and geology, to achieve the optimum WWTPs (Qasim, 2017). However, the design of WWTP includes basic requirements as follows:

- Environmental impact assessment (EIA) and life cycle assessment (LCA)
  EIAs often address neighborhood disruptions, technical concerns, and natural hazards. EIA should be integrated with LCA since different findings were drawn from EIAs with and without LCA input (Larrey-Lassalle et al., 2017). EIA and LCA are two important factors in the design phase. For example, four EIA steps—the environmental comparison of alternatives, the identification of important impacts, the impact assessment, and the impact of mitigation measures—were recognized as potentially benefiting from the application of LCA. Investigating environmental concerns related to the WWTP building phase is crucial, as is examining the functions of various materials in light of their effects on the environment. Transportation and materials for civil works were involved in the two WWTPs' thorough inventory. All effect categories were measured using two methodologies: ReCiPe LCA & EPD 2018. For the assessment, five treatment procedures were taken into consideration: pumping, preliminary treatment (usually physical treatment like sedimentation and filtration), secondary (the main process and usually applied biological-based process) treatment, construction landscape, and sludge line. Concrete and reinforcing steel were found to have equally important roles in the majority of the EPD 2018 effects. More than 90% of the effects were attributed to the considerable effect of reinforcing steel on cancer. Diesel had a 5% greater impact on developing ozone than reinforcing steel. The percentage of glass fiber that contributed to ozone depletion was 70%, which was significantly greater than the combined percentage of reinforcing steel and concrete. Due primarily to the percentage of materials used, primary treatment units only made up 9.5% of the construction effects at one WWTP but up to 43.8% in another WWTP. To put it briefly, thorough data inventories were required in order to assess WWTP's overall environmental effects (Nguyen et al., 2020).
- Influents/effluents characteristics
  For municipal wastewater, identifying the mean values of biochemical oxygen demand (BOD), chemical oxygen demand (COD), total solids (volatile and suspended), heavy metals, and other micropollutants are the main parameters in the raw and treated wastewaters. Additionally, the sources and the physicochemical properties of the main pollutants in the influent wastewater should be identified for industrial wastewater. Additionally, the effluent standards should be set according to the local and international requirement, and environmental standards.

- Identification of unit process and unit operations
  The optimum treatment process (organic oxidation, nitrification/denitrification, adsorption, photocatalytic degradation, ozonation, and other processes) is identified based on bench- and pilot-scale studies. A variety of recent technologies are available that have been shown to quickly and efficiently combine with current treatment procedures. Also, the actual size of the units (reactors, clarifiers, filters) should be identified.
- Sewer segregation efforts
  With an emphasis on the recovery of water, nutrients, and energy, the idea of wastewater treatment is developing from the standpoint of the circular economy. Decentralized wastewater systems may be a suitable option in this situation since they reduce the expenses associated with pumping wastewater and relieve strain on treatment facilities that may be operating at large capacity. However, the management scheme's potential for resource recovery is considerably easier because nearby regions might be fertilized and irrigated using recycled water or biofertilizer flows. The environmental profile of P-recovery technologies used in decentralized and hybrid treatment processes was studied at various scales using the LCA technique. Sludge recovery as fertilizer, wastewater segregation, and reduced energy use all greatly raise the environmental score. The technologies used, however, require particular consideration (Estévez et al., 2023).
- Energy conservation and management
  Energy consumption expenditures represent 20–40% of operational expenses, whereas electrical consumption is 60–90% of total energy consumption in WWTPs. Numerous Supervisory Control and Data Acquisition (SCADA) systems have analytical functionalities aimed at recording the operational and financial consequences of certain energy consumption choices. The ability to combine tactical energy-saving plans and SCADA data within the context of wider strategic data tools that consider other energy-use factors might pave the way for step-change behavior that incorporates water sector energy-optimization techniques. Even yet, the basic cost of energy may not be the sole concern. For example, minimizing the carbon footprint may have an equally major influence on which energy choices are selected. However, there are different sources for green energy that can be applied as an economic and eco-friendly resource for energy in WWTPs such as solar radiation, wind power, and tidal energy in plants near the outfall (Ghoneim et al., 2016).
- Energy recovery
  There are a lot of resources and energy in wastewater and sludge, which should be handled responsibly. Energy and nutrients are viable recycling components. The materials/energy recovery technologies are chemical energy, thermal energy, and biomass and its derivatives. Waste heat from

wastewater is the primary source of heat energy, including hydropower of influent or effluent wastewater, electrical power or heat (from digester biogas or biosolids), utilization of influent or effluent wastewater as a heat or cooling source, and biofuel energy from microbial fuel cell (MFC) (Ghoneim et al., 2016). Heat energy from chemical reactions or their by-products is the source of chemical energy and may be converted into electrical power. Biomass is required as the reaction's carrier for biomass energy. Biofuel from microalgae can store a lot of biomass energy since they can absorb nutrients and store carbon and sun energy. Energy self-sufficiency is made feasible by the fact that the majority of recovered energy is utilized in several WWTP operations (Sun et al., 2019).

- Materials recovery
  The valuable materials to be recovered should be identified, and the required unit operation/process (mechanical evaporators, reverse osmosis, adsorption, membrane systems, etc.) should be optimized. In recent years, scientists have developed a diverse set of technical methods for recovering water, electricity, valuable materials, nutrients, and other goods from different WWTPs. The motivations for wastewater research varied from materials/resources recovery and economic effectiveness to the high energy requirements and significant environmental impacts of present WWTP designs. However, some technologies have been adopted, and the transition from WWTPs to water resource facilities appears to be a long time coming. Much more studies are required for exploring different techniques or modifying the current one and challenges associated with the design of new WWTPs or redesigning the current one from a materials recovery standpoint. Depending on the resource, the supply potential might vary substantially. Second, materials recovery methods explored in literature is thoroughly and deeply investigated. Some nontechnical constraints have been noted in the research that must be addressed before these technologies can be successfully implemented into the design of the WWTPs. The obstacles are connected to value chain growth and economics, the human health and ecosystems, as well as social and policy challenges. Taking market potential, technical breakthroughs, and possible bottlenecks into account in the pre-design stage may help to ease the planning and combining of WWTPs and strengthen the practices of circular economy in wastewater treatment (Kehrein et al., 2020).
- Circular economy
  The implementation of the 10Rs of sustainability in WWTPs should be applied as follows:

  Refuse (R1): Rejection of any harmful process and suggestion of green wastewater technologies. It is also suggested that, if feasible, the chemical compounds functioning as pollutants may be replaced with ecologically friendly substances that are bioaccumulative, not or less persistent, and less hazardous to the environment.

Rethink (R2): This principle can be applied by: (i) the introduction and integration of relevant policies, (ii) implementation of accurate control on wastewater, (iii) monitoring of source/pollutant pathways at the international, national, and local levels, (iv) development of applicable and economic technologies for pollutants management, and (v) establishment of new infrastructure under successful financial programs.

Reduce (R3): Minimize the consumption of water inside the facility along with the chemical used in the WWTPs. Identifying the origins of pollutants (imports and production), manufacturers are required to take responsibility for avoiding or minimizing pollutant discharge to the aquatic bodies. As a result, the environmental concerns associated with the pollutants leakage to aquatic ecosystems should be investigated and properly reported, such as through correct labeling (Kosek et al., 2020).

Reuse (R4): Reusing the wastewater after treatment in other applications with special focus in irrigation of non-edible crops.

Repair (R5): Maintain the wastewater after discharge from further pollution that may affect humans or the environment.

Refurbish (R6): Restoration of wastewater before discharging to the water bodies or application of integrated management of neighboring WWTPs. For example, a study conducted in 2015 shows that building an additional pipeline to connect nearby wastewater systems yields favorable outcomes in the economic assessment. It meets the environmental criteria and yields cost savings of $47 491/year (Morera et al., 2015).

Remanufacture (R7): The possibility to utilize the treated wastewater in another plant/facility.

Repurpose (R8): The treated wastewater could be used inside the plant for another purpose like heating, cooling, pre-washing, etc.

Recycle (R9): Utilize the wastewater after the appropriate treatment in the same process with the same or at a lower efficiency without impacting the overall performance of the plant.

Recover (R10): Design and optimize the processes to recover as much valuable materials and energy as possible.

- Other factors

  There are other structural information required for the WWTPs such as: (i) load design or number of person equivalent (population size), (ii) past and expected future water consumption, (iii) area, location, and topography of WWTP land, and (iv) evaluation of company expansion plans.

10.3 Economics Behind Construction, Maintenance, and Working of WWTP

The primary objective of WWTP design, construction, and operation is to guarantee effluent quality by fulfilling the legally mandated parameter values. However, WWTPs are of high cost owing to expensive construction and high equipment costs, as well as other expenses associated with regular operation, monitoring, and maintenance. Even in wealthy nations, it develops social and economic strain. As a result, engineers, water utility companies, and plant operation personnel are searching for innovative, economical, and ecologically friendly solutions for suitable technology and equipment. While maintenance activities make sure that equipment operates regularly and efficiently to reach sustainable operational goals, operational activities make sure that WWTP provides the required amount and quality of treated water and satisfies requirements. For instance, a modest, straightforward treatment plant with cheap capital costs can have high operating costs, which would raise the overall cost as compared to a different technology. Following the testing phase, when the water utility business and the local community must collect adequate cash to cover high running expenses, WWTPs in undeveloped and emerging nations typically face operational and maintenance issues. That is the first step in reducing the expenses and procedures associated with O&M. The repercussions are frequently severe, have a big impact on the rise in expenses for fixing the damage, and need a lot of work to get a WWTP back to regular functioning (Serdarevic and Dzubur, 2019).

With expanding decarbonization efforts to implement more integrated waste management, there is an increasing need to re-manage waste-related infrastructures in urban contexts. Wastewater treatment facilities are important contributors to increased environmental services but related technological decisions and economic considerations must be considered from a systems approach (Tsui et al., 2022). The comparison of cost categories reveals that the O&M costs greatly outweigh the initial cost. Moreover, the operating cost is dependent on the treatment technology employed because every technology has different capital cost (CC) and operations. For example, the life cycle costing (LCC) of a WTTP at Nashik was carried out for the lifespan of 30 years. The key cost driver in the LCC of WTTP is the O&M cost (72.52% of overall life cycle expenditures). Initial construction costs are the second largest contributor to overall life cycle cost, accounting for 22.17% after O&M costs. The replacement cost accounts for 4.6% of the entire life cycle cost, whereas energy costs make for 1.183%. LCC analysis identifies cost drivers across the life cycle of WTTPs, reducing overall costs and improving economic efficiency. To minimize overall LCC, actions may be implemented to lower the cost of O&M, which accounts for 72.52%. To lower the overall LCC, construction costs should be kept as low as possible. Municipal corporations can utilize the LCC analysis to identify the project's budgetary needs ahead of time and obtain financing for them. This estimate outlines projected expenditures for equipment replacement, O&M, and power, and helps with good scheduling and budget management. LCC analysis can help manage completed WTTPs, choose between competing alternatives, aid in planning, contracting, and budgeting, evaluate the impact of design approaches and support policies, and identify cost drivers for asset support systems (Ambre et al., 2016).

The economic analysis of capital and O&M of WWTPs can optimize the overall performance and cost. For example, an economic analysis of CC and operational cost (OC) for a traditional seawater reverse osmosis (RO), low-pressure reverse osmosis/forward osmosis (LPRO/FO), and an advanced oxidation/reverse osmosis/membrane bioreactor (ARM) for wastewater treatment and reuse. The LPRO/FO system reveals the most relevant variables influencing economic feasibility. The LCCs consider water quality (same feed and generated water and reuse has been conducted), production of 18345.2 gallons/min, materials/chemicals, energy consumption, operation, expenses of modules (FO and RO), and maintenance. Compared to LPRO/FO, the RO systems have a lower CC (21%) and a higher OC (56%) due to fouling control as well as energy conservation. The RO system has a 16% higher cost (cost of water produced) compared to LPRO/FO. Compared to the LPRO/FO, the ARM systems have a higher CC (7%) and lower OC (9%). However, hybrid LPRO/FO systems are economic in terms of water treatment and recovery systems compared to the traditional desalination technologies. Moreover, the total cost of water produced can be achieved via enhancement of water permeability, packing density, and FO membrane modules (Linares et al., 2016). LCC analysis for different hybrid systems in wastewater treatment has been reported (Ilyas et al., 2021). For example, the economic/environmental performance of two small-scale and decentralized WWTPs that are connected to wetlands was assessed. The systems were modeled with a 20-year lifespan in open LCA software using Ecoinvent 3.3 data, and the effect assessment was relied on the ReCIPe technique. Harmful emissions, like nutrients discharged in effluent and greenhouse gas emissions from the septic tank, have been linked to possible implications on climate change, photochemical oxidants, and freshwater eutrophication. Artificial aeration minimizes the area needed for system construction, and power usage accounted for just 7% of the overall climate change effect. However, the environmental performance of the WWTP is promoted using eco-friendly technologies, which are characterized by low airborne and liquid emissions. The examination of the aerobic wetland revealed that the operation period has the most significant effect on the environment, in terms of human toxicity (64%) and freshwater eutrophication (100%). The aerated system had an LCC of treated sewage that was 1.8 times lower than the non-aerated system (Resende et al., 2019). Recent technologies such as microbial electrolysis (MEC), MFCs, and membrane bioreactors (MBRs) were compared to traditional techniques (TT). To evaluate the environmental and economic impacts, economic assessments of total cost (both operating and capital costs) and LCA are used. The assessment of the environmental effect shows that MFC/A_2O and MBR/A_2O hybrid systems are the most eco-friendly during operation and construction periods, respectively. Additionally, the TT/MFC and TT systems have, respectively, the lowest operational and capital costs. TT/MFC has the highest economic impact, including both operational and capital costs over a 20-year period (Rashidi et al., 2018). Moreover, an economic evaluation was conducted using the LCC approach. Hazardous heavy metal emissions, emissions from electricity, and treated effluent are the main contributors to the overall environmental effect. Soil biotechnology (SB) had the lowest environmental effect of any WWTP under consideration, with the exception of the eutrophication effect. Owing to the excessive consumption of chemicals and electricity, aerated lagoon (AL) system produced the poorest performance. Furthermore, as the rate of effluent reuse increases, the toxicity potential decreases (Kamble et al., 2019). Also, LCC theory was used to assess the environmental impact and cost of consumption of various aeration systems (surface aeration and blasting aeration). The research findings revealed that in the 20-year life cycle, the LCC was primarily determined by the cost of energy consumption, regardless of surface aeration or blasting aeration system, whereas the LCC of the latter was influenced by economic losses due to wastewater loss, maintenance cost, and environmental load during the maintenance duration. Generally, the LCC of the surface aeration was higher than that of the blasting aeration system, based on the standard discharge, and the blasting aeration system was estimated to save US$ 8295.89/year in costs when compared to surface aeration (Liai et al., 2017).

Systems analyses and modeling of wastewater treatment process were integrated to determine the parameters under which anaerobic membrane bioreactor (AnMBR) could have lower life cycle environmental emissions and generate more net energy than high rate activated sludge (HRAS)/anaerobic digestion (AD), traditional activated sludge (TAS)/AD, and an AnMBR/AD. Under baseline assumptions of 15 °C for domestic wastewater treatment with medium strength, AnMBR recovered higher energy than HRAS@AD (by 49%). The global warming effect of AnMBR was primarily caused by effluent dissolved methane emissions. AnMBR recovered higher net energy than HRAS@AD (by 15%) for treating domestic wastewater with high strength. Still much more studies are required to increase the flux, control fouling, and lower methane emissions to enhance the performance of AnMBR technology as much as HRAS@AD (Smith et al., 2014).

10.4 Treatment Capacity (Efficiency) and Treatment Methods Used in the WWTP for the Dairy Industry; A Case Study

All aspects of the dairy industry, including milk production products, their storage, packaging, distribution, and marketing, affect the environment (Kushwaha et al., 2010). Dairy industry wastewaters have intermittent flow rates and exhibit high seasonal variations correlated with milk processing volume, which is highest in summer and lowest in winter (Demirel et al., 2005). The properties of dairy industry effluents considerably varied according to the operational system used, due to the production of different dairy products (Vidal et al., 2000). Alkaline and acid cleaners in the dairy industry affect wastewater and yield variation in pH levels. Unfortunately, the data on mean properties of dairy wastewater from large-scale plants is limited. However, a comprehensive study found extensive data on specific properties of dairy wastewater from different large-scale processes (Demirel et al., 2005). Dairy industry wastewater poses challenges due to high organic content (carbohydrates, proteins, and grease/oil) and fluctuations in temperature, pH, phosphate, and nitrate levels. However, the produced wastewater must be properly managed at the level of the dairy industry.

Special techniques are necessary for effective contamination management. Numerous methods, such as phytoremediation, electrocoagulation (EC), anaerobic or/and aerobic processes, water hyacinth, electrochemical, and reverse osmosis processes, can treat wastewater and remove toxicity. Natural coagulants like activated charcoal, rice husk, and rice husk ash are also effective in treating dairy wastewater (Ahmaruzzaman and Gupta, 2011). Chemical coagulants, like alum, PAC, $Al_2(SO_4)_3$, and $FeCl_3$, can be used to treat dairy wastewater with contaminants parameters including COD, BOD, temperature, pH, chlorides, alkalinity, hardness, and nitrates. Elevated concentrations of COD in dairy industry wastewater indicate fluctuating and strong characteristics. The nutrients and organic components in dairy wastewater come mainly from milk and its products, where nitrogen primarily comes from milk proteins, existing as organic nitrogen or ions like NO_2^-, NO_3^-, and NH_4^+. Phosphorus is usually present in inorganic forms like polyphosphate ($P_2O_7^{4-}$) and orthophosphate (PO_4^{3-}) along with some organic forms; high Na concentrations indicate the use of excessive alkaline cleaners at dairy plants. The levels of Cu, Ni, and Zn heavy metals fall within a safe range for biological treatment (Demirel et al., 2005). Dairy wastewater is typically comprised of easily biodegradable carbohydrates such as lactose, with some lipids and proteins. Lactose, the basic carbohydrate in this kind of wastewater, can be used by anaerobic bacteria for methanation, requiring collaboration between acidogens, acetogens, and methanogens (Yu and Pinder, 1993).

Elevated concentrations of carbohydrates in the synthetic dairy wastewater decrease proteolytic enzyme synthesis, leading to reduced protein degradation. Lipids can hinder anaerobic treatment of dairy wastewaters. In degradation, lipids are hydrolyzed to long-chain fatty acids and glycerol, which then undergo b-oxidation to produce hydrogen and acetate. Lipid biodegradation is challenging due to limited bioavailability. Glycerol, from lipid hydrolysis, was non-inhibitory; Long-chain fatty acids (LCFAs) inhibit methanogenic bacteria, as reported. Also, lipids, especially LCFAs, hinder methane production in anaerobic processes by acting as inhibitors (Demirel et al., 2005). Numerous studies have been conducted on the management of dairy wastewater as well as the recovery of some valuable materials from the discharged stream as follows.

10.4.1 Use of Electrocoagulation Process

The effectiveness of EC, an electrolytic treatment, on dairy effluents was assessed. Artificial wastewater made from milk powder solutions was treated using a soluble aluminum anode. Filtration was used to separate the resulting flocks. The results indicated reductions in COD (61%), turbidity (100%), nitrogen (81%), and phosphorus (89%). Interestingly, lactose was not eliminated in the treatment of lactose and phosphate solutions, which may explain the relatively low decrease in COD. Comparing the efficiency of EC with chemical coagulation using aluminum sulfate, both methods showed similar effectiveness. However, the EC technique resulted in lower conductivity and a neutral pH in the treated wastewaters, which suggests the possibility of recycling the water for industrial purposes. Additionally, EC requires fewer reagents compared to chemical coagulation, making it a superior technique. Increasing the COD initial concentration from 1.55 to 19.8 g/L led to a 10% increase in removal efficiency (from 88 to 98%) (Tchamango et al., 2010). EC was also integrated with Fenton reaction and ozone processing for the management of heavily loaded wastewater generated from ice cream manufacturing. The impacts of various operating conditions, like reaction time, applied current density, ozone, iron (Fe^{2+}), and hydrogen peroxide (H_2O_2) dosage, on wastewater management have been studied. At a current density of 5 mA/cm^2, EC achieved a COD removal of 40%. The introduction of the Fenton reaction increased treatment efficiency by approximately 25% with a H_2O_2/Fe^{2+} ratio of 1:1. When combined with ozone, an extra COD removal (30%) was achieved. The findings reveal the effectiveness of the hybrid system for removing contaminants from the industrial wastewater (Tchamango et al., 2010). Another study investigated the efficiencies of stainless steel, Fe, and Al anodes. It was reported that Fe anodes are the most effective in treating mixed wastewaters from a dairy plant, reducing COD in the discharged wastewater from 1.605 to 0.285 g/L, with an equivalent efficiency of 82% (Ghahremani et al., 2012). Fe anodes were also used in batch mode to remove COD. Optimizing the treatment time (60 minutes), pH (4.5), and current density (~36 A/m^2) resulted in the removal of >90% of COD and turbidity and an increment in the effluent pH to 7.5 (Geraldino et al., 2015). Moreover, EC was integrated with adsorption onto natural zeolite (ECA). The response surface methodology (RSM) was used to analyze the process by investigating various operating parameters like electrolysis time, initial pH, zeolite dosage, and current density on the treatment of industrial dairy wastewater. The experimental design comprised 30 runs based on the Central Composite Design (CCD). Key parameters measured included COD removal efficiency, turbidity removal, and energy consumption. Data analysis resulted in regression models showing a high determination coefficient (R^2), showing a robust correlation between predicted and experimental data. Optimizing the operational conditions (zeolite dosage of 183 g/L, electrolysis time of 20 minutes, pH 7, and current density of 38 A/m^2) resulting in achieving as high as 92% COD and 97% turbidity. Comparison with a design involving EC alone showed that the introduction of natural zeolite didn't improve treatment performance in terms of turbidity and COD removal from dairy wastewater (Faraj et al., 2024). In contrast, EC, electro-oxidation, and hybrid processes were integrated to manage wastewater from the dairy industry. The wastewater was characterized based on turbidity, color, COD, and BOD. Comparison of the three techniques revealed that the hybrid process achieved approximately 92–93% color change, 98–99% turbidity removal, and 66–67% COD removal efficiency (Faraj et al., 2024).

10.4.2 Sequencing Batch Reactor

Two aerobic sequencing batch reactors (SBRs) have been studied to manage dairy wastewater discharge and landfill leachate. The reactors operated in a 24-hour cycle with different periods: aerobic filling, aerobic treatment, anoxic treatment, settling, discharging, and idle. The goal was to determine the optimal operational strategy for the removal of COD and nitrogen-based nutrients. The study proved that combined wastewater treatment is applicable, but efficiency depends on factors like phase time, organic loading, and hydraulic retention time (HRT). Scenario I, with 19 hours of aeration and a 2-hour anoxic phase, was found to be the most effective. High organic loading or decreased HRT decreased removal efficiencies (Mohan et al., 2007). Electro Fenton (EF) process (primary stage) was integrated with SBR (main treatment) for treating wastewater from the dairy industry. Optimization of the key factors such as sludge retention time (SRT), HRT, and sludge volume index (SVI) resulting in effectively breaking down difficult-to-degrade compounds through advanced oxidation processes. Laboratory-scale experiments conducted for one year under stable conditions showed impressive removal rates: COD (99%), total phosphorus (95%), and total nitrogen (97%). The hybrid system both outperforms traditional biological treatment and is cost-effective (Heidari et al., 2021). Moreover, the SBR was compared with conventional activated sludge. For example, four bioreactors at pilot scale have been investigated under different temperatures (20, 35, and 45 °C) using synthetic dairy wastewater with specific characteristics. It was reported that the optimum temperature for treatment was 35 °C, with NH_3-N and COD removal efficiency ranging from 89.01% to 93.22% and from 93.52% to 97.79%, across the systems, respectively. The NO_3-N concentration in effluent wastewater varied from 7.56 mg/L to 14.12 mg/L. At 45 °C, oxygen consumption was the greatest, indicating increased microbial activity. Additionally, the sludge settleability improved at lower temperatures. Overall, SBR has been effectively treating dairy wastewater under the optimum conditions, outperforming other treatment systems (Khalaf et al., 2021).

10.4.3 Electrochemical Treatment

There is a growing interest in recovering energy from waste, particularly in the field of green energy production. MFC is a biological–electrochemical hybrid system where the chemical energy of the bio-convertible waste is converted into electrical energy. By using a series of bioelectrochemical reactions, MFCs provide a means of treating waste while also generating energy and reducing excess sludge production. This technology shows promising potential in the research community (Mohan et al., 2010). For example, a non-catalyzed MFC was used with anaerobic mixed consortia as a biocatalyst to treat dairy wastewater and generate power. The MFC demonstrated effective treatment of the wastewater, with COD removal (95.49%), turbidity (99.02%), proteins (78.07%), and carbohydrates (91.98%). As the substrate load increased, the MFC's performance improved, reaching a maximum power production of 1.10 W/m^3 at a substrate load of 4.44 $kgCOD/m^3$. The study evaluated the MFC's power generation potential through sustainable power estimation, cyclic voltammetric analysis, cell potentials, and polarization behavior. This integrated approach shows promise as an alternative to conventional wastewater management routes (Mohan et al., 2010). Also, electrochemical oxidation of aerobically pre-treated dairy wastewaters was explored using IrO_2-Pt coated dimensionally stable anodes. The study revealed that the IrO_2/Ti electrode performed better than Ti@IrO_2-Pt and Ti@Pt at lower current densities, while Ti@Pt showed higher efficiency at a higher current density. The current density was identified as the most significant parameter for process efficiency. With a current density of 0.1 A/cm^2 and the Ti@IrO_2 electrode combined with 0.2 M of NaCl, almost complete removal of 3.7 g/L COD occurred after six hours, while decolorization was reached in less than one hour. The choice of electrolyte also significantly affected the process, with the use of sodium chloride yielding improved COD and color removal. Temperature had a minimal impact on the process, while increasing COD loading led to decreased removal of COD and color (Kushwaha et al., 2010). Also, EF process with an iron electrode was investigated. The study examined the effects of various factors on the process's performance, such as current density, reaction time, pH, and H_2O_2/dairy wastewater and H_2O_2/Fe^{2+} molar ratios. RSM was employed to determine the optimal operating conditions with the fewest experiments. Five variables were carefully optimized: reaction time of 90 minutes, current density 56 mA/cm^2, pH 7.52, H_2O_2/DW volumetric ratio of 0.898 mL/L, and H_2O_2/Fe^{2+} molar ratio of 3.965, resulting in COD removal of 93.93%. Additionally, optimizing current density (55.1 mA/cm^2), reaction time (86 minutes), H_2O_2/Fe^{2+} molar ratio (3.987), H_2O_2/DW volumetric ratio (0.907 mL/L), and pH (7.48) achieved color removal as high as 97.32% (Markou et al., 2017). Also, a full factorial CCD was employed to investigate the effectiveness of the electrochemical method with iron electrode to treat simulated dairy wastewater involving factors such as current density, electrolysis time, pH, and NaCl dosage. Each factor had five levels to optimize COD removal. The optimal values for time, pH, and current density were determined to be 50 minutes, 7.0, and 270 A/m^2, respectively. The highest COD removal percentage achieved was approximately 70%. Analysis of residues (scum and sludge) and iron electrodes was conducted to understand the electrochemical mechanism. The electrochemical process appeared to involve electro-oxidation, electro-floatation, and electrocoagulation for COD removal (Davarnejad and Nikseresht, 2016).

10.5 Summary

In this chapter, the dairy wastewater has been discussed as a case study, and the non-technical and technical concerns that are most frequently addressed in the design and planning of WWTPs, such as EIA and LCA of WWTPs, characteristics of the influent wastewater and treated water, identification of the unit process and unit operations inside the WWTPs, sewer segregation efforts, energy conservation and management, and materials and energy recovery were addressed. Moreover, the implementation of the 10Rs of sustainability in WWTPs was reported. Finally, various methods for treating dairy wastewater were discussed.

Review Questions

1. How can the 10Rs of circular economy be efficiently applied in the planning and design stage of a dairy plant?
2. Suggest a new hybrid treatment for a typical dairy plant.
3. How can fouling be efficiently controlled in membrane-based technologies?

4 Why is lipid biodegradation challenging?
5 How can we convert linear economy to a circular economy for dairy plant?
6 How can we convert from traditional wastewater treatment technologies to green ones in dairy plants?

Acknowledgments

They replied, "Glory be to You! We have no knowledge except what our God taught us. He is truly the All-Knowing, All-Wise."

References

Ahmaruzzaman, M. and Gupta, V.K. (2011). Rice husk and its ash as low-cost adsorbents in water and wastewater treatment. *Industrial & Engineering Chemistry Research*, **50**, 13589–13613. https://doi.org/10.1021/ie201477c

Almukdad, A., Hafiz, M.A., Yasir, A.T. et al. (2021). Unlocking the application potential of electrocoagulation process through hybrid processes. *Journal of Water Process Engineering*, **40**, 101956. https://doi.org/10.1016/j.jwpe.2021.101956

Davarnejad, R. and Nikseresht, M. (2016). Dairy wastewater treatment using an electrochemical method: experimental and statistical study. *Journal of Electroanalytical Chemistry*, **775**, 364–373. https://doi.org/10.1016/j.jelechem.2016.06.016

Demirel, B., Yenigun, O., and Onay, T.T. (2005). Anaerobic treatment of dairy wastewaters: a review. *Process Biochemistry*, **40**, 2583–2595. https://doi.org/10.1016/j.procbio.2004.12.015

Estévez, S., Moreira, M.T., and Feijoo, G. (2023). Chapter 9 – Life-cycle assessment for resource recovery facilities in the wastewater sector. In: *Current Developments in Biotechnology and Bioengineering* (ed. G. Mannina, A. Pandey, and R. Sirohi), 183–220. Elsevier.

Faraj, H., Jamrah, A., Al-Omari, S. et al. (2024). Optimization of an electrocoagulation-assisted adsorption treatment system for dairy wastewater. *Case Studies in Chemical and Environmental Engineering*, **9**, 100574. https://doi.org/10.1016/j.cscee.2023.100574

Geraldino, H.C.L., Simionato, J.I., De Souza Freitas, T.K.F. et al. (2015). Efficiency and operating cost of electrocoagulation system applied to the treatment of dairy industry wastewater. *Acta Scientiarum Technology*, **37**, 401–408. https://doi.org/10.4025/actascitechnol.v37i3.26452

Ghahremani, H., Bagheri, S., Hassani, S.M. et al. (2012). Treatment of dairy industry wastewater using an electrocoagulation process. *Advances in Environmental Biology*, **6**, 1897–1901.

Ghoneim, W.A.M., Helal, A.A., and Wahab, M.G.A. (2016). Renewable energy resources and recovery opportunities in wastewater treatment plants. In: *2016 3rd International Conference on Renewable Energies for Developing Countries (REDEC)*, 1–8. IEEE.

Heidari, M.R., Malakootian, M., Boczkaj, G. et al. (2021). Evaluation and start-up of an electro-Fenton-sequencing batch reactor for dairy wastewater treatment. *Water Resources and Industry*, **25**, 100149. https://doi.org/10.1016/j.wri.2021.100149

Ilyas, M., Kassa, F.M., and Darun, M.R. (2021). Life cycle cost analysis of wastewater treatment: a systematic review of literature. *Journal of Cleaner Production*, **310**, 127549. https://doi.org/10.1016/j.jclepro.2021.127549

Kamble, S., Singh, A., Kazmi, A. et al. (2019). Environmental and economic performance evaluation of municipal wastewater treatment plants in India: a life cycle approach. *Water Science and Technology*, **79**, 1102–1112.

Kehrein, P., Van Loosdrecht, M., Osseweijer, P. et al. (2020). A critical review of resource recovery from municipal wastewater treatment plants–market supply potentials, technologies and bottlenecks. *Environmental Science: Water Research & Technology*, **6**, 877–910.

Khalaf, A.H., Ibrahim, W.A., Fayed, M. et al. (2021). Comparison between the performance of activated sludge and sequence batch reactor systems for dairy wastewater treatment under different operating conditions. *Alexandria Engineering Journal*, **60**, 1433–1445. https://doi.org/10.1016/j.aej.2020.10.062

Kosek, K., Luczkiewicz, A., Fudala-Książek, S. et al. (2020). Implementation of advanced micropollutants removal technologies in wastewater treatment plants (WWTPs) – examples and challenges based on selected EU countries. *Environmental Science & Policy*, **112**, 213–226. https://doi.org/10.1016/j.envsci.2020.06.011

Kushwaha, J.P., Srivastava, V.C., and Mall, I.D. (2010). Organics removal from dairy wastewater by electrochemical treatment and residue disposal. *Separation and Purification Technology*, **76**, 198–205. https://doi.org/10.1016/j.seppur.2010.10.008

Larrey-Lassalle, P., Catel, L., Roux, P. et al. (2017). An innovative implementation of LCA within the EIA procedure: lessons learned from two wastewater treatment plant case studies. *Environmental Impact Assessment Review*, **63**, 95–106. https://doi.org/10.1016/j.eiar.2016.12.004

Liai, C., Hongxun, H.O.U., Weibiao, F.E.I. et al. (2017). Comparative research to surface aeration and blasting aeration system based on LCC Theory. In: *IOP Conference Series: Earth and Environmental Science*, 12029. IOP Publishing.

Linares, R.V., Li, Z., Yangali-Quintanilla, V. et al. (2016). Life cycle cost of a hybrid forward osmosis – low pressure reverse osmosis system for seawater desalination and wastewater recovery. *Water Research*, **88**, 225–234. https://doi.org/10.1016/j.watres.2015.10.017

Markou, V., Kontogianni, M.C., Frontistis, Z. et al. (2017). Electrochemical treatment of biologically pre-treated dairy wastewater using dimensionally stable anodes. *Journal of Environmental Management*, **202**, 217–224. https://doi.org/10.1016/j.jenvman.2017.07.046

Mohan, S.V., Babu, V.L., and Sarma, P.N. (2007). Anaerobic biohydrogen production from dairy wastewater treatment in sequencing batch reactor (AnSBR): effect of organic loading rate. *Enzyme and Microbial Technology*, **41**, 506–515. https://doi.org/10.1016/j.enzmictec.2007.04.007

Mohan, S.V., Mohanakrishna, G., Velvizhi, G. et al. (2010). Bio-catalyzed electrochemical treatment of real field dairy wastewater with simultaneous power generation. *Biochemical Engineering Journal*, **51**, 32–39. https://doi.org/10.1016/j.bej.2010.04.012

Morera, S., Comas, J., Poch, M. et al. (2015). Connection of neighboring wastewater treatment plants: economic and environmental assessment. *Journal of Cleaner Production*, **90**, 34–42. https://doi.org/10.1016/j.jclepro.2014.11.047

Mukhopadhyay, R., Talukdar, D., Chatterjee, B.P. et al. (2003). Whey processing with chitosan and isolation of lactose. *Process Biochemistry*, **39**, 381–385.

Nguyen, T.K.L., Ngo, H.H., Guo, W. et al. (2020). Contribution of the construction phase to environmental impacts of the wastewater treatment plant. *Science of the Total Environment*, **743**, 140658.

Pathak, U., Das, P., Banerjee, P. et al. (2016). Treatment of wastewater from a dairy industry using rice husk as adsorbent: treatment efficiency, isotherm, thermodynamics, and kinetics modelling. *Journal of Thermodynamics*, **2016**, 3746316. https://doi.org/10.1155/2016/3746316

Qasim, S.R. (2017). *Wastewater Treatment Plants: Planning, Design, and Operation.* Routledge.

Rashidi, J., Rhee, G., Kim, M. et al. (2018). Life cycle and economic assessments of key emerging energy efficient wastewater treatment processes for climate change adaptation. *International Journal of Environmental Research*, **12**, 815–827.

Resende, J.D., Nolasco, M.A., and Pacca, S.A. (2019). Life cycle assessment and costing of wastewater treatment systems coupled to constructed wetlands. *Resources, Conservation and Recycling*, **148**, 170–177. https://doi.org/10.1016/j.resconrec.2019.04.034

Ambre, H.P., Saner, A., Aher, P.D. et al. (2016). Life cycle costing of water treatment plant in the Nashik city. *Journal of water pollution & purification research*, **3**(1), 25–30.

Serdarevic, A. and Dzubur, A. (2019). Importance and practice of operation and maintenance of wastewater treatment plants. In: *Advanced Technologies, Systems, and Applications III: Proceedings of the International Symposium on Innovative and Interdisciplinary Applications of Advanced Technologies (IAT)* (vol. 2, pp. 121–137). Springer.

Smith, A.L., Stadler, L.B., Cao, L. et al. (2014). Navigating wastewater energy recovery strategies: a life cycle comparison of anaerobic membrane bioreactor and conventional treatment systems with anaerobic digestion. *Environmental Science & Technology*, **48**, 5972–5981.

Stasinakis, A.S., Charalambous, P., and Vyrides, I. (2022). Dairy wastewater management in EU: produced amounts, existing legislation, applied treatment processes and future challenges. *Journal of Environmental Management*, **303**, 114152.

Sun, Y., Lu, M., Sun, Y. et al. (2019). Application and evaluation of energy conservation technologies in wastewater treatment plants. *Applied Sciences*, **9**, 4501.

Tchamango, S., Nanseu-Njiki, C.P., Ngameni, E. et al. (2010). Treatment of dairy effluents by electrocoagulation using aluminium electrodes. *Science of the Total Environment*, **408**, 947–952. https://doi.org/10.1016/j.scitotenv.2009.10.026

Tsui, T.-H., Zhang, L., Zhang, J. et al. (2022). Methodological framework for wastewater treatment plants delivering expanded service: economic tradeoffs and technological decisions. *Science of the Total Environment*, **823**, 153616. https://doi.org/10.1016/j.scitotenv.2022.153616

Vidal, G., Carvalho, A., Méndez, R. et al. (2000). Influence of the content in fats and proteins on the anaerobic biodegradability of dairy wastewaters. *Bioresource Technology*, **74**, 231–239. https://doi.org/10.1016/S0960-8524(00)00015-8

Yu, J. and Pinder, K.L. (1993). Intrinsic fermentation kinetics of lactose in acidogenic biofilms. *Biotechnology and Bioengineering*, **41**, 479–488. https://doi.org/10.1002/bit.260410412

11

Sewage a Wasting Resource: Need to Rethink

*Olfat Saied[1], Tabbi Wilberforce[2], and Nabila Shehata[1]**

[1]*Environmental Science and Industrial Development Department, Faculty of Postgraduate Studies for Advanced Sciences, Beni-Suef University, Beni-Suef, Egypt*
[2]*Faculty of Natural, Mathematical and Engineering Sciences, King's College London, Strand, London, UK*

*Corresponding author: nabila.shehata@psas.bsu.edu.eg

11.1 Introduction

The energy deficit, the environmental pollution, and the scarcity of drinkable water are some of the major issues the globe is facing and will likely continue to have in the near future. More than 40% of the world's population is predicted to experience stress and water insecurity by 2025 (Wang and Serventi, 2019). Urbanization and population growth are the causes of these issues. Safe drinking water is insufficient and expensive in many developing and impoverished nations (WHO, 2014). Merely 1% of freshwater can be directly consumed by humans. Utilizing freshwater sources for irrigation has decreased due to the use of sewage water. Nowadays, farms are being irrigated with the treated sewage instead of being thrown into freshwater bodies. Particularly in water-scarce and peri-urban settings, sewage water is a good alternative to traditional water sources since it reduces the demand for fertigation and includes a lot of nutrients. In certain regions, this method was beneficial due to its encouraging effects on soil fertility and productivity. Some of the toxins contained in sewage, including heavy metals, are poisonous. Researchers have discovered that sewage water is bad for the environment and crops (Khatri et al., 2022). In addition, a large amount of fuel might be needed to support the energy needs in the upcoming years due to the growing population (Gemar et al., 2021).

Wastewater to Resource Recovery: Applying the Circular Economy Toward Sustainable Development, First Edition. Edited by Arvind Kumar Mungray et al. Published 2026 by John Wiley & Sons, Inc.

British Petroleum's (BP) June 2017 statistics evaluation of world energy found that while worldwide oil output increased by only 0.4 million barrels per day (Mb/d), global oil consumption was 1.6 Mb/d. Furthermore, the estimated global oil reserves in 2016 were 1707 billion barrels, which would be sufficient for 50.6 years of oil production (Tejaswini et al., 2022). New energy sources that can reduce atmospheric greenhouse gas (GHG) emissions—which are mostly caused by industrial processes and fossil fuels—must be used to counteract this decline in energy generation in order to lower the temperature of the planet. All nations are considering their alternatives for sustaining their energy supply over the ensuing decades as a result of growing awareness of global warming and the possibility that fossil fuels, especially natural gas and crude oil, may become more expensive and scarcer. Anaerobic digestion of organic wastes is one of the most notable bioenergy technologies because it can produce substantial amounts of energy from easily accessible feedstocks, such as sewage wastewater, either alone or in combination with other feedstocks, such as cow dung and food waste (Mu et al., 2020). Therefore, this chapter emphasizes the use of sewage for various resources and their applications in sustainable development, accordingly toward changing the mindset for its utilization rather than its discharge.

11.2 Sources and Composition of Sewage

Wastewater classified as sewage contains impurities primarily from human excreta, such as urine and feces, which are transported from the source to the disposal site through a pipe or sewer. Since the sewage's origin has a significant impact, its features may differ from one location to another (Rahman et al., 2016). Based on factors including the initial makeup of the sewage supply and the type of residential or community areas that contribute to the sewage volume, the source of the sewage can change (Gao et al., 2015). In addition to intact living entities (viruses and bacteria), sewage is predominantly a mixture of inorganic and organic materials that combine to form an extremely complicated cocktail. This combination varies according to the time of day, the season, and the type of treatment it undergoes. While some of the more water-soluble components are primarily linked with the solid or particle phase, others are transported with the liquid phase. As a result, tracers could be found in a variety of settings and at various separations from possible sources (Mudge and Ball, 1964).

According to the sources of generation, sewage is a mixture of liquid (or water) that contains waste which has been removed from homes, businesses, and industrial facilities, as well as any surface, groundwater, and storm water that may be present (Zhou and Smith, 2001). Generally speaking, “sewage/community wastewater” refers to the wastewater that is released from residential, commercial, and institutional settings. With 99.9% water and 0.1% solids, it is organic due to the presence of carbon compounds such as paper, vegetable matter, and human waste. In addition to sewage and wastewater from the population, the area has industrial wastewater. In the same manner that sewage is treated physicochemically and/or by microorganisms, many industrial wastes are also organic in composition (NP, 2002). Wastewater’s composition and features are also influenced by its origins. Although sewage and wastewater are frequently used interchangeably, sewage refers to liquid waste, primarily urine and feces, from commercial, industrial, institutional, and residential settings. Enhancing wastewater quality for use in drinking, household, industrial, and agricultural settings is the goal of treatment (Daniels–Azoma et al., 2024). Wastewater is categorized according to where it was generated or originated: (i) domestic sanitary activities (kitchen, toilet, laundry, and bathroom); (ii) industrial activities (water utilized for various industrial processes); and (iii) storm water/surface runoff (rain flood, runoff transporting agricultural waste, municipal runoff) are the main sources of wastewater (Nesterov et al., 2024).

11.3 Chemical Energy Stored in Sewage Water

The amount of energy required to treat wastewater is almost five times that of wastewater itself (Qadir et al., 2020). Because wastewater contains a lot of organic content, it can carry chemical energy. Anaerobic digestion is a process that can transform the organic molecules found in wastewater into biogas that is high in methane. To heat the wastewater treatment reactors and provide energy, certain wastewater treatment facilities use biogas. Wastewater is another source of thermal energy that is largely unexplored, but it needs to be recycled as close to the source as feasible, whereas chemically bonded energy can be carried down the sewage system with minimal loss (Nowak et al., 2015). The prospects for enhancing energy recovery from wastewater are promising. Initially, this will help reduce the primary cost factor—energy—by making the treatment process more energy-efficient. The goal is to achieve an energy-neutral wastewater cycle through on-site renewable energy production, alongside efforts to improve energy efficiency within the treatment systems (Maktabifard et al., 2018). Although a lot has been accomplished in this area, more research on “net-zero” energy and energy-positive technologies is still required.

Investments in energy efficiency and recovery initiatives in wastewater systems can yield high returns when evaluated through life-cycle cost analysis.

There is potential to advance sustainable development by putting current best management practices into practice and including energy considerations through gradual programs, especially in areas and nations where wastewater collection and treatment are not always guaranteed (Lackey and Fillmore, 2017). Methane's caloric value is determined by the anaerobic conversion of organic carbon in wastewater to methane and assuming full energy recovery from wastewater, it is projected that 53.2 billion m^3 of methane, which has a caloric value of 1908 billion MJ, make up the energy value of 380 billion m^3 of wastewater. Taking into account that the average family requires 3350 kWh of electricity. With three to four people living in each household on average, the reported wastewater could supply 474 million to 632 million people's energy needs per capita, or in areas with abundant energy supplies, lessen reliance on less ecologically friendly energy sources (Qadir et al., 2020). By the conclusion of the sustainable development goals (SDGs) timeframe in 2030, wastewater production worldwide might supply electricity to 196 million homes, given the projected rise in wastewater volumes in the upcoming years in relation to the energy embedded in wastewater. Similarly, by 2050, wastewater-driven power might help 239 million homes (Qadir et al., 2020).

11.4 Usage of Sewage as a Source of Energy: Quantification and Assessment

11.4.1 Nutrients Recovery

As municipal wastewater volumes increase and resource recovery technologies advance, there is growing interest in recovering nutrients from waste streams. Among the primary nutrients, wastewater produced globally each year contains 16.6 Tg (Tg = million metric tons) of nitrogen, 3.0 Tg of phosphorus, and 6.3 Tg of potassium (Qadir et al., 2020). The primary goal of nitrogen recovery is using the ammonia and/or ammonium from wastewater to directly/indirectly produce fertilizers. In municipal wastewater treatment plants (WWTPs), biological nitrogen removal (BNR) is commonly achieved through the oxidation of ammoniacal nitrogen to nitrate via nitrification, followed by the partial reduction of nitrate to dinitrogen gas through denitrification (Lee et al., 2021). Sewage sludge from biological wastewater treatment plants serves as a valuable renewable source of readily available nitrogen. Because the discharge of wastewater rich in phosphorus is a main cause of eutrophication and the phosphorus business faces significant challenges due to the lack of natural phosphate rocks, the removal and recovery of phosphorus not only reduce this eutrophication processes but also assist in satisfying the need for fertilizers based on phosphorus. Around 13% of the world's agricultural need for these nutrients would be met if wastewater were fully recovered (Qadir et al., 2020).

Supporting resource recovery from wastewater will require a methodical approach to overcome a number of obstacles in order to provide a high rate of return that directly supports SDGs 6, 7, and 12, as well as other goals like climate change adaptation and initiatives to advance "net zero" energy processes toward a green economy. There are many technologies for nutrients removal and recovery from wastewater conventional and advanced methods. Conventional methods like chemical processes that mainly contain chemical precipitation and adsorption, and biological processes that involve the use of specific bacteria, known as phosphorus-accumulating organisms, which are capable of taking up and storing large amounts of phosphorus under anaerobic conditions. The phosphorus is then released under aerobic conditions and can be removed and recovered. Some advanced technologies like membrane systems, electrodialysis (ED) processes, forward osmosis (FO), and membrane distillation (MD) have been applied for the nutrients recovery (Ye et al., 2020). Nitrogen, phosphorus, and other elements have been extracted from municipal wastewater using photosynthetic bacteria (Cho et al., 2013). Both macroalgae and microalgae have the innate ability to accumulate nutrients (N and P) from various aquatic habitats (Nguyen et al., 2022). Algal processes may be stoichiometrically more efficient in removing and accumulating N and P than traditional bacterial activities, which are hampered by carbon deficiency and an imbalance in the carbon:N:P ratio. From 100 m^3 of settled sewage, this algal sewage treatment and resource recovery system can produce around 1.7 kg of ammonium sulfate and approximately 2.5 kg of struvite, both of which are high-value agricultural fertilizers. With the ability to recover around 31% of nitrogen and 74% of phosphorus from settled sewage for advantageous reuse with minimal energy requirements, this algal pathway presents special prospects to green sewage treatment practices (Munasinghe-Arachchige et al., 2020). Nutrient recovery contributes to environmental sustainability by lowering the concentrations of ammonium and phosphates in the effluent from wastewater treatment facilities. To enhance resource efficiency, it is essential to adopt a circular economy framework that promotes the closed-loop flow of materials and energy, minimizes waste generation, and mitigates environmental pollution (Herzel et al., 2016).

11.4.2 Bio-fuel from Sewage Sludge

11.4.2.1 Bio-oil

Both the digested and undigested dried sewage sludge had high heat values of 8.5–17 MJ/kg and 23 MJ/kg, respectively (Haghighat et al., 2023). Sewage sludge can be turned into energy using a variety of methods. One of the interesting processes is pyrolysis, due to its low cost and excellent efficiency and ease operation. It takes place in the absence of oxygen. The products obtained like bio-oil depend on the pyrolysis operating factors, such as temperature, residence time, heating rate, specifications of the reactor particle size, and the catalyst. High heating rates make it more likely that biomass will decompose into gas components and stop vapors from re-polymerizing, even at low temperatures, increasing the generation of bio-oil (Belotti et al., 2014). Utilizing sewage sludge as a feedstock for the pyrolysis process lowers the cost of producing bio-oil and makes it economically competitive with fossil fuel (Haghighat et al., 2020).

11.4.2.2 Biodiesel

Biodiesel is a fuel generated from both edible and nonedible oils. It is produced by chemically reacting lipids, such as soybean oil, animal fat (tallow), or another vegetable oil, with alcohol to produce methyl, ethyl, or propyl ester (Ahmed and Huddersman, 2022). Lipids and oil from sewage sludge are extracted and then reacted with alcohol (transesterification) to produce biodiesel. The synthesis of biodiesel from sludge is becoming more commercialized since the byproducts are either environmentally safe or have added value (Liu et al., 2021). Figure 11.1 illustrates a typical wastewater treatment plant (Figure 11.1a) and the production of biodiesel from the sewage sludge (Figure 11.1b).

11.4.3 Electricity from Sewage Wastewater

The oxidation of organic materials in wastewater can provide electrical energy using microbial fuel cells (MFCs), but this technology needs to be proven on a wide scale and used to treat actual wastewater (Rossi et al., 2022). By employing various structures, researchers have tried to enhance MFC designs and optimize useful parameters to get around some of their shortcomings and bring this technology to a commercial level (Bazina et al., 2023). MFCs can be classified as membrane-less, single-chambered, or two-chambered. While membrane-less MFCs have been demonstrated to be the least costly, modifications are still necessary to overcome the current efficiency limitations. The working principle is the breakdown of organic materials to produce protons and electrons. Protons diffuse into the cathode chamber through a proton exchange membrane, while electrons are transported to the cathode by an external connection. They mix with oxygen there to produce water. The materials selected for the substrate and electrode have been shown to have a substantial impact on the cost and performance of the MFC. For the production of electrodes, carbon-based materials, including felt, graphite plate, and carbon cloth, are suitable; however, metal ions and stainless steel are

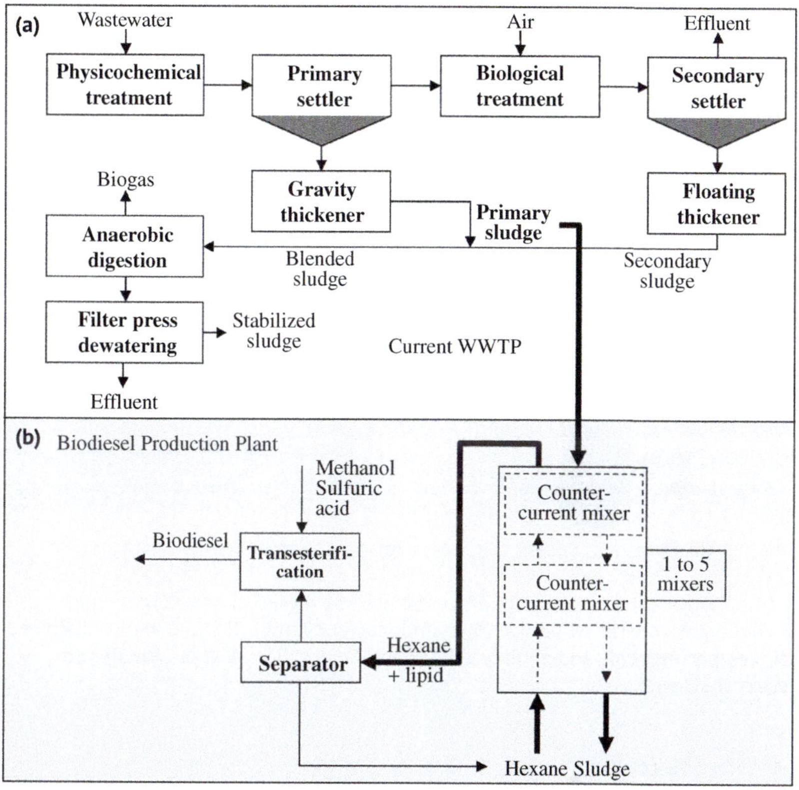

Figure 11.1 A schematic flow diagram for the biodiesel production from sewage sludge. *Source:* Olkiewicz et al. (2016)/with permission of Elsevier.

also being researched. Granular graphite-coated rods activated by nitric acid treatment are being researched as a potential superior alternative to platinum, which is not economically feasible when employed as a catalyst in cathodes. Biocathodes may or may not be beneficial for multifunctional MFC, depending on the substrate. Bazina et al. (2023) reviewed numerous configurations of MFCs. All things considered, MFC technology has a great deal of room to grow, and more research and development should eventually result in a revolutionary alternative energy source. To reduce costs and enhance MFC designs, scientists are attempting to broaden their present body of knowledge and investigate new fields. Different schematics of MFC are shown in Figure 11.2.

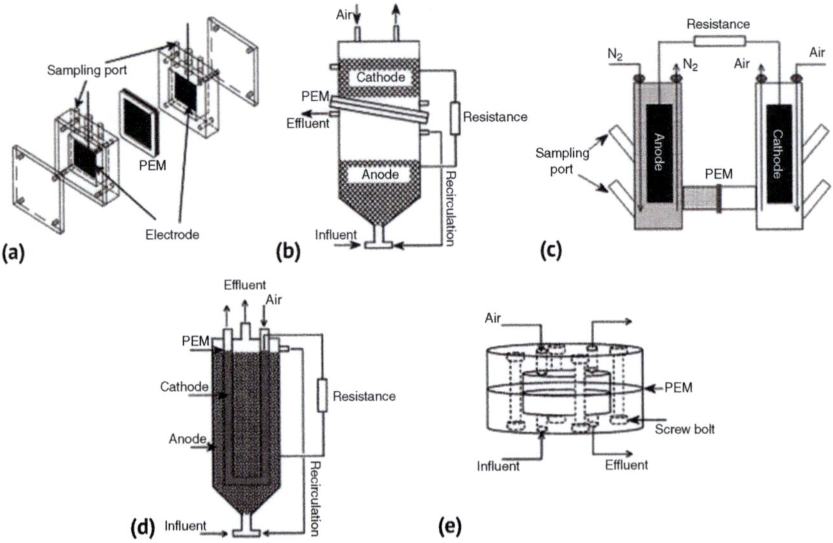

Figure 11.2 Different schematics of a two-compartment MFC in a rectangular (a), upflow configuration with cylindrical (b), cylindrical (c), cylindrical shape with a U-shaped cathodic compartment (d), and miniature (e) shape. *Source:* Bazina et al. (2023)/with permission of Elsevier.

11.4.4 Heat Recovery

One of the renewable energy sources is wastewater heat recovery. Approximately 20% of energy consumption within the domestic sector is linked to water heating for a variety of household applications (Yao and Steemers, 2005). So, the resultant wastewater includes a substantial quantity of heat energy that can be recovered. Heat exchanger and heat pump technologies are used to recover thermal energy from wastewater (Figure 11.3). Heat is extracted from wastewater, then used for heating purposes, and this helps in reducing energy consumption and environmental footprint. Essentially, wastewater heat is CO_2-free and offers a substantial energy transition potential in the heating sector (Alisawi, 2023).

11.4.5 Green Building Materials

Sludge can be used as an environmentally friendly addition to unfired earth bricks. It is possible to produce a new generation of building materials by combining fly ash from burning sewage sludge with a natural raw material, such as clay (Wiśniewski et al., 2024). The valorization of sludge can produce polyhydroxyalkanoates (PHA), which are a class of bioplastics used as

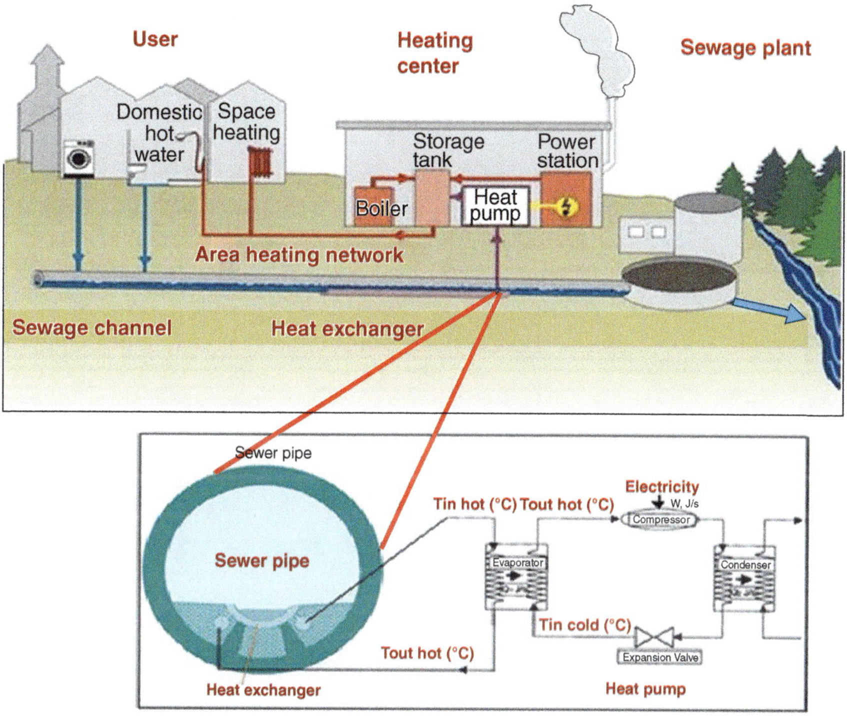

Figure 11.3 A schematic flow diagram for the heat recovery system from wastewater sewage. *Source:* Adapted from Garmsiri et al. (2014).

a building material. Some cement factories use dry sludge as an alternative source of fuel, and this helps to save cost, lower emissions, and promote sustainability (Goh et al., 2022). Treated wastewater can be used for various irrigation purposes like building a landscape and rooftop gardening. This route can lead to zero-liquid discharge (Olabi et al., 2025). Moreover, wastewater was recycled to produce eco-friendly concrete. However, the process depends on the treatment procedure and the treatment quality of the wastewater. For example, industrial effluents contain organic compounds or/and heavy metals need careful management (Zeyad, 2023).

11.4.6 The Recovery of Vermicompost Sewage Sludge in Agriculture

Sludge has plant-necessary nutrients that can boost agricultural yields and enhance soil quality (Figure 11.4). For this reason, sludge can be used in place

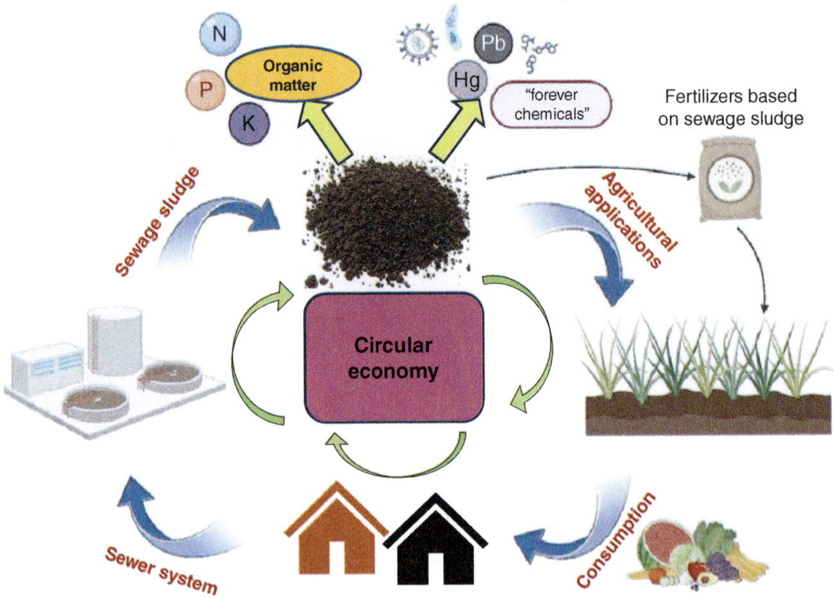

Figure 11.4 A schematic flow diagram for the conversion of sewage sludge into fertilizer. *Source:* Kominoko et al. (2024)/with permission of Elsevier.

of chemical fertilizers, which can contaminate and acidify soil and crops when they are applied. Nitrogen mineralization is accelerated, and organic materials are quickly broken down using the vermicomposting process, which uses earthworms. Adding vermicompost to the soil has been observed to improve plant growth and promote soil microbial activity (Rusănescu et al., 2022).

11.5 Prospects and Challenges

Although numerous technologies for recovering nutrients, energy, fertilizers, and other products from wastewater have been explored in academic research, only a few have been implemented on a large scale due to technical limitations and high energy consumption. Additionally, microorganisms, heavy metals, contaminants, and medications in the sludge or wastewater that need further treatment and extensive monitoring are another challenge, which otherwise would pose a threat to the environment.

The cost of nutrient recovery including energy cost is relatively high. Recovered phosphate is three times more expensive than phosphate made from rock phosphate (Cornel and Schaum, 2009). Sludge must be carefully managed to minimize their environmental footprint so it can be reused. The regulatory laws and frameworks for wastewater and sludge recovery are not fully developed. The successful implementation of wastewater resource recovery requires advanced technologies, reducing its environmental impacts. Also, sludge and wastewater management policies and frameworks require improvement. Governments should support future research and private sectors to develop new applicable and sustainable technologies for wastewater and sludge recovery. The source of the major contaminants in sewage is majorly due to a centralized wastewater treatment system, where industrial, storm wastewater etc., are added along with higher-order microorganisms during the flow in a large sewerage network, and ultimately making resource recovery difficult and costly. For resource recovery, decentralized waste treatment, i.e., onsite treatment i.e., treatment as per the composition of waste seems a better and cost-effective solution.

11.6 Conclusion

Due to growing population and climate change, wastewater reuse is no longer a choice, it's an essentiality. Wastewater can be recycled and reused where water is scarce. Reusing of wastewater and sludge has many advantages: water conservation, reduced pollution, minimized nutrient loads, and strengthened local economy and employment. Nutrient recovery from wastewater to produce fertilizers minimizes the use of chemical fertilizers, which can lead to water pollution. Energy recovered from sludge like bio-oil and biodiesel is considered a renewable energy source and this decreases greenhouse emissions. In general, wastewater reusing contributes to lowering the overall carbon footprint of water management, mitigating climate change, and improving sustainability. Reusing wastewater faces many challenges; to overcome these challenges, we need more research in this sector, governmental support, and public awareness. So, we need to rethink about reusing sewage and enhance population awareness from consumers to businesses about its implementation and the importance of reusing and recycling wastewater and sludge. Future studies should focus on the development of the current recovery processes to be applied both at large scale and onsite with the minimum cost and ecological impact.

Review Questions

1 Why is there a need to rethink wastewater effluents disposal into the environment in the current situation?
2 How can we differentiate sewage and wastewater based on their origin, characteristics, level of different pollutants, etc.?
3 What are the various resources available in the wastewater? Make a list and sources.
4 What are the various contaminants available in sewage (which make resource recovery difficult and costly) at sewage treatment plants and their source of addition?
5 Make a figure of various streams of wastewater generated at the household and finally termed as sewage.
6 How much energy is stored in the sewage and its potential?
7 What is microbial fuel cell (MFC) and its mechanism? How is electricity produced by MFC?
8 What are the various challenges and probable solutions toward the recovery of resources from wastewater?
9 Why is there a need for resource recovery from sewage?
10 Make a prospective flow diagram for the resource recovery of various resources from sewage for sustainable development.

References

Ahmed, R. and Huddersman, K. (2022). Review of biodiesel production by the esterification of wastewater containing fats oils and grease (FOGs). *Journal of Industrial and Engineering Chemistry*, **110**, 1–14.

Alisawi, H.A.O. (2023). Renewable energy of wastewater heat recovery. *Water Science and Technology*, **87**, 2852–2871. https://doi.org/10.2166/wst.2023.167

Bazina, N., Ahmed, T.G., Almdaaf, M. et al. (2023). Power generation from wastewater using microbial fuel cells: a review. *Journal of Biotechnology*, **374**, 17–30.

Belotti, G., de Caprariis, B., De Filippis, P. et al. (2014). Effect of *Chlorella vulgaris* growing conditions on bio-oil production via fast pyrolysis. *Biomass and Bioenergy*, **61**, 187–195.

Cho, S., Lee, N., Park, S. et al. (2013). Microalgae cultivation for bioenergy production using wastewaters from a municipal WWTP as nutritional sources. *Bioresource Technology*, **131**, 515–520.

Cornel, P. and Schaum, C. (2009). Phosphorus recovery from wastewater: needs, technologies and costs. *Water Science and Technology*, **59**, 1069–1076.

Daniels-Azoma, A.E., Ubechu, B.O., Amadi, C.C. et al. (2024). Aspects of wastewater collection, treatment and reuse: review of principles and practices. *International Journal of Geography & Environmental Management (IJGEM)*, **10**(5), 1–37. https://doi.org/10.56201/ijgem.v10.no5.2024.pg1.36

Gao, C., Tang, H., Zhang, M. et al. (2015). Investigation and analysis of college's water system: the case of Northeastern University. *Water Policy*, **17**, 1224–1235.

Garmsiri, S., Koohi, S., and Rosen, M.A. (2014). Recovery of sewer waste heat vs. heat pumps using borehole geothermal energy storage for a small community water heating system: comparison and feasibility analysis. In *The platform for open scholarly exchange and collaboration, Basel, Switzerland*. ISBN9781315366050.

Gemar, G., Soler, I.P., and Sánchez-Teba, E.M. (2021). Waste management: valorisation is the way. *Foods*, **10**, 2373.

Goh, Q.H., Farouk, A.A., and Chew, I.L. (2022). Optimizing the bioplastic chemical building block with wastewater sludge as the feedstock using carbon-hydrogen-oxygen framework. *Resources, Conservation and Recycling*, **176**, 105920.

Haghighat, M., Majidian, N., and Hallajisani, A. (2020). Production of bio-oil from sewage sludge: a review on the thermal and catalytic conversion by pyrolysis. *Sustainable Energy Technologies and Assessments*, **42**, 100870.

Haghighat, M., Majidian, N., Hallajisani, A. et al. (2023). Production and optimization of bio-oil from municipal wastewater sludge by thermal and catalytic pyrolysis process. *Journal of Chemical and Petroleum Engineering*, **57**, 227–247.

Herzel, H., Krüger, O., Hermann, L. et al. (2016). Sewage sludge ash—a promising secondary phosphorus source for fertilizer production. *Science of the Total Environment*, **542**, 1136–1143.

Khatri, N., Tyagi, S., Tharmavaram, M. et al. (2022). Sewage water: from waste to resource–a review. *Environmental Claims Journal*, **34**, 6–33.

Kominko, H., Gorazda, K., and Wzorek, Z. (2024). Sewage sludge: a review of its risks and circular raw material potential. *Journal of Water Process Engineering*, **63**, 105522.

Lackey, K. and Fillmore, L. (2017). *Energy Management for Water Utilities in Latin America and the Caribbean: Exploring Energy Efficiency and Energy Recovery Potential in Wastewater Treatment Plants*. Washington DC: World Bank Group.

Lee, W., An, S., and Choi, Y. (2021). Ammonia harvesting via membrane gas extraction at moderately alkaline pH: a step toward net-profitable nitrogen recovery from domestic wastewater. *Chemical Engineering Journal*, **405**, 126662.

Liu, X., Zhu, F., Zhang, R. et al. (2021). Recent progress on biodiesel production from municipal sewage sludge. *Renewable and Sustainable Energy Reviews*, **135**, 110260.

Maktabifard, M., Zaborowska, E., and Makinia, J. (2018). Achieving energy neutrality in wastewater treatment plants through energy savings and enhancing renewable energy production. *Reviews in Environmental Science and Bio/Technology*, **17**, 655–689.

Mu, L., Zhang, L., Zhu, K. et al. (2020). Anaerobic co-digestion of sewage sludge, food waste and yard waste: synergistic enhancement on process stability and biogas production. *Science of the Total Environment*, **704**, 135429.

Mudge, S.M. and Ball, A.S. (1964). Sewage. In: *Environmental Forensics*, 35–53. Elsevier.

Munasinghe-Arachchige, S.P., Abeysiriwardana-Arachchige, I.S.A., Delanka-Pedige, H.M.K. et al. (2020). Algal pathway for nutrient recovery from urban sewage. *Algal Research*, **51**, 102023.

Nesterov, D., Barrera-Martínez, I., Martínez-Sánchez, C. et al. (2024). Approaching the circular economy: biological, physicochemical, and electrochemical methods to valorize agro-industrial residues, wastewater, and industrial wastes. *Journal of Environmental Chemical Engineering*, **12**, 113335.

Nguyen, L.N., Aditya, L., Vu, H.P. et al. (2022). Nutrient removal by algae-based wastewater treatment. *Current Pollution Reports*, **8**, 369–383.

Nowak, O., Enderle, P., and Varbanov, P. (2015). Ways to optimize the energy balance of municipal wastewater systems: lessons learned from Austrian applications. *Journal of Cleaner Production*, **88**, 125–131.

NP, C. (2002). An overview of water and water treatments. In *Hank Book of Water and Wastewater Treatment Technologies*, 1–60. Butterworth-Heinemann Publication.

Olabi, A.G., Shehata, N., Issa, U.H. et al. (2025). The role of green buildings in achieving the sustainable development goals. *International Journal of Thermofluids*, **25**, 101002. https://doi.org/10.1016/j.ijft.2024.101002

Olkiewicz, M., Torres, C.M., Jiménez, L. et al. (2016). Scale-up and economic analysis of biodiesel production from municipal primary sewage sludge. *Bioresource Technology*, **214**, 122–131. https://doi.org/10.1016/j.biortech.2016.04.098

Qadir, M., Drechsel, P., Jiménez Cisneros, B. et al. (2020). Global and regional potential of wastewater as a water, nutrient and energy source. *Natural Resources Forum*, **44**(1), 40–51. Wiley Online Library.

Rahman, M.M., Hagare, D., and Maheshwari, B. (2016). Use of recycled water for irrigation of open spaces: benefits and risks. In: *Balanced Urban Development: Options and Strategies for Liveable Cities* (eds. B. Maheshwari, B. Thoradeniya, and V.P. Singh), vol. 72, 261–288. Cham: Springer International Publishing. https://doi.org/10.1007/978-3-319-28112-4_17

Rossi, R., Hur, A.Y., Page, M.A. et al. (2022). Pilot scale microbial fuel cells using air cathodes for producing electricity while treating wastewater. *Water Research*, **215**, 118208.

Rusănescu, C.O., Rusănescu, M., Voicu, G. et al. (2022). The recovery of vermicompost sewage sludge in agriculture. *Agronomy*, **12**, 2653.

Tejaswini, M., Pathak, P., and Gupta, D.K. (2022). Sustainable approach for valorization of solid wastes as a secondary resource through urban mining. *Journal of Environmental Management*, **319**, 115727.

Wang, Y. and Serventi, L. (2019). Sustainability of dairy and soy processing: a review on wastewater recycling. *Journal of Cleaner Production*, **237**, 117821.

WHO/UNICEF Joint Water Supply, and Sanitation Monitoring Programme (2014). *Progress on drinking water and sanitation: 2014 update*. World Health Organization.

Wiśniewski, K., Rutkowska, G., Jeleniewicz, K. et al. (2024). Ecologically friendly building materials: a case study of clay–ash composites for the efficient management of fly ash from the thermal conversion of sewage sludge. *Sustainability*, **16**, 3735.

Yao, R. and Steemers, K. (2005). A method of formulating energy load profile for domestic buildings in the UK. *Energy and Buildings*, **37**, 663–671. https://doi.org/10.1016/j.enbuild.2004.09.007

Ye, Y., Ngo, H.H., Guo, W. et al. (2020). Nutrient recovery from wastewater: from technology to economy. *Bioresource Technology Reports*, **11**, 100425.

Zeyad, A.M. (2023). Sustainable concrete production: incorporating recycled wastewater as a green building material. *Construction and Building Materials*, **407**, 133522.

Zhou, H. and Smith, D.W. (2001). Advanced technologies in water and wastewater treatment. *Canadian Journal of Civil Engineering*, **28**, 49–66.

12

Wastewater as a Source of Different Forms of Energy: Resource Recovery, Circular Economy and/or Cleaner Production Perspectives from the Mining and Food Processing Industry

Quinnetta Blessing Clement[1], Abayomi B. C. Grant[2], Abubakar M. Bilyaminu[1], Hector García Hernandez[1], Alka A. Mungray[3], Arvind Kumar Mungray[3], Eldon R. Rene[1], and Anina James[4]**

[1]*Department of Water Supply, Sanitation and Environmental Engineering, IHE Delft Institute for Water Education, Delft, The Netherlands*
[2]*Environmental & Safeguard Officer, Monrovia Metropolitan Climate Resilience Project (MMCRP), Environmental Protection Agency of Liberia (EPA), Monrovia, Liberia*
[3]*Department of Chemical Engineering, S.V. National Institute of Technology, Surat, Gujarat, India*
[4]*J & K Pocket, Delhi, India*

*Corresponding authors: anina_elina@yahoo.co.in; e.raj@un-ihe.org

12.1 Introduction to Resource Recovery from Wastewater

The exponential growth of population and environmentally detrimental methods of production and consumption for increasing economic output is propelling climate change worldwide. The global urban population is projected to increase by ~2.5 billion by the year 2050 (United Nations, 2018), and rising economies are distinguished by escalating consumption needs; therefore, exacerbating the environmental challenges are linked to a linear economy (Durkin et al., 2024). Adoption of circular economic principles, wherein squandered resources are reclaimed and recycled to replace a substantial portion of the primary output could provide relief to the stressed environmental systems. The linear "take-make-waste system" produces substantial waste and exacerbates resource scarcity, since rising demand is satisfied by augmenting the consumption of raw materials and energy. Resource recovery enables the transition from a linear economy to a circular economy (CE) model that extracts value-added goods/products from waste (i.e. as the primary material source). This approach addresses the increasing demands of economic growth in an environmentally sustainable and resource-efficient manner. Reintroducing the recovered resources as feedstocks into the supply chain can diminish the demand for basic resources. Extracting water, energy, and other resources from wastewater would help to alleviate the environmental consequences of industrialization (Keijer et al., 2019).

Conventional wastewater treatment techniques, including flocculation, coagulation, oxidation processes, sedimentation, and filtration, primarily emphasize pollutant elimination while neglecting the recovery of beneficial resources, resulting in resource depletion and environmental damage (Obaideen et al., 2022). Resource recovery seeks to convert wastewater into a useful asset, in accordance with sustainable development principles. This method prioritizes the recovery of nutrients, energy, and water from wastewater while advocating for CE principles (Kumar et al., 2024). For example, phosphorus and nitrogen are recovered as struvite, while algal and lignocellulosic biomass are valorized to produce ethanol, biogas, and biohydrogen as components of sustainable energy. Bio-based soft technology refers to the employment of naturally occurring biological processes, such as those involving microorganisms, plants, or algae, for wastewater treatment and the recovery of valuable resources in an energy-efficient and environmentally sustainable manner (Sharma et al., 2024).

The mining industry wastewater contains high amounts of toxic chemicals such as flotation reagents and metal ions (Matebese et al., 2024). Hence, untreated discharge from mining sites has deleterious effects on organisms and ecology, eliciting public concern. Further, resource recovery from mining industry wastewater is an advantageous proposition since it contains various metal ions. On the other hand, the food manufacturing/processing industry is a water-intensive industry due to the use of water for the following processes: heating, cooling, fermentation, and washing of basic ingredients. In addition, water is employed to sanitize the equipment, pipelines, and various process units in the production chain, thereby guaranteeing adherence to the rigorous sanitation standards that ensure the safety of the food products. The biochemical oxygen demand (BOD) and chemical oxygen demand (COD) are induced by the high concentration of organic components and other substances in wastewater from the food industry, which presents the primary challenge in wastewater treatment. A combination of biological and physicochemical treatment stages is typically employed in the treatment of wastewater in the food industry. This chapter discusses the types of energy contained in wastewater, technological approaches for extracting energy and value-added products from wastewater, the CE framework for energy production, the importance and effects of CE strategies, and case studies from the mining and food processing sectors (Figure 12.1).

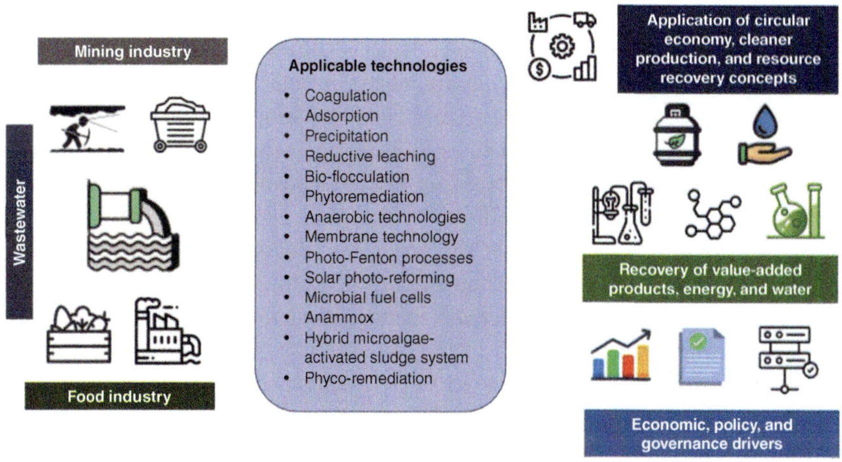

Figure 12.1 Schematic diagram of the different forms of energy, value-added products, the economic, policy, and governance drivers that facilitate the application of circular economy, cleaner production, and resource recovery concepts through different technological interventions for treating mining and food industry wastewater.

12.2 Forms of Energy Stored in Wastewater

Wastewater treatment plants (WWTPs) are considered as large energy-dependent systems incurring up to 40% of their operating costs as energy demand (Panepinto et al., 2016). The major steps involved in the treatment process such as transport of wastewater to the plant, pretreatment techniques applied, biological treatment, and sludge management, all require energy input (Gu et al., 2016). Achieving sustainable treatment and resource recovery from wastewater requires lowering the external energy use of the treatment plants. A suitable proposition would be to extract energy from the WWTPs for a net-zero energy balance. This can be achieved by comprehending the types of energy stored in wastewater. Wastewater can be considered as "resource water" (i.e. treated water for reuse) due to the presence of valuable constituents like water itself, nutrients, and energy. The two most significant energy types are: (i) chemical (bio)energy from organic matter, which can be converted to biofuels; (ii) thermal energy recovered as heat from wastewater. Incineration of sludge from WWTPs also generates thermal energy. The combustion enthalpy of the organics can be used to determine the potential energy that can be extracted/recovered using different technologies, i.e. from the organic carbon fraction present in wastewater (Frijns et al., 2013). Further, H_2 as an energy source can be produced via dark fermentation. Dark fermentation effluents are energy-laden and can further be used as feedstocks for other energy-generating processes such as anaerobic digestion, photo-fermentation, and in bio-electrochemical systems (Boshagh et al., 2025).

During the oxidation of organic matter by microbes, the chemical energy present in the COD is lost as metabolic heat. However, during the digestion of sludge, part of this energy is recovered. Sludge digestion and biogas production are a common practice at several WWTPs worldwide. This generated biogas can be used to heat the digestion reactor as well as to generate energy, i.e. depending on the location of the facilities. Sludge-digesting WWTPs consume ~40% less net energy than non-sludge-digesting plants (Frijns et al., 2013). The primary challenge with direct anaerobic treatment is that the sewage is excessively diluted, hindering optimal digestion, and often a substantial portion of the generated methane is lost (Verstraete et al., 2009). Thus, effectively managed anaerobic treatment and sludge digestion minimize the methane leakage and optimize biogas utilization, thereby dramatically reducing greenhouse gas (GHG) emissions. For instance, the optimization of energy recovery in the production of anaerobic biogas from domestic wastewater can be accomplished by increasing the organic carbon concentration. This can be achieved by implementing two methods: (i) optimizing sludge digestion and increasing the concentration, and (ii) implementing anaerobic digestion and source segregation principles. Thermal energy recovery is a feasible approach for industrial wastewater resource utilization and converting the wastewater to value-added products (Song et al., 2023). Among the industries, food and associated industries have raised concerns due to excessive water consumption and high per capita wastewater generation. Brewery and other agriculture-based industry wastewater with suitable treatment strategies can produce biohydrogen and biogas (Kumar et al., 2022). Recently, Qiao et al. (2024) reported the recovery potential and utilization pathway of chemical energy during wastewater treatment in China. The optimization of anaerobic digestion may be crucial for achieving energy self-sufficiency and resource recovery from both industrial and domestic WWTPs.

The energy inherently stored in chemical compounds by virtue of having bonds of atoms and molecules is called chemical energy. It is a type of potential energy that is quantifiable, and also modified during chemical reactions. To harness energy from wastewater, it is pertinent to understand the types of chemical energy stored in wastewater. The primary source of chemical energy in wastewater is contributed by the organic matter or biomass present in it, containing up to 10 times the energy incurred during the treatment of wastewater (Qiao et al., 2024). The GHGs (e.g. N_2O, CH_4, and indirect CO_2) emitted during the treatment processes also contain substantial amounts of chemical energy (Wang et al., 2022; Tong et al., 2024). Microbial metabolic processes utilize, convert, and/or release the chemical energy present in the constituents of wastewater such as nutrients and organic matter while releasing GHGs (Tong et al., 2024). Photosynthesis is a phenomenon that is being considered nowadays to produce biomass from the nutrients present in the wastewater (Mohammed et al., 2023). The biomass is then used to produce bioenergy.

12.3 Methods to Derive Energy from Wastewater

It is noteworthy to mention that, when the concentration of organics present in the wastewater influent is high, a substantial quantity of energy is readily accessible through treatment and resource recovery-based technologies. For example, the biogas produced in the digester is the primary energy source of a WWTP. The utilization of biogas for digester heating and electricity production is a sustainable method for energy recovery from WWTPs, resulting in a reduction in the input energy required for the operation of the WWTPs and a consequent reduction in sludge volume (Gupta and Khatiwada, 2024). Five WWTPs were studied for energy balance analysis in Catalonia, Spain (Silvestre et al., 2015). The study reported that 67% of the energy contained in the organic matter of wastewater was transferred to the sludge, with anaerobic digestion of the sludge subsequently recovering 52% of this energy through biogas conversion. Hu et al. (2024a) developed a photo-electrochemical anaerobic digestion system (PEADs) wherein the photo-generated carriers of nano-semiconductors caused the photoreduction of CO_2, thus enhancing methanogenesis. The study reported increased efficiency of microbial pathways of anaerobic digestion and resource recovery from a low organic concentration wastewater. According to Islam (2023), the yearly domestic and industrial wastewater produced in Bangladesh is expected to be 4874 million m^3 and 453 million m^3, respectively. The authors indicated that the utilization of up-flow anaerobic sludge blankets (UASB) for resource recovery from home and industrial wastewater possesses an energy recovery potential of 1986 GWh, annually.

Anaerobic ammonium oxidation (anammox) is a technique used to remove nitrogen at reduced aeration rates while ensuring the remanence of adequate organic matter. This improves the energy efficiency of the WWTP (Arora et al., 2021). The methane generation potential can increase up to 47% as no organic carbon is required for the denitrification process (McCarty 2018). Anammox-integrated anaerobic digestion system can remove nitrogen while achieving high phosphorus recovery. The phosphorus obtained as hydroxyapatite ($Ca_5(PO_4)_3OH$) by the addition of calcium sources like lime milk or calcium chloride can be utilized as a beneficial fertilizer, thus promoting the concepts of a CE. Moreover, the sludge produced from this process, which has a lesser volume compared to conventional methods, can be utilized directly as fertilizer or indirectly after incineration, providing a sustainable solution for sludge management (Tran et al., 2024).

In previous studies, it has been shown that electricity is generated through the use of microbial fuel cell (MFC) technology in a number of WWTPs, which include both residential and industrial wastewater treatment facilities (Suransh et al., 2023; Pugazhendi et al., 2022). MFC technology can convert the chemical energy of surplus sludge into electrical energy. In contrast to traditional anaerobic digestion, MFCs produce electricity directly, presenting a significant potential advantage, as the conversion of biogas, via combustion, into electricity yields low efficiency, resulting in the loss of up to 70% of the energy contained in the biogas (Pham et al., 2006). Theoretically, the potential energy that can be recovered from wastewater using MFC can reach 0.65 kWh/m^3 when the coulombic efficiency (the percentage of electron charge that contributes to electricity generation) is 40% and the hydraulic retention time (HRT) of domestic wastewater treatment is 20 hours. Using conventional activated sludge (CAS) technologies, the energy consumption ranges from 0.3 to 0.6 kWh/m^3, which indicates that MFCs have the capacity to achieve energy sufficiency throughout the entire wastewater treatment process (Gu et al., 2017). This calculation illustrates that MFCs have the capacity to generate an adequate amount of energy. MFCs are also energy-saving technologies that can reduce aeration and produce less sludge than CAS (Gu et al., 2017).

Algae or microalgae have the potential to utilize and convert nutrients from wastewater into carbohydrates, proteins, and lipids, contributing to algal biomass. Algal biofilm is a sessile microbial consortium predominantly composed of microalgae, supplemented by various microorganisms, including bacteria, which enhance algal proliferation. The system's or the reactor's ability to recycle nutrients and convert nitrogen and phosphorus into biomass energy is enhanced by the synergistic interaction between microalgae and bacteria in algal biofilms (Liu et al., 2024). This algal biomass can be utilized as a raw material for the production of biofuels, such as bioethanol, biogas, biohydrogen, and biodiesel (Joshi et al., 2024). While biomass derived from microalgae makes it possible to simultaneously produce biopolymers, biofertilizers, and biofuels, the process of extracting these components from microalgae is challenging. This makes it less cost-effective and commercially unviable. It is suggested that biofuel production could be combined with the extraction of high-value components from algae, such as cosmetics, nutraceuticals, and medicines, for cost benefits. Further, if wastewater is used as the feedstock for algae, this utilization of a waste-as-a-value method results in a reduction in the production costs of the algal biofuels. In addition, microalgal biomass could be cultivated on terrain that is not suitable for agriculture, and it demonstrates a higher yield of usable biomass than terrestrial plants. Hence, algae could be an excellent choice for the production of biofuels (Kumar et al., 2021).

Su et al. (2023) reported high growth of microalga *Tribonema aequale* in brewery wastewater (6.45 g/L), which could be harvested by a simple filtration method. The alga removed 79.53% total dissolved nitrogen (TN), 85.39% total organic carbon (TOC), 71.33% total dissolved phosphorus (TP), and 93.38% ammonia nitrogen (NH_3-N) from the brewery wastewater. The treated wastewater complied with the Chinese national wastewater discharge quality. Additionally, the algal biomass was used to extract large amounts of high-value products, eicosapentaenoic acid, chrysolaminarin, and palmitoleic acid.

12.4 Application of Cleaner Production and Circular Economy Approach for Energy Generation

In 1989, the phrase "Cleaner Production" (CP) was introduced to enhance the sustainability of industries and enterprises while mitigating their environmental implications. This was part of the crucial plan for the United Nations Environmental Programme Industry and Environment Programme. The development of CP to produce less waste and reduce harmful emissions has evolved into a concept that improves an industry's environmental and social performance, resource efficiency, and overall cost efficiency. In addition, it also lowers costs of treating waste or pollution by gaining revenue from waste recycling or reuse (Oliveira Neto et al., 2020). Nearly 50% of worldwide wastewater is discharged without treatment into the environment (UN Habitat and WHO, 2021). The increasing demand for renewable energy, in conjunction with the escalating scarcity of resources, has led to a greater emphasis on the principles of CE and sustainable development. The operational status of WWTPs worldwide is a clear indication of the necessity of these principles (Gupta and Khatiwada, 2024).

WWTPs are an essential component of the ecosystem and significant energy users. The wastewater sector, which includes the collection and treatment of wastewater, accounts for 3% of the global GHG emissions and 4% of the global electricity usage. With the rising global population and the escalation of water scarcity, the necessity for the recovery and reuse of wastewater for productive applications is becoming increasingly urgent. Implementing sustainable development and CE principles in WWTPs offers a distinct chance to reclaim important resources, including energy, nutrients, and treated water (Gherghel et al., 2019). As mentioned by the authors, various methods have been proposed for the recovery of energy, nutrients, and other value-added products from wastewater. The Netherlands has introduced the "NEW Factory" concept, proposing that WWTPs can function as facilities for recovering nutrients, energy, and clean water, thereby illustrating a vision for the sustainable operation of WWTPs in the future (Roeleveld et al., 2010). The biorefinery concept views WWTPs as sustainable factories that are analogous to oil refineries, where raw materials (wastewater and sludge) are processed to extract and recover a variety of beneficial products. The primary objective of this approach is to remediate/treat wastewater. In a previous study, it was mentioned that a consensus is forming to reclassify WWTPs as Wastewater Resource Recovery Facilities (WRRF), with resource recovery serving as a primary function in addition to wastewater treatment (Gupta and Khatiwada, 2024). It is imperative to pursue energy self-sufficiency in these facilities in order to establish sustainable wastewater treatment systems, as there is a substantial potential for energy recovery from WWTPs. Additionally, the absence of fossil fuel-derived heat and electricity can considerably reduce GHG emissions through the optimization of energy recovery and utilization in WWTPs (Durkin et al., 2024; Gupta and Khatiwada, 2024).

The recovery of nutrients from wastewater is becoming increasingly common in agricultural applications. In comparison to the effluent, sludge contains a higher concentration of nutrients and organic matter, which results in improved nutrient recovery and processing efficiency. The sludge is classified as a biosolid and may be employed in agricultural or landscaping applications after proper treatment (Lin et al., 2024). In addition, the demand for agricultural fertilizers is increasing globally due to the growing population, and nutrient recovery from sludge and effluent can be used to meet this demand on a local and regional scale (Gupta and Khatiwada, 2024). Wastewater treatment offers a viable opportunity to recover phosphorus from wastewater, which has the potential to replace substantial portions of the global phosphorus requirement (Yan et al., 2024). For instance, Sweden and Switzerland have implemented regulations regarding the recovery of phosphorus from wastewater treatment. These regulations provide preliminary insights into the regulatory framework necessary to realize the global potential for phosphorus recovery (Gupta and Khatiwada, 2024).

12.5 Significance and Impact of Cleaner Production and Circular Approaches on Environment, Climate, and Economics

Sustainable development and CP, which encompass the economic, environmental, and social aspects of development, require the availability of renewable energy, clean water, and efficient resource management. The recovery of resources from wastewater is an opportunity at the waste-nutrient-energy nexus, which can contribute to the advancement of global sustainable development. The implementation of sophisticated WWTPs can directly support the objectives of clean water accessible to everyone. Methane capture in WWTPs reduces its atmospheric leakage, thereby reducing the GHG impact of methane and contributing to climate action (Wechselberger et al., 2025). Nutrient removal during wastewater treatment has been shown to reduce eutrophication/pathogens in downstream aquatic systems, according to several previous studies. According to Soo and Shon (2024), wastewater treatment reduces the pathogens in the effluent and improves the purity of the downstream water. Thus, the requirement for freshwater extraction is reduced, and the burden on overburdened water resources is alleviated by the provision of treated effluent for agricultural irrigation.

The substitution of conventional chemical-based fertilizers with organic biosolids produced as a by-product from the stabilization and treatment of wastewater in WWTPs is also a potential advantage in CP. The necessity for mining value-added products can be reduced by applying resource recovery technologies for wastewater treatment, which serves as, for example, a viable method for extracting phosphorus for fertilizers and other applications (Hu et al., 2024b). The facility's energy requirements are frequently met by the energy recovered from WWTPs, which reduces the need for external electricity sources. Furthermore, the recovered energy resources can provide energy access to individuals residing in the vicinity, particularly in developing nations, if they are repurposed and transported beyond the treatment facility. Biogas is a versatile fuel that has the potential to replace firewood, kerosene, and natural gas in the cooking process. This could improve the quality of life, create economic opportunities, and increase access to renewable energy in energy-deprived areas. This alternative cooking fuel provides a variety of benefits by facilitating cleaner combustion, which in turn improves the purity of the indoor/outdoor air environment (Lohani et al., 2021).

The productivity of agriculture can be improved through the use of water reuse, biosolids, and treated effluent. The mining industry is an appropriate candidate for decarbonization by utilizing electric batteries for its processes and transportation (Pavez et al., 2025). Zhou et al. (2025) reported the high-efficiency recovery of rare earth elements (REEs) from mining wastewater. The mining industry's transition to decarbonization would be facilitated by the utilization of metals recovered for the production of critical components.

On the other hand, advanced WWTPs are required for promoting economic growth and decent work, the development of innovative industries and resilient infrastructure, the mitigation of economic inequities, and the establishment of ethical production and consumption practices (Gupta and Khatiwada, 2024). The widespread implementation of WWTPs in metropolitan areas would foster economic growth, employment opportunities, and the development of new industries by providing a critical utility service with substantial environmental benefits. As a result, there is also a direct correlation between the environment, climate, and economics, as well as the benefits of resource recovery from effluent and sludge and integrated water resource management (IWRM). It is imperative to recognize the contextual relationship between these advantages and local circumstances, such as demography, economic conditions, location, and the institutional and regulatory environment. In order to fully understand the actual benefits that can be achieved through resource recovery systems in WWTPs, it is imperative to conduct empirical analysis and consider the local factors (Gupta and Khatiwada, 2024).

12.6 Case Study: Applying Cleaner Production and Resource Recovery Concepts in the Mining Sector

12.6.1 The Integrated Cleaner Production Approach

The concept of CP emerged as an industrial response to the World Commission on Environment and Development's (WCED) request for sustainable development while protecting the environment. This concept was further explained during Rio Summit 21, where it was based on the three (3) P programs (pollution prevention pays) (Baas, 1995; Hens et al., 2018). CP was therefore defined as "the conceptual and procedural method of production which requires that the entire stages of a product or process life cycle be discussed to prevent or reduce short- and long-term threats to people and the environment" (Baas, 1995). The UNEP developed the commonly quoted definition: "cleaner production (CP) refers to the continuous application of a combined preventive environmental plan to processes, products, and services to increase efficiency and reduce risks to humans and the environment" (Hens et al., 2018). The development of CP to produce less waste and reduce harmful emissions has evolved into a concept that improves an industry's environmental and social performance, resource efficiency, and overall cost efficiency. In addition, it also lowers the costs of treating waste or pollution by gaining revenue from waste recycling or reuse (Hens et al., 2018; Oliveira Neto et al., 2020). This concept was strategically developed to redefine the management approach of industries in terms of long-term requirements based on the principles of sustainable development (Krishna et al., 2020; Mudd, 2007). The regular practice of pollution control is still applied in many companies today. However, some of these companies or government officials have not accepted the current CP recognition that requires new processes and innovative methods (Khalili et al., 2015).

CP is obtained by means of applying new skills and knowledge, the advancement of technology, and the modification of behaviors. The conventional approach to pollution control (end-of-the-pipe treatment) has served as the foundation for the belief that a company's environmental policy and program are always expensive. This belief has been altered by the benefits of implementing pollution prevention from the source, which reveals a better use of resources and energy and reduced dissipation of toxic compounds. The successful implementation of CP in case studies in the United States and Europe since the 1980s proves that ecology and economics can coexist (Baas, 1995; Hens et al., 2018). The approach of CP employs a technique that focuses on identifying and implementing pollution prevention concepts as alternatives or opportunities to waste management. These concepts are planned using a systematic approach, which informs stakeholders within the industry on the practices, processes, and benefits of applying CP. It brings together decision-makers responsible for developing, assessing, and implementing the opportunities for CP (Dieleman, 2007). The UNEP developed framework guides the step-by-step process of applying CP. The procedure consists of four stages: planning and organization, assessment to identify wastes and emissions as well as modifications to be made, analysis or feasibility, and implementation. The performance of CP in industries may vary based on the size and production process of the industry. For mining companies interested in implementing a comprehensive environmental management system (EMS) as a component of CP at their project sites, increased efforts and participation from governments, international environmental organizations, and educational institutions, and the companies are required to enhance the spread of EMS implementation throughout the industry (Hilson and Nayee, 2002).

The goal of the preventive approach of CP is to identify the sources of waste and pollutants generated throughout the production process. When these sources have been discovered, the next stage is to consider all options for eliminating or reducing the waste. Following the generation of a range of future possibilities, the method recommends conducting feasibility studies to analyze the economic and environmental consequences of the choices. Finally, the ideas and techniques that prove economically and financially viable are proposed for adoption (Dieleman, 2007). The process implementation is more dependent on the outcomes of assessments and feasibility. Usually, most companies decide to implement CP practices, and these projects typically begin after a specific issue has been recognized; then the final decision is with companies and their representatives (Dieleman, 2007). Due to the intricacies related to the environmental issues arising from mining (gold and iron ore) and waste management, the concept of CP is used to address the treatment and sustainability of these environmental concerns.

12.6.2 Cleaner Production Alternatives to Gold and Iron Ore Waste Management

The application of scientific knowledge and suitable technologies to recover resources from mining waste is arduous and requires significant financial input. Different procedures have been used over the past several years to create products with market value to offset the treatment costs by the industry (Seadon, 2010). The development of an integrated approach to CP guided by the CE, preventive strategies, or pollution prevention could undoubtedly support the advancement of sustainable mining (Flores et al., 2012). A paradigm shift is necessary to consider employing mine wastes for improved societal, technical, environmental, economic, and regulatory outcomes. Varnish, color and paint, rubber, ceramic enamels, soil restoration and recovery methods, filtration media, and coatings are examples of goods that can be utilized as raw materials in other sectors and help connect the mining industry with end-users. It would also allow the beneficiaries, namely companies, purchasers, and the government, to share a potential mechanism (Lowe, 2005). Social perception and acceptability of those products, in addition to scientific knowledge, are likewise essential and can only be appropriately determined through a multidisciplinary approach (Flores et al., 2012).

Since each mine's pollution potential is unique, the nature and size of the risk involved and the practicability of mitigation strategies will differ from one site to the next. There are no consistent approaches for assessing, measuring, and lowering the acid mine drainage (AMD) risk. The alternative use of waste from the iron ore and gold industries can be easily incorporated by implementing CP concepts (Akcil and Koldas, 2006). In the case of an abandoned colliery mine in the Republic of South Africa, AMD was mitigated or reduced by redirection of surface water that flows in the direction of the contaminated sites, preventing groundwater infiltration into the polluted area, as well as the managed disposal of acid-generating waste (Akcil and Koldas, 2006). With this, further pollution of groundwater was reduced, and the AMD was prevented from leaching to other surface water bodies.

Using iron mining tailings as pigments in the production of sustainable paint is an effective ecological way to handle wastes and promote sustainable paint production (Galvão et al., 2018). For decades, the painting industry has substituted lighter pigments such as hydrous ferric oxides, frequently found in iron tailings, for heavy metals such as lead. However, eco-friendly paint has a lower heavy metal content than traditional paint, which also has a higher range of harmful volatile organic compounds that can cause devastating diseases such as cancer and asthma. Eco-sustainable paint made with Portland cement and lime binders is more resistant to harsh weather and has a longer lifetime. Additionally, sustainable paints have a lower cost of production than conventional paints manufactured with heavy metal pigments (Galvão et al., 2018). Alumina, silica, and calcium oxides, which have similar physicochemical properties to building materials, are found in higher concentrations in iron tailings. As a result, substituting cheaper alternatives, such as silica and alumina, for traditional construction materials is economically and environmentally worthwhile (Krishna et al., 2020). Unfortunately, mining accidents occur, those happening in Sub-Saharan Africa taking devastating life tolls. The most recent tragedy happened in 2019, when the Gbanepea Gold Mine's pit walls crumbled, killing 40 people. Therefore, there is a need to implement an EMS, Corporate Social Responsibility (CSR), and tools of CP in the mining sectors of countries. Utilizing iron ore dry tailings to fill pits beneath the mining shaft is a cost-effective and potentially life-saving method of preventing such disasters (Bing et al., 2018). Iron oxide's applications in the medical, polymer, dermatological, and paint industries have increased due to the alteration of ferrous hydroxy sulfates to hematite in rivers and streams (Singh et al., 2020). The acidity of mining drainages can be neutralized by mixing iron mining tailings, sand, and waste solid debris with Portland cement (Kefeni et al., 2017). Barium sulfate, gypsum, and ferric hydroxide are the end products used as materials in concrete products. Furthermore, AMD has a high proportion of yttrium, a rare element used to make motors, rechargeable batteries, and lighting circuits (Ayora et al., 2016).

Iron ore sludge has a high superficial charge and capacity to function in a wider pH range. It is an effective adsorbent in wastewater treatment, notably for heavy metals removal (Akcil and Koldas, 2006). However, the treatment of AMD is expensive, particularly for countries with a gradually developing economy. Modern technologies such as membrane and magnetic nanotechnology can be used, but the sophisticated treatment of industrial wastewater is not applied in developing countries where they rely heavily on the traditional neutralization process (Wilson et al., 2017). With the severity of addressing AMD, individual gold and iron ore companies must assume their responsibility in monitoring effluents. They must recognize that AMD hazards exist in specific locations and pose specific environmental risks. It is better to modify the response where AMD is unavoidable, requiring site-specific research. Research should be conducted in mining regions to identify ways to prevent AMD. As a result, mine designers and management must be equipped with the information and instruments needed to regulate AMD—especially ways to mitigate AMD's impact on living forms and their support systems. (Akcil and Koldas, 2006). To successfully recover the metal that remains in these trash piles, techniques or a whole process should be created, thereby contributing value to the CE. Therefore, minimizing and recovering waste from watercourses are critical factors to consider when discussing resource efficiency. The CE is a financial concept that focuses on waste reduction, creation, and has vast economic potential. The goal of CE is to retain products and resources in continuous use while also preserving their value (Ellen MacArthur Foundation, 2015).

12.7 Case Study: Treatment of Wastewater and Resource Recovery in the Food Processing Industry

The food processing industry, which encompasses the food packaging industry, produces a variety of contaminated streams, such as sludge, solid debris, wastewater, and odors/gas emissions. The livestock and poultry industry, dairy industry, and seafood, fruit, and vegetable processing facilities are among the most common examples. They are primarily composed of organics, fat, oil, lubricants, nutrients, and other inorganic residues. Typically, a treatment train consisting of various technologies, such as pretreatment, dissolved air flotation, primary treatment, secondary treatment, and tertiary or advanced treatment, is employed to eliminate these contaminants. While the primary objective of wastewater treatment in the food processing industry has been to treat wastewater, there has been a recent trend to recover valuable resources by treating the wastewater and converting it into biogas, biohydrogen, fertilizers, and treated water reuse. In this section, some of the technologies used for the treatment of wastewater and resource recovery in the food processing industry are discussed.

In the context of wastewater treatment from the food processing industry, coagulation and flocculation have been extensively employed due to their low energy consumption, cost-efficiency, and ease of operation (Oladoja, 2015; Balbinoti et al., 2023). Balbinoti et al. (2023) have reported the following in their review paper: (i) plant-based coagulants can be applied efficiently to treat wastewater from the food industry and the partial/complete replacement of synthetic coagulants by plant-based coagulants will reduce the footprint of the effluent treatment, (ii) plant-based coagulants can be applied in the dairy industry, dairy cattle breeding facilities, palm oil milling, cassava processing, tapioca starch, coffee fruit pulping, coffee fermentation units, distillery spent wash, swine farm wastewater, and slaughterhouse wastewater, and (iii) the sources of coagulants are: *Moringa oleifera* seeds, *Cassia obtusifolia* seeds, vegetable tannins, rice starch, okra pods, ipin tree bark, iraodan tree bark, etc. The efficiency of the novel Fuller's earth clay ceramic membrane for treating the actual wastewater generated by the local dairy and palm oil processing industries was assessed in a study conducted by Rani et al. (2025). The following significant findings were reported: for the dairy and palm oil industries, COD removal was 96% and 92.5% at 0.35 bar, TSS and turbidity removal were both >98%, and the total phosphorus (P) concentration was <5 mg/L after filtration. Özmen et al. (2021) investigated the efficiency of a sacrificial Fe anode in a •OH-based electrochemical-Fenton (EF) process in the presence of hydrogen peroxide (H_2O_2) for the treatment of food industry effluent (FIE), i.e. vinegar production. The authors reported that the COD removal was ~99% under the following conditions: a COD concentration of 10 000 mg/L, a current density of 20.00 mA/cm^2, the addition of sodium sulfate (5 mM) as the supporting electrolyte, and an energy consumption of 36.17 kWh/m^3. In addition, it was also reported that the genotoxicity effects of the •OH-based EF were reduced (i.e. below the standard levels) despite the high COD value of the effluent, as demonstrated by the ICT-ANIARA umuC AQ genotoxicity test results. Mateus et al. (2021) investigated adsorption technology for removing nitrates and COD from effluent that is produced during the production of guava bars. The wastewater was pretreated with the coagulation-flocculation-sedimentation processes; subsequently, it was treated with bentonite impregnated with magnetic nanoparticles. The authors reported: (i) the effect of contact time, adsorbent dose, and temperature on the removal of different pollutants, achieving the following removal values—COD ~82%, nitrate ~74%, chloride ~98%, and sulfate ~70%, and (ii) the bentonite-based adsorbent can be reused for up to 10 cycles, and the treated water can be reused as long as the water quality complies with the Reuse Guideline (Res 1207–2014).

Sundaramahalingam and Sivashanmugam (2023) employed oleaginous yeast obtained from the FIE for their biological treatment application. In that study, the microbial oil of the yeast biomass was obtained through the bioaugmentation process, which occurred after the bioremediation step. The following are a few of the most significant findings of this study: (i) the lipid content of the oleaginous yeast (*Rhodotorula mucilaginosa* SML) was 67.95 (*w/w*%) of dry cell biomass, and (ii) the extracted microbial oil was utilized for transesterification, resulting in a 98% conversion at an oil-to-methanol ratio of 5.0, a catalyst concentration of 2.8%, and an operating time of 1.15 hours. The anammox (anaerobic ammonium oxidation) process was the subject of a chapter by Kha et al. (2024), in which the authors explained the mechanism and reviewed its application in the treatment of high BOD and N levels in wastewater from the meat, poultry, and seafood processing industries. For instance, the BOD concentrations in pig meat plants range from 1500 to 2000 mg/L, while they are in the range of 1400–3200 mg/L in cattle/sheep meat plants. The anammox process was also highlighted in the studies of Mai et al. (2021) and Sriariyanun et al. (2015) and it was reported that, using the anammox process, the nitrification step can be avoided in order to reduce the operation cost in industrial-scale treatment plants, and high N removals can be achieved at higher nitrogen loading rates.

In a review by Shrivastava et al. (2022), the authors reviewed the different treatment options for food industry wastewater and it was suggested that, technologies such as dissolved air floatation (DAF), photocatalytic degradation, reverse osmosis with membrane filtration (RO+MF), electrocoagulation (EC) and electrochemical systems can be used to treat food industry wastewater and recovery valuable resources such as methane; N, P, K fertilizers; biopolymers; lipids and biofuels; xanthan gum; and the treated water can be used for agricultural, household, and recreational activities. Neogi and Chatterjee (2021) conducted a review of the currently applied biological treatment technologies and identified their advantages and limitations in the treatment of effluent from the food processing industry. The significance of low pH methanogenesis for food processing industries was emphasized by the authors, who also elucidated the role of hydrolytic microbial consortia, fermentative acidogenic bacteria, acetogenic bacteria, and methanogens. In terms of technological applicability, the authors underscored the significance of considering additional factors when constructing full-scale bioreactors that perform acidic methanogenesis: (i) the use of molecular technology tools and culture methods, (ii) the source of the low pH tolerant inoculum, (iii) the reactor design, (iv) the selection of the appropriate reactor material for long-term operation, and (v) the environmental impact of the treated effluent.

The application of microbial fuel cells (MFCs) for the remediation of wastewaters in the food processing and beverage industry was the subject of a comprehensive chapter by Mathuriya and Pandit (2020). The following aspects, advantages of MFCs, were emphasized by the authors: (i) food waste has high moisture content and rich organics fraction, making it a suitable anolyte in MFCs, (ii) organic wastes can be used as the substrate to achieve sustainable power generation and effective wastewater treatment, and (iii) MFCs have been tested in various food industries and the agricultural sectors, including cassava, chocolate, composite vegetable waste, canteen-based food waste, seafood, cereal, meat packing, dairy, raw dairy manure, whey, molasses, brewery, and distillery industry. Cecconet et al. (2018) evaluated the efficiency of pollutant treatment and the potential for electric generation in MFCs by utilizing dairy effluent over a 75-day period of operation. The authors evaluated the effect of operational parameters on the electrical performance of MFCs and identified prospective strategies for reducing overpotentials. The authors reported a maximal power density of ~27 W/m^3, which was achieved by removing a significant amount of organic matter. Nevertheless, Cecconet et al. (2018) also reported the following operational limitations: (i) identifying strategies to prevent cathodic losses and other internal losses and (ii) minimizing non-exo-electrogenic reactions within the MFC.

Several case studies on the treatment of wastewater from dairy, table olive, and olive mill operations were presented by Tekerlekopoulou et al. (2020), with a focus on biotrickling filter, electrochemical oxidation, and constructed wetlands. CP options, life cycle analysis, biorefineries for food processing waste processing, wastewater treatment, and water reuse in the food industry, etc., among other important aspects, were discussed in the review. In addition, Tekerlekopoulou et al. (2020) provided the following useful suggestions: (i) the effluent discharge limits for the treatment of wastewater from the food industry are exceedingly challenging to achieve with a one-stage or stand-alone treatment system, (ii) to meet the strict regulations, hybrid systems or multiple-stage treatment systems should be implemented on a large scale, and (iii) examples of hybrid systems include biotrickling filter + constructed wetland, which can be used to treat high-strength wastewaters, while constructed wetlands + electrooxidation or biotrickling filter + electrooxidation can be used to treat low to medium-strength wastewaters from the food industry. Valencia-Valero et al. (2025) recently illustrated the utilization of solar photo-reforming process for the treatment of food industry wastewaters and the generation of hydrogen (e.g. >4.5 mmol H_2/g_{cat}. H; simulated sunlight condition). The authors utilized wastewater effluents from a processed food production facility (i.e. mainly fruits) and conducted laboratory-scale studies under simulated sunlight conditions to ascertain the optimum operating conditions of the photocatalytic process. Valencia-Valero et al. (2025) also conducted pre-pilot scale tests with photocatalytic panels utilizing natural sunlight, demonstrating continuous hydrogen production from real food industry wastewaters together with the concurrent breakdown of saccharides and other oxygenates. Consequently, that study illustrated that solar photocatalysis is a practically viable solution for the treatment of food industry wastewater and the recovery of renewable energy.

Another important area of research is the recovery of phosphorus (P) from food industry wastewater, as it is a vital nutrient that can become a pollutant when released in excess into aquatic environments, subsequently causing eutrophication problems. Wastewater treatment facilities can recover and recycle phosphorus, for instance, via chemical precipitation or biological phosphorus removal. Carolina et al. (2011) employed an integrated biological and chemical approach for the remediation of P-containing wastewater from the food industry. The authors evaluated the possibility of P removal in a model wastewater from the dairy industry through simultaneous precipitation in a laboratory-scale activated sludge reactor via the addition of ferric chloride. The authors reported that (i) the introduction of a high dosage of Fe:P (1.9–2.3:1.0) into the reactor's aeration basin resulted in the accumulation of inorganic solids within the sludge, and (ii) the application of the mathematical model and the biological indices ensures proper control of systems utilizing simultaneous phosphorus precipitation via ferric chloride.

In order to address the existing research gaps in cheese whey wastewater (CWW) treatment, and considering the techno-economic feasibility, environmental impact, and design, Tugume et al. (2025) investigated the design and scalability of CWW-based biogas digestion, followed by digestate pyrolysis for enhanced biochar production. The authors also identified the limitations/constraints that impede the implementation of CWW in developing countries. The financial gains from the injection of biogas into the gas grid, the sale of biochar, the low price from pollution reduction, and the acquisition of carbon allowances/credits were demonstrated by the authors to be capable of recouping the initial investment within a reasonable repayment period. The following aspects were highlighted and/or identified by the authors: (i) insufficient/lack of advanced organic matter bioconversion technologies, (ii) ecological emissions from processes, (iii) economic constraints, and (iv) societal implications of sustainable CWW management are the primary obstacles/limitations/barriers that were identified. Therefore, a diverse array of one- or two-stage processes can be employed to efficiently manage the high organic load for treating food industry wastewater, which includes sugars, fats, proteins, nutrients, minerals, solids, and by-products, in order to produce value-added products. Thus, the operational costs, environmental impacts, local bio-CE, and net-zero objective targets would all be either mitigated/achieved by these interventions.

12.8 Conclusion

This chapter focuses on the mining and food processing industries' efforts to transform effluent (wastewater/process water) into a valuable asset in accordance with sustainable development principles, with a particular emphasis on resource recovery and CP. It is clear from the review that the recovery of water, metals, nutrients, and energy from wastewater is feasible, while simultaneously advocating for CE principles. For instance, anaerobic ammonium oxidation (anammox) is a method that is employed to eliminate nitrogen at reduced aeration rates while maintaining an adequate amount of organic matter. This enhances the energy efficiency of the WWTPs. In addition, microbial fuel cells (MFCs) are energy-efficient technologies that can decrease the aeration costs and generate less sludge than the CAS process. It employs a technique that concentrates on the identification and implementation of pollution prevention concepts as alternatives or opportunities to waste management in relation to CP approaches. Anew, various technologies, including DAF, photocatalytic degradation, RO+MF, EC, and electrochemical systems, can be employed to treat wastewater from the food industry. These technologies can recover valuable resources, including methane; N, P, and K fertilizers; biopolymers; lipids; and biofuels, as well as xanthan gum. The treated water can be used for household, agricultural, and recreational purposes.

Acknowledgments

This chapter is part of a Master of Science degree thesis at IHE Delft Institute of Water Education (Delft, the Netherlands), which was sponsored by the Orange Knowledge Programme (NUFFIC scholarship program). Quinnetta Blessing Clement obtained the following funding for this research: (i) MSc study scholarship: OKP scholarship of the Netherlands Government/NUFFIC to participate in the Master Programme 2020–2022; Ref: EB/MRA/ACAD Student number: 1079694, and (ii) Research support for the advanced class: IHE Delft Institute for Water Education, ES programme fund: 100724. Abubakar M. Bilyaminu acknowledges the support received from the Joint Japan World Bank Graduate Scholarship Programme (grant number 583901) for his MSc degree, the Advanced Class in Translating Science into Application (grant number 2023/3711) and the Environmental Science Programme Research Fund (number 1079776) at the IHE Delft Institute for Water Education, the Netherlands.

Review Questions

1. In your home country, what type of mining industries are presently operating and how is their wastewater/process water being treated?
2. In your home country, what are the five major food industries and what process/wastewater treatment technologies do they adopt?
3. Provide some examples wherein extracting water, energy, and other resources from wastewater (industrial) would help to alleviate the environmental consequences of industrialization in your home country?
4. List at least ten (10) conventional industrial wastewater treatment technologies that can be applied in practice for treating mining and food industry wastewater/process water.
5. List at least ten (10) worldwide regulations/policies that facilitate the use of treated industrial wastewater/process water for reuse in different applications.
6. How can the concept of biorefinery be applied in the mining and food industrial sectors? Give five (5) suggestions based on your knowledge/experiences.
7. How can cleaner production principles be applied to reduce risks to humans and the environment?
8. What are the enabling factors and barriers for implementing cleaner production and resource recovery in the field of wastewater engineering? Give some examples from your home country.
9. What are the six (6) most preferred environmental technologies that can be applied in your home country for resource recovery from industrial wastewater? (Note: Please consider your country's present economic, environmental, and climatic conditions.)
10. Is it possible to increase the productivity of agriculture in your region (i.e. in your continent) through the use of water reuse? If yes, write/give some practical examples.

References

Akcil, A. and Koldas, S. (2006). Acid Mine Drainage (AMD): causes, treatment and case studies. *Journal of Cleaner Production*, **14**(12-13 SPEC. ISS.), 1139–1145. https://doi.org/10.1016/j.jclepro.2004.09.006

Arora, A.S., Nawaz, A., Qyyum, M.A. et al. (2021). Energy saving anammox technology-based nitrogen removal and bioenergy recovery from wastewater: inhibition mechanisms, state-of-the-art control strategies, and prospects. *Renewable and Sustainable Energy Reviews*, **135**, 110126.

Ayora, C., Macías, F., Torres, E.et al. (2016). Recovery of rare earth elements and yttrium from passive-remediation systems of acid mine drainage. *Environmental Science and Technology*, **50**(15), 8255–8262. https://doi.org/10.1021/acs.est.6b02084

Baas, L.W. (1995) Cleaner production: beyond projects. *Journal of Cleaner Production*, **3**(1–2), 55–59. https://doi.org/10.1016/0959-6526(95)00042-D.

Balbinoti, J.A., dos Santos Junior, R.E., de Sousa, L.B.F. et al. (2023). Plant-based coagulants for food industry wastewater treatment. *Journal of Water Process Engineering*, **52**, 103525.

Bing, L., Zhongying, Z., Biao, T. et al. (2018). Comprehensive utilization of iron tailings in China. *IOP Conference Series: Earth and Environmental Science*, **199**(4). https://doi.org/10.1088/1755-1315/199/4/042055

Boshagh, F., van Niel, E., and Lee, CJ. (2025). Techno-economic and life-cycle analyses of dark fermentative hydrogen production integrated with anaerobic digestion from coffee-manufacturing wastewater under thermophilic and mesophilic conditions. *Bioresource Technology*, **416**, 131737.

Carolina, D.G., Alejandro, H.C., and Noemí, E.Z. (2011). Application of a combined biological and chemical system for the treatment of phosphorus-containing wastewater from the food industry. *Procedia Food Science*, **1**, 1841–1847.

Cecconet, D., Molognoni, D., Callegari, A. et al. (2018). Agro-food industry wastewater treatment with microbial fuel cells: energetic recovery issues. *International Journal of Hydrogen Energy*, **43**, 500–511.

Dieleman, H. (2007). Cleaner production and innovation theory. Social experiments as a new model to engage in cleaner production. Revista Internacional de Contaminacion Ambiental, **23**(2), 79–94.

Durkin, A., Vinestock, T., and Guo, M. (2024). Towards planetary boundary sustainability of food processing wastewater, by resource recovery & emission reduction: a process system engineering perspective. *Carbon Capture Science & Technology*, **13**, 100319.

Ellen MacArthur Foundation. (2015). Towards a circular economy: business rationale for an accelerated transition. In *Ellen MacArthur Foundation (EMF)*. http://www.ellenmacarthurfoundation.org/case_studies

Flores, R.G., Andersen, S.L.F., Maia, L.K.K. et al. (2012). Recovery of iron oxides from acid mine drainage and their application as adsorbent or catalyst. *Journal of Environmental Management*, **111**, 53–60. https://doi.org/10.1016/j.jenvman.2012.06.017

Frijns, J., Hofman, J., and Nederlof, M. (2013). The potential of (waste)water as energy carrier. *Energy Conversion and Management*, **65**, 357–363.

Galvão, J.L.B., Andrade, H.D., Brigolini, G.J. et al. (2018). Reuse of iron ore tailings from tailings dams as pigment for sustainable paints. *Journal of Cleaner Production*, **200**, 412–422. https://doi.org/10.1016/j.jclepro.2018.07.313

Gherghel, A., Teodosiu, C., and De Gisi, S. (2019). A review on wastewater sludge valorisation and its challenges in the context of circular economy. *Journal of Cleaner Production*, **228**, 244–263.

Gu, Y., Li, Y., Li, X. et al. (2017). The feasibility and challenges of energy self-sufficient wastewater treatment plants. *Applied Energy*, **204**, 1463–1475.

Gu, Y., Dong, Y-N., Wang, H. et al. (2016). Quantification of the water, energy and carbon footprints of wastewater treatment plants in China considering a water–energy nexus perspective. *Ecological Indicators*, **60**, 402–409.

Gupta, A.S. and Khatiwada, D. (2024). Investigating the sustainability of biogas recovery systems in wastewater treatment plants-A circular bioeconomy approach. *Renewable and Sustainable Energy Reviews*, **199**, 114447.

Hens, L., Block, C., Cabello-Eras, J.J. et al. (2018). On the evolution of "Cleaner Production" as a concept and a practice. *Journal of Cleaner Production*, **172**, 3323–3333. https://doi.org/10.1016/j.jclepro.2017.11.082

Hilson, G. and Nayee, V. (2002). Environmental management system implementation in the mining industry: a key to achieving cleaner production. *International Journal of Mineral Processing*, **64**(1), 19–41. https://doi.org/10.1016/S0301-7516(01)00071-0

Hu, A., Fu, T., Ye, J. et al. (2024a). Photoelectrochemical methane production achieved energy recovery from wastewater with low organic concentration. *Chemical Engineering Journal*, **500**, 157113.

Hu, H., Li, P., Tong, Y.W. et al. (2024b). Reutilization recovered nutrients from wastewater by synthesizing eco-friendly slow-release hydrogel fertilizer for sustainable vegetable production. *Chemical Engineering Journal*, **499**, 155943.

Islam, AKMK. (2023). Domestic and industrial wastewater generation and its energy recovery potential in Bangladesh. *Cleaner Energy Systems*, **6**, 100092.

Joshi, J.S., Langwald, S.V., Ehrmann, A. et al. (2024). Algae-based biopolymers for batteries and biofuel applications in comparison with bacterial biopolymers-A review. *Polymers*, **16**(5), 610.

Kefeni, K.K., Msagati, T.A.M., and Mamba, B.B. (2017). Acid mine drainage: prevention, treatment options, and resource recovery: a review. *Journal of Cleaner Production*, **151**, 475–493. https://doi.org/10.1016/j.jclepro.2017.03.082

Keijer, T., Bakker, V., and Slootweg, J.C. (2019). Circular chemistry to enable a circular economy. *Nature Chemistry*, **11**(3), 190–195.

Kha, T.C., Nguyen, C.M., and Le, L.T.P. (2024). Chapter 5 – Microbiology and application of the anammox process for the treatment of wastewater in the food industry. In: *Anammox Process* (ed. Maulin P. Shah), 79–101. Elsevier.

Khalili, N.R., Duecker, S., Ashton, W. et al. (2015). From cleaner production to sustainable development: the role of academia. *Journal of Cleaner Production*, **96**, 30–43. https://doi.org/10.1016/j.jclepro.2014.01.099

Krishna, R.S., Mishra, J., Meher, S. et al.(2020). Industrial solid waste management through sustainable green technology: case study insights from steel and mining industry in Keonjhar, India. *Materials Today: Proceedings*, **33**(Part 8), 5243–5249. https://doi.org/10.1016/j.matpr.2020.02.949

Kumar, A.N., Yoon, J-J., Kumar, G. et al. (2021). Biotechnological valorization of algal biomass: an overview. *Systems Microbiology and Biomanufacturing*, **1**, 131–141.

Kumar, J.A., Sathish, S., Krithiga, T. et al. (2022). A comprehensive review on bio-hydrogen production from brewery industrial wastewater and its treatment methodologies. *Full Length Article*, **319**, 123594.

Kumar, R., Kundu, D., Kormoker, T. et al. (2024). Phycoremediation of potentially toxic elements for agricultural and industrial wastewater treatment: recent advances, challenges, and future prospects. *Desalination Water Treat*, **319**, 100505.

Lin, Z., Price, G.W., Liang, C., et al. (2024). Effects on soil carbon storage from municipal biosolids application to agricultural fields. *Journal of Environmental Management*, **361**, 121249.

Liu, W., Qian, J., Ding, H. et al. (2024). Synergistic interactions of light and dark biofilms in rotating algal biofilm system for enhanced aquaculture wastewater treatment. *Bioresource Technology*, **400**, 130654.

Lohani, S.P., Dhungana, B., Horn. H. et al. (2021). Small-scale biogas technology and clean cooking fuel: assessing the potential and links with SDGs in low-income countries – A case study of Nepal. *Sustainable Energy Technologies and Assessments*, **46**, 101301.

Lowe, E. (2005). Economic solutions. In *Environmental Solutions*. Elsevier Inc. https://doi.org/10.1016/B978-012088441-4/50005-8

Mai, W., Chen. J., Liu. H. et al. (2021). Advances in studies on microbiota involved in nitrogen removal processes and their applications in wastewater treatment. *Frontiers in Microbiology*, **12**, 746293.

Matebese, F., Mosai. A.K., Tutu. H. et al. (2024). Mining wastewater treatment technologies and resource recovery techniques: a review. *Heliyon*, **10**(3), e24730.

Mateus, A., Torres, J., and Marimon-Bolivar, W. (2021). Implementation of magnetic bentonite in food industry wastewater treatment for reuse in agricultural irrigation. *Water Resources and Industry*, **26**, 100154.

Mathuriya, A.S. and Pandit, S. (2020). Chapter 9 – Treatment of food processing and beverage industry wastewaters in microbial fuel cells. In: *Integrated Microbial Fuel Cells for Wastewater Treatment* (eds. R. Abbassi, A.K. Yadav, F. Khan et al.), 99–210. Butterworth-Heinemann.

McCarty, P.L. (2018). What is the best biological process for nitrogen removal: when and why? *Environmental Science & Technology*, **52**(7), 3835–3841.

Mohammed, K., Ahammad, S.Z., Sallis, P.J. et al. (2023). Hybrid microalgae-activated sludge system for carbon-efficient wastewater treatment. *Water Science & Technology*, **88**(3), 586–594.

Mudd, G. M. (2007). Sustainability and Mine Waste Management – A Snapshot of Mining Waste Issues. *Waste Management & Infrastructure Conference - IIR Conferences*, 1–12.

Neogi, S. and Chatterjee, P.K. (2021). Chapter 9 – Acidophilic methanogenesis for treatment of wastewater in food processing industries. In: *Handbook of Advanced Approaches Towards Pollution Prevention and Control* (ed. R.O. Abdel Rahman and C.M. Hussain), 249–273. Elsevier.

Obaideen, K., Shehata, N., Sayed, E.T.et al. (2022). The role of wastewater treatment in achieving sustainable development goals (SDGs) and sustainability guideline. *Energy Nexus*, **7**, 100112.

Oladoja, N.A. (2015). Headway on natural polymeric coagulants in water and wastewater treatment operations. *Journal of Water Process Engineering*, **6**, 174–192.

Oliveira Neto, G.C., de Tucci, H.N.P., Correia, J.M.F. et al. (2020). Assessing the implementation of cleaner production and company sizes: survey in textile companies. *Journal of Engineered Fibers and Fabrics*, **15**, 1558925020915585.

Özmen, F.K., Gerek, E.E., and Koparal, A.S. (2021). Treatment of food industry wastewater by •OH-based electrochemical-Fenton method and toxicity evaluation. *Desalination and Water Treatment*, **233**, 239–252.

Panepinto, D., Fiore, S., Zappone, M. et al. (2016). Evaluation of the energy efficiency of a large wastewater treatment plant in Italy. *Applied Energy*, **161**, 404–411.

Pavez, R., Ripoll, N., Briceño, P. et al. (2025). Review of international standards for the use of green H_2 in Chilean mining industry. *International Journal of Hydrogen Energy*, **103**, 975–986.

Pham, T., Rabaey, K., Aelterman, P. et al. (2006). Microbial fuel cells in relation to conventional anaerobic digestion technology. *Engineering in Life Sciences*, **6**, 285–292.

Pugazhendi, A., Jamal, M.T., Al-Mur, B.A. et al. (2022). Bioaugmentation of electrogenic halophiles in the treatment of pharmaceutical industrial wastewater and energy production in microbial fuel cell under saline condition. *Chemosphere*, **288**, 132515.

Qiao, L.L., Guo, J.S., Fang, F. et al. (2024). The recovery potential and utilization pathway of chemical energy from wastewater pollutants during wastewater treatment in China. *Journal of Environmental Management*, **356**, 120591.

Rani, S.L.S., Satyannarayana, K.V.V., Arthanareeswaran, G. et al. (2025). Treatment of food processing industries wastewaters using a new clay-based inorganic membrane: performance evaluation and fouling analysis. *Journal of the Taiwan Institute of Chemical Engineers*, **166**, 105439.

Roeleveld, P., Roorda, J., and Maarten, S. (2010). *NEWs: the Dutch roadmap for the WWTP of 2030*. Amersfoort.

Seadon, J.K. (2010). Sustainable waste management systems. *Journal of Cleaner Production*, **18**(16–17), 1639–1651. https://doi.org/10.1016/j.jclepro.2010.07.009

Sharma, P., Singh, S., Ramamurthy, P.C. et al. (2024). Regenerative resource recovery from wastewater: state-of-the-art bio-based soft technology. *Current Opinion in Environmental Science & Health*, **43**, 100587.

Shrivastava, V., Ali, I., Marjub, M.M. et al. (2022). Wastewater in the food industry: treatment technologies and reuse potential. *Chemosphere*, **293**, 133553.

Silvestre, G., Fernández, B., and Bonmatí, A. (2015). Significance of anaerobic digestion as a source of clean energy in wastewater treatment plants. *Energy Conversion and Management*, **101**, 255–262.

Singh, S., Sukla, L.B., and Goyal, S.K. (2020). Mine waste & circular economy. *Materials Today: Proceedings*, **30**(Part 2), 332–339. https://doi.org/10.1016/j.matpr.2020.01.616

Song, J., Dong, X., Yang, W. et al. (2023). Environmental co-benefits of energy recovery from wastewater of typical industrial sectors from life cycle perspective: regional potentials in China. *Energy Conversion and Management*, **293**, 117450.

Soo, A. and Shon, H.K. (2024). A nutrient circular economy framework for wastewater treatment plants. *Desalinat*, **592**, 118090.

Sriariyanun, M., Anh, D.H.Q., Tantayotai, P. et al. (2015). Anammox process: the principle, the technological development and recent industrial applications. *Applied Science and Engineering Progress*, **8**, 237–244.

Su, H., Wang, K., Lian, J. et al. (2023). Advanced treatment and resource recovery of brewery wastewater by co-cultivation of filamentous microalga *Tribonema aequale* and autochthonous bacteria. *Journal of Environmental Management*, **348**, 119285.

Sundaramahalingam, M.A. and Sivashanmugam, P. (2023). Concomitant strategy of wastewater treatment and biodiesel production using innate yeast cell (*Rhodotorula mucilaginosa*) from food industry sewerage and its energy system analysis. *Renewable Energy*, **208**, 52–62.

Suransh, J., Jadhav, D.A., Nguyen, D.D. et al. (2023). Scalable architecture of low-cost household microbial fuel cell for domestic wastewater treatment and simultaneous energy recovery. *Science of the Total Environment*, **857**, 159671.

Tekerlekopoulou, A.G., Economou, Ch.N., Tatoulis, T.I. et al. (2020). Chapter 8 – Wastewater treatment and water reuse in the food industry. In: *The Interaction of Food Industry and Environment* (ed. G. Charis), 245–280. Academic Press.

Tong, Y., Liao, X., He, Y. et al.(2024) Mitigating greenhouse gas emissions from municipal wastewater treatment in China. *Environmental Science and Ecotechnology*, **20**, 100341. https://doi.org/10.1016/j.ese.2023.100341

Tran, D.P-H., You, S.J., Bui, X.T. et al. (2024). Anaerobic membrane bioreactors for municipal wastewater: progress in resource and energy recovery improvement approaches. *Journal of Environmental Management*, **366**, 121855.

Tugume, M., Ibrahim, M.G., and Nasr, M. (2025). Valorization of cheese whey wastewater to achieve sustainable development goals. *Renewable and Sustainable Energy Reviews*, **211**, 115273.

United Nations (2018) News: 68% of the world population projected to live in urban areas by 2050, says UN, Department of Economic and Social Affairs. https://www.un.org/development/desa/en/news/population/2018-revision-of-world-urbanization-prospects.html (Accessed in January 2024).

UN Habitat and WHO. (2021). Progress on wastewater treatment – global status and acceleration needs for SDG indicator 6.3.1. Geneva: United Nations Human Settlements Programme (UN-Habitat) and World Health Organization (WHO).

Valencia-Valero, L.C., Simbaña, K.A., Eleraky, J.E. et al. (2025). Demonstration of hydrogen production from food industry wastewaters by solar photoreforming: from the laboratory to outdoors operation in panel reactors. *Chemical Engineering Journal*, **510**, 161636.

Verstraete, W., Van de Caveye, P., and Diamantis, V. (2009). Maximum use of resources present in domestic "used water." *Bioresource Technology*, **100**(23), 5537–5545.

Wang, D., Ye, W., Wu, G. et al. (2022). Greenhouse gas emissions from municipal wastewater treatment facilities in China from 2006 to 2019. *Scientific Data*, **9**(317). https://doi.org/10.1038/s41597-022-01439-7

Wechselberger, V., Hrad, M., Bühler, M. et al. (2025). Assessment of whole-site methane emissions from anaerobic digestion plants: towards establishing emission factors for various plant configurations. *Waste Management*, **191**, 253–263.

Wilson, S.T.K., Wang, H., Kabenge, M., and Qi, X. (2017). The mining sector of Liberia: current practices and environmental challenges. *Environmental Science and Pollution Research*, **24**(23), 18711–18720. https://doi.org/10.1007/s11356-017-9647-4

Yan, X., Zhang, X., Wang, R. et al. (2024). Phosphorus recovery from domestic wastewater via *Candida tropicalis*: performance and mechanism. *Journal of Water Process Engineering*, **68**, 106404.

Zhou, G., Jin, X., and Chen Z. (2025). Biosynthesis of manganese nanoparticles for efficient recovery of rare earth elements from mining wastewater. *Journal of Water Process Engineering*, **70**, 106987.

13

Utilization of Industrial Wastewater as an Input for Resource Recovery

Ahmed Salah Elkomy[1] *and Nabila Shehata*[2*]

[1]*Materials Science and Nanotechnology Department, Faculty of Postgraduate Studies for Advanced Sciences, Beni-Suef University, Beni-Suef, Egypt*
[2]*Environmental Science and Industrial Development Department, Faculty of Postgraduate Studies for Advanced Sciences (PSAS), Beni-Suef University, Beni-Suef, Egypt*

*Corresponding author: nabila.shehata@psas.bsu.ed.eg

13.1 Introduction

The main aim of the conventional wastewater treatment is to remove organic matter and destroy or deactivate pathogens found in the wastewater in order to protect the environment from pollution and stop the spread of disease. On the other hand, these industrial effluents contain high levels of several organic matters such as organics, carbohydrates, oil, inorganic salts, proteins, metals, and chemicals in addition to waste energy. The wastewaters generated from a series of various stages (washing, production, packing, cooling, coating, etc.) contain different waste materials/energy. The effluent wastewater with different levels of pollutants is usually collected and managed on-site or in a traditional sewage treatment plant. However, it is suggested that more effective treatment is required to ensure recovery of the waste materials/energy resulting in water with higher quality and recovery of valuable materials and waste energy.

Some techniques have been applied to recover valuable materials and energy. For example, physicochemical treatment processes are commonly used processes in the field of wastewater treatment as pretreatment. Physicochemical methods of treatment are implemented for the removal of inorganic and organic components, grease, and oil (Shamsan et al., 2023). Biological approaches for

Wastewater to Resource Recovery: Applying the Circular Economy Toward Sustainable Development, First Edition. Edited by Arvind Kumar Mungray et al. Published 2026 by John Wiley & Sons, Inc.

materials recovery are simple, practical, and cost-effective procedures that may be deployed on an industrial scale. Furthermore, it has demonstrated the importance of further research into the recovery of some valuable metals from wastewater (especially Ag, Au, Pd, and Pt), which might be attributed to the high costs and extensive use of these metals. Algae, bacteria, and fungi have generated more interest than other biomass materials utilized for metals biosorption. Ionic liquids (ILs) are significant organics with excellent metal recovery capabilities. Anionic exchange and ionic pairing are key processes of metals recovery. Some ILs, such as [3MOPYR][BF4], [C8mim][BF4], [OPYR][Cl], [OMIM][Cl], [OPYR][Tf2N], and [4MOPYR][TfO], have demonstrated great efficiency for valuable metals recovery in a variety of industrial activities, including liquid/liquid extraction, precipitation, and hollow fiber supported liquid membranes (Taghvaie Nakhjiri et al., 2022). Membrane technologies that include liquid membrane, reverse osmosis (RO), membrane distillation (MD), and forward osmosis (FO) are other promising methods for efficiently recovering valuable materials from waste streams. The FO method has demonstrated good performance in the recovery of various heavy metals from industrial wastewater, an essential pressure-driven membrane technology. RO has some advantages such as low pressure, modest space requirements, good selectivity, and strong separation performance. Recently, hybrid approaches have been regarded as an efficient means of combining biological and chemical processes to reap the benefits of both. Indeed, the applications of hybrid technologies that combine biological and chemical processes in order to improve material recovery efficiencies from various water bodies can be identified as an important outlook for the sustainable supplementation and application of materials (Taghvaie Nakhjiri et al., 2022). Figure 13.1 summarizes the different recovery technologies used in industrial wastewater treatment to recycle materials and energy.

Modern wastewater treatment plants (WWTPs) should contain steps that promote the more sustainable use of existing resources (water, electricity, and raw materials) in order to contribute to natural resource conservation. Numerous valuable metals have been recovered from industrial wastewater such as gold, ruthenium, rhenium, silver, rhodium, palladium, osmium, iridium, and platinum (Taghvaie Nakhjiri et al., 2022). Additionally, nutrients like chemical oxygen demand (COD) can provide a potential for transformation into useful organic compounds such as polyhydroxyalkanoates (PHA), hydrocolloids, and extracellular polymeric substances (EPS). WWTPs may recover vivianite [$Fe_3(PO_4)_2 \cdot 8H_2O$], and struvite ($MgNH_4PO_4 \cdot 6H_2O$) from burned sludge ash, which has a high phosphate concentration (≥90% of influent P) and is easily recoverable. From the perspective of energy recovery, anaerobic digestion of surplus sludge has been viewed skeptically. Incineration of dried sludge has been suggested as an effective method of energy recovery. Incineration may convert chemical energy based on influent

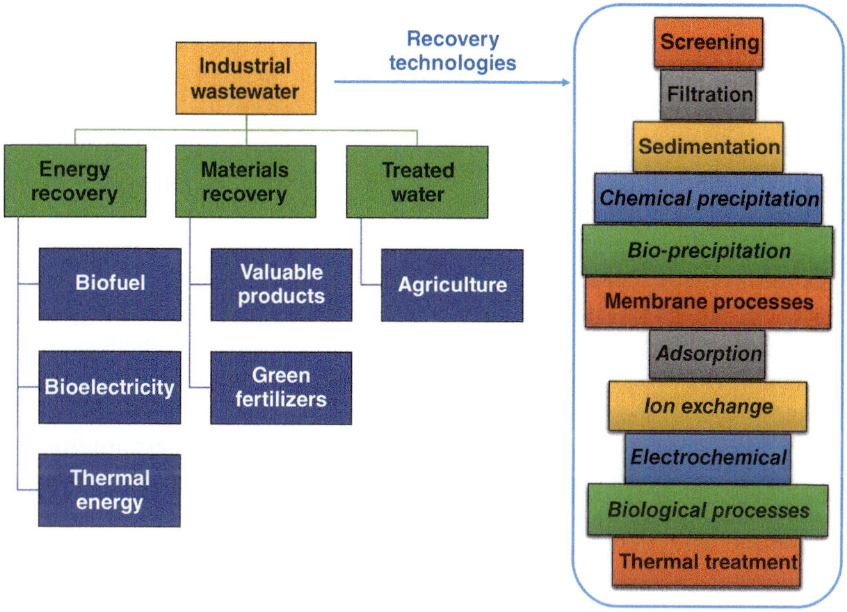

Figure 13.1 Treatment technologies used to recover water, materials, and energy from industrial wastewater.

COD with 32% efficiency, exceeding anaerobic digestion (<15%). Recently, decarbonization of hydrogen synthesized via wastewater effluent has been advocated (Hao et al., 2024). Resource recovery often serves as the foundation for a circular bioeconomy, which aims to conserve natural resources while also reducing the expenses involved with treating industrial wastewater. Hence, the main aim of this chapter is to explore the valuable materials in industrial wastewaters and the current technologies to recover these materials from different effluents and the contribution of the recovered materials to the circular economy.

13.2 Composition of Industrial Wastewater

The following sections list some valuable materials that can be recovered from different industries.

Batteries-based industry: Because of the many chemicals they produce, the battery manufacturing sectors are regarded as hazardous and poisonous. Grid casting, oxide milling, plate parting and curing, paste production, grid pasting, and battery construction are some of the phases in the manufacturing process.

The effluent had heavy metal (Li, Ni, Fe, Ag, Zn, and Pb) levels over the legally permitted discharge limits and had extremely acidic properties. Some sorbents have been used to recover these metals such as commercial resin, mineral dolomite, and residual biosorbents (Ribeiro et al., 2018).

Nuclear power facilities: Radioactive materials are emitted into the air and liquid phases as effluents from nuclear power facilities. Boiler water reactors (BWR) and pressurized water reactors (PWR) also release these effluents. Before being allowed to be released into the sea, liquid radioactive effluents are seen in surface water. Groundwater contamination is the outcome of uncontrollably leaking liquid radioactive effluents caused by fission activity. Radionuclides are used to reduce the amounts of liquid effluents generated by nuclear power reactors. The main isotopes in liquid radioactive waste are Co-60, Co-58, Fe-59, Mn-54, Sb-124, and Cr-51, which are generated by the contour corrosion products and neutron activation of pipes (Burcl, 2013). Radioactive wastewater is treated using biological processes, electrochemical routes, precipitation, adsorption, membranes, and evaporation systems (Kadadou et al., 2023).

Petroleum and petrochemical sectors: In the petroleum industry, waste materials—include ammonia (NH_3), carbon monoxide (CO), CO_2, hydrogen sulfide (H_2S), sulfur oxides (SO_x), nitrogen oxides (NO_x), plastics, oily wastewater, toxic drilling mud, sour water, spent amine solution, desalter wastewater, spent caustic, metal scraps, sewage, empty drums, oil-free wastewater, spent catalyst, charcoal bags, particulate matter, off-gas, oil sludge, oil-contaminated solids, and office and household waste—are all dealt with in the petroleum industry. While managing gaseous pollutants is relatively simpler, effectively treating liquid pollutants is necessary to prevent pollution of the environment. Petroleum industry-produced liquid pollutants are treated in effluent treatment plants (ETPs). Through the use of colloidal particles, biological processes, and refractory organics, the treatment systems remove oil, grease; coarse, suspended, and floating materials; and organic solids. To separate oil and grease, the oil–water separation process is commonly used.

Steel and iron industry: The most polluting source in the steel and iron industry is the by-product from the coke oven plant, which produces wastewater. This wastewater includes ammonia, cyanide, and phenol. Untreated effluent from the steel industry can be poisonous to living things and lower the amounts of dissolved oxygen in the water. The amount of dissolved oxygen in water bodies is reduced when biodegradable organic materials are released. The industry's effluents also have dangerous side effects, like slick development and unsightly appearance caused by trapped oil and grease.

Dairy industry: Large-scale milk processing is the primary source of wastewater with high concentrations of solid organic compounds produced by the dairy industry. The amount of milk processed and the kind of product produced determine the effluent's quality. In the dairy industry, physicochemical techniques such as mechanical, physicochemical, and biological methods are frequently employed for the treatment of wastewater (Kolev Slavov, 2017).

Oil extraction: Wastewater is mostly produced in oil extractive mills during the bleaching, rinsing, and the ETP stage. It contains mineral solids such as potassium and microflora. After undergoing basic treatment, the effluents proceed to secondary processes, including biological and physicochemical ones (Ammar et al., 2005; Hodaifa et al., 2008).

Leather industry: Among the industries with the greatest pollution and the quickest rate of growth worldwide is the leather industry. Unfavorable environmental conditions and the use of chemicals are produced throughout the leather production process. During the preparation, tanning, and crusting stages of the manufacturing process, raw materials undergo different treatments. The wastewater contains high COD/biochemical oxygen demand (BOD) levels, sulfides, trivalent chromium, Mg, Ca, organics, sodium chloride, and other toxic ingredients. Traditionally, the tannery industry's effluent is treated using adsorption, coagulation/flocculation, oxidation using Fenton's reagent, and nanofiltration. However, no treatment procedure has been established to eliminate all types of waste found in leather manufacturing effluent (Mandal et al., 2010).

Pharmaceutical industry: Since the pharmaceutical industry uses almost 99% of water in the process of producing excipients, it is one of the main causes of water contamination. The principal sources are wastewater from healthcare and hospitals, landfill leaks, treatment facilities, home sewage, and sludge. It contains sugar, organic matter, tannins, pharmaceutical residues, acids, and alkali contents. Adsorption, flocculation/coagulation, photolysis, photocatalysis, membrane, photo-Fenton, biological processes, and RO are some of the treatment techniques used to treat the effluent (Eniola et al., 2022).

Agro-industry: The commercial agricultural industry produces a lot of wastewater and uses a lot of water from many sources. Pollutants including organic materials, inorganic matter (dissolved minerals), nutrients (nitrogen, phosphorus, and potassium), hazardous compounds, and pathogens are commonly found in this effluent. Pesticides, BOD, COD, viruses, bacteria, suspended particles, fertilizers, and manure are all removed using different filtration techniques. In the agriculture sector, wastewater treatment includes mechanical techniques such as filtration, sedimentation, separation, flotation, and crystallization; biological techniques such as activated sludge in anaerobic environments; and physicochemical techniques such as bioelectrochemical process, RO, electrodialysis, and ion exchange. Additionally, microbial fuel cells were used to recover energy (Apollon et al., 2024).

Textile industry: One of the biggest and oldest sectors in the world, the textile sector is mostly to blame for pollution. About 70% of the pollution in this industry is caused by chemical processing. The goal of wastewater treatment in the textile industry is to remove suspended materials using techniques such as mechanical flocculation, chemical coagulation, sedimentation, screening, and neutralization. The principal objective of secondary treatment is to eliminate colloidal and dissolved organic components, along with any dyes or color found in wastewater, through the utilization of methods including anaerobic digestion, aerated lagoons, and trickling filtration. Before treated water is discharged into bodies of water, tertiary treatment is applied to boost its effectiveness even more. Membrane technology, electrolytic precipitation, and oxidation processes are used to accomplish this.

Pulp and paper industries: Paper and pulp industries produce a lot of biomass. A number of processes are involved in the production of pulp and paper, including debarking, chipping, pulping, and bleaching. In the paper and pulp industry, secondary biological treatment techniques like aerated lagoons, activated sludge, and anaerobic processes, such as acidogenesis, hydrolysis, methanogenesis, and acetogenesis, are commonly used in conjunction with primary treatment techniques like sedimentation, flotation, and filtration. Membrane processes, such as nanofiltration, ultrafiltration, RO and microfiltration, ozonation, and coagulation/precipitation, are all part of tertiary treatment (Sathya et al., 2022).

Food industry: The food industry also releases wastewater with a high COD and BOD composition, which pollutes the environment. The food industry uses a lot of water to run its operations, and total solids, BOD/COD, and suspended particles in the effluents vary depending on the additives used in different food products. Banana, orange, rice, grape skin pulp extract, wheat, palm biomass wastes, olive pomace, grape marc, oil mill wastewater, sunflower oil, pomegranate peel waste, fermented cheese whey, pomegranate peels, spent coffee ground waste, pineapple cannery waste, rice straw biomass, and empty fruit bunch are just a few of the waste materials that can be used in the recovery process.

Manufacturing and mining industries: Waste materials found in the mining and manufacturing sectors include organic constituents, bases and acid, heavy metal solutions, waste ink solvent, spent solvents, wastewater containing benzene and some hydrocarbons, spent petroleum catalysts, toluene and benzene, waste sludge, metallurgical slags, paint waste, particulate matter, mineral beneficiation tailings, paper mill waste, waste rock, paper sludge, paper/pulp wastewater, and pulp and paper liquor. Sources of polluted mine wastewater from mining operations include runoff from open pit workings, mill tailings, drainage from underground workings, waste rock dumps from mining activities, ore stockpiles, and spent ore from heap leach processing (Brown et al., 2002). Different metals can be recovered from the effluent of quarries and mines. Various methods, including chemical oxidation, neutralization, co-precipitation, and biological treatment, are employed to handle the wastewater discharged by quarries and mines. Moreover, waste materials found in the mining and manufacturing sectors include bases and acids, heavy metal solutions, and spent solvents.

Others: The environment is greatly impacted by the chemical industry, particularly those industries that produce organic chemicals. Many of the organic and inorganic contaminants found in the wastewater released by this industry are poisonous, mutagenic, carcinogenic, and non-biodegradable. Chemical effluents are treated using a variety of techniques, including flocculation, coagulation, de-emulsification, skimming, and dissolved air flotation. Waste products produced by the aluminum industry include spent pot lining, waste rock, red mud, scrap metal, and dross. Particulate matter, off-gas, waste gases (CO_2, SO_x, NO_x), coal fly ash, and coal boiler ash are examples of waste materials in the thermal industry (Srivastava et al., 2023).

13.3 Influence of Industrial Wastewater on Economics of Respective Industry

Impact on health: The uncontrolled release of hazardous materials into the environment is probably caused by an unregulated release of industrial effluent. Wastewater generated by several industries has been found to have negative health effects, such as lung and respiratory ailments, cancer, and immune system problems. Risk assessments are necessary to guarantee safer and more dependable wastewater management. This analysis should include hazard identification, exposure assessment, and risk characterization. Bacteria like *Escherichia coli* and *Salmonella*, which can cause cholera, typhoid fever, and allergy symptoms, are typically found in industrial effluent. Workers in the organic chemical industry are predominantly affected by myeloid leukemia, a disease that is frequently brought on by excessive formaldehyde exposure in the absence of adequate safety measures (Sathya et al., 2022). In conclusion, the improper treatment of industrial wastewater has a direct impact on the health of workers and the surrounding population.

Reducing dependence on fossil carbon: Coal and petroleum products are examples of fossil fuels that are important sources of energy for several sectors. But the amount of these fuels is limited, and continued use could eventually result in a shortage. Researching substitute energy sources that can meet demand in the future without endangering the environment is essential.

This objective gets a step closer with the recovery and utilization of heat from industrial wastewater. A new, naturally renewable energy source is produced by recovering bioenergy from industrial wastewater, which is essential for achieving the objectives of circular bioeconomy.

Production of new substances: Resources recovered from industrial effluent occasionally have distinct chemical properties that result in the development of new substances. These materials may be compounds, polymers, building blocks, or other materials with new properties. New products, including polymers, fertilizers, and biosolids, are produced as a result of the recovery process.

Improving sustainable agriculture: New forms of fertilizers can be used by the agriculture sector to increase output because of continuous advances in technology. Nevertheless, a large number of these fertilizers are harmful to the environment. Under these conditions, recovered biosolids from industrial wastewater (rich in phosphate and nitrogen) can be used as long-term fertilizers. Wastewater from paper, textile, meat, food, and dairy industries is produced in large quantities and contains important nutrients. Biosolids, which are these nutrients, can be recovered from wastewater and put to use on land for agriculture. Biosolids encourage plant development because of their high carbon, nitrogen, and phosphorus contents. As a result, there are no environmental issues with the agriculture sector using recovered biosolids from industrial effluent (Estahbanati et al., 2021).

Providing an ecological carbon source: Significant amounts of carbon are present in the separated sludge from industrial effluent, especially from the food, poultry, paper, and textile industries. These sludges can be used as biomass to produce electricity. This biomass can be converted into biofuels by processes such as hydrothermal liquefaction, thermoconversion, and biochemical conversion. Depending on where the biomass came from, these biofuels can have different calorific values. Biofuels offer a sustainable energy source and are good for the environment. Biofuels aid in the development of a circular bioeconomy by providing a fuel substitute for traditional fuels that doesn't harm the environment (Estahbanati et al., 2021).

Production of energy and electricity: Anaerobic digestion can be used to turn organic waste from municipal wastewater into heat and energy. It also produces humus and compost. This procedure increases power generation capacity, produces fertilizer and biogas, and streamlines trash recycling and landfill management. Anaerobic digestion has been estimated to have the potential to reduce greenhouse gas emissions by 4.36 gigatons of carbon dioxide equivalent when used for biogas generation from affordable waste. This represents 13% of global greenhouse gas emissions from activities like deforestation, evaded emissions management, crop burning, landfill gas, and fertilizer synthesis emissions (Peng et al., 2023).

Several alternative biofuels were investigated for energy recovery. For example, syngas is an efficient fuel that may be produced from wastewater using the supercritical treatment method. Unfortunately, their efficiency is restricted owing to the impurities, which might harm the turbine or fuel cell. As a result, energy recovery from wastewater is favorable since sludge converts into energy in a shorter period. The extreme operational conditions cause precipitation, clogging, and corrosion of the reactor, impeding the operation method. Additionally, nitrogenous fuels, biodiesel, and hydrogen may be used to efficiently recover energy from WWTPs. Photofermentation is used in conjunction with dark fermentation to boost total hydrogen generation. Municipal wastewater includes lipids, which may be accumulated anaerobically by bacteria, and a high output of biodiesel can be achieved by skimming the wastewater surface (Dutta et al., 2021). Phototrophic microalgae may be farmed in ponds and lakes to produce biodiesel. For nitrogenous fuel recovery, sludge incineration by full oxidation produces water, CO_2, and ash. The emitted heat is converted into electricity; a stand-alone combustion plant may generate 25–30% of the energy. Co-combustion with fluidized bed combustion provides energy efficiency of 40% at a low cost. The procedure is limited by the high-water content of waste sludge, which can be decreased to less than 30%. Microbes in bioelectrochemical systems oxidize COD, releasing electrons that are used for energy production (Dutta et al., 2021).

On the other hand, if the volume of treated wastewater is large, producing biogas during wastewater treatment can also save operating costs in addition to allowing the biogas to be recycled as a sustainable energy source. There are several uses for the biogas generated by anaerobic digestion. It can produce heat for heat exchangers and boilers in direct combustion systems. It can also be utilized to produce electricity in fuel cells and gas turbines. Converting biogas to natural gas, which may be used as a fuel for vehicles, is another possibility. Renewable natural gas with a 97% methane content can be developed by eliminating carbon dioxide, hydrogen sulfide, and other elements. Natural gas shortages may be avoided with the use of this procedure.

13.4 Resource Waste During Industrial Usage

This section lists several products that are wasted and can be acquired through resource recovery. Fertilizer, cellulose, biogas, bioproducts, nanoplastic/microplastic, and thermal energy are among these products.

13.4.1 Energy Resources

There are several different kinds of inorganic carbon resources, including carbon dioxide, carbonic acid, bicarbonates, and carbonates. Anaerobic digestion is one way by which industrial effluent with a high organic carbon content might be used as a possible energy source. Achieving sustainable development goals and protecting water resources depend on wastewater management done right (Estahbanati et al., 2021).

Rather than treating organic matter as pollution, it is possible to use it as a source of energy through the recovery of products from wastewater. Biogas is an important product that can be produced during anaerobic digestion from organic materials. The amount of organic matter in anaerobic digestion, for example, affects the composition of biogas, the organic matter's capacity for digestion, the rate at which digestion proceeds, the digester's physicochemical parameters (pH, retention period, temperature, and working pressure), and the existence of additional anions such as nitrates and sulfates. It's vital to remember that microbial habitat is essential to the generation of biogas and the elimination of COD. There are two primary uses for the biogas that is produced, depending on the wastewater flow rate: either upgrading it for other uses or burning it in a flare to use in heating devices (Estahbanati et al., 2021).

In fact, it is not economically possible to reuse biogas in small- to medium-sized treatment facilities with low wastewater flow rates. Pollution from water to air used to be transferred when biogas was released into the environment. Its use has detrimental effects on the environment and human health in addition to increasing greenhouse gas emissions. Consequently, it's crucial to refrain from releasing biogas into the atmosphere directly. In this case, biogas is burned in a flare instead, although this still has negative environmental effects and isn't regarded as an energy source.

The wastewater treatment industry should think about using clean technologies to recover waste thermal energy. One such technology that could be able to meet some of the industry's heating and cooling needs is wastewater heat recovery. This technology may reduce the use of fossil fuels, which will significantly lower operating costs and contribute to a reduction in global warming. Wastewater availability in a particular area and energy prices are two elements that affect wastewater heat recovery's economic viability. Because water has a high heat capacity, high flow rate wastewater can generally be regarded as an excellent source of thermal energy. Heat recovery from wastewater can be applied in three different places: at the WWTP, from the sewer, and inside the industrial plant. However, because of possible issues with wastewater treatment methods, fouling in heat exchangers, and corrosion brought on by the chemical makeup of sewage, heat recovery within industrial plants and sewers is not advised. It is advised to recover heat after treatment since it avoids these issues and offers a dependable energy supply. Recovered thermal energy can be applied to sludge preheating, digester heating, and the wastewater utility building's heating system. To reduce heat loss, excess heat should be used in the vicinity of the wastewater treatment facility. In order to extract heat from sewage, the influent needs to be filtered to get rid of big particles, and the heat exchanger tubes need to be cleaned on a regular basis to get rid of biofilm. Thermal energy recovery minimizes waste generation, saves energy, and lowers greenhouse gas emissions, all of which contributes to a smaller carbon footprint.

13.4.2 Products Recovery

Converting wastewaters into bio-factories is a technology that can potentially solve several problems at once. Water shortage, reducing the consumption of energy and natural resources, and other pressing issues are among these challenges. Wastewater can yield a number of important compounds, including cellulose, iron phosphate, polyhydroxyalkanoates, volatile fatty acids (VFAs), and carbon-chain elongation. Recovery of carboxylates—dissociated organic acids—from wastewater treatment technologies is the first step in the manufacture of VFAs. One potential recovery strategy is to include the carboxylate platform in wastewater. VFAs can be made by hydrolyzing and fermenting various microorganisms with sludge, preventing the growth of methanogens that grow slowly and reducing the amount of time that sludge is retained, or keeping the pH high while fermenting. After being extracted from the fermentation broth, these recovered products can be used as effective starting materials for additional fermentation and electrochemical processes that result in the production of fuels or bioplastics. The VFAs recovered from primary sludge are enhanced by adding sodium dodecylbenzene sulfonate or activated sludge to the fermentation broth, which has a higher pH during fermentation. Since they are biodegradable carbon sources that help with the removal of nitrogen and phosphorus, the liquid products from VFAs fermenter are ideal for the treatment route. Fermenting VFAs also helps with sludge management and disposal and cost reduction (Dutta et al., 2021).

In the circular bioeconomy, resource recovery plays a major role. The waste industry is especially promising in the current environment. Every day, industries produce a lot of wastewaters, which is rich in nutrients including phosphorus, nitrogen, and carbon. It is possible to recover these nutrients and use them to make a variety of products with additional value, such as fertilizers, biosolids, bioenergy, and biofuels. Building a circular bioeconomy depends heavily on the recovery of materials from industrial effluent. Resource recovery adds sustainability, reduces greenhouse gas emissions, increases eco-efficiency, lessens reliance on fossil fuels, produces new kinds of materials, and so on to the circular bioeconomy (Dutta et al., 2021).

13.5 Resource Recovery Methods to Be Applied in Industrial Sector

Some of the most recent developments in treatment procedures are covered in the part that follows, along with the potential and difficulties they present.

13.5.1 Physical Procedures

To remove large or solid particles from the wastewater stream, physical procedures are employed as the first treatment steps. Membrane filtration, sedimentation, screening, and filtering are some of these procedures.

Screening and Filtration: These procedures are used to take out suspended or floating items from wastewater. Various-sized screens are used to filter out different kinds of particles. In contrast, filtration uses particle sizes between 0.1 and 0.5 mm to get rid of things like bacteria, oils, and grease. Nevertheless, this method is sensitive to complex chemicals and ineffective for high metal concentrations.

Sedimentation: The process of lowering suspended solid particles in wastewater is known as sedimentation. To speed up the process, exogenous chemicals might be used. A number of variables, including solid loading, detention duration, and flow rate, affect how effective sedimentation is.

Membrane Filtration: This technique removes metals from mixed metal effluent by passing via a porous or non-porous membrane. The wastewater's composition, the membrane's pore size, and the material utilized all have an effect on membrane performance. Ultrafiltration (UF), RO, nanofiltration (NF), and microfiltration (MF) are examples of this technology. MF efficiently separates suspended or colloidal particles found in wastewater and is mostly employed as a pretreatment procedure. UF can remove viruses, polysaccharides, proteins, and macromolecules because it has a thicker surface coating. NF requires high pressure but effectively separates low molecular weight molecules. RO is frequently used to desalinate wastewater and is effective at removing multivalent ions. The permeability and pore structure of various membrane filtration techniques vary. The technique does have certain drawbacks, though, including expensive initial installation costs, higher running expenses because of fouling, and reduced removal effectiveness when additional metals are present. Membrane filtration uses backwashing and backpulsing to regulate membrane fouling. Backwashing is the process of forcing clean water through the membrane's permeable side, whereas backpulsing is the periodic, brief reversal of the transmembrane pressure. Benefits of backflushing include increased frequency and reduced time requirements. The membrane filtration process doesn't stop during backflushing.

13.5.2 Chemical Processes

One popular conventional method for removing metals from wastewater is chemical precipitation. To help the metals precipitate, several chemicals are applied to the effluent. Chemical precipitation falls into a number of categories, such as chelating, sulfide, and hydroxide precipitation. The process's efficacy is dependent upon various elements, including but not limited to pH variation, wastewater composition, metal concentration, reaction conditions, and the existence of chemicals capable of obstructing the reaction.

Adsorption: Another technique for removing metals from wastewater is adsorption. The wastewater is passed through a solid phase known as an adsorbent in this treatment, which enables the selected metals to be adsorbed onto its surface. The most widely utilized adsorbent is activated carbon, although zeolite, agricultural residues, and carbon nanotubes can also be employed. The process's cost and efficiency are impacted by the adsorbent selection, which makes it important. Nevertheless, certain adsorbents might be weak mechanically, have a low adsorption capacity, or dissolve in acidic environments. The adsorbent can be regenerated by a procedure known as desorption, and the adsorbed metals may be collected individually.

Ion exchange: Exchanging ions between a liquid and a solid phase is called ion exchange. Metals are extracted from wastewater using it. The exchange of ions can be influenced by variables like temperature, pH, anions, sorbate initial concentration and adsorbent mass, and duration of contact. In wastewater treatment, synthetic or natural resins are frequently utilized for ion exchange. Analytical chemistry/main group reactions/reactions in aqueous solutions/precipitation reactions, high cost, low metal extraction efficiency, and the requirement for regular resin replacement are some of the disadvantages of the method.

Electrochemical: Although it is not frequently employed in typical wastewater treatment methods because of their high cost and energy consumption, electrochemical treatments can also be used to remove metals from wastewater. The electrochemical recovery routes include E-oxidation recovery, E-reduction recovery, E-precipitation recovery and E-adsorption recovery. The main process parameters such as current density, electrolysis time and electrolyte type should be precisely controlled to improve the purity of recovered metals (Yang et al., 2021).

Chemical precipitation: Metals can be extracted from wastewater using a standard procedure called chemical precipitation. To help the metals precipitate, several chemicals are applied to the effluent. Chemical precipitation falls into a number of categories, such as chelating, sulfide, and hydroxide precipitation. A number of variables, including pH change, wastewater type, metal concentration, reaction conditions, and the presence of chemicals that can block the reaction, affect how effective the process is. Due to their high cost and energy consumption, electrochemical treatments are not frequently utilized in conventional wastewater treatment methods, but they can be effective in removing metals from wastewater. Microorganisms like bacteria, fungi, and algae are essential for cleaning wastewater and recovering metals in the biological approach. These microbes simplify metallic compounds so that they can be extracted from mixed metal solutions more easily.

13.5.3 Biological Processes

An alternate method for recovering metal from wastewater is through biological processes. Microorganisms, including algae, fungi, and bacteria, are essential for biological wastewater treatment. These microbes aid in the simplification of metallic compounds into simpler forms, which facilitate their separation from the mixtures of metal solutions. Various biological processes have been identified according to these microorganisms' mechanism of action.

Biosorption: Agro-industrial wastes, plant parts, biopolymers, and microorganisms, among others, extract metals from aqueous solutions and attach to their surfaces through a physicochemical process called biosorption. Through chemical interactions with functional groups like carboxyl, hydroxyl, amino, sulfhydryl, ester, phosphate, and carbonyl groups, the metals are bonded to the biosorbent. When it comes to recovering metal from wastewater, the biosorption process has a number of advantages over traditional techniques. It is an easy and economical procedure that doesn't need an extra source of nutrients. High absorption capacity, availability, and regenerability characterize biosorbents. A key distinction between biosorption and other common procedures is that the former does not raise the COD of wastewater. A set amount of time is needed for biosorption to reach equilibrium between the biosorbent (solid phase) and wastewater (liquid phase) containing metals. It then extracts the biosorbent together with the metals bound to it. Between the metals and the biosorbent, there are two steps in this process: physical adsorption and chemical interactions. Transportation through cell walls, ion exchange, complexation, surface adsorption, and precipitation are some of the mechanisms that support biosorption. The participation of microbe metabolism in biosorption determines whether it is considered metabolism-dependent or metabolism-independent. Transport of metals across cell wall is a component of metabolism-dependent biosorption. On the other hand, processes like ion exchange, precipitation, surface adsorption, and complexation are included in the metabolism of biosorption. The site of metal removal also divides biosorption into three categories: transport across cell membranes for extracellular accumulation/precipitation, ion exchange, complexation, precipitation for cell surface adsorption/precipitation, precipitation for intracellular accumulation, and physical adsorption. The biosorption process is affected by a number of variables, such as pH, temperature, time, metal ion concentration, type, specific surface area and mass of the sorbent, and selectivity toward metal. Bacteria are good at binding metals because of their charged polysaccharides and structure on the cell surface. Fungi also have high binding properties because of the components of their cell walls that include chitin, lipids, polyphosphates, and proteins. Yeast, algae, and leftover plant matter have also demonstrated potential for biosorption-based metal recovery.

Bio-precipitation: Biological agents are used in the wastewater treatment process called bio-precipitation to recover valuable components. Bio-precipitation provides an efficient and greener alternative to conventional precipitation, which depends on chemicals and has disadvantages such requiring a lot of chemicals and pH monitoring. Biological agents absorb metals by emitting metabolites that lead to stable compounds such as metal-sulfide precipitation. In the end, this procedure lowers operating expenses by reducing sludge formation and the related disposal issues (Dutta et al., 2021).

13.6 Conclusion

Sustainable development and circular economy have prompted the development of resource recovery. This chapter examined different treatment methods and resources recovered (valuable materials and energy) in current WWTPs. Wasting materials and energy in WWTPs have a negative impact on human health and ecosystems, while the recovery of valuable materials and energy from wastewater reduces the dependence on fossil carbon, develops new substances, enhances sustainable agriculture, provides an ecological carbon source, and acts as renewable energy and electricity. This chapter views industrial wastewater as a natural resource with enormous potential for sustainable development and economic growth. However, successful recovery of materials and energy from industrial wastewaters can only be assessed through life cycle assessment, pilot-scale and full-scale investigations, and feasibility studies.

Review Questions

1. Is there an urgent need to upgrade the current WWTPs?
2. How many new products can be recovered from industrial wastewater?
3. Do the current policies of industrial wastewater management need revisions?
4. What are the current challenges of industrial wastewater management?

5 To what extent does the recovery of valuable products from industrial wastewater impact the related industry?
6 Can the industrial plant achieve zero-liquid discharge?
7 Suggest new practice in the field of energy conservation and energy integration in the industrial plants
8 Does the public awareness contribute to the transition from linear economy to circular economy?
9 Does the recovery of radioactive products from nuclear wastewater have negative impacts?
10 What are the selection criteria for industrial wastewater treatment processes?

References

Ammar, E., Nasri, M., and Medhioub, K. (2005). Isolation of enterobacteria able to degrade simple aromatic compounds from the wastewater from olive oil extraction. *World Journal of Microbiology and Biotechnology*, **21**, 253–259.

Apollon, W., Rusyn, I., Kuleshova, T. et al. (2024). An overview of agro-industrial wastewater treatment using microbial fuel cells: recent advancements. *Journal of Water Process Engineering*, **58**, 104783. https://doi.org/10.1016/j.jwpe.2024.104783

Brown, Melanie, Bob Barley, and Harvey Wood. Minewater treatment. IWA publishing, 2002.

Burcl, R. (2013). Radioactive waste (RAW) categories, characterization and processing route selection. In: *Radioactive Waste Management and Contaminated Site Clean-Up*, 50–72. Elsevier.

Dutta, D., Arya, S., and Kumar, S. (2021). Industrial wastewater treatment: current trends, bottlenecks, and best practices. *Chemosphere*, **285**, 131245. https://doi.org/10.1016/j.chemosphere.2021.131245

Eniola, J.O., Kumar, R., Barakat, M.A. et al. (2022). A review on conventional and advanced hybrid technologies for pharmaceutical wastewater treatment. *Journal of Cleaner Production*, **356**, 131826. https://doi.org/10.1016/j.jclepro.2022.131826

Hao, X., Li, J., Liu, R. et al. (2024). Resource recovery from wastewater: what, why, and where? *Environmental Science & Technology*, **58**, 14065–14067. https://doi.org/10.1021/acs.est.4c05903

Hodaifa, G., Martínez, M.E., and Sánchez, S. (2008). Use of industrial wastewater from olive-oil extraction for biomass production of Scenedesmus obliquus. *Bioresource Technology*, **99**, 1111–1117. https://doi.org/10.1016/j.biortech.2007.02.020

Kadadou, D., Said, E.A., Ajaj, R. et al. (2023). Research advances in nuclear wastewater treatment using conventional and hybrid technologies: towards sustainable wastewater reuse and recovery. *Journal of Water Process Engineering*, **52**, 103604. https://doi.org/10.1016/j.jwpe.2023.103604

Estahbanati, M.R.K, Kumar, S., Khajvand, M. et al. (2021). 5 – Environmental impacts of recovery of resources from industrial wastewater. In: *Biomass, Biofuels, Biochemicals* (eds. A. Pandey, R.D Tyagi, and S. Varjani), 121–162. Elsevier. https://doi.org/10.1016/B978-0-12-821878-5.00003-9

Kolev Slavov, A. (2017). General characteristics and treatment possibilities of dairy wastewater–a review. *Food Technology and Biotechnology*, **55**, 14–28.

Mandal, T., Dasgupta, D., Mandal, S. et al. (2010). Treatment of leather industry wastewater by aerobic biological and Fenton oxidation process. *Journal of Hazardous Materials*, **180**, 204–211. https://doi.org/10.1016/j.jhazmat.2010.04.014

Peng, X., Jiang, Y., Chen, Z. et al. (2023). Recycling municipal, agricultural and industrial waste into energy, fertilizers, food and construction materials, and economic feasibility: a review. *Environmental Chemistry Letters*, **21**, 765–801. https://doi.org/10.1007/s10311-022-01551-5

Ribeiro, C., Scheufele, F.B., Espinoza-Quiñones, F.R. et al. (2018). A comprehensive evaluation of heavy metals removal from battery industry wastewaters by applying bio-residue, mineral and commercial adsorbent materials. *Journal of Materials Science*, **53**, 7976–7995. https://doi.org/10.1007/s10853-018-2150-6

Sathya, K., Nagarajan, K., Carlin Geor Malar, G. et al. (2022). A comprehensive review on comparison among effluent treatment methods and modern methods of treatment of industrial wastewater effluent from different sources. *Applied Water Science*, **12**(4), 70.

Shamsan, A., Fouad, M., Yacoob, W. et al. (2023). Performance of a variety of treatment processes to purify wastewater in the food industry. *Current Chemistry Letters*, **12**, 431–438.

Srivastava, R.R., Rajak, D.K., Ilyas, S. et al. (2023). Challenges, regulations, and case studies on sustainable management of industrial waste. *Minerals*, **13**, 51. https://doi.org/10.3390/min13010051

Taghvaie Nakhjiri, A., Sanaeepur, H., Ebadi Amooghin, A. et al. (2022). Recovery of precious metals from industrial wastewater towards resource recovery and environmental sustainability: a critical review. *Desalination*, **527**, 115510. https://doi.org/10.1016/j.desal.2021.115510

Yang, L., Hu, W., Chang, Z. et al. (2021). Electrochemical recovery and high value-added reutilization of heavy metal ions from wastewater: Recent advances and future trends. *Environment International*, **152**, 106512.

14

Decentralized Systems for Resource Recovery: Water Recovery Strategies from Wastewater

Upexa Patel[1], Ambika Arkatkar[1], Pankaj Pardeshi[2], Alka A. Mungray[3], and Arvind Kumar Mungray[3]*

[1]*Department of Biotechnology, Veer Narmad South Gujarat University, Surat, Gujarat, India*
[2]*Technical & Applied Chemistry Department, Veermata Jijabai Technological Institute (VJTI), Mumbai, India*
[3]*Department of Chemical Engineering, Sardar Vallabhbhai, National Institute of Technology, Surat, Gujarat, India*

*Corresponding author: pmpardeshi@hs.vjti.ac.in

14.1 Introduction

Treatment of wastewater is emerging as a challenge in the world of liquid waste management. Increasing urbanization has added more problems to this scenario. The municipal corporations are struggling to extend the collection network for sewage and maintain the quality of treated wastewater. Problems with wastewater contamination are linked to population expansion. The term "population expansion" describes the gradual rise in the population of a certain area. There is a greater need for shelter, food, and water as populations rise, which raises the demand for a variety of goods and services, and the quantity of wastewater produced rises. Wastewater comprises runoff from many sources, industrial discharges, and domestic sewage from homes (such as from sinks, toilets, and showers).

Global population growth figures show that the population size increases every year, from 2.6 billion to 5.2 billion. It has been forecasted that human population count will grow up to 9 billion by 2050, with 7.8 billion of those living in developing nations. The sanitation analysis reports state that 946 million people worldwide still defecate outside, while UNICEF and the World Health

Wastewater to Resource Recovery: Applying the Circular Economy Toward Sustainable Development, First Edition. Edited by Arvind Kumar Mungray et al. Published 2026 by John Wiley & Sons, Inc.

Organization (WHO) estimate that 2.4 billion people will need access to better sanitation facilities. Conventionally, potable water is used for washing, flushing toilets, industrial processes, and irrigation. In developing nations, wastewater treatment and disposal are common practices, with over 90% of their population contributing to the nation's centralized wastewater treatment systems (WWTS). For, the calculation of budget of WWTS, three fundamental elements are taken into account—collection, treatment, and disposal. The collection network makes up about 60% of the whole budget of WWTS. This component is scaled back and primarily concentrated on wastewater treatment and disposal in decentralized systems (Bernal et al., 2021).

Urban settings are an input source for a centralized wastewater system. In these systems, wastewater is gathered from several sources and sent to a central treatment plant via a network of sewer lines. The conventional centralized wastewater treatment cannot be the only solution for ensuring adequate treatment of contaminated water in the future. In order to achieve current and future sustainability goals, it has been suggested that a new paradigm based on a decentralized approach should be adopted. The collection, treatment based on composition, and reuse of wastewater at or close to the place of generation are known as decentralized wastewater management. Decentralized wastewater treatment systems (D-WWTS) are appropriate for isolated locations, tiny, low-density settlements, or individual properties. These systems feature a smaller conveyance network—sometimes only one pipeline—because they treat wastewater close to the point of generation (Bernal et al., 2021). Thus, it is used to treat or dispose of small volumes of wastewater from single-family homes or groups of homes that are relatively close to each other and are not connected to a regional wastewater treatment plant (WWTP) by a central sewer system. Although it was once thought to be useful for solitary residential areas, its use in new housing developments—such as suburban regions, industrial parks, and small communities—is becoming more and more typical. This solution was chosen initially because it might prevent effluent from being pumped over long distances and overtaxing current treatment facilities (Estévez et al., 2022). The decentralization approach helps in bridging the gap between on-site systems and centralized conventional systems. It creates systems that are more fiscally stable, socially conscious, and environmentally acceptable as compared to centralized conventional systems. Reducing the number of pollutants that water bodies receive after leaving the plant is another benefit of using DWWTS for the environment. Smaller wastewater quantities are handled by DWWTS closer to the source, which aids in more efficient pollution dilution. Chemical oxygen demand (COD), biochemical oxygen demand (BOD), and dissolved oxygen (DO) are three water quality indicators that are less affected by this dilution since it lowers the concentration of dangerous compounds. Essentially, compared to centralized systems handling huge quantities, contaminants are more evenly diffused and their impacts are less severe when wastewater is treated in a decentralized system (Capodaglio, 2017).

The goal of this chapter is to analyze several scenarios—from centralized to decentralized approaches—for the collection and treatment of wastewater from individual houses and small communities. Furthermore, from the standpoint of life cycle evaluation, several treatment plans are contrasted. Lastly, estimates and discussions of the consequences for the recovery of resources like biofertilizers and the reuse of treated wastewater are provided (Devi et al., 2022).

14.2 Need of Decentralized System

Water is the fundamental element required by all living organisms. For the sake of human health and dignity, water is vital. Over the last 250 years, there has been a significant surge in global population and economic activity. A significant amount of social transformation and a sharp rise in the demand for natural resources are the outcomes of this development process. The amount and quality of water resources are significantly impacted by urbanization, industrial growth, and the expansion of agricultural output. Wastewater production increases with the size of cities and the reach of water delivery systems. A significant portion of this wastewater is black water, which is produced when feces are flushed into the sewer line. If black water is utilized untreated, it can harm consumer health, particularly that of the underprivileged. Wastewater is treated and disposed of in metropolitan areas using centralized sewerage networks. Due to their high construction costs as well as ongoing operating and maintenance expenses, these treatment facilities are linked to significant investments.

Conventional wastewater collection and treatment systems need a lot of natural resources in addition to money and energy. Low performance of WWTP is a result of limited financial support in municipal budgets, low local competence, etc. (Gong et al., 2023). The methods used for treating wastewater range from fully onsite decentralized and cluster systems to more traditional centralized systems. The massive quantities of wastewater collected and treated by the centralized systems, which are often owned by the government, require the use of manholes, large pipelines, and extensive excavations. Decentralized onsite systems, on the other hand, handle the wastewater from specific residences and structures. Centralized systems frequently either dispose or reuse the treated wastewater far from its generating point, whereas decentralized systems do all the relevant processes (collection, treatment, and disposal) of treated wastewater at or near its generation point (Sutar and Kulkarni, 2015).

14.3 Decentralized Project Planning and Resource Recovery

Wastewater collection and treatment have a significant effect on the local and global economies as well as the environment. Eco-innovation might be of utmost importance in lessening the environmental effect of these systems as well as increasing their economic, environmental, and social sustainability. Decentralization systems emphasize on-site wastewater treatment as well as local resource recycling and reuse; thus, it seems like a sensible way to address sustainability issues with wastewater management systems (Amoatey et al., 2011).

14.3.1 Planning Infrastructure

The significant problems faced by Indian city inhabitants are the pollution of water resources and the scarcity of drinkable water. The flow of raw sewage in the city drains is considered as one of the prominent causes of contamination in water resources. Effective wastewater management and sanitation are becoming harder to achieve in highly populated places. In many developing nations, centralized sewage and wastewater treatment facilities are only available in big cities, and in heavily populated areas, on-site sanitation is usually insufficient. There is a need for intermediate and complementary solutions.

The planning of infrastructure for decentralized setup can be divided into the initiative taken by the government and non-government sectors. The decentralized systems are implemented by many private companies and educational institutes to achieve the in-campus treatment goals. The small townships, working company campus, and educational campus are treating the wastewater within their vicinity and using the treated water for non-potable purposes (Table 14.1). The decentralized setup is using technology as per the suitability of their treatment requirement. The wastewater is not added to the centralized sewer line; this reduces the burden on the municipal STP.

Decentralized Wastewater Treatment Solutions (DEWATS) have been put into practice in several locations using appropriate treatment techniques that fit the demands of the local population. The limitations of CWTS can be overcome by DEWATS that are generally used to link the communal sanitation centers. These are community-managed anaerobic DWWTS. Its application on the field can potentially enhance sanitation facilities in high-priority areas where the local government does not provide a comprehensive sanitation service. The standard DEWATS normally follows the technical treatment methods like modular fashion settlers, septic tanks, or biodigester in first treatment ponds. Anaerobic/facultative treatments are used as secondary treatment options that may include facultative and anaerobic pond systems,

Table 14.1 Applications of decentralized wastewater treatment solutions (DEWATS).

Place	Technology	Information	Resource recovery	References
Indian Institute of Technology Hyderabad (IITH), Telangana, India	Membrane Bioreactor technology	Two modular sewage treatment plants with a treatment capacity of 650 kiloliters of sewage per module are included in the Master Plan of IIT Hyderabad. Wastewater from the academic buildings, faculty and staff towers, canteen, dining hall, and dormitories is treated. Sewage is collected and sent to the STP using gravity driven through a complex sewer network.	Treated water is utilized in non-potable applications like air conditioning systems, flushing, and gardening. This technique produces sludge, which can be used in gardening.	(Hyderabad, 2023) It's a collaborative work of Prakriti Club, Unnat Bharat Abhiyan, Rural Development center, and climate change—IITH
Indian Institute of Technology Kharagpur (IIT KGP), West Bengal, India	Up-flow anaerobic sludge blanket (UASB) reactor and moving bed biofilm reactor (MBBR)	At the IIT Kharagpur campus, wastewater is gathered from various sources and sent to the treatment facility. Primary sludge is created using flotation or sedimentation techniques. Additional procedures improve the wastewater's ability to remove nutrients and organic debris. Treated wastewater is used for non-potable purposes.	The institution is planning to carry out resource recovery programs to extract important materials from wastewater.	(Kharagpur, 2023)

(Continued)

Table 14.1 (Continued)

Place	Technology	Information	Resource recovery	References
Infosys Bengaluru Campus, India	Advanced membrane bioreactor technology, energy-efficient pumps, Sequential Batch Reactors (SBRs)	Recycled wastewater is used on Infosys campuses for air conditioning cooling, gardening, and landscaping. Furthermore, all buildings have specialized plumbing lines for treated water, which makes it easier to use recovered wastewater for flushing. Infosys and Leibniz University Hannover, Germany, worked together to create automated solar dryers that are used to clean and dry sludge. Wastewater treatment and solar panel cleaning are automated by a robotic system. When irrigating campus plants, a mixture of dried sludge and organic compost is used.	Treated Water and Compost.	(Infosys Limited, 2024)
Jawaharlal Nehru University (JNU), New Delhi, India	Advanced Oxidation Processes (AOP), Membrane Bioreactor (MBR)	Strongly oxidizing hydroxyl radicals (•OH), which are extremely reactive, are produced during AOP. When these radicals come into contact with the organic and inorganic contaminants found in wastewater, they break them down into less toxic and simpler molecules. Inorganic pollutants like cyanides and heavy metals can also be targeted by AOP, which can precipitate or change them into less harmful forms before removal.	Reusing treated wastewater for non-potable uses is possible.	(JNU, 2024)

horizontal gravel filters, anaerobic baffled reactors, and anaerobic filters. The ponds are used after treatment for aerobic polishing. DEWATS usually entails treating wastewater from homes, businesses, or small towns on-site (Geetha Varma et al., 2022).

Saraswati 2.0 is a four-year initiative sponsored jointly by the Department of Science and Technology (DST)/Department of Biotechnology (DBT) of the Government of India and the European Union's Horizon 2020 Research and Innovation program. The project expands on the 2012–2017 implementation of the previous Saraswati initiative, which proved successful.

The project's goals include:

- Saraswati 2.0 seeks to determine which technologies are most suitable for resource recovery and decentralized wastewater treatment in India.
- It offers a flexible solution to problems with water shortage by concentrating on placing water and wastewater treatment facilities at the point of water supply or demand.
- Through the promotion of safe wastewater treatment and effective water use, the initiative aims to improve sustainable practices.

Saraswati 2.0 intends to use decentralized wastewater treatment facilities to improve India's water security. The initiative looks for low-cost, efficient solutions that can minimize pollution, enhance water quality, and advance sustainable water management (EU and DST/DBT, 2019).

14.3.2 Technology Advancement

Around the world, several DWWTS and technologies have been created and put into use.

Nepal's EcoSan Systems: The goal of ECOlogical SANitation (EcoSan) toilets, which have been thoroughly tested in Nepal, is to use human waste as a resource for food security and agriculture (Arora et al., 2016). The goal of EcoSan, or ecological sanitation, is to securely close the sanitation-agriculture loop (Bhandari and Shrestha, 2018). Up till now, Nepal has pushed the Ecosan latrine technology, which operates on a three-system method. Two of its three primary systems handle the excreta in shallow pits; the first system gathers feces in shallow pits and separates them from urine, while the second system prepares feces for composting. Urine and feces are treated independently in the third system, which keeps them apart. The idea behind all Ecosan latrines in Nepal is that adding soil and ashes to feces causes the mixture to break down quickly, producing compost that is beneficial for any farm or garden. As long as the combination is not very moist, it has no smell. In comparison to alternative sanitation systems, Ecosan latrines are inexpensive (limited not just to initial investment) and provide several advantages such as being permanent, manageable, and gradually producing rich compost (Betti, 2008).

Compost toilets: Compost-based sanitation systems are popular under various names like dry toilets, bio-toilets, composting toilets, biological toilets, or waterless toilets. The fundamental principle of these toilets' designs is to convert human waste into compost below the toilet, resulting in a low-cost supply chain need. The closest sewerage system is used to dispose of the pee collected in these toilets (Arora et al., 2016). Because composting toilets use very little to no water, they may be disconnected from the wastewater and water systems. At the system level, these water and wastewater reductions can be substantial. The cycling and transfer of nutrients are another benefit of composting toilets. Similar to those from conventional wastewater treatment facilities, the solids collected from composting toilets can be used as fertilizer. However, if used locally—where compost is made—they won't be polluted by urban runoff and may need less shipping (Anand and Apul, 2014).

Septic tank and absorption fields: The septic tank, a distribution device, and an absorption field are the three main parts of a septic tank system. A septic tank is a sizable, subterranean, leak-proof, and corrosion-resistant receptacle that collects raw sewage from a home's plumbing drains. Anaerobic microorganisms in it partially decompose solids when they are isolated from the raw sewage. The raw sewage and partial breakdown must be retained in the septic tank for a minimum of 48 hours. Undigested solids either settle as sludge at the bottom of the tank or float to the top to form a scum layer. To enable bacterial digestion to continue in the tank, the sludge and scum must be pumped out at least every two to five years, depending on the size of the tank and the volume of sewage. If not, untreated sewage could enter the absorption field through the tank and lead to the collapse of the system.

By passing via the distribution mechanism, the liquid effluent from the septic tank is distributed evenly to every pipe in the absorption field. Receiving the liquid effluent from the distribution device, the absorption field is a subsurface leaching region within the soil that spreads it over a designated area where it is permitted to seep into the soil. After the toxic material in the effluent is treated and disease organisms are eliminated, the soil's filtering effect works in combination with additional bacterial activity to finish the treatment process, allowing the water to be recycled back into the surface or groundwater source (Adegoke and Stenstrom, 2019).

Moving bed biofilm reactor (MBBR) system: The moving bed biofilm reactor (MBBR) is essentially an improved version of the combined activated sludge and biofilter processes, combining the best aspects of each. Utilizing an appropriate sieve system at the reactor exit, the biomass is produced on freely moving carriers known as biocarriers that are positioned inside the reactor (Madan et al., 2022).

The biocarriers are available in many forms; the smallest are hollow plastic cylinders that are widely used, which have a high density of 0.95–0.98 g/cm^3. The agitation generated by the aeration process aids in the movement of the biofilm carrier in aerobic, anoxic, and anaerobic processes. The agitator moves the biofilm carrier physically, as illustrated in Figure 14.1a. Some benefits of the MBBR technology include less head loss, decreased carrier medium clogging, and enhanced volumetric treatment capacity. The entire working tank volume, including the activated sludge, facilitates the growth of biomass. The extra biomass is not removed, thus there is no need for sludge recycling. The ability to evaluate the volume of the biofilm media carrier to be employed in accordance with the needs and treatment quantity is yet another important advantage of employing a moving bed biofilm reactor. The precise biofilm area can therefore

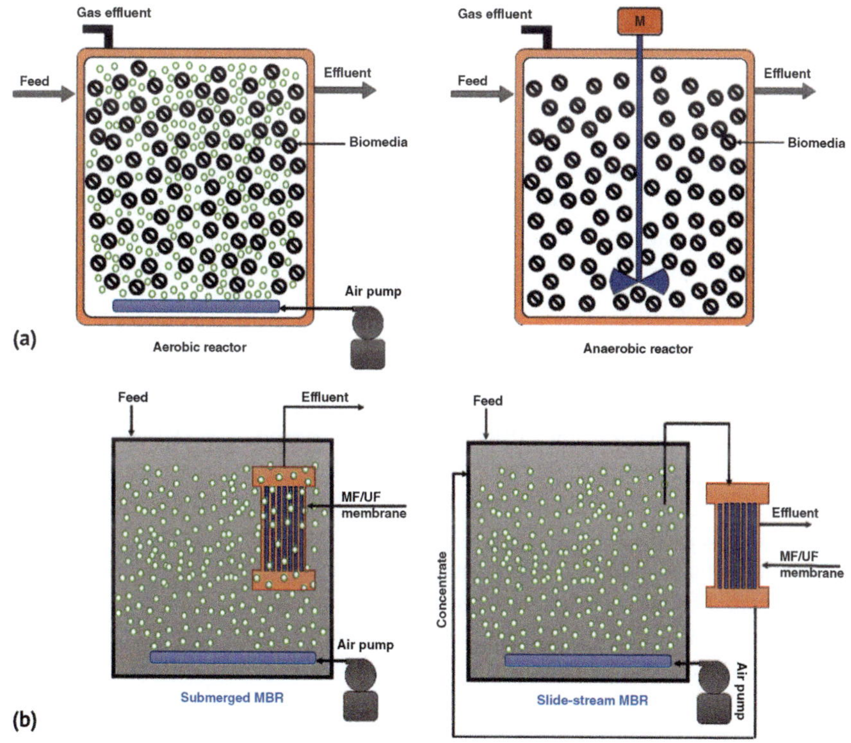

Figure 14.1 Schematic diagrams of (a) aerobic and anaerobic MBBR and (b) submerged and slide-stream MBR process.

be changed appropriately. It is advised that the volume of carriers used should not exceed 70% of the entire reactor volume, nor should it exceed 67% of that amount. For example, the 70% filling criterion is applied with a possible film growth area of 500 m^2/m^3, and since biofilm formation occurs inside the carrier, the specific surface area for biofilm growth will be around 350 m^2/m^3. The fixed biomass in MBBR can reach 10 000–12 000 mg/L, whereas the biomass concentration in the reactor is roughly between 3 and 4 kg SS/m^3, which is comparable to powered sludge. As a result of the high biomass concentration, moving bed processes have a volumetric removal rate that is sufficiently high in comparison to other similar systems, making them far more practical (Lago et al., 2024). The concept of operation of a moving bed biofilm reactor in both aerobic and anaerobic processes is illustrated in Figure 14.1a.

Membrane bioreactor (MBR) system: MBR is a wastewater treatment method that combines membrane technology with biological treatment (aerobic and anaerobic). Unlike conventional biological treatments that use a clarifier for gravity settling, this approach uses microfiltration or ultrafiltration to separate the sludge produced by biological treatments. MBR has a number of advantages over the traditional activated sludge method. While the hydraulic retention time (HRT) in MBR is shorter than in the CAS process, the solid retention time (SRT) in MBR is higher than in CAS. Furthermore, in the case of MBR, sludge separation is more effective. MBR requires less space and is excellent for water reclamation because of its superior effluent quality in terms of suspended particles, turbidity, and BOD. In addition to CAS, MBR can be employed in anaerobic treatments in place of traditional anaerobic digestion (AD) through the use of an anaerobic baffled tank reactor, expanded granular sludge bed (EGSB), or up-flow anaerobic sludge blanket (UASB). By regulating the biomass concentration, the anaerobic membrane bioreactor (AnMBR) may generate high-quality wastewater with a lower COD than the traditional method (Kim et al., 2024).

Membrane separation in MBRs enables the decoupling of SRT and HRT, in addition to rejecting bacteria and larger molecules. Porous, low-pressure-driven membranes are commonly utilized in membrane bioreactors; nevertheless, their effectiveness is primarily restricted to the reuse of wastewater. Many attempts have been made recently to enhance hybrid MBR procedures for wastewater reclamation by combining desalination technology. Reverse osmosis (RO) and nanofiltration (NF) have been used in conjunction with MBRs to enhance effluent quality. The benefits and drawbacks of these processes have been well documented, including issues with fouling management, rejection efficiency, operating energy consumption, and retentate stream recovery. As an alternative, non-pressurized desalination technologies like membrane distillation (MD) and forward osmosis (FO) have been heavily investigated for their low energy consumption and superior solid material rejection efficiency. These technologies could be directly incorporated into MBR processes for the purpose of wastewater reclamation or sometimes even for microbial activity. To put hybrid FO- or MD-MBR processes on the market, a few technological issues still need to be fixed (Kim et al., 2024). A schematic diagram of MBR is shown in Figure 14.1b.

Anaerobic baffled reactor (ABR): The compartments of these reactors that are in charge of the AD of organic wastes have alternating baffle walls that are home to microorganisms. An ABR is made up of a tank and alternating hanging and standing baffles that divide the reactors into compartments and make liquid flow up and down between them.

This allows the fresh wastewater that enters the reactor to have better contact with the residual sludge, which contains the microorganisms that break down the organic pollutants anaerobically. ABRs may remove between 65% and 90% of COD. This is far better than a conventional septic tank. Typically, 50% of the entire volume of TSS is removed in the sedimentation chamber at the start of the ABR. Total suspended solids (TSS) removal of up to 90% may be accomplished because of the unique design, which also enables improved treatment of non-settleable particles. The effluent has a BOD reduction of between 70% and 80% and a high concentration of nutrients. Additionally, the effluent satisfies the requirements for agricultural reuse. A schematic diagram of ABR is shown in Figure 14.2a.

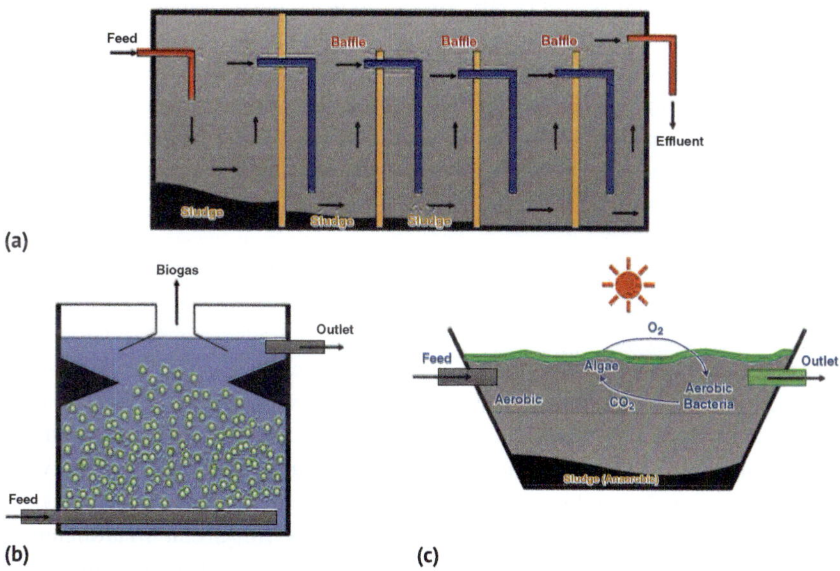

Figure 14.2 Schematic diagrams of (a) anaerobic baffled reactor (ABR), (b) UASB, and (c) waste stabilization pond.

Up-flow anaerobic sludge blanket (UASB): Wastewater is cleaned by a floating sludge blanket as it rises from the tank's bottom. The organic contaminants are anaerobically broken down by the bacteria in the sludge. The granular consistency of the UASB biomass is attributed to the activity of microorganisms, which leads to the production of granular groups that have a larger dimension, compact structure, higher density, and a higher settling capacity than the biomass that is produced in the typical active sludge (CAS), which is generally flock-type. Psychrophilic, mesophilic, or thermophilic operating environments can be achieved using UASB, contingent upon the kind of matrices supplied. Thus, the researchers achieved success in degrading COD that was resistant to degradation at higher temperatures. The case was especially true for industrial wastewaters (WWs). The organic material degrades quickly due to the high microbial variety and biomass concentration (60–100 kgVSS m^{-3}) in the granules. Waste with a high amount of organic matter can also be treated using the UASB method in tiny reactors. A schematic diagram of UASB is shown in Figure 14.2b.

Waste stabilization pond: This treatment technique takes up a sizable amount of space and includes three layers of treatment facilities: facultative, aerobic, and anaerobic. The upper layer receives oxygen from the surroundings while the lower layer is oxygen-depleted. Such a natural waste stabilization pond (WSP) may be added to centralized or semi-centralized sewage systems to treat sludge, septage, and wastewater from homes and businesses. WSPs are inexpensive, easy to use, easy to build, and very efficient. It can be coupled with or utilized alone as a secondary or tertiary treatment unit in a treatment plant. When considering maintenance costs and energy requirements, WSP becomes more cost-effective than other treatment methods because of its algal-bacterial symbiosis, which makes it a fully natural treatment process. WSP wastewater may also be utilized for gardening, farming, road irrigation, car washes, and other purposes. A schematic diagram of a waste stabilization pond is shown in Figure 14.2c.

14.3.3 Probable Resource Recovery

Wastewater has changed over the last few decades from being a waste to a useful resource. By reclaiming water, wastewater not only lessens the impact of water scarcity but it also serves as a medium for the recovery of nutrients and energy, which helps counterbalance the depletion of valuable resources. The resource recovery methods that may be used to recover energy, water, and nutrients from household and municipal wastewater can be divided into categories with an emphasis on implementation size. The system sizes can be divided into three categories: big (scales beyond 3800 m^3/day), medium (design flows of 8–20 000 m^3/day), and tiny (design flows of 17 m^3/day or less). It was noted that concentrated nutrient recovery was found in small-scale systems employing urine source separation methods, and energy recovery was primarily accomplished in large-scale facilities through hydraulic head loss or biosolids management. The key aspects that support larger-scale resource recovery are lower per unit costs and energy consumption for treatment, whereas capacity to recover and reuse resources on-site, which saves up the transportation and distribution costs, is a supportive aspect of small-scale systems. Pharmaceuticals and personal hygiene items pose a limitation for the resource recovery project at all scales, although certain impurities could be effectively removed using unit techniques including RO, nanofiltration, activated carbon, and sophisticated oxidation processes (Diaz-Elsayed et al., 2019).

One of the most essential resources for human existence and productivity is water. Although surface rivers, lakes, and subterranean aquifers provide the majority of the world's fresh water supply, overexploitation of these resources has drawn criticism and is likely exacerbated by urbanization, economic expansion, changing land uses, and climate change. The current state of water scarcity would worsen with an increasing population. A report by the US National Research Council (NRC, 2012) makes reference to the Census Bureau's population projection and notes that, in the absence of the development of significant new resources, per capita water use in the United States is expected to decline for the next 50 years. Treated municipal wastewater offers a climate-independent, high-quality, locally controllable supply of fresh water that is superior to other water obtained from recovery systems like desalinated seawater, collected rainwater, or water from melted icebergs. It also has a large potential to contribute to meeting future water needs. A paradigm change in wastewater management is required to handle the clean water crisis in a sustainable manner. This shift should use wastewater as a renewable resource, and a thorough evaluation of resource recovery technologies is crucial (Diaz-Elsayed et al., 2019).

Tata Steel (in India) is making a concerted effort to maximize the recycling of treated waste effluents inside its operations and minimize the amount of fresh water extracted from rivers. In 2018, the Tata Steel Bara Tertiary Treatment Plant was put into operation. Tata Steel may repurpose the effluent recovered from the Bara WWTP by its Jamshedpur-based facility. Tata Steel has cut its input of fresh water from the Subarnarekha River by 16% by processing and recycling wastewater. Tata Steel gains from this program, and it also helps the area's water security.

One of the creative ways Mahindra & Mahindra has reduced water usage in its offices and industrial facilities is by flushing toilets with the condensate water from its air conditioning units. They followed this technology at their various centers located all over India. Their dedication to sustainability and effective water management is demonstrated by this practice. Mahindra & Mahindra contributes to their overall environmental stewardship aims by minimizing waste and reducing the demand for fresh water through the capture and reuse of AC condensate water (Mahindra Lifespace Developers Ltd, 2024).

One of the best techniques for processing biosolids from industrial and municipal wastewater is AD. Biogas is the term for the methane gas produced during the treatment of wastewater sludge by AD. It has to be viewed not just as a clean energy source but also as one of the most promising answers to the major environmental issues related to CO_2 emissions, waste management, and water contamination. The AD procedure yields biogas, which is usually combustible, moist, and contaminated with caustic substances like hydrogen sulfide. These features must be carefully considered when designing a gas handling system, as improper handling could pose a safety hazard. Compared to an all-fossil fuel-based energy system, biogas is a clean, sustainable energy source and has advantages both on financial and environmental fronts. It is created when organic waste breaks down anaerobically. The economic viability of ADs may be hindered by the diversity of microorganisms that may lead to lower production of biogas or internal utilization of biogas by other microorganisms. The chemical load in the sludge can be one of the factors that hinders biogas production, as it affects the production of microbial enzymes that are required for enzymatic and bioconversion activities. Methane composition is also affected by the procedure and the kind of substrate used. This also determines the energy content or calorific value of biogas. The heating value of biogas typically ranges from 21.5 to 23.5 MJ/m^3. Right now, estimates place the annual amount of electricity produced by biogas at about 7000 MW. Methane composition, which is affected by the procedure and the kind of substrate utilized, determines the energy content of biogas. Although there are a number of elements, including composition and proportion, that affect the biogas's calorific value, the biogas typically has a heating value of 21–23.5 MJ/m^3. One cubic meter of biogas is equivalent to 0.5–0.6 L of fuel and 6 kWh of electricity (Yadav et al., 2024).

The Vellore Institute of Technology (VIT), Vellore, India, campus has set up a biogas plant to process wastewater from its canteens and organic waste. As a component of their sustainability efforts, this biogas plant generates biogas for the campus's use in cooking and other uses. In addition to aiding with trash management, this also lowers carbon emissions and conserves electricity. VIT exemplifies a pragmatic and environmentally conscious strategy for handling organic waste and advancing sustainable energy sources by using wastewater and canteen waste (VIT, 2024).

Microbial fuel cells (MFC) use the catabolic properties of microorganisms to decompose organic materials and liberate electrons, which then travel through an external circuit resulting in energy production. In this technique, high-energy-rich electrons are released during the biochemical process that breaks down the organic compound. These electrons are then taken up by electron acceptors like oxygen. However, because there isn't enough oxygen in the cell, the electrons are moved to the cathode. This external movement of electrons causes electricity production (Bazina et al., 2023).

Green sanitation technology is being advanced by Ghangrekar's bioelectric toilet. It turns human waste into clean water and power using MFCs. By dissolving organic materials, the MFCs release electrons that are then collected to produce energy. The toilet processes wastewater and produces reusable water at the same time, which may be used for flushing or other non-potable uses (Jadhav et al., 2020).

14.4 Case Study

The main issues with wastewater management in Timor-Leste (East Timor) include rapid population increase and inadequate wastewater treatment facilities (da Costa et al., 2024). The government of Timor-Leste has adopted the strategy of separating wastewater from residential, commercial, industrial, and greywater regions into blackwater and greywater categories. In order to create a cost-effective wastewater strategy, three methods were used to gather information about the locations and discharge of grey and black water: an open-source mapping of wastewater discharge locations was used, interviews were carried out with over 130 participants from local authorities and communities, and data collection was also done through a field survey. According to the study's findings, 47.7% of grey wastewater is dumped into open sewers that are directly connected to the ocean. At the grassroots level, many communities lack awareness regarding policies, regulations, and wastewater management. They hardly understand the health risks associated with wastewater. As a result, the majority of towns discharge their wastewater directly into the environment. Due to inadequate wastewater treatment, 73.8% of the children in these villages have experienced diarrhea, and during the rainy season, there is a higher risk of contracting waterborne illnesses. To alleviate the wastewater situation and solve additional issues like biodiversity loss, heat stress, drought, and flooding, low-cost nature-based solutions such as built wetlands and bioswales can be installed with local skills and resources. Given the various natural disasters that might occur in Timor-Leste, including landslides, earthquakes, strong winds, and fluvial and pluvial flooding, these installations can act as coastal protection and are less complicated to reconstruct than large-scale grey infrastructure.

The goal of Mavugara et al. (2024) was to evaluate Zvishavane Urban, Zimbabwe's municipal wastewater system's potential for resource recovery. It justified the reuse of waste, nutrients (nitrogen, phosphorus, and potassium), and energy recovery. This analysis used wastewater data and material flow analysis. The relationship between the quantities of resources to be recovered and physicochemical parameters of wastewater was found to be mathematically linear in nature. The Resource Value Mapping online tool was used for data input and analysis, and descriptive statistics were used to show the results. The study's conclusions showed that 103 660 m^3 of water could be reused, which would cover 22%

of Zvishavane Urban's water needs for irrigation of green spaces and local agriculture. Furthermore, the findings show that the valorization of sewage sludge has the potential to provide up to US$1 739 142.80 in revenue annually, 92940 Nm^3 of biogas, and 1557 tonnes of solid fuel. The study showed that the Mabula wastewater treatment facility can yield 11 861 tonnes of digestate and 3613 tonnes of compost in terms of nutrient recovery. Regarding the nutritional composition, it is possible to retrieve 589 tonnes of phosphorus, 663.2 tonnes of nitrogen, and 294 tonnes of potassium. The potential income from recovering nutrients was US$1 413 600. The study shows how municipal wastewater streams in Zvishavane have the potential for recover resources. Studies conducted in underdeveloped nations on the circular economy frequently exclude this information. Wastewater treatment for resource recovery is in line with both local and global priorities as outlined in the Sustainable Development Goals, as well as Zimbabwe's National Development Strategy. This study attests to the substantial resources that are incorporated in and recoverable from Zvishavane Urban wastewater. The findings of this study are significant because they shed light on the advantages of resource recovery for planners and decision-makers in Zvishavane Urban and Zimbabwe. This is an essential step in the right direction for providing efficient wastewater management and outlining the best course for recovering resources from urban wastewater. These plans and strategies will not be limited to Zimbabwe but can be applicable to other developing nations in the area and beyond.

Swan et al. (2023)'s investigation in Yangon, Myanmar, revealed that the city's rapidly increasing urbanization and deteriorating infrastructure significantly impeded wastewater management. Even with initiatives like the Greater Yangon Project to improve conditions, funding limitations and limited land availability may prevent advancement. To influence long-term infrastructure planning, it was essential to understand the performance of the CWTS and the decentralized wastewater treatment system (DEWATS). The technological, social, economic, and environmental elements of these systems were thoroughly examined in this study. The study's key finding was that DEWATS, as opposed to the costly and time-consuming CWTS master plan, is a feasible option for the city. They did the analysis using mixed methods like surveys, interviews, and weighted sum techniques involving citizens and municipal authorities. Conversations revealed that residents are more in favor of DEWATS due to its affordability and environmental advantages, but they are less satisfied with CWTS due to its high operating and maintenance (O&M) costs and unreliability. Over the course of 10 days, DEWATS consistently met acceptable levels in the wastewater quality analysis, demonstrating stable performance and effective pollutant removal. However, CWTS often exceeds permitted limits in order to ensure consistent adherence to standards. DEWATS also surpassed CWTS by employing Weighted Sum Methods, scoring 102 across three sustainability scenarios.

DEWATS is a superior option in the context of resource conservation. As compared to CWT it provides efficient treatment and pollution reduction. The implementation of a CWTS's requires a handsome budget and land area. Its reliability and flexibility become the responsibility of both decision-makers and end-users. These findings corroborate the findings of previous studies conducted in other developing countries, which demonstrated the superiority of DEWATS. The limitations of the in-depth study stem from people's perceptions on the costs and advancements of technology. Further study was suggested in order to fully comprehend wastewater management methods. Not just Yangon, but other developing cities with similar problems can also benefit from these findings.

In Abu Dhabi University's Design and Economic Evaluation of Recycling Greywater, Madhuranthakam et al. (2023) investigated the decentralization of wastewater management. In order to manufacture fresh water derived from the greywater generated at the Abu Dhabi University campus in the United Arab Emirates, they investigated the economic viability of doing so and built a decentralized plant. A compact design of filtration unit, chemical treatment, and disinfection processes will be required for the production of treated wastewater. This treated water can be utilized for horticulture on and around the nearby campus or stored and supplied for irrigation. Throughout the process, a number of parameters were measured and tracked, including total suspended solids, COD, and biological oxygen demand. These parameters were controlled by the proper operations carried out for each unit. Lab-scale studies utilizing a factorial three-level design were conducted to treat greywater from Abu Dhabi University. The results indicated that treated water suitable for horticulture could be achieved through the use of coagulation, flocculation, and chlorination stages. Greywater can be treated with alum (30 mg/L), PAM (0.6 mg/L), and chlorine (1 mg/L) (sodium hypochlorite) to lower TDS, TSS, BOD, and COD levels while keeping the ideal pH range. With the goal of processing 90 000 tons of water, a conceptual modular greywater treatment unit was designed using the experimental results. This plant can be built to the size of two parking lots with a speculative fixed capital investment of US$8 million and a discounted rate of return of 34%, according to detailed design and economic calculations. The primary constraint on this research would be the constant fluctuation in the parameters of greywater. However, the ideal dosages of sodium hypochlorite, flocculant, and coagulant can be related to various beginning greywater parameters using an empirical model. In addition to lessening the strain on the standard CWTS, the suggested decentralization of greywater treatment will also cut treatment costs and provide a sustainable means of closing the water consumption and wastewater production.

Small towns and the majority of rural settlements in the Kingdom of Saudi Arabia (KSA) store household wastewater in home septic tanks and transport it to the closest centralized wastewater treatment facility, according to the case study examined by Alresheedi et al. (2023). The lack of a sanitary sewerage system presents the locals with a number of socioeconomic and environmental issues, including the creation of disagreeable gases, leaking septic tanks, and sewage collection vehicles. For a small community of 1300 members (160 homes) in Qassim (KSA), this study created a resident perception-based technique to evaluate the sustainability of a low-cost ceramic filter bioreactor-type decentralized wastewater treatment system (DWWTS). Six demographic and nine other parameters evaluated the opinions of the locals regarding the current and proposed wastewater management systems. The three elements of sustainability were assessed through the use of four environmental, nine social, and three economic indicators in a hierarchical sub-index system. The study area's 34 respondents were asked questions based on the indicators. The relationship between the responses and the projected DWWTS's willingness to accept was evaluated statistically. The responses were converted into performance scores using a subjective evaluation system, and the scores were then combined into sub- and top-level indices using a fuzzy-based technique. The highest level of the hierarchy demonstrated a strong correlation between the sustainability of DWWTS and the residents' perceptions. The study discovered that a somewhat high readiness to reuse treated wastewater and WTA the decentralized system was a result of citizens' awareness of the environment and resource conservation. The analysis also demonstrated that a DWWTS's economic feasibility maintained a reasonable performance level because of its cheap monthly waste disposal costs. The results of the study show great promise for long-term, government-funded, community-maintained DWWTS. Decision-makers in ministries responsible for agriculture, water resources, and environmental conservation can more easily assess the sustainability of DWWTS for small towns in arid areas with the help of this suggested methodology.

14.5 Economic Feasibility

Worldwide, WWTPs have generally been used on a large scale. The amount of money spent on building, designing, and acquiring land is all considered capital costs. According to the report, system size directly affects total cost, but plant size has little bearing on performance. There is a dearth of information on the economic analysis of decentralized wastewater management systems in underdeveloped nations, but some data is accessible for industrialized nations. Developing nations typically adopt anaerobic and mixed treatment systems for small communities. This method works in a smaller zone and produces higher-quality organics and nutrient removal. Conversely, natural systems are less effective and require a bigger area, especially when it comes to removing nutrients. A thorough wastewater treatment approach evaluation and selection must take into account the life cycle costs associated with three fronts: design and construction, operation and maintenance, and repair and replacement. The main reason why cost estimates are difficult to determine nationally and generally is that the conditions in each location vary, including local expenses, land costs, and population density. According to a study, decentralized systems are typically less expensive in rural areas than urban areas. Some studies have been reviewed in terms of economic feasibility of decentralized systems as follows.

Rodríguez et al. (2020) conducted a social and economic analysis of greywater treatment pilot projects that were installed in remote schools in Chile's Coquimbo Region. The treatment systems' initial investment, ongoing operating and maintenance expenses, and environmental costs were calculated for this study. The advantages were measured in terms of water savings and the advantages that green spaces have for the environment. The treatment systems can be advantageous for these communities on both social and environmental fronts. Additionally, a sensitivity analysis was added to the study that considered investment cost, discount rate, amount of greywater treated per capita, green area expenses, and frequency of system maintenance. As a consequence of the initial investment expenditures above US$5200, which are not compensated by the water savings. Therefore, the results indicated that the adoption of these systems would not be economically feasible. The assessment of treatment systems should, therefore, be given top priority to additional advantages such as improved life quality, temperature regulation, and the sense of wellbeing and contentment experienced by teachers and pupils.

This kind of project should be evaluated with greater consideration, though, given the area's increasing water shortage and the socioeconomic benefits that come from more water being available in arid and semi-arid regions. Green spaces can be irrigated thanks to greywater, which is particularly useful in regions with limited water supplies and little rainfall. More green space in schools is associated with improved output and performance as well as higher levels of happiness. Additionally, it helps the area's green space availability per resident to rise, which is below the WHO-recommended threshold. Thus, it is important to assess greywater initiatives from a social perspective and improve funding by means of public policies that encourage increased sustainability in educational institutions. In order to support sustainable education, the state should encourage laws that permit financial assistance and encourage partnerships with private organizations.

The techno-economic assessment of extreme decentralization of water and wastewater was conducted by Garrido-Baserba et al. (2022). The researchers studied the techno-economic viability of establishing independent building-scale decentralized systems. The study combined the rainfall collection, potable water production, wastewater treatment and recycling of water. The data were collected from six major building types that included single-family houses and even high-rise complexes. Five different treatment approaches were evaluated in five different climatic scenarios for each type of structure. The plans considered nearly all types of toilet (vacuum, urine diverting, and conventional), and various levels of source separation (i.e. black, grey, yellow, brown, and combined wastewater) along with the required pipelines and pumping. A detailed economic analysis that considered both capital expenditure (CapEx) and operating expenditure (OpEx) indicated that the cost of installing on-site water harvesting and recycling systems would increase the overall construction costs. The increase in cost will be about 6% for multi-family buildings and about 12% for single-family homes. In case of a building or water system in which we have more than 300 occupants, the anticipated total cost of on-site water provision (including harvesting, treatment, recycling, and monitoring) ranged from $1.5/m^3 to 2.7/m^3. Compared to the normal rates collected by utilities in the United States and Western Europe, this is a huge reduction. When buildings are able to avoid connecting to centralized supplies for the disposal of potable water and sewage, their water bills may be greatly reduced. Urine-diversion is the least tested and has the most uncertainty in the cost analysis, yet it can result in the least priced option. There was less cost uncertainty and greater competition for more proven systems, including membrane bioreactors. The findings of this study indicated that economically viable solutions that eliminate or drastically reduce the need for centralized water and sewage collection systems as well as substantially reduce the demand on centralized utilities can be constructed.

14.6 Future Prospects and Conclusion

The majority of cities rely on centralized systems for treating wastewater, which come with substantial operating and maintenance expenses in addition to hefty capital requirements. A decentralized wastewater management system could be a useful way to reduce pollution and recover valuable recourses given the speed at which urbanization is occurring. Furthermore, a decentralized system for treating small amounts of wastewater needs to offer advanced waste treatment, just like conventional wastewater treatment procedures. It also needs to be affordable, simple to use, reliable, and easy to maintain. Decentralized approaches can typically satisfy regional regulations for water use and reuse. Locally treated water, for instance, can be added to supplies of drinking water or utilized to boost agricultural production. Further evaluations are necessary due to the difficulties and potential future prospects of treating wastewater using decentralized systems such as:

- One of the biggest obstacles to putting such projects into action is frequently the high expense of wastewater treatment and management.
- The concerns of governments in developing nations, such as those pertaining to food supply, healthcare, and conflict resolution, are more urgent than

wastewater management. Oftentimes, wastewater management is not given high priority.

- Political meddling in environmental issues, including site selection and other building and operation-related matters, is a problem for many developing nations.
- To ensure optimal effectiveness, even the most advanced technology needs to be supported by the relevant institutions and laws that are enforced.
- International organizations and developed nations must provide financial help, but in order to fully utilize any aid, local conditions must be taken into account. Funding these kinds of projects would be pointless otherwise.
- Inappropriate technology adoption and disregard for the local conditions of the targeted community lead to project failure, which is frequently attributed to a lack of funding and technical expertise.
- Millions of dollars are occasionally spent on building, while only a few pennies are used to acquire accurate design data. It is advantageous to replicate successful projects, but the system should be modified to account for local circumstances, particularly the climate. The least expensive technology is typically selected without further thought.
- One funded project is insufficient to fulfill the existing and future sanitation needs of rural communities in developing countries. Sustainable wastewater management necessitates a comprehensive, long-term plan with significant phases of planning and implementation.
- By fostering an environment that is favorable for learning and encouraging technicians to participate, the government may support the development of a decentralized wastewater treatment system. In order to encourage the usage of these frameworks, governments might also create new rules and laws. At the policy level, the government ought to encourage and raise awareness of the potential use of decentralized systems in rural areas.

Review Questions

1. What is the difference between centralized and decentralized wastewater treatment systems?
2. Resource recovery from decentralized system or from centralized system? Which is preferred? Give logical answer with some data in support.
3. Why there is a need for decentralized systems toward resource recovery?
4. What are the risks in decentralized wastewater treatment systems?
5. Give some examples of decentralized wastewater treatment systems and accordingly generated resources and their economics.

6 Give some examples of centralized wastewater treatment systems and accordingly their resources and economics.
7 Give some best examples of decentralized wastewater treatment systems as case study.
8 What are the various challenges toward implementation of decentralized systems?
9 Justify the need of self-sustainable wastewater treatment system.
10 For achieving Sustainable Development Goals, which one is necessary: decentralized or centralized wastewater treatment systems. Give proper examples.

References

Adegoke, A.A. and Stenstrom, T.-A. (2019). Septic systems. In: *Water and Sanitation for the 21st Century: Health and Microbiological Aspects of Excreta and Wastewater Management (Global Water Pathogen Project)*. Michigan State University. https://doi.org/10.14321/waterpathogens.59

Alresheedi, M.T., Haider, H., Albuaymi, A.M. et al. (2023). Sustainability of a low-cost decentralized treatment system for wastewater reuse: resident perception-based evaluation for arid regions. *Water (Switzerland)*, **15**(19), 3458. https://doi.org/10.3390/w15193458

Amoatey, P. and Bani, R. (2011). Wastewater management, Chapter 20. *Waste Water—Evaluation and Management*. InTech. https://www.researchgate.net/publication/221911472

Anand, C.K. and Apul, D.S. (2014). Composting toilets as a sustainable alternative to urban sanitation – a review. *Waste Management*, **34**(2), 329–343. https://doi.org/10.1016/j.wasman.2013.10.006

Arora, J., Yadav, A., and Saroj, D. (2016). Potential of decentralised wastewater treatment systems applicable to India. *Current World Environment*, **11**(2), 338–350. https://doi.org/10.12944/cwe.11.2.01

Bazina, N., Ahmed, T.G., Almdaaf, M. et al. (2023). Power generation from wastewater using microbial fuel cells: a review. *Journal of Biotechnology*, **374**, 17–30. https://doi.org/10.1016/J.JBIOTEC.2023.07.006

Bernal, D., Restrepo, I., and Grueso-Casquete, S. (2021). Key criteria for considering decentralization in municipal wastewater management. *Heliyon*, **7**(3), e06375. https://doi.org/10.1016/j.heliyon.2021.e06375

Betti, M. (2008). Ecological sanitation latrines: the experience of Nepal WaterAid in Nepal. *Water Aid*. **1**, 1–20

Bhandari, B.C. and Shrestha, P. (2018). *Understanding to Ecological Sanitation (EcoSan): Research and Implementation-a Review*. www.enpho.org

Capodaglio, A.G. (2017). Integrated, decentralized wastewater management for resource recovery in rural and peri-urban areas. *Resources*, **6**(2), 22. MDPI AG. https://doi.org/10.3390/resources6020022

da Costa, Z.X., Boogaard, F.C., Ferreira, V. et al. (2024). Wastewater management strategy for resilient cities—case study: challenges and opportunities for planning a sustainable Timor-Leste. *Land*, **13**(6), 799. https://doi.org/10.3390/land13060799

Devi, K., Singh, D.M., Kumar, P. et al. (2022). Household domestic water consumption patterns in a rural semi-arid village: a case study of Kumbha village in Hisar district, Haryana. *International Journal of Geography, Geology and Environment*, **4**(2), 1–5. https://doi.org/10.22271/27067483.2022.v4.i2a.105

Diaz-Elsayed, N., Rezaei, N., Guo, T. et al. (2019). Wastewater-based resource recovery technologies across scale: a review. *Resources, Conservation and Recycling*, **145**, 94–112. https://doi.org/10.1016/j.resconrec.2018.12.035

Estévez, S., González-García, S., Feijoo, G. et al. (2022). How decentralized treatment can contribute to the symbiosis between environmental protection and resource recovery. *Science of the Total Environment*, **812**, 151485. https://doi.org/10.1016/J.SCITOTENV.2021.151485

EU's, H. 2020 and DST/DBT, I. (2019). Sarswati2.0.pdf. *Annual Report*, 1–9. https://09ggn.mjt.lu/nl3/_0bK4wdzjrtOj5bvL3el9w?hl=en

Garrido-Baserba, M., Barnosell, I., Molinos-Senante, M. et al. (2022). The third route: a techno-economic evaluation of extreme water and wastewater decentralization. *Water Research*, **218**, 118408. https://doi.org/10.1016/j.watres.2022.118408

Geetha Varma, V., Jha, S., Himesh Karthik Raju, L. et al. (2022). A review on decentralized wastewater treatment systems in India. *Chemosphere*, **300**, 134462. https://doi.org/10.1016/j.chemosphere.2022.134462

Gong, F., Wang, S., Li, H. et al. (2023). Sustainable, decentralized water treatment system fabricated from domestic waste materials and ultrafast-reduced graphene oxide. *Journal of Environmental Chemical Engineering*, **11**(5), 110480. https://doi.org/10.1016/j.jece.2023.110480

Hyderabad, I. (2023). *IIT Hyderabad.pdf*. Waste Management. https://sustainability.iith.ac.in/zeroliq.html

Infosys limited. (2024). *Infosys.pdf*. Waste Management. https://www.infosys.com/industries/waste-management/insights/zero-wastewater.html

Jadhav, D.A., Das, I., Ghangrekar, M.M. et al. (2020). Moving towards practical applications of microbial fuel cells for sanitation and resource recovery. *Journal of Water Process Engineering*, **38**, 101566. https://doi.org/10.1016/j.jwpe.2020.101566

JNU. (2024). *JNU.pdf*. Waste Management. https://www.jnu.ac.in/content/technologies-water-treatment-and-reuse-advanced-oxidation-processes-aop

Kharagpur. (2023). *IIT KHARAGPUR.pdf*. Waste Management. https://www.iitkgp.ac.in/department/CE/faculty/ce-ghangrekar

Kim, J., Wu, B., Jeong, S. et al. (2024). Recent advances of membrane-based hybrid membrane bioreactors for wastewater reclamation. *Frontiers in Membrane Science and Technology*, **3**, 1–14. https://doi.org/10.3389/frmst.2024.1361433

Lago, A., Rocha, V., Barros, O. et al. (2024). Bacterial biofilm attachment to sustainable carriers as a clean-up strategy for wastewater treatment: a review. *Journal of Water Process Engineering*, **63**, 105368. Elsevier Ltd. https://doi.org/10.1016/j.jwpe.2024.105368

Madan, S., Madan, R., and Hussain, A. (2022). Advancement in biological wastewater treatment using hybrid moving bed biofilm reactor (MBBR): a review. *Applied Water Science*, **12**(6), 141. Springer Science and Business Media Deutschland GmbH. https://doi.org/10.1007/s13201-022-01662-y

Madhuranthakam, C.M.R., AbuZaid, M., Chaalal, O. et al. (2023). Sustainable water management with design and economic evaluation of recycling greywater at Abu Dhabi University—a case study on decentralization. *Sustainability (Switzerland)*, **15**(23), 16208. https://doi.org/10.3390/su152316208

Mahindra Lifespace Developers Ltd. (2024). *mahindra.pdf*. Waste Management. https://www.mahindralifespaces.com/sustainability-impacts/water/

Mavugara, R., Matsa, M., and Defe, R. (2024). Potential recovery assessment of resources embodied in the organic component of municipal wastewater: the case of Zvishavane Urban, Zimbabwe. *Physics and Chemistry of the Earth, Parts A/B/C*, **135**, 103627. https://doi.org/10.1016/j.pce.2024.103627

Rodríguez, C., Sánchez, R., Rebolledo, N. et al. (2020). Cost–benefit evaluation of decentralized greywater reuse systems in rural public schools in chile. *Water (Switzerland)*, **12**(12), 1–14. https://doi.org/10.3390/w12123468

Sutar, S. and Kulkarni, M.G.S. (2015). Review paper on decentralized wastewater treatment system. *International Journal of Innovative Research and Creative Technology*, **1**, 396. www.ijirct.org

Swan, K.H., Surinkul, N., and Visvanathan, C. (2023). Investigation of decentralized wastewater treatment system in urban wastewater management: case study in Yangon, Myanmar. *Sustainability (Switzerland)*, **15**(24), 16756. https://doi.org/10.3390/su152416756

VIT. (2024). *VIT.pdf*. Waste Management. https://vit.ac.in/center/facilities/co2

Yadav, P., Singh, R.P., Singh, S.K. et al. (2024). Chapter Fifteen – Biogas production using wastewater: methodologies and applications. In: *Recent Advancements in Wastewater Management: Nano-based Remediation* (eds. A. Kumar, M. Bilal, and L.F.R. Ferreira) (Vol. 10, pp. 347–361). Elsevier. https://doi.org/10.1016/bs.apmp.2023.08.005

15

Conversion Strategies for Utilization of Wastewater as a Fertilizer

Sudeeptha Girijan[1], *K. Hasim Suhaib*[2], *M. Devasena*[1], *and Praveena Gangadharan*[1,2]*

[1]*Global Sanitation Center of Excellence, IIT Palakkad, Kerala, India*
[2]*Department of Civil Engineering, IIT Palakkad, Kerala, India*

*Corresponding author: praveenag@iitpkd.ac.in

15.1 Introduction

The term "wastewater" refers to the impure water generated after domestic or industrial use and is termed as sewage or industrial wastewater, respectively. Globally, 1.04×10^6 MLD of wastewater is generated (Qadir et al., 2020). With increasing population, urbanization, and economic prosperity, wastewater generation is expected to increase further by at least 24% by the year 2030 (Gowd et al., 2022; Qadir et al., 2020). In India, urban sewage generation alone is estimated to exceed 80 000 MLD by 2025 (Mishra et al., 2023). Wastewater is generally discharged into a water body after complete/partial/no treatment. However, partially treated or untreated wastewater may contain pathogens and contaminants. Consequently, the water bodies become unfit for potable and any other secondary uses. Therefore, the major objective of wastewater treatment is the removal of constituents that cause these effects. One of the major constituents that is sought to be removed during treatment is nutrients, since wastewater contains a considerable concentration of nutrients like nitrogen (N), phosphorus (P), and potassium (K). Nutrient discharge into surface or groundwater can cause eutrophication (algal growth and subsequent depletion of dissolved oxygen). Additionally, nitrite, nitrate, and high P concentrations can adversely affect human health.

Wastewater to Resource Recovery: Applying the Circular Economy Toward Sustainable Development, First Edition. Edited by Arvind Kumar Mungray et al. Published 2026 by John Wiley & Sons, Inc.

Essential plant nutrients are usually supplied in excess (as fertilizers) for sustaining agricultural practices. Among the nutrients, N, P, and K are considered the major nutrients that are required in larger quantities. Other essential nutrients such as Na, Ca, Mg, Zn, Cu, etc., which are required in smaller quantities, are also present in wastewater (Shilpi et al., 2018). If the nutrients are effectively recovered from wastewater, they could be beneficially used in agriculture to boost the growth and yield of crops. This is also economically favorable, especially in the context of depleting P reserves (Cordell, 2013; Zhu et al., 2023). Wastewater from dairy farms, abattoirs, wineries, etc., has been demonstrated to positively influence the biomass yield of crops (Shilpi et al., 2018). Thus, nutrient recovery measures not only offset fertilizer requirements of agriculture but also reduce adverse effects (such as eutrophication) on the environment.

This chapter discusses wastewater as a potential fertilizer source, detailing the nutrient composition of wastewater and the interaction with soil and crops once applied. Having realized the potential of wastewater, it is necessary to understand how nutrients can be recovered from wastewater. Thus, the second section discusses various nutrient recovery technologies, which are based on physicochemical/ biological/ or hybrid technologies. Finally, the application of the recovered product is discussed, focusing on centralized and decentralized approaches.

15.2 Wastewater as a Source of Resources

N and P are the major nutrients sought for plant fertilization. N and P are essential for the synthesis of essential biomolecules [like amino acids, nucleic acids, and adenosine triphosphate (ATP)] and cellular material; therefore, they are absorbed by living things. N and P are cycled naturally in the environment through which they are made available to plants and animals. However, human interventions have caused an imbalance in these cycles, owing to the demand for nutrients. Figure 15.1 represents how human activities interfere with nutrient concentrations in different environmental matrices and how nutrient recovery can effectively reduce the adverse impact on the environment.

In the environment, atmospheric N is fixed by microbes into plant-available forms and absorbed by plants. Phosphate rocks are the initial sources of P, which reach the soil through weathering and are absorbed by plants. Subsequently, humans consume these plants or plant-consuming animals and release these nutrients back into the environment through excretion or decay. The nutrient content in human excreta (untreated or treated) generally ends up in water bodies.

Conventionally, N-fertilizer production starts with the conversion of atmospheric N to ammonia (Haber–Bosch process), and P-fertilizer production starts

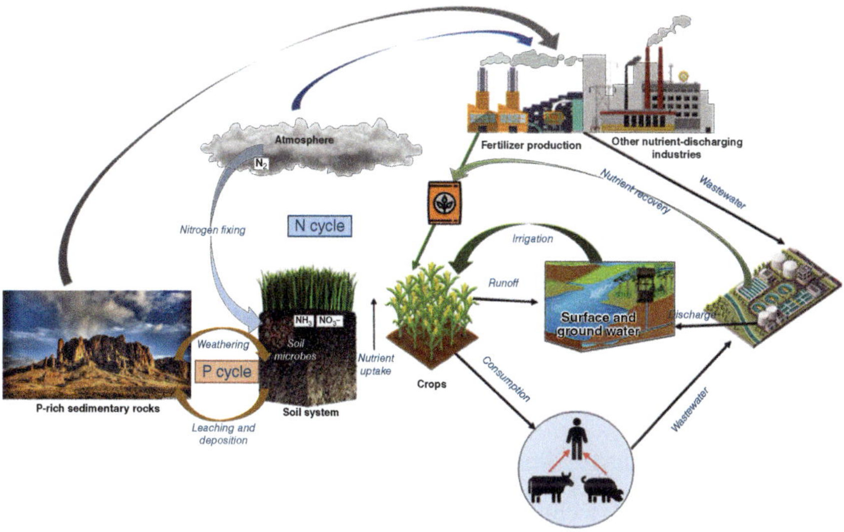

Figure 15.1 Schematic diagram showing interrelationship between nutrient cycles, fertilizer production, and nutrient recovery systems. *Source:* WallpapersWide.com/Public domain.

by the reaction of phosphate rocks with H_2SO_4. Commercial fertilizers have high mobility and also result in discharge into water bodies through excessive leaching. Thus, water bodies act as a sink for these nutrients. Moreover, fertilizer production contributes to considerable greenhouse gas emissions. Through nutrient recovery, the nutrient discharge into water bodies can be reduced and diverted to land application. This ensures that the nutrients are made available to plants and animals, thereby completing the cycle. This also reduces the demand for commercial fertilizer production, improving the sustainability quotient of wastewater treatment.

15.2.1 Source of Nutrients in Wastewater

Domestic wastewater is mainly composed of excreta, other body exudates, cleaning agents, and wash waters. They primarily contain food waste, detergents, and products of human metabolism, which are rich in nutrient content and are also present in wastewater. Similarly, industries generating nutrient-rich wastewater include fertilizer, paper and pulp, chemical, petrochemical, microbiological, textile, and food (fermentation, abattoirs, etc.) (Nie et al., 2018; Tymchuk et al., 2020). The nutrient content in domestic wastewater could be attributed to the plants and animals consumed by humans. Consequently, the major source of nutrients in industries that use plant and animal products as raw materials (such as distilleries, abattoirs, olive mills, paper and pulp, etc.) can also be attributed to the nutrient composition of plants and animals. In the case of petrochemical industries, the crude oil contains nitrogenous compounds, which are released as ammonia/ammonium during the distillation and scrubbing process, leading to their presence in the wastewater from the corresponding processing units (Yang, 2020). In textile industries, N and P are introduced through compounds like ammonium phosphate and urea during the dyeing process (Sarayu and Sandhya, 2012).

In addition to nutrients, wastewater may also contain heavy metals, pathogens, and other constituents that hinder its application to agriculture. Therefore, treatment to remove these contaminants is necessary. This is usually imposed through standards, with more stringent regulations for edible-crop irrigation (Kretschmer et al., 2002).

15.2.2 Nutrient Composition of Wastewater

Most nutrient recovery technologies can perform efficiently only if the influent wastewater has an adequate nutrient concentration. Therefore, it is necessary to understand the nutrient composition of wastewater to justify nutrient recovery from it. Differences in the type/source of wastewater lead to differences in its nutrient composition (Table 15.1). As mentioned above, the wastewater could be broadly classified as municipal and industrial wastewater, based on its composition. In terms of volume, municipal wastewater is considered to make larger contributions to nutrient loads than industrial wastewater. However, non-point sources included, the major contributor to nutrient pollution is agriculture (Nie et al., 2018; Pehlivanoglu-Mantas and Sedlak, 2006). For municipal wastewater, protein consumption, sewerage connection, treatment efficiency, and socio-economic status are some of the key factors that influence its nutrient composition (van Puijenbroek et al., 2019).

The quality of wastewater also differs considerably between municipal and industrial wastewater and among the different industrial wastewater types. In the case of industrial wastewater, food-related industries generally show a higher concentration of nutrients (Kretschmer et al., 2002). Moreover, municipal wastewater tends to be more prone to contamination since it is unregulated.

15.2.3 Forms of Nutrients Applied as Fertilizers

In wastewater influents, nutrients are present in different forms. Nitrogen-containing compounds are usually present as oxidized forms (as nitrate and nitrite) and Kjeldahl N (organic N and ammonia). In domestic wastewater, N is mainly present as organic N (protein, amino acids, and urea) and ammonia (Pehlivanoglu-Mantas and Sedlak, 2006). During treatment, organic N may

Table 15.1 Nutrient composition of different types of nutrient-rich wastewater.

Source	COD mg/L	BOD mg/L	N mg/L	P mg/L	K mg/L	Other nutrients mg/L	Location	Reference
Olive mill	420–97000	230–30600	13–532	4–182			Jordan	Ammary (2004)
Paper and pulp	4160						India	Gowd et al. (2022)
Winery	320	122	7.8	28.2	271	Na: 60.1 Ca: 123 Mg: 17.8	Australia	Shilpi et al. (2018)
Abattoir	365	125	359	52.9	67.2	Na: 112 Ca: 48.7 Mg: 17.9	Australia	Shilpi et al. (2018)
	6185–44122	3000–3500	1050–1982	520			India	Gowd et al. (2022)
Dairy	370	120	422	27.8	112	Na: 137 Ca: 235 Mg: 144	Australia	Shilpi et al. (2018)
	989–1100	500–540	215–305	46–54			India	Gowd et al. (2022)
			267	70			United States	Pang et al. (2020)

(Continued)

Table 15.1 (Continued)

Source	COD mg/L	BOD mg/L	N mg/L	P mg/L	K mg/L	Other nutrients mg/L	Location	Reference
Municipal	91	35	9.72	0.07	44.2	Na: 244 Ca: 45.1 Mg: 30	Australia	Shilpi et al. (2018)
			43.7	7.8	16.5		Global	Qadir et al. (2020)
	500–1200	230–560	30–100	6–25			Global	Henze and Comeau (2008)
			8–98					Beckinghausen et al. (2020)
Distillery	70000–98000	45000–60000	1000–1200	500–1500			India	Gowd et al. (2022)
	100000–150000	35000–50000	1660–4200	225–308	9600–15475	Ca: 2300–2500 Mg: 220–250 Fe: 1550–1800	Global	Mikucka and Zieli (2020)
Textile	736	56	69.5				India	Gowd et al. (2022)
	8000	237	506	410			India	Brar et al. (2019)
	150–12000	80–6000	70–80		30–50		Global	Yaseen and Scholz (2019)
Tannery	2512	1484	1200				India	Gowd et al. (2022)
Industrial			96–3500				Global	Beckinghausen et al. (2020)

be converted to ammonia, which could subsequently be oxidized to nitrite and nitrate. Therefore, the effluent composition depends on the type of treatment employed.

Untreated/partially treated wastewater (which retains some nutrient concentration) can be directly applied to agriculture by diverting it to irrigated lands instead of surface waters. This is especially significant for countries like India, where nutrient removal/recovery is limited (Qadir et al., 2020). The solids settled out during treatment may be further processed (dewatered, dried, composted, or pyrolyzed) to produce biosolids; subsequently, they could also be directly applied to land. Different types of wastewater and biosolids may also be combined and then applied for agriculture (Mikucka and Zieli, 2020).

Furthermore, technologies recover nutrient resources in various forms. They could be directly recovered as liquid fertilizers or extracted as solid products (Gowd et al., 2023). P is commonly recovered as mineral salts, such as struvite, hydroxyapatite, and calcium phosphate (Tymchuk et al., 2020), which are suitable for direct land application. N could also be recovered as struvite. Additionally, N is recovered as ammonium in acid solutions, ammonium hydroxide, ammonium sulfate, and ammonium chloride (Batstone et al., 2015; Beckinghausen et al., 2020). These ammonium salts are subsequently distilled to obtain ammonia.

Plants satisfy their N demands through the uptake of ammonium and nitrate, with preferences varying between ammonia and nitrate depending upon the soil and climatic conditions (Muratore et al., 2021; Wang and Macko, 2011). Similarly, they can absorb P only as orthophosphates (i.e. $H_2PO_4^-$ and HPO_4^{2-}). In the case of K, plants directly take up the K^+ cation that is adsorbed on soil particles or available in the soil water (Sardans and Peñuelas, 2021).

15.2.4 Improvement in Crop Growth with Recovered Fertilizer Application

Previous studies have demonstrated that wastewater application can improve crop growth. In the studies, crop growth using the application of wastewater was compared with the application of conventional irrigation sources such as tap water, groundwater, and surface water. An excessive amount of nutrients in wastewater may prove detrimental to the crops. Moreover, wastewater may contain heavy metals, salts, emerging contaminants, and melanoidins that cause detrimental effects (Mikucka and Zieli, 2020; Shilpi et al., 2018). Other important wastewater characteristics like pH, TDS, and EC also affect the improvement in crop growth due to the application of nutrient-rich wastewater/product (Shilpi et al., 2018). Nutrient recovery strategies are more favored over direct application of wastewater, since recovered products help in retaining better control over the application process.

15.3 Technologies for Simultaneous Wastewater Treatment and Resource Recovery

In recent years, various methodologies for nutrient recovery that also achieve wastewater treatment have been investigated and commercially implemented. These approaches encompass a range of techniques, including physical methods like adsorption and ion exchange (Oudad et al., 2022), chemical processes such as precipitation (Xia et al., 2020), and biological treatments like constructed wetlands, vermifilters, and microalgae cultivation (Adhikari et al., 2015; Miito et al., 2022; Rossi et al., 2022). Additionally, membrane technologies and thermal treatment methods are advancing. Physicochemical processes employ techniques such as electrostatic adsorption, cation exchange, and chemical precipitation to effectively separate and recover nutrients like ammonia and P. Electrochemical methods use oxidation, reduction, electrocoagulation (EC), and electrodialysis (ED) to facilitate nutrient recovery through electrochemical processes (Liu et al., 2021; Zhang et al., 2012).

Microbial electrochemical technologies, including electrolysis cells, microbial fuel cells (MFCs), and recycling cells, and integrated microbial processes with electrochemical systems, enhance nutrient retrieval (Deng et al., 2023; Mukherjee et al., 2021). Biological approaches harness microorganisms, earthworms, algae, and plants in systems such as constructed wetlands, vermifilters, and algal treatments to promote higher nutrient recovery. Membrane separation techniques utilize various membranes including reverse osmosis (RO), forward osmosis (FO), microfiltration, and nanofiltration (NF) to concentrate and separate nutrients from wastewater (Xiang et al., 2020). Technologies like incineration, pyrolysis, hydrothermal carbonization, and gasification transform sludge into ash and recover concentrated N and P (Kwapinski et al., 2021). Tables 15.1 and 15.2 provide a comprehensive summary, detailing nutrient removal efficiencies achieved by these technologies and their evaluations in terms of energy and economic viability, environmental sustainability, and carbon footprint. This compilation serves as a valuable resource for engineers and researchers aiming to identify optimal technologies for wastewater treatment that aim for nutrient recovery, facilitating informed decisions in technology selection and implementation.

15.3.1 Physicochemical Technology

Ammonia stripping, adsorption, and ion exchange are the physicochemical technologies mainly employed for nutrient recovery from wastewater.

Table 15.2 Summary of nutrient removal by various technologies.

Technology	Wastewater	Influent characteristics (mg/L)	Nutrient recovery (%)	Form of nutrient recovered	References
Air stripping	Compost leachate	NH_4^+- N 400	N 94	Ammonium sulfate	Oudad et al. (2022)
Adsorption	Secondary effluent	P 20	P 80	Biochar	Xia et al. (2020)
Ion-exchange	Domestic wastewater	PO_4-P 270 NH_4-N 120	N 97 P 98	Struvite	Williams et al. (2015)
Struvite precipitation	Swine wastewater	NH_4^+-N 1013–1426 PO_4^{3-} 55.1–139.7 TKN, 1381–2001	PO_4^{3-} 94 NH_4^+-N 87	Struvite	Zhang et al. (2012)
Electro hydro modulation	Municipal wastewater	PO_4^{3-} 3.07 NH_3 24.07	TN 89 TP 97	Amorphous calcium phosphate	Perera and Englehardt (2020)
Photosynthetic MFC	Swine wastewater	COD 7786 TN 2048 TP 124	NH_4^+-N 99 TP 98	Microalgal biomass	Li et al. (2021)
Forward osmosis	Digested Swine wastewater	PO_4^{3-} 166 NH_4^+ 413	NH_4^+-N 93 PO_4^{3-} 99	Struvite	Wu et al. (2018)
Microalgal treatment	Piggery wastewater	NH_4^+-N 195±55 PO_4^{3-} 19 ±14	NH_4^+-N 90 PO_4^{3-}-P 90	Microalgal biomass	Rossi et al. (2022)
Constructed wetland	Dairy farm wastewater	TN 30–118 TP 2.9–25.2	TN 47.88 TP 50.35	Duckweed biomass	Adhikari et al. (2015)
Vermifilter	Dairy wastewater	TN 857 TP 92	TN 77 TP 48	Vermicompost	Miito et al. (2022)

15.3.1.1 Ammonia Stripping

Ammonia stripping is one robust physicochemical technology typically implemented at centralized wastewater treatment plants to remove/recover ammonia from nutrient-rich wastewater. Ammonia stripping can be achieved through air stripping or steam stripping methods. Air stripping typically employs wastewater with an ammonia concentration exceeding 2000 mg/L (Kundu et al., 2022b). During this process, ammonia is transferred into the air phase and subsequently absorbed into an acid solution to produce an ammonium compound that can be utilized as a fertilizer. For instance, air-stripped ammonia could be made to react with H_2SO_4 to yield ammonium sulfate, a valuable fertilizer compound. In contrast, steam stripping involves extracting ammonia at high temperatures using steam, followed by condensation into liquid form. This liquid then transforms into ammonium sulfate.

The removal rate of ammonia via the ammonia stripping method primarily depends on the pH and temperature of the wastewater. Other factors affecting ammonia stripping include the initial concentration of ammonia in the wastewater, hydraulic loading rate (HLR), and the rate of airflow. According to Norddahl et al. (2006), maintaining the pH range between 11 and 12 is advantageous for achieving optimal recovery of ammonia, incurring minimum costs. Oudad et al. (2022) successfully extracted ammonium sulfate fertilizer from compost leachate using the air stripping method (Table 15.1). They recovered 1.5 g ammonium sulfate from 1 L compost leachate and achieved 94% removal of ammonium N.

The feasibility and sustainability of ammonium stripping technology are evaluated based on energy demand, capital and operation cost, and carbon emissions, as tabulated in Table 15.2. The energy requirements and costs associated with nutrient recovery through ammonium stripping primarily depend on the aeration/steam supply for the process and the addition of chemicals. De Vrieze et al. (2016) reported that economic feasibility is realized for ammonium stripping only when ammonium-N concentrations exceed 1000–1500 mg/L.

15.3.1.2 Adsorption

Adsorption is the best available demonstrated technology for the removal and recovery of nutrients from wastewater. The effectiveness of nutrient recovery by adsorption relies significantly on the appropriate selection of adsorbent and optimization of operational parameters such as pH, temperature, initial nutrient concentrations, hydraulic retention time (HRT), and dosage ratio of adsorbent used. Hydrotalcite, aluminum oxide, iron oxide, zirconium oxide, zeolite, resin, biochar, activated carbon, activated red mud, activated alumina, and blast furnace slag are commonly employed as adsorbents for recovering nutrients from various kinds of wastewater (Kundu et al., 2022b).

Xia et al. (2020) reported the utilization of iron-enriched sludge derived from a municipal wastewater treatment plant as an adsorbent for P recovery from secondary effluent. They determined that the modified sludge displayed a maximum adsorption capacity of 22 mg-P/g within a pH range of 5–8. This treatment effectively reduced the P concentration in secondary effluent below 0.1 mg/L, achieving an 80% removal rate. The P-loaded adsorbent was further pyrolyzed to biochar and reused as a soil conditioner to enhance the growth of wheat plants.

15.3.1.3 Ion Exchange

Ion exchange is a well-developed technique for concentrating nutrients for subsequent recovery at the source and centralized facilities. Ion exchange processes are effective in nutrient removal and recovery from various types of wastewater; they are characterized by lower costs, reduced greenhouse gas emissions, and nutrient recovery benefits, which promote the circular economy (Huang et al., 2020). This process involves the selective removal of N and P using ion exchange resins or materials. Research has identified clinoptilolite resin as highly efficient for ion exchange, particularly for removing and recovering ammonium ions from wastewater, even at low N loading rates. Williams et al. (2015) utilized clinoptilolite and Dow-FeCu resins in an ion exchange column to remove N and P from diluted domestic wastewater and to recover struvite. Their investigation demonstrated impressive removal efficiencies, achieving 98% removal of phosphate P and 97% removal of ammonium N. The recovered struvite contained 13% P and 1.6% N.

Further, Huang et al. (2020) assessed the feasibility and viability of the ion exchange process for nutrient recovery by evaluating its performance and economics on a larger scale. Their study found that the whole life cost of the ion exchange process (calculated over a period of 40 years), combined with traditional activated sludge or anaerobic membrane processes, is 17–27% lower than that of conventional biological nutrient removal plants.

15.3.1.4 Struvite precipitation

Precipitation is an effective technique for recovering struvite from nutrient-rich industrial and agricultural wastewater, as well as landfill leachate (Kabdaşlı and Tünay, 2018). Struvite is a white-colored crystalline powder composed of magnesium, ammonium, and phosphate in equimolar concentration, offering significant potential as a fertilizer. Several studies have demonstrated struvite precipitation from wastewater such as that from the fertilizer industry (Hutnik et al., 2013), swine wastewater (Ichihashi and Hirooka, 2012), municipal wastewater (Mavhungu et al., 2021), and urine (Etter et al., 2011). Zhang et al. (2012) proposed struvite precipitation as an effective pretreatment method for anaerobic digestion of swine wastewater, for pollutant removal (Table 15.2).

Studies indicate that struvite crystallization is primarily influenced by the initial concentration of P, temperature, pH, and mixing speed (Huang et al., 2011; Sangeetha et al., 2022). Pastor et al. (2010) and Kundu et al. (2022a) reported that effective struvite precipitation occurs when the phosphorus concentration exceeds 100 mg/L. A low concentration of P reduces struvite precipitation and requires a longer reaction time, significantly diminishing the economic feasibility of struvite recovery by precipitation.

15.3.2 Electrochemical Technologies

Electrochemical technologies are being extensively implemented for wastewater treatment and nutrient recovery owing to their superior treatment effectiveness, adaptability, simplified maintenance and flexibility, absence of chemical additives, minimal sludge generation, and automated operation (Liu et al., 2021). Electrochemical oxidation (EO), Electrochemical reduction (ER), EC, ED, and capacitive deionization (CDI) are the core electrochemical technologies generally employed in wastewater treatment and nutrient removal/recovery (Liu et al., 2021). ED and CDI are employed for concentrating and recovering nutrients from low-strength wastewater. The precipitation of struvite, hydroxyapatite, and amorphous calcium phosphate from electrochemical systems is valuable for P recovery from wastewater (Perera and Englehardt, 2020).

Electrochemical stripping combined with the acid trap is an effective method for the recovery of N as ammonia from wastewater (Ye et al., 2018). Tarpeh et al. (2018) reported a N recovery rate of 93% from source-separated urine in the form of ammonium sulfate using an electrochemical stripping method. Building on this approach, Gao et al. (2020) developed an integrated electrochemical process incorporating bipolar membrane electrodialysis (BMED) and membrane capacitive deionization (MCDI) systems. In their study, BMED facilitates struvite precipitation, achieving 88% P recovery rate. Subsequently, MCDI was employed to recover ammonia, achieving a recovery efficiency of 77%.

15.3.3 Microbial Electrochemical Technologies

MFCs, microbial electrolysis cells (MECs), and microbial recycling cells (MRCs) represent advanced microbial electrochemical technologies developed for energy generation and nutrient recovery from wastewater, supporting the circular economy and carbon neutrality. In these technologies, organic matter present in wastewater produces electrons while undergoing degradation with the simultaneous generation of electricity. Simultaneously, this microbial activity helps in treating wastewater and nutrient recovery.

MFCs are considered a promising technology to recover nutrients and electricity from wastewater (Mukherjee et al., 2021). MFCs are composed of anode and cathode chambers, separated by a cation exchange membrane. MFCs with single chambers, double chambers, and multiple chambers have been designed for the recovery of nutrients from wastewater. The design of the MFC and the nature of bacterial growth may affect the treatment efficiency and nutrient recovery from wastewater. The major mechanisms behind the removal/recovery of nutrients from MFCs are the precipitation of struvite, cattite, or vivianite, nutrient assimilation by microalgae, air stripping, nitrification, and denitrification (Ye et al., 2019).

The efficiency of nutrient removal and power generation from MFCs is influenced by various parameters, including temperature, pH, electrode and membrane types, organic loading rate (OLR), HRT, light intensity, initial ammonium concentration, dissolved oxygen levels, and electrical resistance. Ye et al. (2019) conducted a feasibility study of double-chamber MFC with three types of separators (cation exchange membrane [CEM], FO, and nonwoven [NW]) on nutrient recovery from municipal wastewater. The study found that the conventional MFC with a CEM separator achieved more than 97% ammonium-N and more than 95% phosphate P removal, followed by MFC with NW and MFC with FO. The removal/recovery of these nutrients is primarily attributed to the chemical precipitation occurring in the cathodic chamber.

Another novel technology receiving increased research attention is the microalgal-based photosynthetic MFC; in this technology, the algal biomass assimilates nutrients. The algae can subsequently be used for biodiesel production, CO_2 sequestration, and recovery of nutrients and other value-added products (Arun et al., 2020; Wang et al., 2019). Li et al. (2021) used mixotrophic cultivation of *Chlorella vulgaris* in airlift-type photosynthetic MFCs for energy and nutrient recovery from swine wastewater. The study found that photosynthetic MFCs are a sustainable technology for swine wastewater treatment and energy recovery. The investigations found that the system removed 96% COD, 99% ammonium-N, and 98% total P with an energy recovery of 1.92 kWh/m^3. However, the commercial application of microalgae-based technology is limited by its considerable space requirements (Gao et al., 2016).

MECs and MRCs represent an evolution of MFCs, designed to electrochemically oxidize organic compounds. In MECs, biofilms in the anodic chamber facilitate the generation of protons and electrons by microorganisms, which are then employed in oxidation-reduction reactions. On the other hand, MRCs utilize easily recyclable and biocompatible components for the anode, cathode, and separator. Finally, the recovered products along with the electrochemical-cell components can be recycled to produce fertilizer and soil conditioners.

15.3.4 Biological Method

15.3.4.1 Constructed Wetland and Vermifilter Treatment

Constructed wetlands and vermifilters are low-cost, environmentally sustainable wastewater treatment technologies developed for the treatment of domestic and industrial wastewater and the recovery of nutrients. In constructed wetlands, the concentration of N is primarily influenced by ammonification, nitrification, denitrification, and plant and microbial uptake processes, whereas P concentrations depend on sorption, precipitation, plant and microbial uptake, and peat or soil accretion (Vymazal, 2007). Adhikari et al. (2015) investigated the capability of duckweed-constructed wetlands to treat and recover nutrients from high-strength manure-containing dairy wastewater (Table 15.2). The results showed that COD removal was 81%, with an average recovery of 22.4 g $N/m^2/yr$ and 7.4 $P/m^2/yr$ achieved through harvesting duckweed from the treatment system.

A vermifilter operates as a biofilter that exploits the symbiotic relationship between earthworms and microbial communities within the vermibed (Suhaib and Bhunia, 2022). Through their combined activities, earthworms significantly enhance the removal of organic matter and nutrients from wastewater by ingesting and digesting contaminants. This biological technology not only facilitates efficient wastewater treatment but also enables the recovery of nutrients via the production of vermicompost. In vermifiltration, N removal primarily occurs through nitrification and denitrification processes. In contrast, P removal relies entirely on the sorption capacity, surface area, and size of the vermifilter bed material. Malal et al. (2024) evaluated the efficacy of the vermicompost, derived from vermifilters, in enhancing soil fertility and squash crop yields using field-level experiments. The study revealed that the application of vermicompost significantly increased total soil carbon, total N, and the activity of C-N-P enzymes (particulate organic carbon and organic nitrogen increased by 93%, compared to control). These findings underscore vermifiltration as a sustainable approach to nutrient recovery and soil fertility enhancement.

15.3.4.2 Micro-algal Treatment

Microalgal treatment is a polishing step for wastewater treatment and has been implemented for various kinds of wastewater such as municipal wastewater, industrial wastewater, and agricultural wastewater. Algae-based treatment technology provides significant benefits over conventional wastewater treatment by assimilating nutrients into algae biomass. Nutrient-rich algal biomass can be used as a fertilizer or can be converted into value-added products such as biochar by thermochemical methods. Biochar is a good fertilizer that has a high potential to enrich the fertility of the soil and nutrient uptake rate by crops and their yield.

The efficiency of the treatment and production of algal biomass may vary depending on the selected algal species, C/N and N/P ratios, wastewater characteristics, and environmental conditions. Rossi et al. (2022) utilized multi-culture of *Chlorellaceae, Scenedesmaceae*, and *Chlamydomonadaceae* for the treatment of wastewater from piggery farms (500 L) and found a 90% removal of N and P with algal biomass productivity of 10.7 $g/m^2/day$. Xie et al. (2022) reported 85% N and 30% P removal with 1.4 kg/day biomass productivity while treating Swain wastewater (3000L) by *Chlorella vulgaris* species.

Generally, wastewater with low nutrient levels such as municipal wastewater is more suitable for microalgal growth. High-strength wastewater, such as agricultural wastewater (animal manure containing wastewater) with an excess amount of turbidity, suspended solids, nutrients, and COD concentration, might hinder the growth of algae and is inappropriate for algal-based wastewater treatment. There are a few species of algae employed in animal wastewater treatment. Generally, diluted animal wastewater is used for the higher recovery of the nutrients by using some species of algae. Zhu et al. (2013) diluted piggery wastewater to below 2000 mg/L COD concentration to achieve its optimal recovery of nutrients. Algal-based nutrient removal and recovery studies have been done in pilot-scale and full-scale systems. Their implementation in the industrial field is much less due to challenges in scaling up the nutrient recovery and maintaining the stability of the treatment efficiency. Moreover, its extensive spatial requirements will impede the implementation or integration of algal-based wastewater treatment in urban and other space-constrained areas.

15.3.5 Membrane Separation Technology

Membrane separation technology is one of the most effective and viable technologies to concentrate and recover nutrients from wastewater. Generally, technologies such as RO, FO, ultrafiltration (UF), NF, membrane distillation (MD), and ED are used to recover nutrients from both domestic and industrial wastewater. Technologies such as RO, NF, and UF are conventional pressure-driven membrane technologies widely used for nutrient recovery from various type of wastewater and urine. For example, Maurer et al. (2006) recovered 70% ammonium, 73% phosphate, and 71% K from source-separated urine using RO. However, the challenges remain in its economic application due to the requirements of energy for pressure, low nutrient recovery efficiency, and membrane fouling (Ugwuanyi et al., 2024).

MD, ED, and FO are the three emerging technologies developed to overcome the limitations of conventional membrane filtration technology. The driving force behind the mass transfer in FO arises from the osmotic pressure difference developed across two solutions with different concentrations separated by a semipermeable membrane. In the case of MD, the mass transfer is driven by the development of a thermal gradient across the membrane. However, in the case of ED, the mass transfer is due to the generation of a direct current field due to the movements of ions toward the cathode and anode. These technologies were used effectively in nutrient recovery from various waste streams such as wastewater, urine, and sewage sludge (Ge et al., 2012; Hau et al., 2014; Zhang et al., 2012, 2014).

Wu et al. (2018) investigated the recovery of struvite from digested swine wastewater using FO. Their study demonstrated that the FO system achieved a high level of ammonium recovery (>93%) and phosphate recovery (>99%). Compared to pressure-driven membrane technologies, FO exhibits better resistance to membrane fouling and higher fouling reversibility. Membrane fouling in MD could be attributed to dissolved organic matter and colloids in wastewater streams, and fouling in the ED membrane could depend on the charge of the membrane.

15.3.6 Thermal Treatment

In contrast to P recovery from the liquid phase through the precipitation method, P in wastewater sludge can also be extracted by a thermal process. Thermal treatment represents a promising technology for sewage sludge management, enabling the production of char and ash, which can subsequently be utilized to recover P. Generally, incineration, pyrolysis, hydrothermal carbonization, and gasification methods are employed for P recovery (Kwapinski et al., 2021). These thermal processes facilitate the recovery of P from wastewater sludge in various forms, including calcium phosphate, calcium pyrophosphate, and hydroxyapatite, which can be immediately used as fertilizer. While thermal treatment is suitable for all types of sludge and allows for nutrient and energy recovery simultaneously, it is a cost-intensive technique.

15.4 Decentralized and Centralized Approaches for Utilization of Wastewater as Fertilizer

A centralized system refers to a system that collects wastewater from a large area and treats it at a centralized facility. It is characterized by extensive sewerage networks, service areas encompassing cities/municipalities, and high treatment volumes. Contrarily, decentralized systems generally service individual households/communities/enterprises. Centralized and decentralized approaches to wastewater management are generally understood from the focus point of treatment, since it has been the major objective of wastewater management strategies. However, with increasing focus on resource recovery and reuse, the suitability of these approaches needs to be evaluated in a new light. Taking an informed decision on whether to adopt centralized/decentralized/hybrid strategies is crucial, especially during the initial phase of implementing reuse approaches. Some of the major factors to be considered are listed below, and both systems are discussed under each head.

15.4.1 Costs of Wastewater Reuse

Centralized approaches are large-scale systems benefited by the economy of scale, whereas decentralized approaches present a higher cost per MLD. However, decentralized system cost is influenced only by the design flow rate (Wang et al., 2008). Centralized systems require extensive pipeline networks for wastewater collection, whereas decentralized systems do not require these as they rely on on-site treatment and reuse. High initial costs of centralized systems are unfavorable where there is instantaneous demand for wastewater reuse (Wilderer and Schreff, 2000). Moreover, distributed reuse of treated wastewater from a centralized facility would require more infrastructure in the form of a separate pipeline network. In the case of remote generators that need to be linked to the collection system, centralized systems become more expensive than decentralized ones. Thus, the economy of centralized systems is subject to factors such as pipe length and elevation difference.

Wang et al. (2008) defined a critical distance (from a centralized sewerage) as a decision-making tool for selecting centralized/decentralized approaches. In addition to the capital costs, the O&M of these complex centralized systems also incurs higher costs compared to decentralized systems (Chen et al., 2017). Consequently, in centralized systems, the cost of treated water ends up being greater than that of potable water, which discourages treated water use for non-potable uses (Wang et al., 2008). Considering reuse, decentralized strategies have been reported as more economically sustainable, especially for urban uses (Chen and Wang, 2009; Ghafoori et al., 2021).

15.4.2 Technology and Operation

Centralized systems are flexible to employ even advanced technologies that produce high-quality treated water (with low solids and pathogen content). However, the choice of technology in a decentralized system is heavily influenced by the geography of the location, the intended use, and the available technical and financial resources of the community/household (Chen and Wang, 2009). In most cases, this limits the technology to simpler systems such as wetlands and planted gravel filters. The agricultural reuse potential of simpler technologies such as lagoons and wetlands has been reported to be higher than that of conventional technologies usually employed in centralized systems (Bernal, 2021).

Generally, the operation of centralized systems is smooth with well-defined protocols for daily operation, exigencies, maintenance, and monitoring. The employees have appreciable technical expertise and skills. This results in a uniform and high-quality product (Hocaoglu et al., 2018). In the case of decentralized systems, the operators usually have low technical expertise. Effective operation requires considerable community participation; maintenance and monitoring are limited to easily doable tasks (Capodaglio, 2017). Furthermore, owing to the closeness of decentralized systems to residences, decentralized systems tend to be more sensitive to operational problems, such as odor. The impact of variation in influent load and quality and operational parameters on the effluent is higher in decentralized systems than in centralized systems.

From the above discussion, we observe that decentralized systems suffer in terms of technical competence, operation, and maintenance; however, this scenario could improve with advances in remote monitoring/operation, specialized services, and economic manufacturing of advanced small-scale treatment systems (Capodaglio, 2017; Wilderer and Schreff, 2000). Furthermore, characteristics of different waste streams define their reusability; for example, urine/yellow water streams are the most efficient in recovering nutrient-rich products. Such separation of waste streams into black and grey water and black water further into urine and feces (urine diversion technologies) is more favorable (in terms of economy, energy, GHG emissions, etc.) in decentralized systems (Chipako and Randall, 2020; Estévez et al., 2022; Hocaoglu et al., 2018).

15.4.3 Distribution for Reuse

For fertilizer reuse applications, distribution strategies for the treated water/recovered product should be established. Centralized systems require a distribution network, since the reuse application site is usually away from the treatment facility. This becomes unsustainable in reuse applications such as urban gardening, where decentralized approaches are favored. Agricultural lands are generally accumulated in one area, and only those decentralized facilities that are close to such areas can benefit from agricultural reuse. Thus, agricultural reuse generally favors centralized reuse approaches as the treatment facility could be located close to the reuse site (Oron et al., 1999). Centralized systems also face challenges in terms of capacity and safety of water/product stored during non-agricultural periods (Estévez et al., 2022).

Marketing and generating demand for product reuse is key to sustaining reuse practices. The market for decentralized systems is limited to local demands. Consequently, the supply (wastewater generation) and demand (irrigation requirement) must be balanced locally, which leads to lower water-saving potential in decentralized systems. Centralized systems can achieve a better supply-demand balance by diverting excess supply to high-demand areas (Hocaoglu et al., 2018). Contrarily, centralized systems provide a platform for expansion of market opportunities beyond the local scale through integration of waste streams, wider product range, etc. (Kehrein et al., 2020).

15.4.4 Sustainability

This discussion is confined to the sustainability of centralized and decentralized nutrient recovery, and differences attributable to the variation in technology, infrastructure requirement, etc., are not elaborated. Life cycle analysis (LCA) studies are a major source of evaluating the sustainability of centralized and decentralized systems. Economic sustainability is a major parameter monitored and is already discussed in Table 15.3. Furthermore, resource use is another important factor considered; studies have highlighted that electricity consumption (using fossil fuels) is the major driver of unsustainability in the investigated systems, and infrastructure development plays a major role in it (Mehmeti and Canaj, 2022; Risch et al., 2021). In this regard, centralized wastewater reuse relying on extensive pipeline networks for reuse fares worse than decentralized systems.

In addition to resource use and cost, public health and social sustainability are crucial factors to be considered. In decentralized nutrient recovery, higher community participation, community employment, and local landscaping lead to better social sustainability (Opher et al., 2019). In terms of public health, pathogen quality and level of contact determine the sustainability of systems. In comparison to centralized systems, decentralized systems generate lower-quality water with higher pathogen content (due to a lack of tertiary treatment processes). Household/community-level reuses tend to have lower safety precautions and higher proximity to residences, increasing the level of disease risk.

The above discussion reiterates the fact that no single system can be stated as the best-performing. Both systems have their inherent advantages and disadvantages. Depending on the geography, recovered product, application site, demand, capacity of the implementing body, etc., the best-suited system, which minimizes economic, social, and environmental costs, should be selected.

Table 15.3 Comparison of the reviewed technologies.

Technology	Energy feasibility	Economic feasibility	Environmental sustainability	Carbon footprint	References
Ammonium stripping	High energy demand for air stripping and steam stripping	High operation cost	Highly sustainable	Medium to high	Kar et al. (2023); Kundu et al. (2022a); Ye et al. (2018)
Adsorption	Low to moderate energy demand	Low capital and operational costs	Moderate to high	Low	Wu and Vaneeckhaute (2022); Xia et al. (2020)
Ion-exchange	Low to moderate	High capital cost	Moderate to high	Low	Huang et al. (2020); Saliu et al. (2023)
Struvite precipitation	Low to moderate energy demand	Moderate to high capital cost	Moderate to highly sustainable	Low to moderate	Sena and Hicks (2018)
Electrochemical oxidation	High	High capital cost	Moderately sustainable	High	Xiang et al. (2020)
Microbial fuel cell	High	High initial cost, energy recovery can offset operation cost	Highly sustainable	Low	Gowd et al. (2023); Saliu et al. (2023); Shah et al. (2019)
Constructed wetlands	Low energy demand	Low capital, operation, and maintenance costs	High sustainability	Low	Flores et al. (2020); Parde et al. (2021)

Vermifilter	Low energy demand	Low capital, operation cost, and maintenance costs	High sustainability	Low	Dey Chowdhury et al. (2022); Miito et al. (2022)
Microalgal treatment	Low energy demand	Low capital and operation cost	High sustainability	Low	Abdelfattah et al. (2023); Li et al. (2023); Rossi et al. (2022); You et al. (2022)
Membrane filtration	Moderate to high energy demand	High capital and maintenance costs	Moderate to high	High	Xiang et al. (2020); Yadav et al. (2021)
Thermal treatment	High energy demand	High capital cost	–	High	Kwapinski et al. (2021); Sylwan et al. (2019)
Forward osmosis	Low energy demand	High capital cost	Moderately to highly sustainable	Low	Saliu et al. (2023)
Capacitive deionization	Low to moderate energy demand	Low to moderate operation cost	Moderate to highly sustainable	Low to moderate	Xiang et al. (2020); Yu et al. (2016)

15.5 Summary

Using wastewater as a fertilizer incorporates sustainability into wastewater management. It reduces the demand for fertilizer production and associated environmental impacts, including resource exploitation and environmental pollution. Moreover, it offsets the adverse effects of nutrient pollution. If we understand wastewater/nutrient-based products as a saleable commodity, the principle of the circular economy is also evident. However, the conventional wastewater treatment and management systems are not effective in optimally utilizing the nutrient potential of wastewater. Innovations in nutrient recovery technologies can bridge this gap. It is evident that promising research is being conducted in this area, and several technologies have been implemented at the field scale.

Recovery efficiencies above 90% have been achieved by advanced technologies for N and P. Moreover, the nutrient-based product "struvite" is gaining traction, and several technologies are based on nutrient recovery in this form. It has been directly applied to the soil as a fertilizer. Apart from technical feasibility, the cost of installation and operation of these technologies is an important factor to be considered. Furthermore, technologies need to be evaluated in the context of a centralized or decentralized scheme of reuse. A developing country like India is likely to lean toward decentralized reuse, since the existing wastewater management systems are decentralized. Decentralized systems cannot afford highly sophisticated systems without a degree of remote monitoring. The reuse would also be restricted by local demand. However, the conditions vary with each scenario, and there is no one technology that suits all conditions. Invariably, for the widespread adoption of these technologies, establishing a market for nutrient-rich wastewater/products and policy engagement is necessary. It is proposed that the future steps toward wastewater reuse as a fertilizer should have these activities at their core.

Review Questions

1. What are the various challenges of effective wastewater treatment?
2. What are the major objectives of a conventional wastewater treatment system?
3. What are the different inorganic components available in wastewater? Are they considered as pollutants or resources?
4. Give the quantification of N, P, and K available in wastewater.
5. What are nutrient cycles in the environment and their flow?
6. Prepare a closed-loop concept of nutrients showing interrelationship between biotic and abiotic components.

7 What are the various forms of nutrients available in wastewater?
8 Explain various technologies for nutrient removal from wastewater.
9 Explain emerging technologies (MD, ED, and FO) to overcome the limitations of conventional membrane filtration technology.
10 Wastewater as a fertilizer incorporates sustainability into wastewater management. Comment on this and justify.

References

Abdelfattah, A., Ali, S.S., Ramadan, H. et al. (2023). Microalgae-based wastewater treatment: mechanisms, challenges, recent advances, and future prospects. *Environmental Science and Ecotechnology*, **13**, 100205. Editorial Board, Research of Environmental Sciences. https://doi.org/10.1016/j.ese.2022.100205

Adhikari, U., Harrigan, T., and Reinhold, D.M. (2015). Use of duckweed-based constructed wetlands for nutrient recovery and pollutant reduction from dairy wastewater. *Ecological Engineering*, **78**, 6–14. https://doi.org/10.1016/j.ecoleng.2014.05.024

Ammary, B.Y. (2004). Nutrients requirements in biological industrial wastewater treatment. *African Journal of Biotechnology*, **3**, 236–238.

Arun, S., Sinharoy, A., Pakshirajan, K. et al. (2020). Algae based microbial fuel cells for wastewater treatment and recovery of value-added products. *Renewable and Sustainable Energy Reviews*, **132**, 110041. Elsevier Ltd. https://doi.org/10.1016/j.rser.2020.110041

Batstone, D.J., Hülsen, T., Mehta, C.M. et al. (2015). Platforms for energy and nutrient recovery from domestic wastewater: a review. *Chemosphere*, **140**, 2–11. https://doi.org/10.1016/j.chemosphere.2014.10.021

Beckinghausen, A., Odlare, M., Thorin, E. et al. (2020). From removal to recovery: an evaluation of nitrogen recovery techniques from wastewater. *Applied Energy*, **263**, 114616. https://doi.org/10.1016/j.apenergy.2020.114616

Bernal, D. (2021). Key criteria for considering decentralization in municipal wastewater management. *Heliyon*, **7**, e06375. https://doi.org/10.1016/j.heliyon.2021.e06375

Brar, A., Kumar, M., Vivekanand, V. et al. (2019). Phycoremediation of textile effluent-contaminated water bodies employing microalgae: nutrient sequestration and biomass production studies. *International Journal of Environmental Science and Technology*, **16**(12), 7757–7768. https://doi.org/10.1007/s13762-018-2133-9

Capodaglio, A.G. (2017). Integrated, decentralized wastewater management for resource recovery in rural and peri-urban areas. *Resources*, **6**(2) 22. https://doi.org/10.3390/resources6020022

Chen, R. and Wang, X.C. (2009). Cost-benefit evaluation of a decentralized water system for wastewater reuse and environmental protection. *Water Science and Technology*, **59**(8), 1515–1522. https://doi.org/10.2166/wst.2009.156

Chen, Z., Wu, Q., Wu, G. et al. (2017). Centralized water reuse system with multiple applications in urban areas: lessons from China's experience. *Resources, Conservation & Recycling*, **117**, 125–136. https://doi.org/10.1016/j.resconrec.2016.11.008

Chipako, T.L. and Randall, D.G. (2020). Investigating the feasibility and logistics of a decentralized urine treatment and resource recovery system. *Journal of Water Process Engineering*, **37**, 101383. https://doi.org/10.1016/j.jwpe.2020.101383

Cordell, D. (2013). Peak phosphorus and the role of P recovery in achieving food security. In: *Source Separation and Decentralization for Wastewater Management* (eds. T.A. Larsen, K.M. Udert, and J. Lienert). IWA Publishing.

De Vrieze, J., Smet, D., Klok, J. et al. (2016). Thermophilic sludge digestion improves energy balance and nutrient recovery potential in full-scale municipal wastewater treatment plants. *Bioresource Technology*, **218**, 1237–1245. Elsevier Ltd. https://doi.org/10.1016/j.biortech.2016.06.119

Deng, S., Wang, C., Ngo, H.H. et al. (2023). Comparative review on microbial electrochemical technologies for resource recovery from wastewater towards circular economy and carbon neutrality. *Bioresource Technology*, **376**, 128906. Elsevier Ltd. https://doi.org/10.1016/j.biortech.2023.128906

Dey Chowdhury, S., Bhunia, P., and Surampalli, R.Y. (2022). Sustainability assessment of vermifiltration technology for treating domestic sewage: a review. *Journal of Water Process Engineering*, **50**, 103266. Elsevier Ltd. https://doi.org/10.1016/j.jwpe.2022.103266

Estévez, S., González-García, S., Feijoo, G. et al. (2022). How decentralized treatment can contribute to the symbiosis between environmental protection and resource recovery. *Science of the Total Environment*, **812**, 151485. https://doi.org/10.1016/j.scitotenv.2021.151485

Etter, B., Tilley, E., Khadka, R. et al. (2011). Low-cost struvite production using source-separated urine in Nepal. *Water Research*, **45**(2), 852–862. https://doi.org/10.1016/j.watres.2010.10.007

Flores, L., García, J., Pena, R. et al. (2020). Carbon footprint of constructed wetlands for winery wastewater treatment. *Ecological Engineering*, **156**, 105959. https://doi.org/10.1016/j.ecoleng.2020.105959

Gao, F., Li, C., Yang, Z.H. et al. (2016). Continuous microalgae cultivation in aquaculture wastewater by a membrane photobioreactor for biomass production and nutrients removal. *Ecological Engineering*, **92**, 55–61. https://doi.org/10.1016/j.ecoleng.2016.03.046

Gao, F., Wang, L., Wang, J. et al. (2020). Nutrient recovery from treated wastewater by a hybrid electrochemical sequence integrating bipolar membrane electrodialysis and membrane capacitive deionization. *Environmental Science: Water Research and Technology*, **6**(2), 383–391. https://doi.org/10.1039/c9ew00981g

Ge, Q., Wang, P., Wan, C. et al. (2012). Polyelectrolyte-promoted Forward Osmosis-Membrane Distillation (FO-MD) hybrid process for dye wastewater treatment. *Environmental Science and Technology*, **46**(11), 6236–6243. https://doi.org/10.1021/es300784h

Ghafoori, S., Hassanpour Darvishi, H., Mohamadvali Samani, H. et al. (2021). Enhancing the method of decentralized multi-purpose reuse of wastewater in urban area. *Sustainability (Switzerland)*, **13**(24), 1–12. https://doi.org/10.3390/su132413553

Gowd, S.C., Ramakrishna, S., and Rajendran, K. (2022). Wastewater in India: an untapped and under-tapped resource for nutrient recovery towards attaining a sustainable circular economy. *Chemosphere*, **291**, 132753. https://doi.org/10.1016/j.chemosphere.2021.132753

Gowd, S.C., Ramesh, P., Vigneswaran, V.S. et al. (2023). Life cycle assessment of comparing different nutrient recovery systems from municipal wastewater: a path towards self-reliance and sustainability. *Journal of Cleaner Production*, **410**, 137331. https://doi.org/10.1016/j.jclepro.2023.137331

Hau, N.T., Chen, S.S., Nguyen, N.C. et al. (2014). Exploration of EDTA sodium salt as novel draw solution in forward osmosis process for dewatering of high nutrient sludge. *Journal of Membrane Science*, **455**, 305–311. https://doi.org/10.1016/j.memsci.2013.12.068

Henze, M. and Comeau, Y. (2008). Wastewater characterization. In *Biological Wastewater Treatment: Principles Modelling and Design* (eds. M. Henze, M.C.M. van Loosdrecht, G.A. Ekama et al.) **27**, 33–52. ISBN: 9781843391883. London, UK: IWA Publishing.

Huang, H.M., Song, Q.W., and Xu, C.L. (2011). The mechanism and influence factors of struvite precipitation for the removal of ammonium nitrogen. *Advanced Materials Research*, **189–193**, 2613–2620. https://doi.org/10.4028/www.scientific.net/AMR.189-193.2613

Huang, X., Guida, S., Jefferson, B. et al. (2020). Economic evaluation of ion-exchange processes for nutrient removal and recovery from municipal wastewater. *NPJ Clean Water*, **3**(1), 7. https://doi.org/10.1038/s41545-020-0054-x

Hutnik, N., Kozik, A., Mazienczuk, A. et al. (2013). Phosphates (V) recovery from phosphorus mineral fertilizers industry wastewater by continuous struvite reaction crystallization process. *Water Research*, **47**(11), 3635–3643. https://doi.org/10.1016/j.watres.2013.04.026

Ichihashi, O. and Hirooka, K. (2012). Removal and recovery of phosphorus as struvite from swine wastewater using microbial fuel cell. *Bioresource Technology*, **114**, 303–307. https://doi.org/10.1016/j.biortech.2012.02.124

Kabdaşlı, I. and Tünay, O. (2018). Nutrient recovery by struvite precipitation, ion exchange and adsorption from source-separated human urine–a review. *Environmental Technology Reviews*, **7**(1), 106–138. Taylor and Francis Ltd. https://doi.org/10.1080/21622515.2018.1473504

Kar, S., Singh, R., Gurian, P.L. et al. (2023). Life cycle assessment and techno-economic analysis of nitrogen recovery by ammonia air-stripping from wastewater treatment. *Science of the Total Environment*, **857**, 159499. https://doi.org/10.1016/j.scitotenv.2022.159499

Kehrein, P., Van Loosdrecht, M., Osseweijer, P. et al. (2020). A critical review of resource recovery from municipal wastewater treatment plants-market supply potentials, technologies and bottlenecks. *Environmental Science: Water Research and Technology*, **6**(4), 877–910. https://doi.org/10.1039/c9ew00905a

Kretschmer, N., Ribbe, L., and Gaese, H. (2002). Wastewater reuse for agriculture. *Technology Resource Management & Development- Scientific Contributions for Sustainable Development*, **2**, 37–64.

Kundu, D., Dutta, D., Samanta, P. et al. (2022a). Valorization of wastewater: a paradigm shift towards circular bioeconomy and sustainability. *Science of the Total Environment*, **848**, 157709. Elsevier B.V. https://doi.org/10.1016/j.scitotenv.2022.157709

Kundu, S., Pramanik, B.K., Halder, P. et al. (2022b). Source and central level recovery of nutrients from urine and wastewater: a state-of-art on nutrients mapping and potential technological solutions. *Journal of Environmental Chemical Engineering*, **10**(2), 107146. Elsevier Ltd. https://doi.org/10.1016/j.jece.2022.107146

Kwapinski, W., Kolinovic, I., and Leahy, J.J. (2021). Sewage sludge thermal treatment technologies with a focus on phosphorus recovery: a review. *Waste and Biomass Valorization*, **12**(11), 5837–5852. https://doi.org/10.1007/s12649-020-01280-2

Li, L., Gao, K., Yang, M. et al. (2023). Challenges and potential solutions of microalgae-based systems for wastewater treatment and resource recovery. *Frontiers in Bioengineering and Biotechnology*, **11**, 1210228. Frontiers Media S.A. https://doi.org/10.3389/fbioe.2023.1210228

Li, M., Zhou, M., Tian, X. et al. (2021). Enhanced bioenergy recovery and nutrient removal from swine wastewater using an airlift-type photosynthetic microbial fuel cell. *Energy*, **226**, 120422. https://doi.org/10.1016/j.energy.2021.120422

Liu, Y., Deng, Y.Y., Zhang, Q. et al. (2021). Overview of recent developments of resource recovery from wastewater via electrochemistry-based technologies. *Science of the Total Environment*, **757**, 143901. Elsevier B.V. https://doi.org/10.1016/j.scitotenv.2020.143901

Malal, H., Romero, V.S., Horwath, W.R. et al. (2024). Vermifiltration and sustainable agriculture: unveiling the soil health-boosting potential of liquid waste vermicompost. *Frontiers in Sustainable Food Systems*, **8**, 1383715. https://doi.org/10.3389/fsufs.2024.1383715

Maurer, M., Pronk, W., and Larsen, T.A. (2006). Treatment processes for source-separated urine. *Water Research*, **40**(17), 3151–3166. Elsevier Ltd. https://doi.org/10.1016/j.watres.2006.07.012

Mavhungu, A., Foteinis, S., Mbaya, R. et al. (2021). Environmental sustainability of municipal wastewater treatment through struvite precipitation: influence of operational parameters. *Journal of Cleaner Production*, **285**, 124856. https://doi.org/10.1016/j.jclepro.2020.124856

Mehmeti, A. and Canaj, K. (2022). Environmental assessment of wastewater treatment and reuse for irrigation: a mini-review of LCA Studies. *Resources*, **11**(10), 94. https://doi.org/10.3390/resources11100094

Miito, G.J., Ndegwa, P.M., Alege, F.P. et al. (2022). Efficacy of a vermifilter at mitigating greenhouse gases and ammonia emissions from dairy wastewater. *Journal of Environmental Quality*, **51**(4), 644–655. https://doi.org/10.1002/jeq2.20367

Mikucka, W. and Zieli, M. (2020). Mikucka-Zielińska2020_Article_DistilleryStillageCharacterist.pdf. *Applied Biochemistry and Biotechnology*, 770–793.

Mishra, A., K, A., Anand, S. et al. (2023). *Reuse of Treated Wastewater in Urban/Peri-Urban Agriculture in India*. NITI Aayog.

Mukherjee, A., Zaveri, P., Patel, R. et al. (2021). Optimization of microbial fuel cell process using a novel consortium for aromatic hydrocarbon bioremediation and bioelectricity generation. *Journal of Environmental Management*, **298**, 113546. https://doi.org/10.1016/j.jenvman.2021.113546

Murat Hocaoglu, S., Erzi, I., and Ergenekon, Ş. (2018). Centralized versus decentralized wastewater reuse: a case study for a touristic area. *Uludağ University Journal of The Faculty of Engineering*, **23**(2), 309–322. https://doi.org/10.17482/uumfd.421941

Muratore, C., Espen, L., and Prinsi, B. (2021). Nitrogen uptake in plants: the plasma membrane root transport systems from a physiological and proteomic perspective. *Plants*, **10**(4), 681. MDPI AG. https://doi.org/10.3390/plants10040681

Nie, J., Feng, H., Witherell, B.B. et al. (2018). Causes, assessment, and treatment of nutrient (N and P) pollution in rivers, estuaries, and coastal waters. *Current Pollution Reports*, **4**(2), 154–161. Springer. https://doi.org/10.1007/s40726-018-0083-y

Norddahl, B., Horn, V.G., Larsson, M. et al. (2006). A membrane contactor for ammonia stripping, pilot scale experience and modeling. *Desalination*, **199**(1–3), 172–174. https://doi.org/10.1016/j.desal.2006.03.037

Opher, T., Friedler, E., and Shapira, A. (2019). Comparative life cycle sustainability assessment of urban water reuse at various centralization scales. *International*

Journal of Life Cycle Assessment, **24**(7), 1319–1332. https://doi.org/10.1007/s11367-018-1469-1

Oron, G., Campos, C., Gillerman, L. et al. (1999). Wastewater treatment, renovation and reuse for agricultural irrigation in small communities. *Agricultural Water Management*, **38**(3), 223–234. https://doi.org/10.1016/S0378-3774(98)00066-3

Oudad, M.A., Kumar, P., Chaali, M. et al. (2022). Optimized ammonium sulphate recovery by stripping-scrubbing sequence system from compost leachate at mesophilic temperatures. *Case Studies in Chemical and Environmental Engineering*, **5**, 100198. https://doi.org/10.1016/j.cscee.2022.100198

Pang, N., Bergeron, A.D., Gu, X. et al. (2020). Recycling of nutrients from dairy wastewater by extremophilic microalgae with high ammonia tolerance. *Environmental Science and Technology*, **54**(23), 15366–15375. https://doi.org/10.1021/acs.est.0c02833

Parde, D., Patwa, A., Shukla, A. et al. (2021). A review of constructed wetland on type, treatment and technology of wastewater. *Environmental Technology and Innovation*, **21**, 101261. Elsevier B.V. https://doi.org/10.1016/j.eti.2020.101261

Pastor, L., Mangin, D., Ferrer, J. et al. (2010). Struvite formation from the supernatants of an anaerobic digestion pilot plant. *Bioresource Technology*, **101**(1), 118–125. https://doi.org/10.1016/j.biortech.2009.08.002

Pehlivanoglu-Mantas, E. and Sedlak, D.L. (2006). Wastewater-derived dissolved organic nitrogen: analytical methods, characterization, and effects – a review. *Critical Reviews in Environmental Science and Technology*, **36**(3), 261–285. https://doi.org/10.1080/10643380500542780

Perera, M.K. and Englehardt, J.D. (2020). Simultaneous nitrogen and phosphorus recovery from municipal wastewater by electrochemical pH modulation. *Separation and Purification Technology*, **250**, 117166. https://doi.org/10.1016/j.seppur.2020.117166

Qadir, M., Drechsel, P., Jiménez Cisneros, B. et al. (2020). Global and regional potential of wastewater as a water, nutrient and energy source. *Natural Resources Forum*, **44**(1), 40–51. https://doi.org/10.1111/1477-8947.12187

Risch, E., Boutin, C., and Roux, P. (2021). Applying life cycle assessment to assess the environmental performance of decentralised versus centralised wastewater systems. *Water Research*, **196**, 116991. https://doi.org/10.1016/j.watres.2021.116991

Rossi, S., Pizzera, A., Bellucci, M. et al. (2022). Piggery wastewater treatment with algae-bacteria consortia: pilot-scale 2 validation and techno-economic evaluation at farm level. *Bioresource Technology*, **351**, 127051. https://ssrn.com/abstract=4028802

Saliu, T.D., Oladoja, N.A., and Sauvé, S. (2023). Eco-friendly and sustainability assessment of technologies for nutrient recovery from human urine—a review. *Frontiers in Sustainability*, **4**, 1338380. Frontiers Media SA. https://doi.org/10.3389/frsus.2023.1338380

Sangeetha, V., Devasena, M., Nambi, I.M. et al. (2022). Crystallization of struvite family crystals from cow urine: analysis, characterization, and effects of crystallization method, retention time, rate of mixing, and competing ions. *Biomass Conversion and Biorefinery*, **14**, 2357–2368. https://doi.org/10.1007/s13399-022-02452-x

Sarayu, K. and Sandhya, S. (2012). Current technologies for biological treatment of textile wastewater-a review. *Applied Biochemistry and Biotechnology*, **167**(3), 645–661. https://doi.org/10.1007/s12010-012-9716-6

Sardans, J. and Peñuelas, J. (2021). Potassium control of plant functions: ecological and agricultural implications. *Plants*, **10**(2), 419. https://doi.org/10.3390/plants10020419

Sena, M. and Hicks, A. (2018). Life cycle assessment review of struvite precipitation in wastewater treatment. *Resources, Conservation and Recycling*, **139**, 194–204. Elsevier B.V. https://doi.org/10.1016/j.resconrec.2018.08.009

Shah, S., Venkatramanan, V., and Prasad, R. (2019). Microbial fuel cell: sustainable green technology for bioelectricity generation and wastewater treatment. In: *Sustainable Green Technologies for Environmental Management* (eds. S. Shah, V. Venkatraman, and R. Prasad), 199–218. Singapore: Springer. https://doi.org/10.1007/978-981-13-2772-8_10

Shilpi, S., Seshadri, B., Sarkar, B. et al. (2018). Comparative values of various wastewater streams as a soil nutrient source. *Chemosphere*, **192**, 272–281. https://doi.org/10.1016/j.chemosphere.2017.10.118

Suhaib, K.H. and Bhunia, P. (2022). Impact of hydraulic loading rate on the removal performance and filter-bed clogging of horizontal-subsurface-flow macrophyte-assisted vermifilter treating dairy wastewater. *Journal of Hazardous, Toxic, and Radioactive Waste*, **26**(3), 1–11. https://doi.org/10.1061/(asce)hz.2153-5515.0000698

Sylwan, I., Zambrano, J., and Thorin, E. (2019). Energy demand for phosphorus recovery from municipal wastewater. *Energy Procedia*, **158**, 4338–4343. https://doi.org/10.1016/j.egypro.2019.01.787

Tarpeh, W.A., Barazesh, J.M., Cath, T.Y. et al. (2018). Electrochemical stripping to recover nitrogen from source-separated urine. *Environmental Science and Technology*, **52**(3), 1453–1460. https://doi.org/10.1021/acs.est.7b05488

Tymchuk, I., Shkvirko, O., Sakalova, H. et al. (2020). Wastewater a source of nutrients for crops growth and development. *Journal of Ecological Engineering*, **21**(5), 88–96. https://doi.org/10.12911/22998993/122188

Ugwuanyi, E.D., Nwokediegwu, Z.Q.S., Dada, M.A. et al. (2024). The role of algae-based wastewater treatment systems: a comprehensive review. *World Journal of Advanced Research and Reviews*, **21**(2), 937–949. https://doi.org/10.30574/wjarr.2024.21.2.0521

van Puijenbroek, P.J.T.M., Beusen, A.H.W., and Bouwman, A.F. (2019). Global nitrogen and phosphorus in urban waste water based on the shared socio-economic pathways. *Journal of Environmental Management*, **231**, 446–456. https://doi.org/10.1016/j.jenvman.2018.10.048

Vymazal, J. (2007). Removal of nutrients in various types of constructed wetlands. *Science of the Total Environment*, **380**(1–3), 48–65. https://doi.org/10.1016/j.scitotenv.2006.09.014

Wang, L. and Macko, S.A. (2011). Constrained preferences in nitrogen uptake across plant species and environments. *Plant, Cell and Environment*, **34**(3), 525–534. https://doi.org/10.1111/j.1365-3040.2010.02260.x

Wang, X.C., Chen, R., Zhang, Q.H. et al. (2008). Optimized plan of centralized and decentralized wastewater reuse systems for housing development in the urban area of Xi'an, China. *Water Science and Technology*, **58**(5), 969–975. https://doi.org/10.2166/wst.2008.456

Wang, Y., Lin, Z., Su, X. et al. (2019). Cost-effective domestic wastewater treatment and bioenergy recovery in an immobilized microalgal-based photoautotrophic microbial fuel cell (PMFC). *Chemical Engineering Journal*, **372**, 956–965. https://doi.org/10.1016/j.cej.2019.05.004

Wilderer, P.A. and Schreff, D. (2000). Decentralized and centralized wastewater management: a challenge for technology developers. *Water Science and Technology*, **41**(1), 1–8. https://doi.org/10.2166/wst.2000.0001

Williams, A.T., Zitomer, D.H., and Mayer, B.K. (2015). Ion exchange-precipitation for nutrient recovery from dilute wastewater. *Environmental Science: Water Research and Technology*, **1**(6), 832–838. https://doi.org/10.1039/c5ew00142k

Wu, H. and Vaneeckhaute, C. (2022). Nutrient recovery from wastewater: a review on the integrated physicochemical technologies of ammonia stripping, adsorption and struvite precipitation. *Chemical Engineering Journal*, **433**, 133664. Elsevier B.V. https://doi.org/10.1016/j.cej.2021.133664

Wu, Z., Zou, S., Zhang, B. et al. (2018). Forward osmosis promoted in-situ formation of struvite with simultaneous water recovery from digested swine wastewater. *Chemical Engineering Journal*, **342**, 274–280. https://doi.org/10.1016/j.cej.2018.02.082

Xia, W.J., Xu, L.Z.J., Yu, L.Q. et al. (2020). Conversion of municipal wastewater-derived waste to an adsorbent for phosphorus recovery from secondary effluent. *Science of the Total Environment*, **705**, 135959. https://doi.org/10.1016/j.scitotenv.2019.135959

Xiang, S., Liu, Y., Zhang, G. et al. (2020). New progress of ammonia recovery during ammonia nitrogen removal from various wastewaters. *World Journal of Microbiology and Biotechnology*, **36**(10), 1–20. Springer. https://doi.org/10.1007/s11274-020-02921-3

Xie, D., Ji, X., Zhou, Y. et al. (2022). Chlorella vulgaris cultivation in pilot-scale to treat real swine wastewater and mitigate carbon dioxide for sustainable biodiesel production by direct enzymatic transesterification. *Bioresource Technology*, **349**, 126886. https://doi.org/10.1016/j.biortech.2022.126886

Yadav, G., Mishra, A., Ghosh, P. et al. (2021). Technical, economic and environmental feasibility of resource recovery technologies from wastewater. *Science of the Total Environment*, **796**, 149022. https://doi.org/10.1016/j.scitotenv.2021.149022

Yang, C. (2020). *Petrochemical Wastewater and Its Treatment*. Centria University of Applied Sciences Environmental Chemistry and Technology.

Yaseen, D.A. and Scholz, M. (2019). Textile dye wastewater characteristics and constituents of synthetic effluents: a critical review. *International Journal of Environmental Science and Technology*, **16**(2), 1193–1226. Springer Berlin Heidelberg. https://doi.org/10.1007/s13762-018-2130-z

Ye, Y., Ngo, H.H., Guo, W. et al. (2019). Effect of organic loading rate on the recovery of nutrients and energy in a dual-chamber microbial fuel cell. *Bioresource Technology*, **281**, 367–373. https://doi.org/10.1016/j.biortech.2019.02.108

Ye, Y., Ngo, H.H., Guo, W. et al. (2018). A critical review on ammonium recovery from wastewater for sustainable wastewater management. *Bioresource Technology*, **268**, 749–758. Elsevier Ltd. https://doi.org/10.1016/j.biortech.2018.07.111

You, X., Yang, L., Zhou, X. et al. (2022). Sustainability and carbon neutrality trends for microalgae-based wastewater treatment: a review. *Environmental Research*, **209**, 112860. https://doi.org/10.1016/j.envres.2022.112860

Yu, T.H., Shiu, H.Y., Lee, M. et al. (2016). Life cycle assessment of environmental impacts and energy demand for capacitive deionization technology. *Desalination*, **399**, 53–60. https://doi.org/10.1016/j.desal.2016.08.007

Zhang, D.M., Chen, Y.X., Jilani, G. et al. (2012). Optimization of struvite crystallization protocol for pretreating the swine wastewater and its impact on subsequent anaerobic biodegradation of pollutants. *Bioresource Technology*, **116**, 386–395. https://doi.org/10.1016/j.biortech.2012.03.107

Zhang, J., She, Q., Chang, V.W.C. et al. (2014). Mining nutrients (N, K, P) from urban source-separated urine by forward osmosis dewatering. *Environmental Science and Technology*, **48**(6), 3386–3394. https://doi.org/10.1021/es405266d

Zhu, F., Cakmak, E.K., and Cetecioglu, Z. (2023). Phosphorus recovery for circular economy: application potential of feasible resources and engineering processes in Europe. *Chemical Engineering Journal*, **454**, 140153. Elsevier B.V. https://doi.org/10.1016/j.cej.2022.140153

Zhu, L., Wang, Z., Shu, Q. et al. (2013). Nutrient removal and biodiesel production by integration of freshwater algae cultivation with piggery wastewater treatment. *Water Research*, **47**(13), 4294–4302. https://doi.org/10.1016/j.watres.2013.05.004

16

Contribution of Modern Biological Sewage Resource Recovery Technologies in Circular Economy of Water

Minh-Ky Nguyen[1], *Quoc-Minh Truong*[2], *The-Anh Luu*[3], *Van-Giang Le*[3], *Hoang-Lam Nguyen*[4], *Minh-Thuan Pham*[5], *and D. Duc Nguyen*[6*]

[1]*Faculty of Environment and Natural Resources, Nong Lam University, Ho Chi Minh City, Vietnam*
[2]*Faculty of Management Science, Thu Dau Mot University, Binh Duong, Vietnam*
[3]*Central Institute for Natural Resources and Environmental Studies, Vietnam National University (CRES-VNU), Hanoi, Vietnam*
[4]*Department of Civil Engineering, McGill University, Montreal, Canada*
[5]*Center for Environmental Toxin and Emerging-Contaminant Research, Cheng Shiu University, Kaohsiung, Taiwan*
[6]*Department of Civil & Energy System Engineering, Kyonggi University, Suwon, South Korea*

*Corresponding author: nguyensyduc@gmail.com

16.1 Introduction

The modern industry is fundamental to a transformative movement toward promoting transformation for the development of circular economies. To protect human health and the earth's safety, innovative waste treatment models are applied to promote beneficial resource recovery from waste as a valuable resource (Le et al., 2024b; Lin et al., 2013; Puyol et al., 2017). They rely on material flow analysis and cutting-edge concepts of sustainable development. The novel concepts and ideas of the circular economy (CE) are groundbreaking as they simultaneously embrace the wider understanding of industrial ecology and place it within an economic framework to promote interest and engagement among individuals, societies, and communities (Nielsen, 2017). The CE aims to efficiently manage resources by promoting the reuse of waste as feedstock for further production processes. The CE envisions achieving worldwide sustainable development by establishing a self-regenerating production system and converting waste into

Wastewater to Resource Recovery: Applying the Circular Economy Toward Sustainable Development, First Edition. Edited by Arvind Kumar Mungray et al. Published 2026 by John Wiley & Sons, Inc.

raw materials. The need to ensure sustainable development is pushing existing industrial schemes toward self-recycling, replacing the current 3Rs model—"*reduce, reuse and recycle*"—with a further effective archetype (Puyol et al., 2017).

Technological innovations are transforming wastewater treatment by enhancing its efficiency and sustainability. As treatment methods evolve, they enable industries to adopt cleaner, more advanced practices, leading to improved environmental management (van Loosdrecht and Brdjanovic, 2014). Approximately 50–100% of wasted resources are found within wastewater. Consequently, economic and environmental experts, as well as corporate interests, are motivating efforts to retrieve and restore these compounds. Countries worldwide have invested significant resources in the bioeconomy to develop the progress of the modern economy in parallel with sustainable development. This process also includes the development of an orientation of wastewater treatment plants (WWTPs), in which organic or biological matter and other useful resources are extracted and utilized instead of being wasted. The driving force behind this is not only the desire to lower costs and conserve resources, especially energy, but also the global depletion of non-renewable macronutrients like easily accessible phosphorus (P), nitrogen (N), and the necessity to reduce human-caused impacts on the earth's nitrogen cycle (Batstone et al., 2015).

There has been recent interest in microbial communities, which are various interacting populations of microorganisms, in order to carry out complex functions and make them more resistant to changes in the environment. Although numerous novel approaches are aiding in the task of extracting resources from wastewater, biotechnologies show the most potential for effectively recovering valuable minerals from low-concentration streams (Johnson, 2018; Kuhad and Singh, 2013). Examples encompass rapidly proliferating bacteria, marine algae, and terrestrial plants that aim to recover organic substances (Barsanti and Gualtieri, 2022; Pinnaka and Tanuku, 2019). Additionally, highly specialized organisms capable of reducing and oxidizing metals are employed for metal recovery. Accumulative bacteria can create biopolymers like polyhydroxyalkanoates and alginates by absorbing complicated organic substances. Microorganisms, which are the most varied organisms on Earth, play a crucial role in maintaining the cycles of phosphorus, sulfur, and different elements that support the functioning of our planet (Wu and Yin, 2020). Microorganisms are the primary drivers of all ecosystem activities, and many processes rely exclusively on microorganisms. Understanding the fundamental processes that drive microbial biodiversity is crucial for predicting how ecosystems will respond to climate change and improving biological processes, such as sewage treatment and/or resource recovery technologies. The currently preferred type of reactor is the activated sludge, although alternative varieties may sometimes be more prevalent than these traditional schemes (Tran et al., 2022b). However, the fundamental bacteria involved in the process are typically the same, regardless of the kind of reactor. Despite having some understanding of microorganisms, the majority of them remain poorly understood, and their functions or roles are still unclear. Furthermore, the wastewater treatment industry is currently facing many difficulties due to strict discharge regulations and the presence of new types of pollutants. These pollutants include micropollutants such as phthalates (Tran et al., 2022c), organic chemical sunscreens, pharmaceuticals, per- and polyfluoroalkyl substances (Le et al., 2024a), heavy metal(loid)s, micro/nano-plastics, and others (Minh Ky et al., 2022; Nguyen et al., 2024; Tran et al., 2022a; Tran et al., 2024).

Approximately 80% of municipal wastewater in developed nations undergoes treatment via the activated sludge technique. Aeration treatment accounts for most of the required energy (>60%). This energy need is affected by factors like the capacity of the treatment plant, the treatment scheme, and the profiles of the wastewater (Ghimire et al., 2021). The incorporation of energy and resource recovery from wastewater into the process of producing clean water has become a fundamental aspect of the CE sustainability movement. The accomplishments thus far encompass the successful execution of experiments at various scales, ranging from small-scale laboratory or pilot-scale projects to widespread industrial applications (Kaszycki et al., 2021). After a century of the rapid growth of the activated sludge procedure, new ideas have emerged that recognize sewage as used water and focus on purifying the aqueous solution and recycling energy. The ultimate discharges can be effectively and securely reused for many applications, such as in agriculture, which is the sector that consumes the most freshwater (Nunes, 2021). Shifting the way people consume toward a climate-neutral and CE, as well as recovering waste, has made us consider the importance of not only liquid as a valuable reserve but also the matter present in waste streams. This transformation has turned sanitation from an expensive service into a self-sustaining one that contributes to the economic framework (Cerruti et al., 2021).

It is urgent to concede the rising importance of microbes in the monitoring system of WWTPs, and how understanding the structure of bacterial communities can aid in the improved design and operation of these treatment stations to cope with various ecological variations. Understanding the fundamental microbial features of water remediation system plans and procedures can assist in studying the microbiomes and identifying the microbes, their metabolisms, and interactions in a more detailed and efficient manner. This knowledge can benefit efforts to protect and restore environmental resources and can aid professionals such as microbiologists and environmental scientists in developing collaborative research approaches. The upcoming version of WWTPs aims to achieve carbon neutrality and recover energy, precious metals, nutrients (specifically nitrogen and phosphorus), etc. There is a growing trend to extract valuable products from various types of waste and wastewater, including those generated by industrial manufacturing and mining activities. Further, biological technologies have the ability to recover resources from these contaminants. The biological methods offer significant potential for reclaiming and reusing valuable resources, contributing to the protection of ecological environments.

16.2 Wastewater and Circular Economy

Water issues play a crucial role in a CE, promoting sustainable and balanced management of raw resource materials, such as water/wastewater and waste streams (Smol, 2022). One approach to transitioning the water resources and treatment sector toward the CE is to adapt existing tools to meet the specific needs of this sector (Figure 16.1). For example, WWTPs offer significant opportunities within the framework of a CE (Hernández-Chover et al., 2023). Wastewater not only serves as an alternative water source but also supplies valuable resources like water and energy to other sectors. Designing sustainable WWTPs should prioritize ecological protection while minimizing resource and energy consumption (Ghimire et al., 2021). Additionally, the nutrient sources in wastewater are precious for agricultural activities, considering they could be applied as valuable fertilizers (Hernández-Chover et al., 2023). Wastewater sludge contains rich organic matter, phosphorus (P), and nitrogen (N), which could be regenerated or recovered and utilized as biogas and/or fertilizers, offering a significant benefit (Ky et al., 2022; Mavhungu et al., 2020). In this context, there is a strong demand for innovative WWTPs that generate useful energy while effectively removing nutrient pollutants.

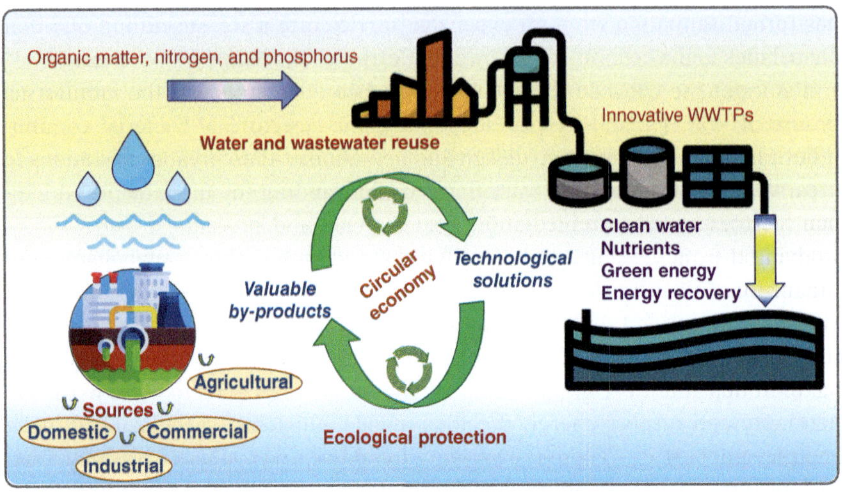

Figure 16.1 Circular economy approaches for water and wastewater reuse.

Water is used as the primary and most critical resource that could be recovered and reused from these operations. Wastewater also includes significant amounts of valuable energy, which can be transformed into heat or electrical potential. Most WWTPs use aerobic biological approaches to treat wastewater, converting organic matter into biomass and carbon dioxide. Recently, there has been increased focus on the energy utilization of these systems, given that they are known to be notably energy-intensive operations. Henriques and Catarino (2017) suggest that WWTPs have a significant potential for improvement in this regard, estimating potential reductions in energy consumption from 20% to 40%. Based on a dry weight basis, wastewater sludge reaches about 60% organic matter, consisting of 50–55% organic carbon that is mostly biodegradable, 10–15% N-nitrogen, and 1–3% P-phosphorus (Du et al., 2019; Sarpong et al., 2019). Especially, the energy content in the wastewater's nutritional elements, e.g. nitrogen and phosphorus, is up to 0.79 kWh/m^3 (Chae and Kang, 2013). This is crucial for establishing methods to transform the potential energy into a useful resource from wastewater and assessing energy recovery performance from wastewater.

A CE model or closed-loop system aims to maximize the utilization of materials and related products by transforming waste into resources, enhancing process efficiency, and prolonging the useful life of materials and products. Recycled wastewater is recognized as a substitute water resource that supports water accessibility while alleviating influence on natural water sources, thereby contributing to their sustainability (Sfez et al., 2019). On the other hand, valuable by-products can be obtained as significant results related to wastewater treatment. The nitrogen recovered from WWTPs has significant potential for agricultural use, as it is eliminated in a form that mimics nitrogenous fertilizers produced through highly energy-intensive industrial processes. From the CE perspective, WWTPs represent a valuable opportunity. WWTPs can adopt principles of the CE, which emphasize retaining materials and/or products in circulation for as long as practicable through recycling and reuse, applicable at both large and small scales. For instance, the nutrients found in wastewater streams are highly valuable for agricultural fields as they could serve as efficient fertilizers (Hernández-Chover et al., 2023). Microalgae-based biofuel production from wastewater is also being explored to enhance the energy sustainability of WWTPs (Otondo et al., 2018). Through the processing of organic waste materials, including FOG (i.e. fats, oils, and grease), agricultural wastes, and food waste, WWTPs could convert these wastes into beneficial products, e.g. biosolids and methane, which have various beneficial uses (Ghimire et al., 2021). However, the potential for energy recovery in a WWTP is contingent upon several factors, including wastewater profiles, the operating process of the biological treatment, the capacity of the plant, and the chosen energy recovery method (Sarpong et al., 2019).

16.3 Biological Technology Used for Resource Recovery

The shift in societal production systems from waste treatment to resource recovery is significantly influenced by the potential of biotechnological processes. These processes, with their cost-effective and flexible nature, play a crucial role in converting resources from wastewater and waste into significant products. This conversion is essential for the development of an eco-effective bio-based economy, highlighting the importance of biotechnological processes in this transition (Yadav et al., 2021).

Table 16.1 highlights critical technologies for resource recovery throughout the wastewater processing cycle. Bioenergy, including biohydrogen (generated through photo fermentation and dark fermentation) and biogas (produced through anaerobic digestion), has traditionally been a key focus. Additionally, the direct conversion of lipidic biomass into biodiesel is being considered. This approach is more sustainable than earlier biofuel concepts, as it enables the production of next-generation biofuels and additional products from waste biomass (Ali et al., 2021). The synthesis of high-value biopolymers from hydrogen, organic acids, and methane, such as those used in bioplastics, presents another avenue for carbon recovery (Kong et al., 2021). The carbon and nutrient recovery can also be obtained through organic fertilizer manufacturing or the generation of single-cell protein, which can serve as feed additives, feed, probiotics, novel fertilizers, etc. Furthermore, chemical redox reactions and bioelectrochemical systems could recover inorganic compounds or organic products beyond the capabilities of microbial metabolic pathways (Meena et al., 2019).

While many novel technologies are advancing wastewater-based resource recovery, biological approaches have the greatest capacity for efficiently recovering significant resources from dilute streams. These methods, which include the use of fast-growing bacteria, algae, and specialized species, are highly effective in recovering resources (Craggs et al., 1997). These biological methods reveal their potential to recover resources from industrial and domestic wastewater and wastes.

The subsequent formation of domestic WWTPs has also been used for energy neutrality and complete nutrient recovery, e.g. nitrogen and phosphorus (Yetilmezsoy and Sapci-Zengin, 2009). There is an increasing drive to recover significant products from various waste streams and wastewater. Biological technologies can recover resources like metals, precious elements, fertilizers, radioactive, and emerging contaminants such as pharmaceuticals, hormones, enzymes, and (bio)plastics (Colica et al., 2012; Mannina et al., 2020; Puyol et al., 2017). Although there have been efforts to recover these important resources, there remains a necessity to improve and consolidate biological solutions for reclaiming and reusing these essential materials.

Table 16.1 Overview of valuable products recovered from waste.

Type of wastewater	Process	Microorganism involved	Product type	Applications	References
Lead–zinc mine tailings (China)	Bioaccumulation/ biomineralization	*Bacillus cereus* (340 mg-Pb/g)	$Ca_{2.5}Pb_{7.5}(OH)_2(PO_4)_6$	In situ restoration of Pb(II)-contaminated	Chen et al. (2016)
Industrial wastewater containing Au, Cu, Zn, and Ni (Italy)	Biosorption	*Cyanothece sp.* (164 mg-Au/g), *R. palustris* (201 mg-Cu/g)	Cyanobacterial biomass	Recovering the metal from the biomass	Colica et al. (2012)
Potato processing plant wastewater (United Kingdom)	Co-culture of green microalgae and methanotrophs	*C. sorokiniana, M. capsulatus*	Single-cell protein	Human food and animal feed	Rasouli et al. (2018)
Raw municipal wastewater (Japan)	Anaerobic membrane bioreactor	Anaerobic microorganisms	Biogases (N_2, CH_4 and CO_2)	Bioenergy electricity	Kong et al. (2021)
Swine wastewater (South Korea)	Anaerobic digester and microbial fuel cell	Anaerobic microorganisms	Biogases (CH_4, CO_2), K, and NH_4	Bioenergy, electricity, and fertilizer	Kim et al. (2015)
Poultry manure wastewater (Turkey)	Up-flow anaerobic sludge blanket (UASB)	Anaerobic microorganisms	$MgNH_4PO_4$	Slow-release fertilizer	Yetilmezsoy and Sapci-Zengin (2009)

(Continued)

Table 16.1 (Continued)

Type of wastewater	Process	Microorganism involved	Product type	Applications	References
Soybean processing plant wastewater (China)	Bioaccumulation	*Chlorella pyrenoidosa*	Cleaner water and biomass	Bioenergy, electricity, and reuse of water	Hongyang et al. (2011)
Copper-silver ores leaching (Germany)	Mesophilic bioleaching	*Lysinibacillus sphaericus, Bacillus sp.*	Copper	Recovery of Cu from ores	Kostudis et al. (2015)
Drilling fluids or additives wastewater (United States)	Microbial cell lysis	Microalgae	Microbial biomass	Lubricants	Go et al. (2019)
Wastewater treatment plants (Italy)	Bacterial fermentation	*E. coli* and *Azotobacter vinelandii*	Polyhydroxyalkanoates	Bioplastics	Mannina et al. (2020)

16.4 Contribution of Biological Technology in Circular Economy

The circular bioeconomy has gained significant global motivation in recent years. Central to its vision is the "3Rs" principle—Reduce, Reuse, and Recycle—which aims to minimize excessive and unnecessary waste within a circular economic framework. The circular bioeconomy focuses on maximizing the efficient use of all available bioresources by reducing waste during product development, recycling waste that is generated, and reusing valuable by-products and residues. Additionally, biological technology significantly contributes to the CE in various ways (Ilmasari et al., 2022; Kaszycki et al., 2021; Kusumo et al., 2022). For example, biotechnology plays a key role in optimizing resource use, making biological agents and bioprocesses critically important in this effort (Dahiya et al., 2022). The use of specific strains in modern agricultural practices and industries could enhance the productive yield of food and feed over time. The vision of biotechnology in the bioeconomy emphasizes the importance of biotechnology research, focusing on the efficient utilization or commercialization of products derived from its effective application across various sectors (Bugge et al., 2016). This includes fertilizing arid soils, converting industrial by-products, and utilizing agricultural waste. It facilitates waste valorization through processes like fermentation and bioconversion, transforming waste materials into valuable products and reducing landfill use while promoting resource recovery (Li and Yu, 2011; Lu et al., 2019; Molla et al., 2002).

Wastewater treatment is increasingly aligned with the circular sustainability campaign by incorporating resource recovery and energy production into the process of generating safe water (Nielsen, 2017). Microbial-aided biotechnology is crucial for advancing the CE in wastewater treatment by integrating resource recovery and various energy generation with the process of generating clean water (Nielsen, 2017). Microalgae cultivated on wastewater represent a promising circular pathway (Catone et al., 2021). The most suitable microalgae for cultivation in wastewater are primarily *Chlorella* and *Scenedesmus*, owing to their rapid growth rates and high-level resistance to contaminants (Catone et al., 2021). The use of microalgae for a variety of products has demonstrated its potential as an eco-friendly renewable bioresource that supports the circular biological economy (Moreira et al., 2023). This approach increases the utilization of algal biomass for biological remediation of organic waste while simultaneously reducing environmental pollution resulting from improper discharge of organic residues in wastewater streams (Bulgariu and Bulgariu, 2017; Le et al., 2024a). These approaches will contribute to reaching Sustainable Development Goal 6 "*SDG 6: Ensure availability and sustainable management of water and sanitation for all*" while also supporting Goal 7 "*SDG 7: affordable and clean energy*" (Nielsen, 2017; Obaideen et al., 2022). Therefore, the significance of this goal should be highlighted to various organizations and governments to enhance the role of WWTPs in the Sustainable Development Goals (SDGs).

The key principles of a CE include recycling by-products and end-products, promoting sensible resource use, and minimizing waste. This approach ultimately contributes positively to the global economy (Didenko et al., 2018). Further, bioprocesses that utilize microorganisms can facilitate the recycling of industrial waste and significantly contribute to the circular bioeconomy. Microbial-assisted remediation is an effective and cost-efficient strategy for the biotreatment of pollutants (Tiwari and Lata, 2018). This process involves various mechanisms, including the uptake, absorption, and accumulation of metals, as well as the extracellular precipitation of metals and the oxidation and reduction of micropollutants via enzymatic activities. Certain microbes could be employed in the "*in vitro*" enzyme synthesis, which is crucial for applications in various industries. For instance, it is possible to isolate a significant quantity of beneficial microbes from our planet (Dahiya et al., 2022).

Additionally, biological technologies aid in bioremediation, helping to clean up contaminated environments by breaking down pollutants and restoring ecosystems, thus minimizing environmental impact (Azubuike et al., 2020). They also enable sustainable production of biofuels, bioplastics, and other eco-friendly materials, providing alternatives to fossil fuels and traditional manufacturing processes. Furthermore, processes like composting and anaerobic digestion promote nutrient recycling from organic waste back into the soil, supporting sustainable agriculture (Cucina, 2023; Nguyen et al., 2022; Sharma et al., 2019). Utilizing microorganisms for wastewater treatment not only allows for energy recovery but also contributes to clean water production (Munoz-Cupa et al., 2021; Puyol et al., 2017). Overall, by integrating biological technology into various industrial processes, organizations can enhance resource efficiency and reduce waste, aligning with the principles of a CE. Certain biological technologies, such as algae cultivation, can capture carbon dioxide from the atmosphere, helping to mitigate climate change.

16.5 Summary

The importance of developing resource-efficient WWTPs cannot be ignored. Integrating these solutions is crucial to maximize energy recovery potential and to foster sustainable wastewater treatment practices. Reusing nutrients recovered from wastewater streams has a profoundly positive environmental effect, as it prevents the discharge of excess nutrients into water bodies, thereby mitigating eutrophication. Additionally, it alleviates pressure on conventional nitrogen and phosphorus sources.

Resources recovered from wastewater are progressively recognized as a strategy to address challenges such as regional and national resource performance and the financial sustainability of wastewater treatment. Implementing a CE model helps minimize resource extraction from nature-based systems while decreasing waste and environmental contamination. The function of the CE is particularly critical in the water sector, where water is vital for human beings and ecosystems, and crucial for numerous economic sectors.

Review Questions

1. Mention a comparative approach for linear and circular economy approaches for water and wastewater reuse.
2. Report and explain various biological techniques for resource recovery.
3. Make a comparative approach toward aerobic and anaerobic mechanisms for wastewater treatment toward resource recovery.
4. How bio-hydrogen is produced from wastewater?
5. What is the potential of biogas as an energy from wastewater? How can it be refined?
6. Mention a circular economy or closed-loop model for resource recovery from wastewater.
7. What are the various applications of micro-algae and their role toward resource generation?
8. Discuss the role of technological innovations toward transforming wastewater into resources.
9. What are the driving forces behind resource recovery from wastewater?
10. Compare biological systems like ASP and UASB in terms of energy requirements and energy generation during wastewater treatment.

References

Ali, S., Peter, A.P., Chew, K.W. et al. (2021). Resource recovery from industrial effluents through the cultivation of microalgae: a review. *Bioresource Technology*, **337**, 125461.

Azubuike, C.C., Chikere, C.B., and Okpokwasili, G.C. (2020). Bioremediation: an eco-friendly sustainable technology for environmental management. *Bioremediation of Industrial Waste for Environmental Safety: Volume I: Industrial Waste and Its Management*, 19–39. Springer.

Barsanti, L. and Gualtieri, P. (2022). *Algae: Anatomy, Biochemistry, and Biotechnology*. CRC press.

Batstone, D.J., Hülsen, T., Mehta, C.M. et al. (2015). Platforms for energy and nutrient recovery from domestic wastewater: a review. *Chemosphere*, **140**, 2–11.

Bugge, M.M., Hansen, T., and Klitkou, A. (2016). What is the bioeconomy? a review of the literature. *Sustainability*, **8**, 691.

Bulgariu, L. and Bulgariu, D. (2017). Sustainable utilization of marine algae biomass for environmental bioremediation. In *Prospects and Challenges in Algal Biotechnology*, 179–217. https://link.springer.com/chapter/10.1007/978-981-10-1950-0_6

Catone, C.M., Ripa, M., Geremia, E. et al. (2021). Bio-products from algae-based biorefinery on wastewater: a review. *Journal of Environmental Management*, **293**, 112792.

Cerruti, M., Guo, B., Delatolla, R. et al. (2021). Plant-wide systems microbiology for the wastewater industry. *Environmental Science: Water Research & Technology*, **7**(10), 1687–1706.

Chae, K.-J. and Kang, J. (2013). Estimating the energy independence of a municipal wastewater treatment plant incorporating green energy resources. *Energy Conversion and Management*, **75**, 664–672.

Chen, Z., Pan, X., Chen, H. et al. (2016). Biomineralization of Pb(II) into Pb-hydroxyapatite induced by Bacillus cereus 12-2 isolated from Lead–Zinc mine tailings. *Journal of Hazardous Materials*, **301**, 531–537.

Colica, G., Caparrotta, S., Bertini, G. et al. (2012). Gold biosorption by exopolysaccharide producing cyanobacteria and purple nonsulphur bacteria. *Journal of Applied Microbiology*, **113**(6), 1380–1388.

Craggs, R.J., McAuley, P.J., and Smith, V.J. (1997). Wastewater nutrient removal by marine microalgae grown on a corrugated raceway. *Water Research*, **31**(7), 1701–1707.

Cucina, M. (2023). Integrating anaerobic digestion and composting to boost energy and material recovery from organic wastes in the circular economy framework in Europe: a review. *Bioresource Technology Reports*, **24**, 101642.

Dahiya, D., Sharma, H., Rai, A.K. et al. (2022). Application of biological systems and processes employing microbes and algae to Reduce, Recycle, Reuse (3Rs) for the sustainability of circular bioeconomy. *AIMS microbiology*, **8**(1), 83.

Didenko, N.I., Klochkov, Y.S., and Skripnuk, D.F. (2018). Ecological criteria for comparing linear and circular economies. *Resources*, **7**(3), 48.

Du, X., Tao, Y., Li, H. et al. (2019). Synergistic methane production from the anaerobic co-digestion of Spirulina platensis with food waste and sewage sludge at high solid concentrations. *Renewable energy*, **142**, 55–61.

Ghimire, U., Sarpong, G., and Gude, V.G. (2021). Transitioning wastewater treatment plants toward circular economy and energy sustainability. *ACS Omega*, **6**(18), 11794–11803.

Go, L.C., Fortela, D.L.B., Revellame, E. et al. (2019). Biobased chemical and energy recovered from waste microbial matrices. *Current Opinion in Chemical Engineering*, **26**, 65–71.

Henriques, J. and Catarino, J. (2017). Sustainable value–an energy efficiency indicator in wastewater treatment plants. *Journal of Cleaner Production*, **142**, 323–330.

Hernández-Chover, V., Castellet-Viciano, L., Fuentes, R. et al. (2023). Circular economy and efficiency to ensure the sustainability in the wastewater treatment plants. *Journal of Cleaner Production*, **384**, 135563.

Hongyang, S., Yalei, Z., Chunmin, Z. et al. (2011). Cultivation of Chlorella pyrenoidosa in soybean processing wastewater. *Bioresource Technology*, **102**(21), 9884–9890.

Ilmasari, D., Kamyab, H., Yuzir, A. et al. (2022). A review of the biological treatment of leachate: available technologies and future requirements for the circular economy implementation. *Biochemical Engineering Journal*, **187**, 108605.

Johnson, D.B. (2018). The evolution, current status, and future prospects of using biotechnologies in the mineral extraction and metal recovery sectors. *Minerals*, **8**(8), 343.

Kaszycki, P., Głodniok, M., and Petryszak, P. (2021). Towards a bio-based circular economy in organic waste management and wastewater treatment–the polish perspective. *New Biotechnology*, **61**, 80–89.

Kim, T., An, J., Jang, J.K. et al. (2015). Coupling of anaerobic digester and microbial fuel cell for COD removal and ammonia recovery. *Bioresource Technology*, **195**, 217–222.

Kong, Z., Wu, J., Rong, C. et al. (2021). Large pilot-scale submerged anaerobic membrane bioreactor for the treatment of municipal wastewater and biogas production at 25 °C. *Bioresource Technology*, **319**, 124123.

Kostudis, S., Bachmann, K., Kutschke, S. et al. (2015). Leaching of copper from Kupferschiefer by glutamic acid and heterotrophic bacteria. *Minerals Engineering*, **75**, 38–44.

Kuhad, R.C. and Singh, A. (2013). *Biotechnology for Environmental Management and Resource Recovery*. Springer.

Kusumo, F., Mahlia, T.M.I., Pradhan, S. et al. (2022). A framework to assess indicators of the circular economy in biological systems. *Environmental Technology & Innovation*, **28**, 102945.

Ky, N.M., Nhut, H.T., Hiep, N.T. et al. (2022). Investigation of Nitrogen and Phosphorus recovery from Swine wastewater by struvite crystallization. *Journal of the Faculty of Agriculture Kyushu University*, **67** (1), 65–74. https://catalog.lib.kyushu-u.ac.jp/opac_download_md/4772412/6701_p065.pdf

Le, V.-G., Nguyen, M.-K., Nguyen, H.-L. et al. (2024a). Ecotoxicological response of algae to contaminants in aquatic environments: a review. *Environmental Chemistry Letters*, **22**(2), 919–939.

Le, V.G., Luu, T.A., Tran, H.T. et al. (2024b). Recovery of lithium from industrial Li-containing wastewater using fluidized-bed homogeneous granulation technology. *Minerals*, **14**(6), 603.

Li, W.-W. and Yu, H.-Q. (2011). From wastewater to bioenergy and biochemicals via two-stage bioconversion processes: a future paradigm. *Biotechnology Advances*, **29**(6), 972–982.

Lin, C.S.K., Pfaltzgraff, L.A., Herrero-Davila, L. et al. (2013). Food waste as a valuable resource for the production of chemicals, materials and fuels. Current situation and global perspective. *Energy & Environmental Science*, **6**(2), 426–464.

Lu, H., Zhang, G., Zheng, Z. et al. (2019). Bio-conversion of photosynthetic bacteria from non-toxic wastewater to realize wastewater treatment and bioresource recovery: a review. *Bioresource Technology*, **278**, 383–399.

Mannina, G., Presti, D., Montiel-Jarillo, G. et al. (2020). Recovery of polyhydroxyalkanoates (PHAs) from wastewater: a review. *Bioresource Technology*, **297**, 122478.

Mavhungu, A., Masindi, V., Foteinis, S. et al. (2020). Advocating circular economy in wastewater treatment: struvite formation and drinking water reclamation from real municipal effluents. *Journal of Environmental Chemical Engineering*, **8**(4), 103957.

Meena, R.A.A., Yukesh Kannah, R., Sindhu, J. et al. (2019). Trends and resource recovery in biological wastewater treatment system. *Bioresource Technology Reports*, **7**, 100235.

Minh Ky, N., Hadi, M., Lin, C. et al. (2022). Microplastics in sewage sludge: distribution, toxicity, identification methods, and engineered technologies. *Chemosphere*, **308**, 136455.

Molla, A.H., Fakhru'l-Razi, A., Abd-Aziz, S. et al. (2002). A potential resource for bioconversion of domestic wastewater sludge. *Bioresource Technology*, **85**(3), 263–272.

Moreira, J.B., Santos, T.D., Duarte, J.H. et al. (2023). Role of microalgae in circular bioeconomy: from waste treatment to biofuel production. *Clean Technologies and Environmental Policy*, **25**(2), 427–437.

Munoz-Cupa, C., Hu, Y., Xu, C. et al. (2021). An overview of microbial fuel cell usage in wastewater treatment, resource recovery and energy production. *Science of the Total Environment*, **754**, 142429.

Nguyen, M.-K., Lin, C., Bui, X.-T. et al. (2024). Occurrence and fate of pharmaceutical pollutants in wastewater: insights on ecotoxicity, health risk, and state–of–the-art removal. *Chemosphere*, **354**, 141678.

Nguyen, M.K., Lin, C., Hoang, H.G. et al. (2022). Evaluate the role of biochar during the organic waste composting process: a critical review. *Chemosphere*, **299**, 134488.

Nielsen, P.H. (2017). Microbial biotechnology and circular economy in wastewater treatment. *Microbial Biotechnology*, **10**(5), 1102–1105.

Nunes, O.C. (2021). The challenge of removing waste from wastewater: let technology use nature! *Microbial Biotechnology*, **14**(1), 63–67.

Obaideen, K., Shehata, N., Sayed, E.T. et al. (2022). The role of wastewater treatment in achieving sustainable development goals (SDGs) and sustainability guideline. *Energy Nexus*, **7**, 100112.

Otondo, A., Kokabian, B., Stuart-Dahl, S. et al. (2018). Energetic evaluation of wastewater treatment using microalgae, Chlorella vulgaris. *Journal of Environmental Chemical Engineering*, **6**(2), 3213–3222.

Pinnaka, A.K. and Tanuku, N.R.S. (2019). Marine microbial diversity for sustainable development. *Microbial Diversity in Ecosystem Sustainability and Biotechnological Applications*: Volume 1. *Microbial Diversity in Normal & Extreme Environments*, 117–158. Springer.

Puyol, D., Batstone, D.J., Hülsen, T. et al. (2017). Resource recovery from wastewater by biological technologies: opportunities, challenges, and prospects. *Frontiers in Microbiology*, **7**, 2106.

Rasouli, Z., Valverde-Pérez, B., D'Este, M. et al. (2018). Nutrient recovery from industrial wastewater as single cell protein by a co-culture of green microalgae and methanotrophs. *Biochemical Engineering Journal*, **134**, 129–135.

Sarpong, G., Gude, V.G., and Magbanua, B.S. (2019). Energy autarky of small scale wastewater treatment plants by enhanced carbon capture and codigestion—a quantitative analysis. *Energy Conversion and Management*, **199**, 111999.

Sfez, S., De Meester, S., Vlaeminck, S.E. et al. (2019). Improving the resource footprint evaluation of products recovered from wastewater: a discussion on appropriate allocation in the context of circular economy. *Resources, Conservation and Recycling*, **148**, 132–144.

Sharma, B., Vaish, B., Monika. et al. (2019). Recycling of organic wastes in agriculture: an environmental perspective. *International Journal of Environmental Research*, **13**, 409–429.

Smol, M. (2022). Chapter 1 – Circular economy approach in the water and wastewater sector. In: *Circular Economy and Sustainability* (eds. A. Stefanakis and I. Nikolaou), 1–19. Elsevier.

Tiwari, S. and Lata, C. (2018). Heavy metal stress, signaling, and tolerance due to plant-associated microbes: an overview. *Frontiers in Plant Science*, **9**, 452.

Tran, H.-T., Dang, B.-T., Thuy, L.T.T. et al. (2022a). Advanced treatment technologies for the removal of organic chemical sunscreens from wastewater: a review. *Current Pollution Reports*, **8**(3), 288–302.

Tran, H.T., Lesage, G., Lin, C. et al. (2022b). Chapter 3 – Activated sludge processes and recent advances. In: *Current Developments in Biotechnology and Bioengineering* (eds. X.-T. Bui, D.D. Nguyen, P.-D. Nguyen et al.), 49–79. Elsevier.

Tran, H.T., Lin, C., Bui, X.-T. et al. (2022c). Phthalates in the environment: characteristics, fate and transport, and advanced wastewater treatment technologies. *Bioresource Technology*, **344**, 126249.

Tran, T.-K., Huynh, L., Nguyen, H.-L. et al. (2024). Applications of engineered biochar in remediation of heavy metal(loid)s pollution from wastewater: current perspectives toward sustainable development goals. *Science of The Total Environment*, **926**, 171859.

van Loosdrecht, M.C.M. and Brdjanovic, D. (2014). Anticipating the next century of wastewater treatment. *Science*, **344**(6191), 1452–1453.

Wu, G. and Yin, Q. (2020). Microbial niche nexus sustaining biological wastewater treatment. *NPJ Clean Water*, **3**(1), 33.

Yadav, G., Mishra, A., Ghosh, P. et al. (2021). Technical, economic and environmental feasibility of resource recovery technologies from wastewater. *Science of The Total Environment*, **796**, 149022.

Yetilmezsoy, K. and Sapci-Zengin, Z. (2009). Recovery of ammonium nitrogen from the effluent of UASB treating poultry manure wastewater by MAP precipitation as a slow release fertilizer. *Journal of Hazardous Materials*, **166**(1), 260–269.

17

Adopting Conventional and Modern Water Treatment Technologies to Achieve Circular Economy by Resource Recovery

Hebatullah H. Farghal[1], Asmaa Ali[1,2], Nabila Shehata[2], and Mayyada El-Sayed[1]**

[1]*Department of Chemistry, School of Sciences and Engineering, The American University in Cairo, New Cairo, Egypt*
[2]*Environmental Science and Industrial Development Department, Faculty of Postgraduate Studies for Advanced Sciences, Beni-Suef University, Beni Suef, Egypt*

*Corresponding authors: nabila.shehata@psas.bsu.edu.eg; mayyada@aucegypt.edu

17.1 Introduction

Water is one of the most vital resources for human survival and production processes. Although rivers, lakes, and subterranean aquifers provide the majority of the world's fresh water supply today, many concerns are raised about the overexploitation of these resources due to urbanization, economic expansion, changing climate conditions, and increased land usage (Lin et al., 2022). Water is also essential for living organisms' survival and growth. On 28 July 2010, water and sanitation were recognized as human rights by the UN General Assembly. In addition, water is crucial for many facets of life and in various sectors of the economy (aprobada por la Asamblea, 2010).

The growing pressures of environmental sustainability and resource depletion have brought the circular economy (CE) concept to the forefront of current global efforts on waste management and responsible industrial processing. Recently, the CE concept has gained specific popularity promoting for the extension of the lifespan of products and materials, including raw materials, and the reduction

Wastewater to Resource Recovery: Applying the Circular Economy Toward Sustainable Development, First Edition. Edited by Arvind Kumar Mungray et al. Published 2026 by John Wiley & Sons, Inc.

of energy consumption by adopting recycling and reuse approaches while enhancing process efficiency (Rashidi et al., 2015). Waste, in this context, is regarded as a secondary raw material that can be processed, recycled, and reused. Thus, the CE approach contrasts with the linear economy model, which applies a "take-make-use-dispose" approach, where waste typically represents the final stage of a product's life cycle (Ghisellini et al., 2016). CE can be employed in wastewater treatment systems (Shehata et al., 2023) where, for instance, reclaiming and reusing wastewater presents a promising opportunity to enhance water resources while simultaneously minimizing their possible hazardous effects on the environment (Salgot and Folch, 2018) by utilizing this water for agricultural, industrial, and recreational purposes. However, myriad opportunities arise when considering substances contained within wastewater effluents. Beyond the well-known recovery of nutrients and energy from wastewater, other options include the extraction of biomolecules, metals, and various organic and inorganic compounds. Additionally, CE is applicable to other fields of water management, such as the valorization of sewage sludge, which is traditionally reused for agricultural purposes (Salgot and Folch, 2018; Maryam and Büyükgüngör, 2019).

This chapter delves into the practical applications of integrating CE in wastewater treatment systems, examining the current technological advancements and their economic implications. An overview of the conventional and modern wastewater treatment technologies that adopt CE is presented, illustrating how CE can transform wastewater treatment by optimizing resource use, mitigating environmental impact, and potentially reducing the incurred costs.

17.2 Conventional Technologies for Wastewater Treatment

Conventional treatment can be achieved chemically, physically, or biologically through various means that target the removal of solid particles, organics, and/or nutrients from raw water. Chemical treatment methods involve chemical precipitation such as coagulation and flocculation, ion exchange, neutralization, adsorption, and disinfection via chlorination, ozone treatment, or UV irradiation.

17.2.1 Coagulation

The coagulation process involves rapid dissolution and even distribution of a chemical, such as alum, in water. Successful coagulation is contingent upon many factors, such as the type and dosage of this chemical, the pH, the initial turbidity of the feed water, and the characteristics of the existing contaminants. Additionally, pre-treatments such as oxidation can affect the effectiveness of the coagulation

(Alexander et al., 2012). There are two classes of coagulants: primary coagulants and coagulant aids. The former neutralizes the electrical charges of suspended solids in water, whereas the latter works on forming denser and stronger flocs to prevent shearing during mixing and settling. Coagulant chemicals can be either inorganic metallic salts or polymers (El-taweel et al., 2023). Furthermore, advancements in coagulation technologies would potentially optimize the use of coagulants, and this would eventually reduce chemical consumption and associated environmental impacts. This optimization enhances the sustainability of the treatment process. This could be achieved by applying natural coagulants, which can be modified to enhance efficiency, or combining with other wastewater treatment techniques such as membrane filtration or biological treatment. Also, CE investigates the potential of reusing the applied coagulants (Ang and Mohammad, 2020). Therefore, this treatment method is thought of as a necessary step in wastewater treatment that aligns well with the principles of CE. By enabling water reuse, recovering valuable resources, and reducing waste, coagulation supports sustainable water management practices.

17.2.2 Flocculation

Another means of chemical precipitation is flocculation, whereby organics and other pollutants of lower density compared to water are removed by forming flocs. This treated water can be further treated and can be supplied to nearby industries for additional uses (Arias et al., 2021). Also, the metals and other compounds that are removed from wastewater during flocculation can be reused in manufacturing processes. Flocs are formed by the aggregation of particles such as polymeric or small organic materials with varying surface charges. The flocculation process depends on factors such as sediment volume, rate of flocs, turbidity, and the water recovery (Lee et al., 2014). Therefore, it can be said that flocculation remains an indispensable technology in both conventional and advanced water treatment systems, contributing to the provision of clean and safe water for many applications.

17.2.3 Filtration

Filtration is a process of utmost importance in wastewater treatment, which entails eliminating suspended solids, microorganisms, and other particulate contaminants from the water. This process enhances the purification and quality of the treated water. Filtration typically follows sedimentation and flocculation, providing an additional level of clarification. During the passage of water through a filter, the majority of fine suspended solids present in water are removed, where the chemical characteristics of water are changed for possible water reclamation (Hube et al., 2020). This filtered water can be recovered and reused for many purposes including agriculture, irrigation, industrial processes, and even as potable water for drinking after further treatment, thus decreasing the demand for fresh water and conserving water resources. Moreover, the recovered water can be reused in cooling towers and water boilers, hence promoting the circular approach to water use (Czuba et al., 2021).

Several filter types could be applied in water treatment: slow sand filters (SSF), rapid gravity filters, and pressure filters. SSF is a crucial method in water treatment, known for its simplicity and effectiveness in mechanical filtration. It is regarded as one of the most facile techniques for water purification, effectively removing turbidity and various insoluble substances. Because of its efficiency and cost-effectiveness, SSF remains an excellent method for providing potable water for poor countries (Abdiyev et al., 2023). For rapid gravity filters, they are used to capture suspended solids and other impurities from wastewater. Hence, after filtration, backwashing needs to be done to maintain filter efficiency (Brandt et al., 2017). In pressure filters, the water is forced through the filter media at high flow rates by applying pressure. Both could be partially applied in small-scale plants. All these filtration methods are driven by physical forces; however, they differ in their pressure requirement, flow rate, and technical specifications of the operating vessels (Cescon and Jiang, 2020).

17.2.4 Adsorption

Adsorption is regarded as a simple and cost-effective method for the removal of various types of contaminants from water and wastewater via physical or chemical interactions. Efficiency of the adsorbent to uptake contaminants is assessed by calculating the maximum adsorption capacity obtained from applying adsorption isotherms, such as Langmuir and Freundlich, among others. These isotherms are plotted as the milligrams of adsorbed molecules per gram of the adsorbent against the equilibrium contaminant concentration.

Ideally, an adsorbent should possess sufficient binding sites to effectively adsorb pollutants. Some of the popular conventional adsorbents for removing pollutants are biosorbents, silica, alumina, activated carbon, clay, and metal oxides. Biosorbents include wastes from agriculture, wood, and fisheries. They are an excellent source for adsorbents due to their capabilities to uptake contaminants, such as heavy metals (Madeła and Skuza, 2021). One of the adsorbents that showed excellent adsorption capacities is activated carbon. Recently, research efforts have been directed toward preparing activated carbon from agricultural wastes, aiming to implement the concepts of CE. Thus, biowastes could be used as adsorbents or converted to activated carbon, which would decrease the harmful influence of these wastes and potentially apply it in wastewater treatment (Tibor and Grande, 2022). Chemical, physical, and biological techniques can regenerate the adsorbent to recover its adsorption capacity and reuse it in the treatment process. For instance, activated carbon can be regenerated and reused many times to reduce raw material consumption and minimize operational costs. Adsorption can also be deployed for capturing nutrients like phosphorus and nitrogen from wastewater. These nutrients can be recovered and used as fertilizers in agriculture, incorporating the principles of CE in nutrient treatment (Wong et al., 2020; Wadhawan et al., 2020). The US Environmental Protection Agency (USEPA) has recognized adsorption as one of the most effective and superior techniques for wastewater treatment, as the adsorption efficiency can approach 100% (Anil et al., 2020). To date, the development of novel adsorbents and advancements in material science continue to enhance the performance and sustainability of the adsorption technology. This could be achieved by using green adsorbent materials, regeneration of the exhausted adsorbents for additional water treatment, and using the exhausted adsorbents in other applications like green construction materials.

17.2.5 Aerobic Digestion

Aerobic digestion is a biological process used to decompose organic matter that exists in wastewater in the presence of oxygen. The process is necessary for reducing the amount of organic contaminants, stabilizing sludge, and reducing odors, resulting in cleaner effluent suitable for discharge or further treatment. Microorganisms such as bacteria feed on organic matter to decompose it into carbon dioxide, water, and biomass. Aerobic digestion is best applied for the stabilization of the thickened activated sludge rich in nitrogen and phosphorus. This biosolid could be deployed as a fertilizer in agriculture, hence returning essential nutrients to the soil and minimizing the application of synthetic fertilizers. During aerobic digestion, pH levels should be monitored as they could decline by nitrification, leading to the inhibition of microbial metabolism and the deterioration of the effluent quality, noting that a pH of 6.5 yields the highest decay coefficient value. The effluent from aerobic digestion can undergo additional treatment and be reused for various applications, including irrigation, industrial processes, or even as potable water after advanced pre-treatment. This approach decreases the demand for fresh water and helps conserve water resources (D'Antonio, 1983; Bernard and Gray, 2000).

17.2.6 Anaerobic Digestion

Anaerobic digestion is a treatment method that entails the degradation and stabilization of organic materials in the absence of air using microbial organisms to generate biogas (carbon dioxide and methane mixture) along with microbial biomass, and thus it reduces operational costs and environmental impacts. The obtained biogas could be applied for direct heating or electricity generation, or alternatively upgraded to biomethane to be used as a vehicle fuel. This treatment technique is one of the most promising methods in the management and treatment of waste and wastewater. Encompassing the four steps of hydrolysis, acidogenesis, acetogenesis, and methanogenesis, it emerges as among the most prevalent and widely utilized methods for treating and stabilizing waste-activated sludge (WAS). This process not only stabilizes organic matter but also reduces the total volume of biosolids requiring disposal by approximately 40% (Qadri et al., 2022). Therefore, anaerobic digestion can promote sustainable urban development by implementing integrated waste management in the energy and agricultural sectors.

17.3 Case Studies on Conventional Technologies Adopting CE

Conventional technologies for wastewater treatment are still applied for the possibility of producing water suitable for other applications. The approaches for incorporating CE in water treatment follow one of two pathways. One pathway is for reusing the material applied in water treatment and/or recovering the contaminants from this material, and another pathway is for reusing the treated water as such or after further treatment.

Several recent studies were concerned with the first pathway of reusing the adsorbent or membrane materials. For adsorbent and catalytic materials, metal organic framework (MOF) adsorbent was applied in the removal of Ag(I) from water, then completely regenerated by 1% HNO_3, or thermally treated at 800 °C to form Ag/C/TiO_2, which was successfully used as a catalyst for methylene blue degradation in the presence of H_2O_2, achieving a degradation efficiency of >90% in 80 minutes with 0.2 g/L catalyst and at pH 5.9, initial concentration of 10 mg/L, and a hydrogen peroxide dose of 10 μL. This preparation method preserves the catalytic activity of titania compared to the regular method of preparation of Ag-decorated TiO_2, which suffers aggregation and decline in the catalytic performance. Herein, silver has a role in inhibiting the recombination of electrons and holes. The Ag/C/TiO_2 photocatalyst exhibited improved catalytic performance in visible light than in UV light and showed better performance than C/TiO_2. The photocatalyst was reused for three cycles while almost maintaining its stability and its catalytic efficiency at 90% after 90 minutes (Ren et al., 2022). Moreover, iron and copper-modified clays were regenerated after applying them in a Photo–Fenton process for removing indigo blue dye, where the removal percentage of the dye remained 90% after four cycles. However, the continuous decomposition of H_2O_2 and loss of iron ions, as well as sludge formation, were reported as the major drawbacks of this method (Shahadat and Isamil, 2018). Biochar—obtained from lime waste—was almost completely regenerated by acetonitrile after adsorbing sulfamethoxazole and methyl paraben. However, this method leads to the formation of another source of contamination, which is the contaminant recovery solution. Therefore, the Fenton reaction was applied to degrade the contaminants and potentially offer a greener solution. However, this reaction affected the removal efficiency of the adsorbent, which decreased to 60% after the fifth cycle. Applying the electro-advanced process enhanced the regeneration. Electro-Fenton proved successful in the complete degradation of the contaminants. After adsorption of 50 ppm of the contaminants using 0.1 g/L adsorbent, electro-Fenton was applied at 300 mA using CBH (carbon black modified carbon felt electrode) as a cathode at Fe^{2+} concentration of 0.29 mM. Almost 100% removal efficiency was obtained after six cycles, but slightly decreased to about 80% in the seventh cycle (Acevedo-García et al., 2020).

For reusing filtration membranes, a carbonaceous nanofibrous membrane was successful in removing methylene blue from water and was regenerable after six cycles without any deterioration in its adsorption/desorption capabilities. Adsorption was conducted at a feed concentration of 20 mg/L, flow rate of 0.25 mL/min at pH 7, and desorption was carried out using 1 M HCl in 80% methanol (Liang et al., 2011). Membranes used for oil separation were also efficiently regenerated. KOCH carbon MF and KOCH XM50 polymeric UF membranes applied for the removal of oils in a circular loop were regenerated with hot water at different temperatures. Modeling predictions concluded that by increasing the temperature from 25 to 126 °C, the optimal cleaning time decreased, while the effective flux increased from 173.8 to 296.4 L/(m^2 h). The chemical treatment at 25 °C, on the other hand, required an optimum cleaning time of 7.4 hours, yielding an effective flux of 143.3 L/m^2 h^{-1}. It was concluded that regeneration with hot water is better than chemical treatment as it could be performed in-line while chemical treatment requires a separate loop (Peng and Tremblay, 2008). Polyvinylidene fluoride (PVDF)-Proanthocyanidins-FeOOH membrane was also successful in treating water from oil and methylene blue simultaneously. The membrane was subjected to washing by deionized water and H_2O_2 under xenon lamp irradiation to remove any methylene blue retained on the membrane by Photo–Fenton catalysis. The membrane was reusable for five cycles with a methylene blue removal efficiency of 99.9% (Yang et al., 2024).

In addition to its importance in reusing the adsorbent or membrane material, the regeneration process was also applied for the purpose of recovering the contaminants in water. Several reports studied the extraction of heavy metals from the adsorbent materials. Bio-based adsorbents were regenerated to recover the adsorbed heavy metals. For example, the biomass-based baker's yeast and *Phragmites australis* shoot-based biosorbents were regenerated using 0.1 M HCl for sixth and third cycles to recover 95% and 98.5% Cd (II), and 95% and 90.5% Pb (II), respectively. Also, 75% Cr(VI) was recovered from *Oedegonium hatei*-based biosorbent using 0.1 M NaOH for five cycles, whereas 99% of the same heavy metal was recovered by *Mucor hiemalis*-based biosorbent using the same desorbing agent for the same number of cycles (Gautam et al., 2014). Heavy metals were also recovered from chitosan-based biopolymer adsorbents. Glutaraldehyde cross-linked carboxymethyl chitosan was regenerated after Cu(II) adsorption to recover 99.93% of the heavy metal after the sixth cycle using 0.1 M HCl, while chitosan-*Laminaria japonicum* was regenerated to recover 36.3% and 31.53% of adsorbed Ni (II) after the fifth cycle, using HCl or HNO_3, respectively. HCl was also used to recover 89.9% As(III) from chitosan/iron oxide nanocomposites after the fifth cycle. Additionally, Pb(II) was recovered from epichlorohydrin-cross-linked chitosan fibers using 0.5 M thiourea and 1 M HCl, with a desorption efficiency of 95.6% after the third cycle. The same metal along with Cd(II) was recovered from ethylene glycol-bis(2-aminoethylether)-*N*,*N*,*N*,*N*-tetraacetic acid-chitosan using 2 M HNO_3 with desorption efficiencies of 99.91% and 99.18%, respectively, after the tenth cycle (Vakili et al., 2019).

The second pathway of incorporating CE in water treatment entails using the treated water for industrial or agricultural purposes. Treated water containing RR141 dye was successfully applied in the dyeing process of a fabric cloth. The treated water resulted from the adsorption of 28.8 mg/g of the dye onto pyrolyzed textile sludge at 25 °C. The benefit of using the treated water was to conduct the dyeing process at lower energy than that normally required if recycled untreated dye water was used (Leal et al., 2018). In a different study, textile wastewater was obtained from Sun Tekstil Inc., Izmir, Turkey, after coagulation and biological treatment and was subjected to an advanced oxidation process followed by an adsorption process to produce treated water suitable for irrigation. The advanced oxidation process took place via Photo–Fenton-like experiments using a perovskite-type $BiFeO_3$ on activated carbon support, while adsorption was performed using the same activated carbon support only, which was prepared from corn cobs after being activated by $ZnCl_2$. The levels of total organic carbon (TOC), total suspended solids (TSS), turbidity, dissolved oxygen, and color in the resulting treated water were within the range of irrigation water standards; however, the conductivity and salinity were out of range. The wastewater source had high concentrations of these, due to their application heavily in the dyeing process, and their removal was not successful (Yılmaz et al., 2023). Similarly, walnut or rice husks were converted to activated carbon after activation using $ZnCl_2$. The activated carbon was used as a support for $BiBO_3$ catalysts in the Photo–Fenton treatment step. This step was followed by an adsorption step using the activated carbon to remove the products of catalysis.

After this two-step treatment, the quality of the produced treated water was suitable for irrigation despite its high salt concentration (sodium adsorption ratio, SAR = 23.37). The treated water was applied in the irrigation of barley, cotton, and sugar beet (Bayrakdar et al., 2021). A two-step water treatment process was applied to real wastewater systems taken from three different points of the Zamora River, Ecuador. First, the iron form of a clay adsorbent was used for the removal of phosphates and ammonium salts, and then citrus lemon or *eucalyptus* essential oil was applied to remove bacteria and total coliforms, to produce water suitable for agricultural purposes. The adsorbent removed phosphates by 65.8% and ammonium salts by 72.3%, along with sulfates, nitrates, bicarbonates, and chlorides. Since the desorption of phosphates from the adsorbent was insufficient, while the composition of the adsorbent was close to that of soil, it was recommended to use the nutrient-enriched adsorbent for soil amendment. In the second step, total coliforms were removed at high percentages (approx. 100%) by both essential oils (Guaya et al., 2023). In another study, hemodialysis wastewater was treated by electrocoagulation/electrooxidation and subsequently by adsorption on activated carbon, resulting in water with low salinity (SAR 1.78 mEq/L). The treated water was tested for the germination of maize and alfalfa plants and showed a much greater germination percentage after seven days, compared to salty water or the original hemodialysis wastewater (Jallouli et al., 2022). Using treated wastewater in agriculture is a double-blade knife in terms of both the soil and water quality. Applying treated wastewater in irrigation could supply the soil with the necessary macro- and micronutrients for cultivation. For example, nitrogen is important for the growth of stems and leaves, and in the uptake of phosphorus by plants. Also, treated wastewater is an important source of organic carbon that is necessary for soil stability and nutrient holding. However, wastewater increases the salinity of the soil, which in turn raises the osmotic stress affecting the growth of plants and causes plant toxicity. This salinity also causes a reduction in the soil structural porosity and leads to the destruction of soil structure through the swelling of clay minerals. From the water quality viewpoint, the use of treated wastewater will prevent discharging wastewater into surface water streams and reduce the required amounts of fertilizers that might leach into natural water resources. On the other hand, treated wastewater still contains traces of xenobiotics and harmful chemical and biological pollutants that might contaminate natural waters, leading to endocrine disruption in aquatic organisms, in addition to eutrophication, which causes an increase in algal growth and subsequent oxygen depletion, thus affecting fish survival (Ofori et al., 2021). Figure 17.1 shows the main water treatment processes used to recover different metals, dyes, essential nutrients, and pharmaceutical products (Singh et al., 2021; Maged et al., 2023).

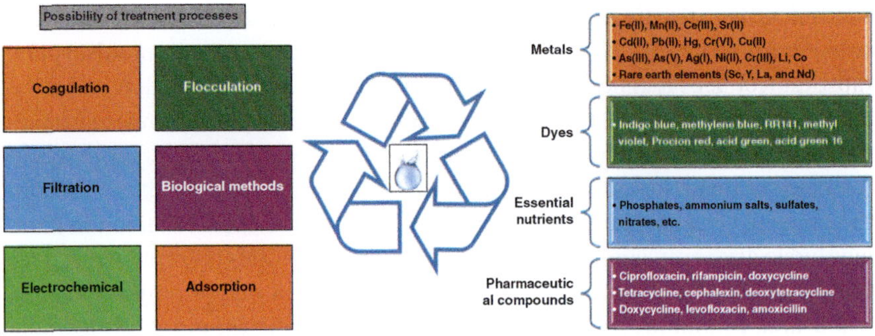

Figure 17.1 Schematic representation of water treatment processes used to recover numerous materials from wastewater. *Source:* Krafla/Adobe Stock Photos.

17.4 Modern Technologies Incorporating CE for Wastewater Treatment

The world today is facing major issues of food insecurity and environmental pollution. One of the recent technologies that arose to overcome these issues is the aquaculture process. Aquaculture was designed to provide fish as food for human consumption and to compensate for the depletion of wild fish. Aquaculture of shellfish produce large amounts of waste, which could be transformed into char by means of slow or fast pyrolysis or hydrothermal carbonization. These chars were proven successful in adsorbing several heavy metals such as Pb(II), Cd(II), Cr(VI); antibiotics such as tetracycline, sulfamethazine, and chloramphenicol; and dyes such as malachite green (Mahari et al., 2022). The disadvantage of aquaculture is the generation of wastewater containing dissolved and particulate organic matter and nutrients (nitrogen and phosphorus), which, when disposed of in rivers, would enhance the metabolic activity of aerobic bacteria, leading, in turn, to O_2 depletion and thus affecting the aquatic life in the receiving waters. To avoid such issues, recirculating aquaculture systems (RAS) were designed to treat wastewater generated by these aquaculture farms for a more sustainable process. In RAS, wastewater is treated and recirculated again into the fish tanks to mechanically remove the solids and convert ammonia and nitrites into nitrates. This method is not very efficient, since it still needs a backup of fresh water to be mixed with the treated water to minimize the effect of the generated harmful nitrates that accumulate in the treated water. Further modification to the RAS was implemented by incorporating wetlands in the recirculation system. This is efficient in minimizing the concentration of ammonia and nitrites in water, and it consumes less energy; however, it is less effective in removing nitrates and phosphates, and requires large areas of wetlands.

Petroleum-based polyolefins and biodegradable polymers known as polyhydroxyalkanoates (PHAs) are comparable in properties, making PHAs a viable substitute for petroleum-based plastics (Guerra-Rodríguez et al., 2020). PHAs production may be combined with sludge management and urban water treatment in WWTPs (Morgan-Sagastume et al., 2014). This suggests a four-stage process: (i) taking out easily biodegradable carbon from wastewater in conjunction with the growth of selective biomass that has the ability to store PHAs; (ii) fermenting sludge acidically to develop a liquid stream that is high in volatile fatty acids (VFAs); (iii) utilizing the grown enriched biomass to accumulate PHAs from the liquid stream of large VFAs concentration; and (iv) recovering and characterizing PHAs (Figure 17.2) (Guerra-Rodríguez et al., 2020; Tyagi et al., 2009).

In search of zero waste management approaches, aquaponics systems recently emerged. They depend on growing plants in the fish tanks, so that a symbiotic interaction takes place. The plants use the fish excreta as manure for their growth, meanwhile purifying the water in these tanks. However, this method consumes high energy to ensure temperature stability all year round and hence avoids fluctuations that may harm the components of the aquaponics system (Goddek et al., 2015). Microalgae-assisted aquaculture can also be applied to convert the waste into biomass necessary for the immunity of fish, and to provide oxygen to avoid its depletion. However, consequences such as algal blooms could be mitigated by harvesting them using coagulation/flocculation, centrifugation, or filtration (Tom et al., 2021). Algal biomass is also beneficial as a biostimulant for the growth of plants, such as *Triticum aestivum* (wheat) and *Nasturtium officinale* (watercress). *Chlorella vulgaris* at 0.5 g/L stimulated the growth of watercress at a germination index of 275.2%, while *Scenedesmus obliquus* at 0.2 g/L enhanced the growth of wheat at a 197% germination index, compared with the control (100%).

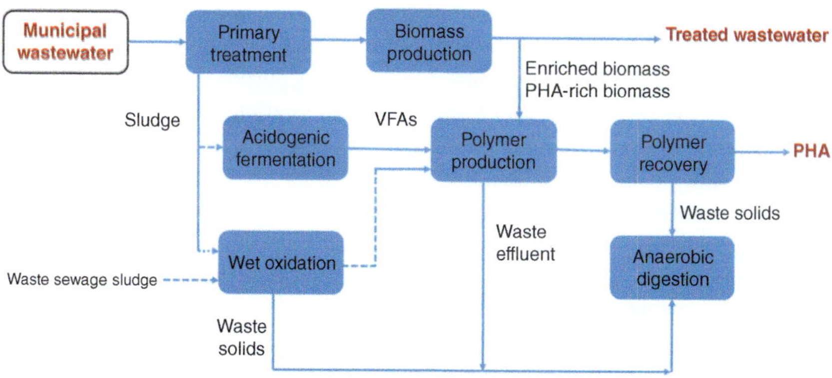

Figure 17.2 A schematic flow diagram for the production of polyhydroxyalkanoates (PHAs) combined with wastewater and sludge treatment.

Torrefaction of biomass was also performed, and the yielded biochar was applied successfully as a fertilizer for wheat and watercress. It was also utilized efficiently to adsorb methylene blue; particularly, the *Scenedesmus* biochar manifested an adsorption efficiency of 62.9% after 48 h, using 100 mg/L methylene blue concentration and 5 g/L adsorbent dose (Viegas et al., 2021). A modification to the microalgae-assisted aquaculture system was implemented by applying a mussels/microalgae/bacteria ecosystem for the treatment of aquaculture wastewater. Mussel breeding density was first optimized to be 4 mussels/m^3 to avoid high production of mussel waste. Then, *Bacillus subtilis, Bacillus licheniformis*, and *Chlorella vulgaris* were added, and the system proved to be very efficient in treating the aquaculture wastewater. The microalgae helped in reducing nitrogen, phosphorus, and chemical oxygen demand (COD); the mussels aided in removing the organic debris and making the water transparent, while the bacteria enhanced the pectin and amylase activity of the mussels. The process was scaled up to be applied on the Kulv River in China, maintaining concentrations of 0.3, 0.8, 0.3, and 30 mg/L for ammonia nitrogen, total nitrogen, total phosphorus, and COD, respectively, which hence promotes the application of this water for industrial engineering and entertainment purposes (Geng et al., 2022).

17.5 Summary

This review highlights the distinct current developments in industrial and municipal water resource recovery routes, as well as their significant contribution to CE. Different water treatment processes, such as coagulation, flocculation, filtration, biological, electrochemical, and adsorption, have been implemented in the water treatment sector to recover different resources from various wastewaters, polluted with contaminants such as heavy metals, dyes, nutrients, and pharmaceutical compounds. The challenges toward the application of such technologies are the lack of economic and feasibility studies required for upscaling purposes. Moreover, the characteristics of the wastewater vary from location to location in terms of volumetric flow rate, composition, pH, temperature, and the presence of other constituents (pollutants). Lastly, it should not be overlooked that the commitment of governments and the awareness of the public are important for the CE to become a reality.

Review Questions

1. What are the benefits of applying CE in wastewater treatment?
2. What are the criteria for selecting the appropriate wastewater treatment method that incorporates CE principles?
3. Suggest methods by which exhausted water treatment materials can be reused?

4 Discuss methods by which the quality of the treated water can be enhanced to become suitable for agriculture?
5 What are the main elements of the aquaponics system? Discuss the advantages of this system over the RAS.

References

Abdiyev, K., Azat, S., Kuldeyev, E. et al. (2023). Review of slow sand filtration for raw water treatment with potential application in less-developed countries. *Water*, **15**, 2007.

Acevedo-García, V., Rosales, E., Puga, A. et al. (2020). Synthesis and use of efficient adsorbents under the principles of circular economy: waste valorisation and electroadvanced oxidation process regeneration. *Separation and Purification Technology*, **242**, 116796.

Alexander, J.T., Hai, F.I., and Al-Aboud, T.M. (2012). Chemical coagulation-based processes for trace organic contaminant removal: current state and future potential. *Journal of Environmental Management*, **111**, 195–207.

Ang, W.L. and Mohammad, A.W. (2020). State of the art and sustainability of natural coagulants in water and wastewater treatment. *Journal of Cleaner Production*, **262**, 121267.

Anil, I., Gunday, S.T., Bozkurt, A. et al. (2020). Design of crosslinked hydrogels comprising poly (vinylphosphonic acid) and bis [2-(methacryloyloxy) ethyl] phosphate as an efficient adsorbent for wastewater dye removal. *Nanomaterials*, **10**, 131.

Aprobada por la Asamblea, O.R. (2010). General el 28 de julio de 2010. 64/292. *El derecho humano al agua y el saneamiento*, **660**, 9–11.

Arias, B.G., Merayo, N., Millán, A. et al (2021). Sustainable recovery of wastewater to be reused in cooling towers: towards circular economy approach. *Journal of Water Process Engineering*, **41**, 102064.

Bayrakdar, M., Atalay, S., and Ersöz, G. (2021). Efficient treatment for textile wastewater through sequential photo Fenton-like oxidation and adsorption processes for reuse in irrigation. *Ceramics International*, **47**, 9679–9690.

Bernard, S. and Gray, N. (2000). Aerobic digestion of pharmaceutical and domestic wastewater sludges at ambient temperature. *Water Research*, **34**, 725–734.

Brandt, M.J., Johnson, K.M., Elphinston, A.J. et al. (2017). Specialized and advanced water treatment processes. In: *Twort's Water Supply*, 407–473. Butterworth-Heinemann.

Cescon, A. and Jiang, J.-Q. (2020). Filtration process and alternative filter media material in water treatment. *Water*, **12**, 3377.

Czuba, K., Bastrzyk, A., Rogowska, A. et al. (2021). Towards the circular economy—A pilot-scale membrane technology for the recovery of water and nutrients from secondary effluent. *Science of The Total Environment*, **791**, 148266.

D'Antonio, G. (1983). Aerobic digestion of thickened activated sludge: reaction rate constant determination and process performance. *Water Research*, **17**, 1525–1531.

El-Taweel, R.M., Mohamed, N., Alrefaey, K.A. et al. (2023). A review of coagulation explaining its definition, mechanism, coagulant types, and optimization models; RSM, and ANN. *Current Research in Green and Sustainable Chemistry*, **6**, 100358.

Gautam, R.K., Mudhoo, A., Lofrano, G. et al. (2014). Biomass-derived biosorbents for metal ions sequestration: adsorbent modification and activation methods and adsorbent regeneration. *Journal of Environmental Chemical Engineering*, **2**, 239–259.

Geng, B., Li, Y., Liu, X. et al. (2022). Effective treatment of aquaculture wastewater with mussel/microalgae/bacteria complex ecosystem: a pilot study. *Scientific Reports*, **12**, 2263.

Ghisellini, P., Cialani, C., and Ulgiati, S. (2016). A review on circular economy: the expected transition to a balanced interplay of environmental and economic systems. *Journal of Cleaner Production*, **114**, 11–32.

Goddek, S., Delaide, B., Mankasingh, U. et al. (2015). Challenges of sustainable and commercial aquaponics. *Sustainability*, **7**, 4199–4224.

Guaya, D., Cuenca, G., Mendoza, E. et al. (2023). Wastewater treatment focused on the reuse and provision of agricultural supplies: combination of phosphate adsorption and microbial reduction. *Industrial Crops and Products*, **197**, 116600.

Guerra-Rodríguez, S., Oulego, P., Rodríguez, E. et al. (2020). Towards the implementation of circular economy in the wastewater sector: challenges and opportunities. *Water*, **12**, 1431.

Hube, S., Eskafi, M., Hrafnkelsdóttir, K.F. et al. (2020). Direct membrane filtration for wastewater treatment and resource recovery: a review. *Science of the Total Environment*, **710**, 136375.

Jallouli, S., Chouchene, K., Ben Hmida, M. et al. (2022). Application of sequential combination of electro-coagulation/electro-oxidation and adsorption for the treatment of hemodialysis wastewater for possible reuse. *Sustainability*, **14**, 9597.

Leal, T.W., Lourenco, L.A., Scheibe, A.S. et al. (2018). Textile wastewater treatment using low-cost adsorbent aiming the water reuse in dyeing process. *Journal of Environmental Chemical Engineering*, **6**, 2705–2712.

Lee, C.S., Robinson, J., and Chong, M.F. (2014). A review on application of flocculants in wastewater treatment. *Process Safety and Environmental Protection*, **92**, 489–508.

Liang, H.W., Cao, X., Zhang, W.J. et al. (2011). Robust and highly efficient free-standing carbonaceous nanofiber membranes for water purification. *Advanced Functional Materials*, **21**, 3851–3858.

Lin, L., Yang, H., and Xu, X. (2022). Effects of water pollution on human health and disease heterogeneity: a review. *Frontiers in Environmental Science*, **10**, 880246.

Madeła, M. and Skuza, M. (2021). Towards a circular economy: analysis of the use of biowaste as biosorbent for the removal of heavy metals. *Energies*, **14**, 5427.

Maged, A., El-Fattah, H.A., Kamel, R.M. et al. (2023). A comprehensive review on sustainable clay-based geopolymers for wastewater treatment: circular economy and future outlook. *Environmental Monitoring and Assessment*, **195**, 693.

Mahari, W.A.W., Waiho, K., Azwar, E. et al. (2022). A state-of-the-art review on producing engineered biochar from shellfish waste and its application in aquaculture wastewater treatment. *Chemosphere*, **288**, 132559.

Maryam, B. and Büyükgüngör, H. (2019). Wastewater reclamation and reuse trends in Turkey: opportunities and challenges. *Journal of Water Process Engineering*, **30**, 100501.

Morgan-Sagastume, F., Valentino, F., Hjort, M. et al. (2014). Polyhydroxyalkanoate (PHA) production from sludge and municipal wastewater treatment. *Water Science and Technology*, **69**, 177–184.

Ofori, S., Puškáčová, A., Růžičková, I. et al. (2021). Treated wastewater reuse for irrigation: pros and cons. *Science of The Total Environment*, **760**, 144026.

Peng, H. and Tremblay, A. (2008). Membrane regeneration and filtration modeling in treating oily wastewaters. *Journal of Membrane Science*, **324**, 59–66.

Qadri, R.Z., Ali, M., Rajpal, A. et al. (2022). Anaerobic wastewater treatment for energy recovery and water reclamation. In: *Clean Energy and Resource Recovery*. Elsevier.

Rashidi, H., Ghaffarianhoseini, A., Ghaffarianhoseini, A. et al. (2015). Application of wastewater treatment in sustainable design of green built environments: a review. *Renewable and Sustainable Energy Reviews*, 49, 845–856.

Ren, X., Wang, C.-C., Li, Y. et al. (2022). Ag (I) removal and recovery from wastewater adopting NH2-MIL-125 as efficient adsorbent: a 3Rs (reduce, recycle and reuse) approach and practice. *Chemical Engineering Journal*, **442**, 136306.

Salgot, M. and Folch, M. (2018). Wastewater treatment and water reuse. *Current Opinion in Environmental Science & Health*, **2**, 64–74.

Shahadat, M. and Isamil, S. (2018). Regeneration performance of clay-based adsorbents for the removal of industrial dyes: a review. *RSC Advances*, **8**, 24571–24587.

Shehata, N., Egirani, D., Olabi, A. et al. (2023). Membrane-based water and wastewater treatment technologies: issues, current trends, challenges, and role in achieving sustainable development goals, and circular economy. *Chemosphere*, **320**, 137993.

Singh, S., Kumar, V., Anil, A.G. et al. (2021). Adsorption and detoxification of pharmaceutical compounds from wastewater using nanomaterials: a review on mechanism, kinetics, valorization and circular economy. *Journal of Environmental Management*, **300**, 113569.

Tibor, S.T. and Grande, C.A. (2022). Industrial production of activated carbon using circular bioeconomy principles: case study from a Romanian company. *Cleaner Engineering and Technology*, **7**, 100443.

Tom, A.P., Jayakumar, J.S., Biju, M. et al. (2021). Aquaculture wastewater treatment technologies and their sustainability: a review. *Energy Nexus*, **4**, 100022.

Tyagi, R., Surampalli, R.Y., Yan, S., et al. (2009). *Sustainable Sludge Management: Production of Value Added Products*. Reston, Va: American Society of Civil Engineers.

Vakili, M., Deng, S., Cagnetta, G. et al. (2019). Regeneration of chitosan-based adsorbents used in heavy metal adsorption: a review. *Separation and Purification Technology*, **224**, 373–387.

Viegas, C., Gouveia, L., and Gonçalves, M. (2021). Aquaculture wastewater treatment through microalgal. Biomass potential applications on animal feed, agriculture, and energy. *Journal of Environmental Management*, **286**, 112187.

Wadhawan, S., Jain, A., Nayyar, J. et al. (2020). Role of nanomaterials as adsorbents in heavy metal ion removal from waste water: a review. *Journal of Water Process Engineering*, **33**, 101038.

Wong, S., Ghafar, N.A., Ngadi, N. et al. (2020). Effective removal of anionic textile dyes using adsorbent synthesized from coffee waste. *Scientific Reports*, **10**, 2928.

Yang, H., Zhang, Y., Huang, C. et al. (2024). Antifouling multi-functional membrane for oil/dye wastewater treatment: regeneration while treating mixed pollutants. *Separation and Purification Technology*, **332**, 125804.

Yılmaz, F.G.T., Tekin, G., Ersöz, G. et al. (2023). Reclamation of real textile wastewater by sequential advanced oxidation and adsorption processes using corn-cob based materials. *Environmental Pollution*, **335**, 122196.

18

Resource Recovery from Urine and Fecal Material at On-site Treatment and Its Economic Evaluation: A Case Study

Zavin Gajera[1], *Monali Chhatbar*[1], *Ambika Arkatkar*[2], *Alka A. Mungray*[1], *and Arvind Kumar Mungray*[1*]

[1]*Department of Chemical Engineering, Sardar Vallabhbhai National Institute of Technology Surat, Gujarat, India*
[2]*Department of Biotechnology, Veer Narmad South Gujarat University, Surat, Gujarat, India*

*Corresponding author: akm@ched.svnit.ac.in

18.1 Introduction

The combination of an expanding human population and urbanization demands improved sustainable wastewater management systems to operate effectively. The current centralized wastewater treatment structure encounters three primary obstacles, which include elevated running expenses and substantial energy usage, together with detrimental environmental effects. On-site and decentralized treatment systems gain interest because they focus on recovering resources from human urine and fecal waste. Resource recovery from human waste is a developing technology that transforms waste items into useful resources, specifically from urine and feces. The processing of generated waste at its location becomes essential when centralized sewage infrastructure does not exist in rural and remote areas. The systems enable recovery of nutrients that become fertilizer while generating biogas energy and making non-drinking-quality water available.

The mixture of human waste contains vital agricultural compounds like nitrogen, phosphorus, and potassium, which prove essential for farming success. Modern sanitation traditional methods consider waste products from human bodies through disposal methods that produce environmental pollution

Wastewater to Resource Recovery: Applying the Circular Economy Toward Sustainable Development, First Edition. Edited by Arvind Kumar Mungray et al. Published 2026 by John Wiley & Sons, Inc.

and waste nutrient content. The recovery of these resources helps in decreasing dependency on synthetic fertilizers because these have negative environmental consequences related to their manufacturing process, resource depletion, and greenhouse gas emissions. Studies prove that recovering nutrients facilitates food security in areas where local populations cannot afford chemical fertilizer options (Rose et al., 2015).

The systems that treat wastewater on-site matter most in settlements that do not have basic sanitation systems, including rural and distant areas. Low-technology systems such as composting toilets combined with urine diversion systems serve as on-site solutions that perform waste treatment directly where waste is produced. Most people globally depend on on-site systems for sanitation because improved facilities linked to centralized treatment reach only 38% of the worldwide population (Kelova et al., 2021).

The chapter evaluates the application of different on-site treatment systems in turning urine and fecal materials into valuable resources. The study evaluates how well urine diversion methods, together with composting and anaerobic digestion systems, retain nutrients as they produce energy. A financial evaluation helps to determine the practicality and affordability of such systems, specifically aimed at areas without centralized infrastructure access. Through the combined evaluation of technical elements and financial considerations, this chapter's effort delivers essential data that supports sustainable sanitation development work done by policy creators and researchers, as well as engineers.

18.2 Urine and Fecal Matter as a Material for Resource Recovery

The waste material from excretions contains substantial organic matter along with essential nutrients, including nitrogen, phosphorus, and potassium, despite many current sanitation systems discarding it as useless material. The recovery of elements from fecal matter serves two major functions in agriculture: it replaces chemical fertilizers, which harm the environment through pollution and deplete natural resources. The recovery of nutrients from fecal matter supports food security according to research studies that focus on areas where chemical fertilizers are scarce. Domestic sewage wastewater comprises black water and grey water. The black water is the combination of fecal matter and human urine. An average individual may produce 128 g/cap/day fecal mass (Rose et al., 2015). It contains approximately 80% water and 20% dry material (Houck et al., 2012). On the other hand, urine contains 70–90% nitrogen (N), 45–80% phosphorus (P), and 70–95% potassium (K) (Perez-Mercado et al., 2024; Lentner, 1981; Vinnerås, 2002).

Human urine only contributes ~1% to the wastewater that is reaching the sewage treatment plants (STPs) for treatment (Larsen et al., 2013). The nitrogen is present as mainly urea (75–90%), and the remainder can be as ammonium ions. Phosphorus (95–100%) is mostly present as inorganic phosphate ions (Mihelcic et al., 2011). The nutrients in urine are in the form of plant-available chemical forms, so it can be considered as a possible substitute for chemical fertilizer (Jönsson et al., 2004).

The recovery and reuse of these nutrients are essential to close the nutrient cycle as there is a looming crisis of wastewater management. The recovery of these nutrients will prevent excessive loading of N, P, and K in the natural water bodies, in turn, reducing the algal bloom. Human urine has a very low number of pathogens, and the chances of disease transfer is very low, especially while using as fertilizer. Only undesirable micro-pollutants present in human urine are pharmaceuticals and steroids, which possibly can be removed using the latest technologies, but they are energy intensive (Escher et al., 2006; Jönsson et al., 2004). The geospatial data is crucial in understanding the nutrient recovery. Urine and fecal matter can be considered as affordable fertilizers. Sustainable approaches based on urine and fecal matter collections serve both waste disposal objectives and agricultural uses. Urine containing N, P, and K can be used through struvite precipitation methods, nitrification processes, and direct application. When nutrient loops become circular, these techniques promote agricultural sustainability, thus decreasing industries' dependency on manufactured fertilizers.

18.3 Limitations of Centralized Treatment System for Urine and Fecal Matter

The centralized sewage treatment facilities were established by considering human urine and fecal matter as "waste" that needs to be removed from the environment. In centralized systems, the domestic wastewater is transported to far away treatment facility leading to high capital, operation, transportation, and maintenance costs. A total of 80–90% cost is contributed by the collection system, and the dilution of human waste leads to higher treatment costs (Maurer et al., 2005). The centralized or conventional wastewater treatment facilities are designed to remove chemical oxygen demand (COD) but the nutrient recovery is not the focus here, which leads to nutrient contamination in the environment (Carpenter et al., 1998; Larsen et al., 2007). They are energy-intensive and can be responsible for the emission of greenhouse gases (GHGs) (Xue et al., 2019). The dispositional quantity of urine and fecal matter leads to energy-intensive nutrient removal with low efficiency in the case of traditional facilities treating a large amount of the influx.

In centralized sewage treatment, the fecal matter and human urine are mixed first and diluted by water flushing, leading to an unnecessary increase in the overall volume of the wastewater supposed to be treated. The centralized treatment facilities are highly dependent on electrical supply for operation and potable water to clean the whole system periodically (Bakir, 2001; Maurer et al., 2005; Wilderer and Schreff, 2000). The centralized facilities can cater highly populated areas, but they do not fulfill the new age water recycle and nutrient recovery demands. These systems are not equipped with nutrient recovery and nutrients are diluted due to mixing of waste streams. The effective removal of potential micro-pollutants like pharmaceuticals, heavy metals, and pathogens is also a huge challenge in the centralized sewage treatment. The collection of large volumes at a single location may lead to public health risk if there is any vulnerability like pipe leakage or system breakdown. The issues of a centralized facility can be resolved by implementing a decentralized sewage treatment facility to treat the waste on location, reducing the transportation cost. Decentralized systems can promote water recycling and nutrient recovery. Small-scale systems are easy to operate and maintain.

18.4 Methods for Urine and Fecal Treatment

18.4.1 Physical Methods

The physical treatment of fecal matter and urine can be carried out by various methods such as evaporation, drying, pyrolysis, adsorption, and precipitation. Evaporation is used to remove water, mainly from urine, leaving the concentrated and hygienized liquid or solid fertilizer. Testing of the method has been made within pilot projects in Germany, Vietnam, and Canada, and has frequently been used in space applications for recycling water (Antonini et al., 2012; Bethune et al., 2015; Pahore et al., 2012). Although evaporation involves high energy consumption and loss of ammonia (up to 93% nitrogen volatilization), it is therefore subject to intensive study. To improve these problems, acidification with hydrochloric, acetic, sulfuric, or phosphoric acid, ammonia stripping, and solar heat utilization have been practiced (Moharramzadeh et al., 2022; Saetta et al., 2020). Photo-thermal conversion materials have also been explored by researchers to improve the solar evaporation efficiency to as high as 87% (Zhang et al., 2023).

Another widely used method includes struvite, calcium phosphate, and potassium struvite precipitation. Struvite precipitation is achieved with appropriate reaction of phosphate ions with Mg salt under alkaline conditions to produce struvite ($MgNH_4PO_4 \cdot 6H_2O$), a slow-release fertilizer, and can be sometimes costly (Pathy et al., 2021). However, calcium being present, the final product is nearly always 30–50% pure and therefore further refinement is necessary (Zhang et al., 2022). Another efficient method of recovery of phosphorus is calcium phosphate precipitation, which contributes to the recovery of both potassium and phosphorus where phosphorus recovery rate could achieve up to 100% (Wilsenach et al., 2007). It has also been shown by researchers that sodium struvite (Na-struvite) formation is possible, thereby enlarging options for nutrient recovery.

Pyrolysis involves the thermal decomposition of black water at a relatively high temperature in an oxygen-reduced environment producing biochar (Krueger et al., 2021). Waste is further converted to hydrochar and bio-crude oil through related processes like hydrothermal carbonization (HTC) and hydrothermal liquefaction (HTL) (Fakkaew et al., 2018; Hossain et al., 2022). However, because black water has high water content, HTL is very suitable since water itself can simply be used as the solvent. The heating efficiency can be improved by microwave technology, electricity-driven methods, and catalyst. These thermochemical processes remain still in the research phase, but they are promising for large-scale applications.

Drying is used to remove moisture via natural, mechanical, or convective drying with hot air (Ling et al., 2022). This method greatly decreases the volume and mass of waste by exposing pathogens to heat. The use of the energy source for drying can be made with renewable energy sources (wind and solar power), which makes it an economic solution (Bennamoun, 2012). Convective drying of fecal sludge at 40–80 °C can dry the sludge at rates of 1–40 $g/min/m^2$ and moisture content of 2.4–3.2 g/g db (Pocock et al., 2022). Models of solar and wind energies for human feces drying have been developed, and a solar drying system was able to reduce blackwater from 70% to 6.82% moisture content for safe discharge and resource recovery for non-sewered sanitation (Seodigeng et al., 2022). Being rich in phosphorus, it can be used as an organic fertilizer or as fuel, the heating value being very important.

18.4.2 Biological Methods

The biological treatment methods are preferred for black water management systems because they provide economic advantages along with their potent ability to recover nutrients and break down organic materials and destroy pathogens. Anaerobic digestion functions as a main treatment method to sanitize black water and produce both biogas and nutrients. The choice of anaerobic digesters includes anaerobic baffled reactors (ABRs) and up-flow anaerobic sludge blanket (UASB) reactors together with continuous stirred tank reactors depending on the waste composition. The use of low-flush or vacuum toilets proves advantageous because it lowers dilution rates that enhance digestion effectiveness (Mang and Li, 2009). Digesting anaerobically allows the retention of agricultural nutrients that generate renewable biogas energy. Blackwater contains high amounts of nitrogen that causes ammonia inhibition, which negatively affects microbial activity and leads to reduced methane production (Fuchs et al., 2018). The process of digestion receives enhancement through the implementation of biochar addition, granular activated carbon, and zero-valent iron supplementation. The joint application of black water waste with kitchen waste generates additional benefits for biogas production. A fecal sludge processing facility located in Ouagadougou, Burkina Faso handles daily volumes of 400 tons and generates 5000 m^3 of biogas (Wang et al., 2023).

The biological process of composting produces stable compost as well as CO_2 and water, while high temperatures reaching 70 °C eliminate pathogens (Zhou et al., 2022). The composting process gains efficiency through the addition of sawdust or wood chips, which control moisture content and maintain the right carbon-to-nitrogen ratio. Organic waste managers have studied various types of bulking substances to determine their effects on nutrient depletion, pathogen elimination rates, and methane production measurements (Oarga-Mulec et al., 2019). Full-scale agricultural operations benefit significantly when composting black water solids because they produce better soil conditions. The utilization of greenhouses in cold climate regions provides stable composting conditions.

In the biological process of nitrification, ammonia is oxidized to nitrite in partial nitrification before completing the conversion to nitrate. Partial nitrification happens in urine because its low buffer capacity allows the process only until ammonium stabilization but avoids nitrogen losses. Pilot testing has proved that nitrification procedures when used with distillation methods can generate concentrated liquid fertilizer products (Fumasoli et al., 2016).

Biological treatment is also partially related to urine storage, in that urea is converted enzymatically to ammonia, increasing pH and inactivating pathogens. However, research also suggests that storing urine at 20 °C for six months is sufficient and safe for agricultural use (Vinnerås et al., 2008). Other countries that have implemented urine storage for fertilizer use are Sweden, Germany, and China (Alemayehu et al., 2020).

Microalgae-based treatment is an evolving biological process where microalgae convert nitrogen and phosphorus from blackwater into biomass that is high in lipids, suitable for biofuels. However, different microalgae species differ in the uptake efficiency of nutrients, being 23–100% for nitrogen and 20–100% for phosphorus (Khan and Yoshida, 2008). Systems of algal bacteria have been developed such that the bacteria decompose organic matter while the microalgae become the oxygen producer through photosynthesis to make a complete wastewater treatment cycle. There have been recent advancements, such as algal-bacterial aerobic granular sludge, which both enhance nutrient removal and biomass settleability, and the ability to capture the CO_2 when the biomass is being degraded (Li et al., 2018).

18.4.3 Bioelectrochemical Systems

The microbial fuel cell (MFC) can be a potential candidate to treat fecal matter and urine in a decentralized manner as an eco-friendly alternative with bioenergy production as shown in Figure 18.1 (Santoro et al., 2017). MFCs use microorganisms as a catalyst to degrade wastewater and convert chemical energy to electrical energy without any external energy supply (Falcucci et al., 2017). The dual functions of MFCs can address two major aspects: waste-to-energy and sustainable waste treatment. 85% COD removal and 39% lower phosphorus output by means of a standard septic tank, while the same septic tank, upgraded to a microbial electrolysis cell (MEC)-septic tank, enhanced biogas production 5 times and reduced H_2S by 2.5 times compared to a standard septic tank. Energy recovery in wastewater treatment was improved with efficient conversion of electricity input to extra biogas (Zamalloa et al., 2013). The system integrated in an ABR-MFC-MEC combined system was capable of simultaneously removing nitrogen and carbon from fecal sewage with complete energy recycling. Startup time was reduced and hydrogen production was increased to 3.42×10^{-3} m^3-$H_2/m^3/d$ when applying 0.5 V. COD removal rates reached 99.2% (single chamber) and 98.9% (double chamber), which were proof of the high shock resistance of this system (Liu et al., 2020). A 71% TCOD, 88% soluble COD, and 44% NH_4^+ removal in dual chamber MFCs treating human waste, along with 70.8 mW/m^2 power production, were obtained while a cubical MFC produces 251 mW/m^2 (Fangzhou et al., 2011; Li et al., 2009). The performance of a stacked air-cathode MFC system was 142 mW/m^2, and a 20 L MFC treating septage removed 99.8% TSS, 86.4% BOD, and 82.8% nitrate (Yazdi et al., 2015).

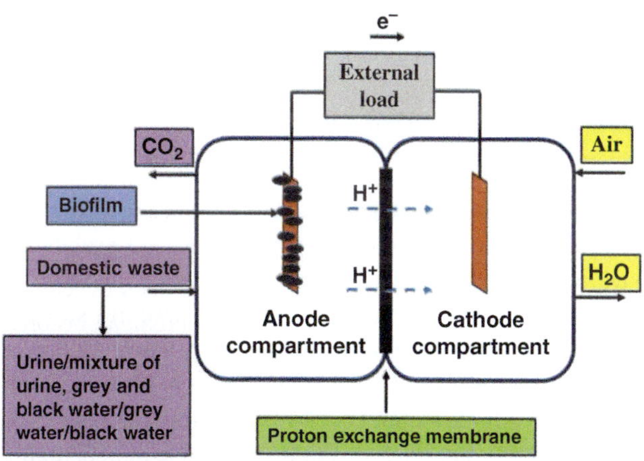

Figure 18.1 General schematic diagram of bioelectrochemical system.

While these progress have been done, low power output is still a pending issue in large-scale applications.

Nutrient dilution leads to the inefficient use of energy from sewage, but urine-powered MFCs achieve COD removal between 50% and 95% and a maximum 2500 mW/m^2 power density (Wang et al., 2017). Successful applications of these MFCs have been driving mobile phones, LED bulbs, electronic gadgets, and sensors, emphasizing their applications for energy solutions in sustaining the environment. The real-world application of urine-fed MFC stacks is to charge mobile phones, and this has been achieved successfully. A 24-unit MFC stack was utilized to charge a 1000 mAh Li–ion battery at 3.7 V in 30 hours (Ieropoulos et al., 2013). The system was scaled up by Walter et al. (2017) using 6 MFC modules to charge a Samsung GT E2121B in 42 hours and a Samsung Galaxy S I9000 in 68–82 hours according to usage (Walter et al., 2017). In the first such demonstration of their power, MFCs were used to charge smartphones, a clear example of an off-grid power solution suitable for remote locations. To investigate urine-based power generation for emergency locator transmitters (ELT), Winfield et al. (2015) developed origami tetrahedron MFC (TP MFC) using photocopier paper. The operation of MFCs immersed in buffer solution was tested without cathode addition in a six-unit abiotic MFC stack, which produced a working voltage within 3 hours and 15 minutes for operating the management system. Fresh urine was able to reactivate dried anodes within 35 minutes after 4 weeks. Parallel TP-MFCs powered 85 radio signals in tests; series TP-MFCs transmitted 238 signals for more than 24 hours, and both could be urine-activated as life rescue systems that are lightweight and low cost (Winfield et al., 2015).

18.5 Economic Analysis of Centralized and Decentralized Treatment of Urine and Fecal Matter

18.5.1 Comparative Analysis of Centralized and Decentralized Systems

This section provides key metrics to evaluate centralized and decentralized systems, including capital costs, operational costs, transportation, and scalability. Centralized systems have a high initial capital investment that is justified in urban areas due to economies of scale (Verma et al., 2024). These are suitable for processing large volumes of waste in an efficient manner. However, lower capital investment is required for decentralized systems, and it is ideal for rural or small-scale applications if the infrastructure cost for the centralized systems is prohibitively high (Jung et al., 2018). They are also more adaptable in geographic and demographic diversity. Automation and integration of centralized systems lead to operational costs that can be achieved with cost-efficient large-scale operations. Nevertheless, there are high energy and maintenance costs involved with their extensive sewer networks that may negate these efficiencies (Eggimann et al., 2016). On the other hand, decentralized systems do not incur transport and network-related expenses and thus provide lower operational costs in case of waste treatment on site (Wang, 2014).

In high-density urban areas, economies of scale allow for centralized systems to justify their significant infrastructure expenses. In addition, these systems also have excellent quality in terms of the advanced treatment capability, i.e. nutrient recovery and biogas generation, resulting in high resource utilization efficiency (Naik and Stenstrom, 2016). Centralized systems, however, are quite vulnerable to disruptions such as mechanical failures or power outages. Reliance of such facilities on a single facility makes these failures have far-reaching consequences (Woods et al., 2013). While decentralized systems are quite better at dealing with localized failures and possess certain benefits as shown in Figure 18.2. This is by design as their distributed nature means that the failure of one unit does not affect the entire network. In rural or water-scarce regions, especially, centralized systems could be impractical or too cost-prohibitive, and less suitable (Zhang and Tan, 2010). Decentralized systems are restricted in their use due to their reliance on community engagement in maintaining and operating properly. Most often, educational initiatives and training are needed to achieve long-term success (Massoud et al., 2009). Furthermore, they are not scalable in densely populated areas since urban settings could produce too much waste for small-scale systems to handle (Gupta, 2007).

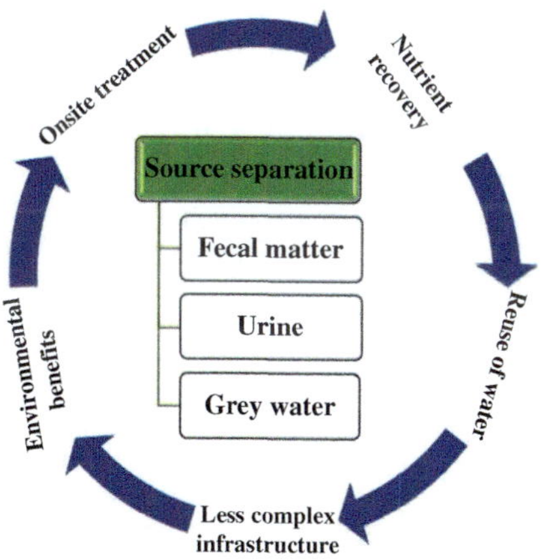

Figure 18.2 Decentralized system characteristics.

Both systems are economically viable only in local conditions and are compared in Table 18.1. Municipalities with robust funding and infrastructure tend to need less flexibility for maintaining their pensions and other systems better addressed using centralized systems, though this comes with a lot of upfront effort. Because of their size and scale and the availability of their advanced technologies, they are economically efficient in urban environments (Eggimann et al., 2016). However, decentralized systems can reach financial break-even faster since their capital and operational costs are lower. This is very advantageous in rural or remote regions where there is a shortage of funding and infrastructure (Naik and Stenstrom, 2016). Sensitivity analyses show that decentralized systems tend to cost less than centralized ones, particularly when scaled for the target population and geographical area at which the system will operate (Jung et al., 2018).

Table 18.1 Comparison of centralized and decentralized systems.

Parameter	Decentralized system	Centralized system
Nutrient recovery	High due to low dilution	Low due to large volume generated by dilution
Water usage	Low due to waterless system	High due to flushing
Cost	Low due to less complex system that does not need a piping network	High due to complex piping and pumping network to handle a huge amount of mixed waste streams
Environmental impact	Low as nutrients are recovered and not wasted as runoff	High as wastage of nutrients may result in eutrophication
Operation	High community involvement	Centrally handled by technically trained staff
Economic aspect	Cost-effective at community or rural level	Cost-effective at large scale
Energy need	Low and may generate energy via biogas	High due to pumping for transporting waste from all the places to a single point for treatment
Resilience	Failure impacts a smaller community area or a single unit	Failure can disrupt the entire urban infrastructure
Reuse	Nutrients can be recovered as bio-fertilizer from urine or by composting, biogas generation	Reuse is possible only by using advanced technologies

18.6 Case Study for Economic Analysis

18.6.1 Case Study 1: Techno-economic Analysis of Nutrient Recovery from Urine – Centralized vs. Decentralized Approaches

This case study describes the implementation of nutrient recovery processes in Université Laval, Canada. A techno-economic assessment was performed over a 30-year project period to evaluate the feasibility of centralized and decentralized urine treatment options. Approximately 15.3 m^3/day urine is generated from the university campus of 47 000 students and 7240 employees. Both methods were analyzed to determine the financial and technical viability of each in recovering nutrients efficiently and economically. The centralized treatment approach includes the use of advanced techniques such as thermal ammonia stripping, Na-chabazite adsorption, and struvite precipitation for hydrolyzed urine processing. Because hydrolyzed urine has such a high ammonium concentration, it needs a lot of treatment before it can be repurposed effectively. Ammonia removal from the hydrolyzed urine is accomplished by ammonia stripping, and the resulting ammonia is converted into ammonium sulfate for use as a fertilizer. Potassium is captured by Na-chabazite adsorption and phosphorus by struvite precipitation to recover most of the valuable nutrients efficiently (Wu et al., 2024).

On the contrary, the decentralized approach takes the fresh urine before it goes through hydrolysis. In this method, urea and potassium are adsorbed from fresh urine, treated with Na-chabazite and biochar. Urea hydrolysis follows this process, converting nitrogen into ammonium. After treatment, the urine is further treated for nitrogen and phosphorus recovery as struvite by thermal stripping. The decentralized approach has the advantage that it reduces the investment costs of the initial investment by integrating adsorption and hydrolysis processes at the source and is more economically viable than the centralized approach.

The nutrient recovery efficiencies of both treatment methods were evaluated by the technical evaluation. In the centralized approach, hydrolyzed urine was stripped of ammonium using ammonia stripping at 60 °C. The potassium was recovered by the Na-chabazite adsorption process, and the phosphorus was extracted by the struvite precipitation at the end of the process. It was found that the centralized approach recovered 74.3% of the nitrogen and 76.8% of the phosphorus. However, the decentralized approach performed better in nutrient recovery. Initially, the fresh urine was treated with Na-chabazite and biochar to adsorb urea and potassium. The transformation of nitrogen into ammonium was facilitated by urea hydrolysis and subsequently thermal stripping and struvite precipitation. The efficiency of the decentralized method was 89.4% for nitrogen recovery and 98.7% for phosphorus recovery, which was superior to the centralized method. These results suggest the possibility of decentralized treatment for optimizing nutrient recovery from urine, and that it is a more suitable option for large-scale implementation.

Both treatment approaches were assessed in terms of cost-effectiveness using a financial assessment. The centralized method had a CAPEX of 2.93 million CAD and an annual OPEX of 303 290 CAD. Annual gross profit was 62 690 CAD from selling recovered ammonium sulfate, potassium-enriched Na-chabazite, and struvite fertilizer. Despite that, the centralized method was not able to break even within the 30-year project period, and so it was not financially feasible. On the other hand, the CAPEX of 2.28 million CAD was 22% lower than the centralized approach. While its OPEX was higher at 158 630 CAD due to biochar procurement costs, the annual gross profit was much higher at 158 630 CAD. As a result, the decentralized method was projected to break even in the twenty-first year, making it a more viable long-term investment. Decentralized urine treatment offers lower initial investment and higher revenue generation than centralized urine treatment for nutrient recovery.

To determine the possible effect of variables that fluctuate, such as urine inflow rate, nitrogen concentration, and urea-enriched biochar pricing, a sensitivity analysis was carried out. It was discovered that increasing the urine inflow rate by 20% resulted in a 35% increase in the net present value (NPV) and improved economic feasibility. Likewise, ammonium sulfate production was increased 50% by increasing nitrogen concentration, which bolstered the economic viability of the treatment methods even more. Increasing the price of urea-enriched biochar by 20% also increased NPV by 30%, reinforcing the decentralized approach as the more financially viable option.

Nutrient recovery from urine has environmental benefits far beyond economic ones. Diverting nutrients at the source significantly reduces both centralized and decentralized approaches' burden on wastewater treatment plants. By doing this, it reduces energy use, reduces nutrient pollution, and decreases greenhouse gases. In particular, the decentralized method conforms well with circular economy principles as it incorporates biochar, a renewable material, in the nutrient recovery process. Further enhancement of the feasibility of these technologies would be possible through future policies supporting the development and implementation of urine-based fertilizers. Regulatory frameworks, incentives, and public acceptance play an important role in allowing for the adoption of large-scale urine nutrient recovery systems. Use of recovered nutrients in agriculture can also help governments and environmental organizations promote sustainable sanitation practices. Financial incentives for decentralized treatment facilities can be further provided through policies.

The decentralized nutrient recovery from urine is economically more viable and environmentally sustainable than centralized approach. The research recommends that the decentralized water resource recovery facilities (WRRFs) be scaled up in urban and semi-urban settings for providing a model for sustainable sanitation and nutrient management systems on a global scale. Cities can decrease their dependency on synthetic fertilizers, apply the circular economy principles, and support the environmental conservation efforts by investing in the decentralized nutrient recovery technologies. Such systems are implemented in a transformative move toward realizing sustainable wastewater management and recovery of nutrients on a global scale.

18.6.2 Case Study 2: Decentralized Wastewater Treatment at Aravind Eye Hospital (AEH), Puducherry

AEH in Puducherry is a leading specialty eye hospital in India that has adopted decentralized wastewater treatment as a part of its sustainability initiatives. Considering the hospital's daily operations, a considerable amount of wastewater is produced, and thus, a suitable and eco-friendly treatment option is needed. As a result, the hospital implemented a nature-based treatment system to address wastewater treatment in an efficient manner while conserving and reusing water. The decentralized wastewater treatment system of AEH is an anaerobic baffled reactor (ABR), a planted gravel filter, and a polishing pond. A consortium of environmental engineering organizations designed and built the system. The design was meant to be simple, inexpensive, and ecologically sustainable. The whole treatment facility occupies an area of 1774 square meters and daily treatment capacity of 307 m^3 of wastewater. It goes through a six-stage wastewater treatment process, including a grease trap, settler, anaerobic baffled reactor, anaerobic filter, horizontal planted gravel filter, and a final polishing pond. With this setup, the treatment efficiency is 90–95% and is sufficient to meet environmental discharge standards. The water is primarily reused for gardening and toilet flushing, saving an annual of 100 000 m^3 (Rath et al., 2020).

The construction of the decentralized wastewater treatment plant at AEH cost approximately $170 000 (₹11.16 million). The operational cost per cubic meter of treated water is $0.10 (₹6.31), which is much lesser than that of conventional wastewater treatment methods. Besides, the hospital also saves an estimated ₹44–50 million in 25 years by reusing treated water instead of depending on municipal facilities. AEH shows that a successful model for bringing sustainable wastewater treatment into institutional settings is by reducing freshwater consumption and minimizing wastewater discharge.

18.6.3 Case Study 3: Fecal Sludge Management in Devanahalli, Bengaluru, Karnataka

A fast-growing town in Karnataka, Devanahalli had no central sewerage system and faced a lot of problems in managing fecal sludge. To overcome this, the town-built India's first dedicated Fecal Sludge Treatment Plant (FSTP) with the goal of sustainable and cost-effective sanitation management. It was in 2015 when the Devanahalli FSTP was constructed with the capacity of processing 6000–8000 liters of fecal sludge per day. The plant uses a combination of anaerobic digestion, unplanted sludge drying beds, and a DEWATS-based water treatment system. The FSTP occupies a total land area of 650 square meters, which is small and efficient for urban sanitation. The sludge collection, transport, and treatment processes are integrated in the FSTP. Fecal sludge is collected by registered desludging operators from households and community toilets and conveyed to the plant where it is treated systematically. The effluent is treated and released into the environment, and the sludge is dried and co-composted for agricultural use (Rath et al., 2020).

The cost of building the FSTP was ₹6.7 million and annual operation and maintenance cost of about ₹600 000. It is maintained under a public–private partnership (PPP) model for financial sustainability. A part of the operational cost is covered by households paying ₹1200–1500 per desludging service. This initiative has increased sanitation coverage greatly in Devanahalli, ending open defecation and environmental contamination. In addition, it also gives rise to an additional revenue stream from its use as soil conditioner and organic fertilizer.

18.6.4 Case Study 4: Fecal Sludge Management in Leh, Ladakh

Wastewater and sludge management become difficult in a high-altitude town such as Leh in Ladakh due to extreme climatic conditions. In 2017, the process of constructing a Fecal Sludge Treatment Plant (FSTP) using public–private partnership (PPP) mechanisms was initiated to ensure safe and sustainable sanitation services. A DEWATS-based water treatment system and planted drying beds are used to treat 12 000 liters of fecal sludge per day in the Leh FSTP. Since the treatment system was to operate in sub-zero winter temperatures, it was designed to operate efficiently in harsh climatic conditions. A desludging service was introduced on a scheduled basis to ensure that all septic tanks are cleaned annually. The plant includes unplanted drying beds, planted gravel filters, and a polishing pond to allow treated water to be safely reused for landscaping. The total investment in the Leh FSTP was ₹7.8 and ₹2.3 million for procurement of desludging trucks. The plant works on PPP model where the facility is managed by a private operator and municipal government enforces the desludging service at regular intervals. Revenue is generated to sustain operations through ₹3500 per desludging service paid by households and businesses. The plant has treated more than 2.5 million liters of fecal sludge since it came into operation and has significantly reduced groundwater pollution and improved hygiene standards in Leh (Rath et al., 2020).

18.6.5 Case Study 5: Public–Private Partnership for Decentralized Sewage Treatment in New Delhi

To alleviate water shortages and pollution problems, New Delhi adopted a decentralized sewage treatment strategy to enhance the reuse of sewage for urban landscaping. Small-scale STPs were installed by the New Delhi Municipal Corporation (NDMC) under a public–private partnership (PPP) model to meet local wastewater effectively. Three STPs were set up as a pilot initiative with the treatment capacity of 1 MLD to 1.1 MLD. The two technologies used by the plants are Soil Biotechnology (SBT) and Membrane Bioreactor (MBR). These systems include local treatment of sewage without large networks of sewers; the treated water is utilized for landscaping and irrigation (Rath et al., 2020).

The STPs are built, operated, and maintained by private entities under the PPP model while NDMC buys treated water from them at a price of ₹31–37 per kiloliter. This approach will minimize the capital expenditure of the government and will ensure operational efficiency. NDMC treats and reuses the wastewater hence reducing the dependency on fresh water sources, and the estimate amount water saved annually is over 1 billion liters. A scalable solution to other urban centers facing water scarcity and pollution is the decentralized STP model.

These case studies present the successful practice of decentralized wastewater and fecal sludge management in India. All projects show that low-cost, sustainable treatment technologies are viable under local conditions. These initiatives have significantly improved urban sanitation, water reuse, and environmental sustainability through leveraging PPP models and nature-based treatment technologies in combination with decentralized infrastructure. These projects are economically feasible due to their low cost of construction and operation, and revenue generation from the resource recovered and improved public health outcomes. The replicable blueprint implemented by these case studies will be a development opportunity for other regions that also aim to develop efficient and sustainable wastewater and fecal sludge management.

18.6.6 Case Study 6: Solid Waste Management in Rural Maharashtra

Solid and wastewater management, especially in rural and semi-urban areas, is a severe challenge for India. In rural area, the country produces about 0.3–0.9 kg of waste per capita per day, more than half of which is organic. In recent years, the Government of India has taken initiatives to promote sustainable solid waste treatment technologies through the Swachh Bharat Mission (SBM). Composting and anaerobic digestion are among the best methods for treating and reducing organic waste, while depending on landfills and enhancing the quality of soil. Analysis of design and socioeconomic impacts of decentralized solid waste and wastewater management is made in this case study with both approaches ranging from centralized models to decentralized models (Patwa et al., 2020).

For instance, in Maharashtra, in villages like Mahalung and Dummanpur, solid waste management was a big issue because of open dumping and poor collection system. In 2017, a decentralized waste management strategy was introduced, where there was community participation and building of composting units. The goal of the project was to process 70% of the organic waste produced in these villages, the current figure being 2.36 tons per year per 500 households. The windrow composting and NADEP composting were based on the waste management system. At the household level, waste segregation was mandated to reduce contamination of organic waste. The NADEP composting units were 3.5 m × 1.5 m with 2 tons organic waste per cycle retention period of three months. However, the composting in windrow was done in larger open spaces and turned every 10 days to maintain aerobic degradation. The cost of the project was about ₹2.4 million, which was a cheaper solution to the problem of rural waste management. Compost generated an income of ₹200 000 per year that helped to offset operational costs. This, in addition, methane emission was reduced by 25%, keeping 14 metric tons of CO_2 equivalent emissions per year from entering the atmosphere annually. Replication of the project was done in 35 villages in Maharashtra.

18.6.7 Case Study 7: Sewage Treatment Systems in Kampala

As with many rapidly urbanizing cities in low- and middle-income countries, Kampala, Uganda, seeks to address severe challenges in its sanitation infrastructure and wastewater treatment. The city's sanitation systems are under extreme pressure due to its high urban population growth rate of 5.2% per year. On-site sanitation systems are used by the population of 3.2 million, which is about 1.3% of the city that is connected to the sewer network. This paper studies the infrastructure investments and operating costs for both fecal sludge management (FSM) systems and sewage treatment systems in Kampala, Uganda. This highlights large cost differences, service efficiency problems, and policy recommendations for future improvement (McConville et al., 2019).

Bugolobi WWTP with a capacity of 45 000 m^3/day and Lubigi WWTP with a capacity of 5000 m^3/day are the two major wastewater treatment plants (WWTPs) in Kampala. The sewer network is 180 km long, most of which was installed between 1940 and 1969. The system now serves only 41 000 people (1.3% of the metropolitan population), and industrial and commercial entities contribute about 56% of the total wastewater flow. It is very capital-intensive to expand and maintain Kampala's sewerage system. The Bugolobi WWTP upgrade cost requirement was about $55 million, and the Lubigi WWTP needed $18 million. The total estimated capital investment in the sewage regime per capita per year is $304, or 23 times higher than that of the FSM system. Sewage treatment costs are also much higher than for FSM. Operating cost per capita for the sewage regime is $74, 101 times higher than that of FSM ($0.73). With these high costs, though, the expenses paid for by the sanitation fees collected from sewer customers ($205.98 per capita per year) do not cover the cost of operations. Financial sustainability problems are exacerbated by maintenance inefficiencies and high energy costs.

18.6.8 Case Study 8: Fecal Sludge Management in Kampala

With only 1% of the city covered by sewerage, 99% of the population of Kampala are serviced by on-site sanitation systems such as septic tanks and pit latrines. Lubigi Fecal Sludge Treatment Plant is the main FSM facility with a capacity of 400 m^3/day, but it is insufficient to cater for the city's growing sanitation needs. Less than 35% of the fecal sludge generated is collected, transported, and treated. The remaining 65% is discharged into the environment or inadequately contained, causing severe public health risks and environmental pollution. The sewage system is much more expensive than the FSM regime. The annual capital investment in FSM infrastructure is $13 per capita, vs. $304 per capita for sewage treatment. Household sanitation units ($12 per capita), vacuum trucks ($0.12 per capita), and treatment facilities ($0.66 per capita) are the primary capital costs (McConville et al., 2019).

For FSM, operating costs are $0.73 per capita per year, and households pay $3500 per desludging service. Currently, there are 89 vacuum trucks in Kampala, of which 49 are small trucks (<5 m^3) and 40 are large trucks (>5 m^3). At Lubigi, the fecal sludge collected is processed; the dried sludge is sold to farmers as fertilizer. But that valorization only generates a small portion of the revenue, equal to 1% of the operational costs. Compared to sewage treatment, FSM costs have poor financial sustainability and efficiency. The cost per capita for the total annualized sewage treatment is much greater at $186 than at $14 for FSM. This also applies to capital costs, as FSM requires $13 per capita for sewage treatment, whereas sewage treatment costs $304 per capita.

The contrast is further emphasized by operating costs: sewage treatment costs $74 per capita per year, while FSM is operated at a very low cost of $0.73 per capita. Even though sewage treatment is costly, sanitation fees are not sufficient to cover the expenses. Fees of $205.98 per capita annually from the sewage system are made, but the system is still financially loss-making. On the other hand,

FSM services are cheaper by far but are mostly dependent on household funding as opposed to structured fee-based revenue. However, the sewage treatment system theoretically offers higher treatment efficiency but only serves 1.3% of the population. On the other hand, FSM can treat 35% of the population's sludge safely but fails to treat or mismanage 65% of the sludge. Areas that depend on FSM have significantly higher public health risks because of lack of sludge containment and untreated discharges to water bodies.

To achieve this, \$350 million is needed to upgrade all on-site systems to a minimum of Ventilated Improved Pit (VIP) latrines and increase capacity to treat fecal sludge. Such a strategy would raise FSM coverage to 100% of the population using on-site sanitation at an annualized capital cost of \$9 per capita. The current sewage system is operating far below its capacity. If 130 000 more users are added to the sewer connection, it would cost the annual per capita cost of \$186–96; thus the system will be 50% cheaper. Nevertheless, FSM is seven times cheaper than the sewage system, even with full capacity utilization. The sewage system is currently subsidized by public funds, while its reach is limited. In FSM, the per capita public investment in sewage treatment is \$160, but FSM receives only \$6.7 million per year, which is 35% of the sludge management needs. FSM's budget could be reallocated from public funds to an amount of \$7.4 million annually, which would allow for expanding the treatment facilities and service coverage to the underserved communities.

The findings indicate that FSM is a much more cost-effective and scalable sanitation solution for Kampala than centralized sewage treatment. Investments in FSM would have more immediate and widespread impact on only 1.3% of the population connected to sewers. Optimizing service efficiencies and expanding infrastructure in some cases will enable Kampala to achieve a higher level of coverage while easing the financial burden to the government and households. Future sanitation planning should be done considering sanitary impacts, service provision accessibility, and cost-effectiveness for decision-makers.

18.6.9 Case Study 9: Al-Azraqain Landfill, Sana'a, Yemen

The rapid population growth, urbanization, economic instability, and lack of infrastructure in Yemen make solid waste management (SWM) an important challenge for the country. In urban areas, the country produces about 0.6 kg per capita per day, while in rural areas, it is about 0.35 kg per capita per day, and the annual MSW generation is about 3%. Landfilling is the primary method of disposal and is highly risky to the environment and public health since greenhouse gases are released, leachate contaminates, and there is inadequate waste treatment. In this study, the challenges and sustainable solutions of SWM in Sana'a, Yemen are analyzed. Sana'a's 500 000 tons per year of solid waste is dumped there at Al-Azraqain landfill, the main wasteland site. The landfill does not have proper engineering designs like gas collection systems, leachate treatment, and efficient segregation of waste, which results in large quantities of methane emissions and contamination of groundwater. Frequent reports of open dumping and waste fires have been reported that release toxic gases such as sulfur dioxide (SO_2) and nitrogen dioxide (NO_2), which pose a severe health risk to nearby communities (Al-Dailami et al., 2022).

The landfill operates with almost no financial support, and the waste collection efficiency has fallen from 85% to 75% after 2015 due to the war and an economic crisis. About 500 000 metric tons of CO_2 equivalent are annually released from methane emissions from the landfill, which are an important source of climate change. Twelve water sources within a 10 km radius of the incinerator have been contaminated with leachate, which has rendered them unsafe for human consumption. Investing about $10 million over the next five years would be necessary for cleanup and environmental restoration efforts. Sana'a's waste collection is mainly through municipal services and informal waste pickers. Fuel shortages have severely impacted the fleet of collection trucks, driving efficiency of 80–40% in the last decade. Due to the waste pileups, exceeding 3000 tons in Sana'a streets, the average number of waste collection shifts in Sana'a decreased from two per day to one per day.

The annual operational cost of the Sana'a waste collection system is $20 million; the revenue from the municipal fees accounts for $5 million leaving $15 million grant gap each year. Spare parts and maintenance have gotten more and more expensive because of import restrictions and the cost of maintaining the waste collection fleet has risen to 300%.

18.7 Conclusion

The nutrients present in human urine and fecal material provide opportunities for recycling of nutrients and sustainable waste treatment. The comparison of decentralized treatment systems with centralized systems demonstrates that certain critical differences exist in the overall efficiency, feasibility, and economic performance of these systems. Because source separation is inherently necessary in some decentralized systems, these systems preserve nutrient concentrations due to no dilution and allow for the specific type of treatment. With these systems, it is possible to achieve high nutrient recovery, energy saving, and adaptability to local conditions. Urine diverting, composting, and high solids anaerobic digestion are especially suitable for decentralized application. Additionally, decentralized systems also do not need expensive and complex sewage infrastructure, thereby reducing capital and operational costs. Nevertheless, centralized systems can process large amounts of waste, but construction and operation costs are very high. The nutrients present in the fecal and urine are diluted in the centralized system, reducing the nutrient recovery efficiency, and the pumping cost is quite high for the centralized system. Furthermore, the cost of sophisticated and expensive equipment for nutrient recovery technologies in the centralized facilities may be higher too. From the economic point of view, decentralized systems are more affordable if a specific area is sparsely populated or infrastructure development is restricted. The decentralized system can survive only with community involvement and appropriate and timely maintenance for smooth operation. Systems are centralized in urban areas, but not feasible for nutrient recovery except upon high investments in advanced treatment and recovery technologies. In general, decentralized treatment of urine and fecal material is well suited for efficient nutrient recovery, and it can be more cost-effective in the long run, especially in the context of rural areas. Based on these component perspectives, a balance pathway is found by using a strategic combination of decentralized approaches with selective centralized processing.

Review Questions

1. Why can domestic waste be considered a nutrient resource rather than waste?
2. Which nutrients are present in the fecal matter and human urine? Why can urine be a superior source for nutrient recovery?
3. What are the crucial aspects that can limit the nutrient recovery from fecal matter?
4. Compare centralized and decentralized systems in terms of treatment efficiency, nutrient recovery, and sustainability. If you are assigned to establish a waste treatment facility in a rural area, which one would you choose and why?
5. For a community-based waterless system, which wastewater treatment method would be suitable?
6. Will source separation of human urine and fecal matter be of any benefit to a city to reduce the overall treatment cost of wastewater?
7. Why are nutrients diluted in the centralized system and difficult to recover at a large scale?
8. Which sustainable waste treatment methods can be beneficial for a rural area or a small community at a nominal cost without complex technical requirements?
9. Is it possible to recover nutrients from the urine, and what makes them an alternative to synthetic chemical fertilizers?
10. If centralized systems are to be integrated with the decentralized technology to meet the demands of expanding modern cities, which integrated strategy will you propose?

References

Al-Dailami, A., Ahmad, I., Kamyab, H. et al. (2022). Sustainable solid waste management in Yemen: environmental, social aspects, and challenges. *Biomass Conversion and Biorefinery*, **12**, 1–27. Springer. https://doi.org/10.1007/S13399-022-02871-W

Alemayehu, Y.A., Asfaw, S.L., and Terfie, T.A. (2020). Nutrient recovery options from human urine: a choice for large scale application. *Sustainable Production and Consumption*, **24**, 219–231. https://doi.org/10.1016/J.SPC.2020.06.016

Antonini, S., Nguyen, P.T., Arnold, U. et al. (2012). Solar thermal evaporation of human urine for nitrogen and phosphorus recovery in Vietnam. *Science of the Total Environment*, **414**, 592–599. https://doi.org/10.1016/j.scitotenv.2011.11.055

Bakir, H.A. (2001). Sustainable wastewater management for small communities in the Middle East and North Africa. *Journal of Environmental Management*, **61**(4), 319–328. https://doi.org/10.1006/jema.2000.0414

Bennamoun, L. (2012). Solar drying of wastewater sludge: a review. *Renewable and Sustainable Energy Reviews*, **16**(1), 1061–1073. https://doi.org/10.1016/j.rser.2011.10.005

Bethune, D.N., Chu, A., and Ryan, M.C. (2015). Passive evaporation of source-separated urine from dry toilets: prototype design and field testing using municipal water. *Journal of Water, Sanitation and Hygiene for Development*, **5**(3), 392–401. https://doi.org/10.2166/washdev.2015.158

Carpenter, S.R., Caraco, N.F., Correll, D.L. et al. (1998). Nonpoint pollution of surface waters with phosphorus and nitrogen. *Ecological Applications*, **8**(3), 559–568. https://doi.org/10.1890/1051-0761(1998)008[0559:NPOSWW]2.0.CO;2

Eggimann, S., Truffer, B., and Maurer, M. (2016). Economies of density for on-site waste water treatment. *Water Research*, **101**, 476–489. https://doi.org/10.1016/j.watres.2016.06.011

Escher, B.I., Pronk, W., Suter, M.J.F. et al. (2006). Monitoring the removal efficiency of pharmaceuticals and hormones in different treatment processes of source-separated urine with bioassays. *Environmental Science & Technology*, **40**(16), 5095–5101. https://doi.org/10.1021/es060598w

Fakkaew, K., Koottatep, T., and Polprasert, C. (2018). Faecal sludge treatment and utilization by hydrothermal carbonization. *Journal of Environmental Management*, **216**, 421–426. https://doi.org/10.1016/j.jenvman.2017.09.031

Falcucci, G., Nastro, R.A., Minutillo, M. et al. (2017). Microbial fuel cells in solid waste valorization: trends and applications. *Modelling Trends in Solid and Hazardous Waste Management*, 159–171. Springer. https://doi.org/10.1007/978-981-10-2410-8_9

Fangzhou, D., Zhenglong, L., Shaoqiang, Y. et al. (2011). Electricity generation directly using human feces wastewater for life support system. *Acta Astronautica*, **68**(9–10), 1537–1547. https://doi.org/10.1016/j.actaastro.2009.12.013

Fuchs, W., Wang, X., Gabauer, W. et al. (2018). Tackling ammonia inhibition for efficient biogas production from chicken manure: status and technical trends in Europe and China. *Renewable and Sustainable Energy Reviews*, **97**, 186–199. https://doi.org/10.1016/j.rser.2018.08.038

Fumasoli, A., Etter, B., Sterkele, B. et al. (2016). Operating a pilot-scale nitrification/distillation plant for complete nutrient recovery from urine. *Water Science and Technology*, **73**(1), 215–222. https://doi.org/10.2166/wst.2015.485

Gupta, S.K. (2007). Decentralized wastewater treatment technology: a promise for the future. In: *Sustainable Development of Energy, Water and Environment Systems*, 322–330. World Scientific. https://doi.org/10.1142/9789812771285_0029

Hossain, M.R., Khalekuzzaman, M., Kabir, S.B. et al. (2022). Enhancing faecal sludge derived biocrude quality and productivity using peat biomass through co-hydrothermal liquefaction. *Journal of Cleaner Production*, **335**, 130371. https://doi.org/10.1016/j.jclepro.2022.130371

Houck, M.M., Crispino, F., and McAdam, T. (2012). Biological evidence. *The Science of Crime Scenes*, 239–258. Academic Press. https://doi.org/10.1016/B978-0-12-386464-2.00021-5

Ieropoulos, I.A., Ledezma, P., Stinchcombe, A. et al. (2013). Waste to real energy: the first MFC powered mobile phone. *Physical Chemistry Chemical Physics*, **15**(37), 15312–15316. https://doi.org/10.1039/C3CP52889H

Jönsson, H., Stintzing, A.R., Vinnerås, B. et al. (2004). *Guidelines on the Use of Urine and Faeces in Crop Production*. EcoSanRes Programme.

Jung, Y.T., Narayanan, N.C., and Cheng, Y.-L. (2018). Cost comparison of centralized and decentralized wastewater management systems using optimization model. *Journal of Environmental Management*, **213**, 90–97. https://doi.org/10.1016/j.jenvman.2018.01.081

Kelova, M.E., Ali, A.M., Eich-Greatorex, S. et al. (2021). Small-scale on-site treatment of fecal matter: comparison of treatments for resource recovery and sanitization. *Environmental Science and Pollution Research International*, **28**(45), 63945. https://doi.org/10.1007/S11356-021-12911-Z

Khan, M. and Yoshida, N. (2008). Effect of L-glutamic acid on the growth and ammonium removal from ammonium solution and natural wastewater by Chlorella vulgaris NTM06. *Bioresource Technology*, **99**(3), 575–582. https://doi.org/10.1016/j.biortech.2006.12.031

Krueger, B.C., Fowler, G.D., Templeton, M.R. et al. (2021). Faecal sludge pyrolysis: understanding the relationships between organic composition and thermal decomposition. *Journal of Environmental Management*, **298**, 113456. https://doi.org/10.1016/j.jenvman.2021.113456

Larsen, T.A., Maurer, M., Udert, K.M. et al. (2007). Nutrient cycles and resource management: implications for the choice of wastewater treatment technology. *Water Science and Technology*, **56**(5), 229–237. https://doi.org/10.2166/wst.2007.576

Larsen, T., Udert, K., and Lienert, J. (2013). *Source Separation and Decentralization for Wastewater Management*. Iwa Publishing.

Lentner, C. (1981). Units of Measurement, Body Fluids, Composition of the Body, Nutrition. Switzerland: Ciba-Geigy. https://books.google.co.in/books/about/Units_of_Measurement_Body_Fluids_Composi.html?id=zQOKwiBj9gMC&redir_esc=y

Li, Y., Wang, Y., Gao, Y. et al. (2018). Seawater toilet flushing sewage treatment and nutrients recovery by marine bacterial-algal mutualistic system. *Chemosphere*, **195**, 70–79. https://doi.org/10.1016/j.chemosphere.2017.12.076

Li, Z., Liu, H., Kong, L. et al. (2009). Preparatory research of microbial fuel cells capable of using human deces in the space base. *Habitation*, **12**(1), 87–96. https://doi.org/10.3727/154296610X12686999887409

Ling, W., Xing, Y., Hong, C. et al. (2022). Methods, mechanisms, models and tail gas emissions of convective drying in sludge: a review. *Science of the Total Environment*, **845**, 157376. https://doi.org/10.1016/j.scitotenv.2022.157376

Liu, H., Lv, Y., Xu, S. et al. (2020). Configuration and rapid start-up of a novel combined microbial electrolytic process treating fecal sewage. *Science of the Total Environment*, **705**, 135986. https://doi.org/10.1016/j.scitotenv.2019.135986

Mang, H. and Li, Z. (2009). Biogas sanitation for blackwater, brown water, or for excreta treatment and reuse in developing countries. *Technology Review* (ed. Dr. Elisabeth von Münch), GTZ. Program "Sustainable Sanitation–Ecosan" of Deutsche Gesellschaft für Technische Zusammenarbeit (GTZ) GmbH.

Massoud, M.A., Tarhini, A., and Nasr, J.A. (2009). Decentralized approaches to wastewater treatment and management: applicability in developing countries. *Journal of Environmental Management*, **90**(1), 652–659. https://doi.org/10.1016/j.jenvman.2008.07.001

Maurer, M., Rothenberger, D., and Larsen, T.A. (2005). Decentralised wastewater treatment technologies from a national perspective: at what cost are they competitive? *Water Science and Technology: Water Supply*, **5**(6), 145–154. https://doi.org/10.2166/ws.2005.0059

McConville, J.R., Kvarnström, E., Maiteki, J.M. et al. (2019). Infrastructure investments and operating costs for fecal sludge and sewage treatment systems in Kampala, Uganda. *Urban Water Journal*, **16**(8), 584–593. https://doi.org/10.1080/1573062X.2019.1700290

Mihelcic, J.R., Fry, L.M., and Shaw, R. (2011). Global potential of phosphorus recovery from human urine and feces. *Chemosphere*, **84**(6), 832–839. https://doi.org/10.1016/j.chemosphere.2011.02.046

Moharramzadeh, S., Ong, S.K., Alleman, J. et al. (2022). Stabilization and concentration of nitrogen in synthetic urine with peracetic acid and progressive freeze concentration. *Journal of Environmental Chemical Engineering*, **10**(3), 107768. https://doi.org/10.1016/j.jece.2022.107768

Naik, K.S. and Stenstrom, M.K. (2016). A Feasibility analysis methodology for decentralized wastewater systems–energy-efficiency and cost. *Water Environment Research*, **88**(3), 201–209. https://doi.org/10.2175/106143016X14504669767337

Oarga-Mulec, A., Hanssen, J.F., Jenssen, P.D. et al. (2019). A comparison of various bulking materials as a supporting matrix in composting blackwater solids from vacuum toilets. *Journal of Environmental Management*, **243**, 78–87. https://doi.org/10.1016/j.jenvman.2019.05.005

Pahore, M.M., Ushijima, K., Ito, R. et al. (2012). Fate of nitrogen during volume reduction of human urine using an on-site volume reduction system. *Environmental Technology*, **33**(2), 229–235. https://doi.org/10.1080/09593330.2011.560192

Pathy, A., Ray, J., and Paramasivan, B. (2021). Challenges and opportunities of nutrient recovery from human urine using biochar for fertilizer applications. *Journal of Cleaner Production*, **304**, 127019. https://doi.org/10.1016/j.jclepro.2021.127019

Patwa, A., Parde, D., Dohare, D. et al. (2020). Solid waste characterization and treatment technologies in rural areas: an Indian and international review. *Environmental Technology & Innovation*, **20**, 101066. https://doi.org/10.1016/j.eti.2020.101066

Perez-Mercado, L.F., Simha, P., Moreira, A.P. et al. (2024). Circular fertilisers combining dehydrated human urine and organic wastes can fulfil the macronutrient demand of 15 major crops. *Science of the Total Environment*, **951**, 175655. https://doi.org/10.1016/j.scitotenv.2024.175655

Pocock, J., Septien, S., Makununika, B.S. et al. (2022). Convective drying kinetics of faecal sludge from VIP latrines. *Heliyon*, **8**(4), e09221. https://doi.org/10.1016/j.heliyon.2022.e09221

Rath, M., Schellenberg, T., Rajan, P. et al. (2020). *Decentralized Wastewater and Fecal Sludge Management: Case Studies from India*. Asian Development Bank Institute: Tokyo, Japan. https://www.adb.org/sites/default/files/publication/634586/adbi-cs2020-2.pdf

Rose, C., Parker, A., Jefferson, B. et al. (2015). The characterization of feces and urine: a review of the literature to inform advanced treatment technology. *Critical Reviews in Environmental Science and Technology*, **45**(17), 1827. https://doi.org/10.1080/10643389.2014.1000761

Saetta, D., Zheng, C., Leyva, C. et al. (2020). Impact of acetic acid addition on nitrogen speciation and bacterial communities during urine collection and storage. *Science of the Total Environment*, **745**, 141010. https://doi.org/10.1016/j.scitotenv.2020.141010

Santoro, C., Arbizzani, C., Erable, B. et al. (2017). Microbial fuel cells: from fundamentals to applications. a review. *Journal of Power Sources*, **356**, 225–244. https://doi.org/10.1016/j.jpowsour.2017.03.109

Seodigeng, R., Kabuba, J., and Rutto, H. (2022). Modelling the drying characteristics of human faeces using thin-layer drying models and calculation of mass transfer properties at ambient conditions. *Environmental Challenges*, **9**, 100648. https://doi.org/10.1016/j.envc.2022.100648

Verma, G., Sharma, S., and Vishard, N. (2024). A comparative study on environmental and economic benefits of centralized versus decentralized solid waste management systems. *International Journal for Research in Applied Science and Engineering Technology (IJRASET)*, Vol. 12. IJRASET. https://doi.org/10.22214/ijraset.2024.63416

Vinnerås, B. (2002). Possibilities for sustainable nutrient recycling by faecal separation combined with urine (No. 353). ISSN 1401-6249 ISBN 91-576-6167-7. https://www.susana.org/knowledge-hub/resources?id=1685#

Vinnerås, B., Nordin, A., Niwagaba, C. et al. (2008). Inactivation of bacteria and viruses in human urine depending on temperature and dilution rate. *Water Research*, **42**(15), 4067–4074. https://doi.org/10.1016/j.watres.2008.06.014

Walter, X.A., Stinchcombe, A., Greenman, J. et al. (2017). Urine transduction to usable energy: a modular MFC approach for smartphone and remote system charging. *Applied Energy*, **192**, 575–581. https://doi.org/10.1016/j.apenergy.2016.06.006

Wang, X., Chen, J., Li, Z. et al. (2023). Nutrient recovery technologies for management of blackwater: a review. *Frontiers in Environmental Science*, **10**, 1080536. https://doi.org/10.3389/fenvs.2022.1080536

Wang, Y.K., Geng, Y.K., Pan, X.R. et al. (2017). In situ utilization of generated electricity for nutrient recovery in urine treatment using a selective electrodialysis membrane bioreactor. *Chemical Engineering Science*, **171**, 451–458. https://doi.org/10.1016/j.ces.2017.06.002

Wang, S. (2014). Values of decentralized systems that avoid investments in idle capacity within the wastewater sector: a theoretical justification. *Journal of Environmental Management*, **136**, 68–75. https://doi.org/10.1016/j.jenvman.2014.01.038

Wilderer, P.A. and Schreff, D. (2000). Decentralized and centralized wastewater management: a challenge for technology developers. *Water Science and Technology*, **41**(1), 1–8. https://doi.org/10.2166/wst.2000.0001

Wilsenach, J.A., Schuurbiers, C.A.H., and Van Loosdrecht, M.C.M. (2007). Phosphate and potassium recovery from source separated urine through struvite precipitation. *Water Research*, **41**(2), 458–466. https://doi.org/10.1016/j.watres.2006.10.014

Winfield, J., Chambers, L.D., Rossiter, J. et al. (2015). Urine-activated origami microbial fuel cells to signal proof of life. *Journal of Materials Chemistry A*, **3**(13), 7058–7065. https://doi.org/10.1039/C5TA00687B

Woods, G.J., Kang, D., Quintanar, D.R. et al. (2013). Centralized versus decentralized wastewater reclamation in the Houghton area of Tucson, Arizona. *Journal of Water Resources Planning and Management*, **139**(3), 313–324. https://doi.org/10.1061/(ASCE)WR.1943-5452.0000249

Wu, H., Hernández, E.M., and Vaneeckhaute, C. (2024). Techno-economic analysis of nutrient recovery from urine: centralized treatment of hydrolyzed urine vs. decentralized treatment of fresh urine. *Science of the Total Environment*, **956**, 177295. https://doi.org/10.1016/j.scitotenv.2024.177295

Xue, X., Cashman, S., Gaglione, A. et al. (2019). Holistic analysis of urban water systems in the greater Cincinnati region: (1) life cycle assessment and cost implications. *Water Research X*, **2**, 100015. https://doi.org/10.1016/j.wroa.2018.100015

Yazdi, H., Alzate-Gaviria, L., and Ren, Z.J. (2015). Pluggable microbial fuel cell stacks for septic wastewater treatment and electricity production. *Bioresource Technology*, **180**, 258–263. https://doi.org/10.1016/j.biortech.2014.12.100

Zamalloa, C., Arends, J.B.A., Boon, N. et al. (2013). Performance of a lab-scale bio-electrochemical assisted septic tank for the anaerobic treatment of black water. *New Biotechnology*, **30**(5), 573–580. https://doi.org/10.1016/j.nbt.2013.01.009

Zhang, B., Tian, S., and Wu, D. (2022). Phosphorus harvesting from fresh human urine: a strategy of precisely recovering high-purity calcium phosphate and insights into the precipitation conversion mechanism. *Water Research*, **227**, 119325. https://doi.org/10.1016/j.watres.2022.119325

Zhang, D.Q. and Tan, S.K. (2010). Decentralized wastewater management and its application in an Urban area of Beijing, China. In: *2010 4th International Conference on Bioinformatics and Biomedical Engineering*, 1–4. IEEE. https://doi.org/10.1109/ICBBE.2010.5517527

Zhang, L., Liu, J., Yang, L. et al. (2023). Interfacial solar evaporation toward efficient recovery of clean water and concentration of nutrients from urine with polypyrrole-based photothermal conversion films. *Resources, Conservation and Recycling*, **188**, 106645. https://doi.org/10.1016/j.resconrec.2022.106645

Zhou, Y., Xiao, R., Klammsteiner, T. et al. (2022). Recent trends and advances in composting and vermicomposting technologies: a review. *Bioresource Technology*, **360**, 127591. https://doi.org/10.1016/j.biortech.2022.127591

19

Technologies for Treatment of Wastewater: Scale-up Plant

*Nabila Shehata**

Environmental Science and Industrial Development Department (ESID), Faculty of Postgraduate Studies for Advanced Sciences (PSAS), Beni-Suef University (BSU), Beni-Suef, Egypt

*Corresponding author: nabila.shehata@psas.bsu.edu.eg

19.1 Introduction

Improper wastewater management has emerged as a critical environmental concern that requires rapid attention from the scientific community. The growing occurrence of emerging contaminants (ECs), which are new synthetic substances discovered in trace amounts across a variety of environmental matrices, is a source of great worry. Recent publications highlight the existence of ECs in wastewater as a result of uncontrolled discharges of pesticides, medicines, endocrine disruptors, and other contaminants. The extensive distribution and effect of ECs emphasize the need for effective detection and eradication. Despite advances in Wastewater Treatment Technologies (WWTTs), current approaches have effectiveness limits. WWTTs have been extensively studied in laboratories for the elimination of various contaminants. However, the commercial application of the treatment procedures has only been assessed in a small number of papers. Adsorption and membrane technologies, while efficient against many ECs that employ nanoscale materials to increase contamination removal, are hampered by physicochemical properties (Ealias et al., 2024). Despite their efficiency, coagulation/flocculation methods have challenges in decolorization and sludge formation. Chemical processes such as electrochemical and photo-electrochemical oxidation, Fenton's oxidation, and ozonation reveal advanced oxidation mechanisms capable of reducing contaminants.

Wastewater to Resource Recovery: Applying the Circular Economy Toward Sustainable Development, First Edition. Edited by Arvind Kumar Mungray et al. Published 2026 by John Wiley & Sons, Inc.

Nonetheless, issues such as energy costs and by-product development necessitate constant development. Biological treatments have great removal efficiency, but they need longer retention durations and sludge disposal issues (Ealias et al., 2024). Biofilm applications have the potential to treat sewage water by highlighting microorganisms' adaptability and enzymatic activity. This emphasizes the critical need for further research and technology breakthroughs to solve the worldwide issue of water shortage and water pollution. It underlines the necessity of including sustainable, cost-effective, and environmentally friendly wastewater treatment methods in global water management strategies, in order to ensure a resilient future for both conventional pollutants and CEs (Shekho and Hassan, 2024).

The studies show that a lot of treatment routes have promising performances; nevertheless, more full- and pilot-scale research is required to verify treatment efficacy for pollutants of rising concern. To inform prospects for full-scale adoption, more research is needed to develop inexpensive yet efficient materials/energy, address the negative environmental impacts, lower operational expenses, and overcome the pragmatic difficulties in designing large-scale reactors for use in pilot-scale systems. Although some studies are inapplicable due to the difficulty in properties of materials, separation of catalysts, cost of regeneration, harsh conditions, hydraulic loading, composition of real wastewater, etc., well-designed systems will boost the overall performance of reactors and improve the hydrodynamics inside the reactors. Nevertheless, industrial issues, including glue, seal, spacer, and feed distributor compatibility, also affect the design and construction of reactors/units themselves. Improvements in the efficiency of the materials/catalysts, such as chemical resistance, durability, selectivity, temperature/pressure resistance, and reduced materials cost, have been made in an effort to increase the treatment performance. Fouling is thought to be the main barrier to membrane-based applications among those restrictions, causing a sharp drop in flux. Fouling also affects capital investment since it necessitates the addition of pretreatment equipment, as well as the energy, materials, and chemicals needed to combat the fouling (Wenten et al., 2016). To overcome the limitations of each of these treatment technologies, the research has been extended to hybrid systems. Hybrid/integrated technologies have recently evolved to overcome each technology's restrictions while maintaining great flexibility and efficiency. Such integrated technologies primarily comprise four categories: (i) biological–chemical, (ii) chemical–physical, (iii) biological–physical, and (iv) physical–physical technological connections (Mumtaz et al., 2024). While such paired technologies have great adaptability, efficiency, and economic promise for treating wastewaters, they also face several future hurdles including tuning the operational settings to optimize the synergistic effects of the integrated technologies while minimizing total treatment costs, recycling huge amounts of adsorbents/catalysts, and preventing secondary contamination from used chemicals, adsorbents, and biomaterials, utilizing process optimization and integrating techniques for treating wastewaters from various sources and development of cost-effective nano, chemical, and biological materials (Mumtaz et al., 2024).

Aside from water, nutrient, and material recoveries, recovery efficiency and energy recovery can lower the operational cost of wastewater treatment plants (WWTPs) significantly. Also, the economic value of recovered materials has a considerable impact on determining the viability of deploying innovative and revolutionary technologies for commercial upscaling, as industrial precious metals are more sensitive to cost. Biological approaches appear to have numerous applications across both types of wastewater, with the possibility of bridging the gap between municipal and industrial WWTPs. Scaling up reconcentrating biomasses and biological recovery methods for high-energy/nutrient recovery continues to be challenging. Then there's the direct recovery of energy, water, and valuable materials from wastewater. Throughout the research, this wastewater has been heavily focused on volume reduction, biological, and membrane. Sewage showed great promise for energy recovery, including biogas and biofuel, in addition to other renewable energetic materials. Energy recovery from urine is performed by microalgae's synthesis of biomass, glucose, and lipids, which require less energy; nevertheless, reconcentrating microalgae as a resource for renewable energy generation is difficult. Biological approaches in highly nutritious wastewater streams are more compatible and sustainable than industrial uses, with phytoremediation being noted as successful in nutrient extraction, carbon sequestration, and toxin removal (Soo et al., 2024).

Hence, the main aims of this chapter are to explore the current technologies in wastewater treatment, focusing on their underlying mechanics, performance, limits, and possibilities, and to discuss the technical requirements for pilot-scale wastewater treatment, and the main parameters affecting the scaling-up of the treatment technology. Finally, a case study from lab to scale-up system is provided. Moreover, this article highlights the importance of the contribution of the circular economy in WWTPs on a large scale.

19.2 Technologies Used for Lab-scale Treatment of Wastewater

Specific technologies have been employed successfully at lab scale such as biological technologies in addition to advanced technologies like advanced oxidation, membrane, and adsorption. However, their application to current continuous-flow systems remains difficult. The following sections list these technologies.

19.2.1 Biological Wastewater Treatment Processes

These processes are among the most complicated microbial systems ever designed for a specific use. Several kinds of heterotrophs and autotrophs may now coexist to perform various activities, such as nitrification, denitrification, and removal of biological phosphorus and organic carbon. They can be of suspended growth and attached growth processes, with the development of bacteria/microorganisms/biofilm in both aerobic and anaerobic environments.

The main processes investigated in the lab are as follows:

1. Rotating biological contactor (RBCs)
   The bacteria develop a film on an inert support medium. The sequential disc support medium is immersed and spins gently along a horizontal axis in a tank, allowing air and wastewater to flow (Chan et al., 2009). Overall, RBCs are a popular choice for wastewater treatment because of their simple design, resilience, low energy needs, and high efficiency. However, their limited application in specific scenarios, high initial cost, limited capacity to retain sludge, and risk of clogging are all key aspects to be considered when selecting a specific technology for a wastewater management system. RBCs are commonly employed in wastewater management because of their low maintenance and operating costs, high efficiency, and ease of use. The optimization of the operational conditions, performance as well as design of RBCs has been investigated, and new materials including carbon nanotubes and graphene have been investigated for potential application in RBCs. Also, this process can be used with other processes, including membrane filtration, electrochemical processes, and enhanced oxidation to improve the overall performance of the treatment process. For example, it can be paired with membrane biological reactors (MBRs) to provide hybrid systems that produce water with high quality and contain lower solids, where it can be used as a primary stage to remove big particles, and the MBR performs the ultimate contaminant removal and filtering. Moreover, these units can be integrated with anaerobic digestion to manage organic materials (soluble) and minimize the AS load, resulting in enhanced biogas output and higher treatment efficiency (Waqas et al., 2023b).
2. Sequencing batch reactors (SBRs)
   The SBR technique is essentially a fill-and-draw biological wastewater management method that functions similarly to an AS process (Fadali et al., 2014). According to the scope of operation, the SBR systems, along with their modifications and hybrids, may have one or more tanks, each with five fundamental operating cycles: fill, operation, settle, discharge, and idle. They are highly efficient in the elimination of chemical oxygen demand (COD), phosphorus, and nitrogen, whereas new anaerobic-aerobic-anoxic SBRs provide effective nutrient elimination techniques. Hybrid systems, notably

photocatalytic SBRs, can remove persistent pollutants, demonstrating the need for improved oxidation processes. However, research gaps remain, such as the need for further investigations into various SBR designs, as well as complete assessments of long-term performance, economic sustainability, and environmental variability (Askari et al., 2024). For example, two types of SBRs—up-flow and airlift reactors—were used to test the probability of quick granule development, the effectiveness of NH_3 retention, organic carbon removal, and the composition of microbial communities. The operation supplied with industrial wastewater indicated that the airlift reactor caused faster granule formation than the up-flow reactor, due to its ability to maintain NH_3 retention stability up to around 85%, in addition to the high shear force. During operation, variations in the settling time, the hydraulic retention time (HRT), and the exchange ratio impacted the composition of the microbial community and the granular particle size. The reduced settling time and HRT increased granulation while inhibiting the action of nitrifiers. The efficacy of granulation for NH_3 conversion/retention promotes the transition from NH_3 removal to NH_3 recovery from industrial effluent (Duan et al., 2024).

3. Up-flow anaerobic sludge bed reactor (UASBR)

   It involves the up-flow of wastewater via a thick sludge bed with strong microbiological activity. The UASBR requires thick granules (1–4 mm) produced by anaerobic bacteria to function properly. Protein, cellulose, and fat suspended in wastewater can negatively impair UASB reactor efficiency, prohibiting high organic loading rates (OLRs) and deteriorating microbial activity (Sikosana et al., 2019). UASBRs are high-rate systems capable of performing anaerobic reactions at a lower HRT than standard digesters. In addition to conventional anaerobic treatment of biodegradable and soluble substrates, recent research is now focusing on the removal of slow degradable matrices and refractory compounds, to better explore reactor hydrodynamics and microbial community composition, as well as the use of advanced modeling tools. There are promising options such as using a UASBR to remove nitrogen as well as producing volatile fatty acids and bio-hydrogen. The digestion of different biomass present in the same area is being widely researched in order to maximize biogas generation and ensure smooth ongoing processes in a circular economy context. There is increasing interest in decentralized treatment, which may deliver energy and heat to local customers while also integrating with other green energies. Efficient implementation of pretreatment boosts biogas output, but further posttreatment is necessary to fulfill effluent regulations, both toxicologically and otherwise. Increasing applications of UASB technology at a large scale are favorable for achieving the sustainability goals and circular economy, including efficient biogas exploitation, lowering greenhouse gas emissions, and achieving renewable energy targets (Mainardis et al., 2020).

4. Fluidized-bed bioreactor
Fluidization is used to improve the performance of treatment in biochemical- and chemical-based industrial processes, although it is not commonly employed in wastewater treatment at full-scale. The circulating fluidized-bed bioreactor (CFBBR) is invented by the University of Western Ontario to treat wastewater, where the particles act as carriers for useful microorganisms with a special focus on bacteria. Fluidization's outstanding mixing and mass transport qualities make it ideal for treating both industrial and municipal wastewater. The pilot- and lab-scale systems studies reported that the CFBBR can eliminate 80% of nitrogen and >90% of the influent organic matter. Because of its promising performance, CFBBRs were utilized to manage heavily organic wastewater, which is considered a challenge for traditional techniques due to long HRT; this technology may also be utilized to minimize footprint and system size. Furthermore, it outperforms current systems in terms of managing dynamic loadings (changing influent concentrations and volume). However, this technology represented promising results in wastewater treatment, capable of treating higher quantities of wastewater with a shorter contact period and a lower reactor capacity. Furthermore, its small form allows for remote and localized wastewater treatment units (Nelson et al., 2017).
5. Fixed film bioreactor
The anaerobic digestion process in a two-stage (methanogenesis and acidogenesis) anaerobic fixed-film bioreactor may effectively remove personal care products and pharmaceuticals from water/wastewater. The methanogenic/acidogenic bioreactor removed over 95% of acetaminophen, naproxen, methylparaben, sulfamethoxazole, and propylparaben, and outperformed the methanogenic bioreactor (single stage) in the elimination of diclofenac, ibuprofen, ciprofloxacin, metoprolol, and carbamazepine (Carneiro et al., 2023). Moreover, an up-flow anaerobic sludge-fixed-film reactor (UFASFF) is a granular sludge reactor designed to convert organic matter to biogas quickly using an aggregated microbial community. The biggest issue with traditional UASB reactors is the lengthy starting phase. Hence, the UFASFF reactor with tubular flow behavior was designed to reduce the start-up time at low HRT. For example, the reactor was operated at 36 °C for 36 and 48 hours (Najafpour et al., 2008).
6. Continuous stirred tank reactor (CSTR)
Biohydrogen can be generated while wastewater is treated in a CSTR with mixed wastewater as a substrate in the presence of mixed anaerobic sludge. For example, a CSTR inoculated with preheat-treated anaerobic digested sludge was operated with an HRT ranging from 8 to 24 hours and an OLR of 2.470–3.330 tons of COD/d/m^3. After 20 days of continuous operation at

temperatures ranging from 35 to 70 °C, the CSTR bioreactor achieved steady fermentation. On the sixtieth day of operation, the OLR was raised to a maximum of 29.87 kg COD/d/m^3, enhancing the fermentation-dominated microflora. The HRT at temperatures of 35, 55, and 70 °C resulted in the highest H_2 production (1577, 2608, and 4161 mL/d), COD removal percentage (50%, 67%, and 83%), VFA/alkalinity ratio (1.0, 0.4, and 0.3), and VSS (5.64, 6.28, and 7.5 g/L). The CSTR reactor demonstrated improved microbial activity and stability, leading to higher substrate utilization rates and hydrogen generation capabilities (Anantharaj et al., 2020).

7. Aerobic membrane bioreactor (MBR)
   It integrates membrane filtration and biodegradation, where liquid–solid separation is achieved through sieving. It is one of the most successful commercial uses of membrane technology where the membrane works as a barrier, allowing permeate to flow through but rejecting solid particles. The ultimate product is a high-quality permeate with extremely low suspended solids. Aerobic MBRs have new uses, such as removing ECs, thickening sludge, reduction, and dewatering. However, future advancements in aerobic MBRs include increased energy efficiency, hybrid aerobic MBR processes for nutrient recovery, high BOD/COD removal, and the development of novel membrane materials (Zhang et al., 2020).

19.2.2 Adsorption

Among all available wastewater treatment systems, adsorption is the simplest, sustainable, cost-effective, and environmentally benign. However, the integrated process requires more optimization, research and development, and numerous practical applications. Many adsorbents and processes for removing pollutants from wastewater have been reported, including biomass materials (Shehata et al., 2021), industrial waste and byproducts (Amer et al., 2024), and agricultural and natural materials. The selection of an adsorbent for removing pollutants is primarily determined by the concentration and type of the pollutant in the wastewater, as well as the adsorption tendency of the sorbent and efficacy of the pollutant. Moreover, adsorbents should be freely accessible, cost-effective, non-toxic, and quickly regenerated. Despite improvements, there is still a need to produce such adsorbents that can handle a wide range of contaminants, especially on a large scale. Wastewater treatment research should be focused on assessing the potential of waste to synthesize cost-effective sorbents and use them on a large scale in practical applications (Rashid et al., 2021). Activated carbon is the most often used adsorbent for eliminating numerous contaminants from water/wastewater owing to the applicability of surface functionalization and its microporous structure. However, separating activated carbon from water after treatment has proven challenging, and its expensive cost has prevented widespread use. Recently, the appearance of many innovative materials has demonstrated their competitiveness in pollutant removal. These prospective new materials have various remarkable properties, including a huge surface area, chemical inertness, and strong mechanical strength. However, there are several constraints to the cost efficacy and efficiency of innovative materials, including MXenes (Tyagi et al., 2024), metal organic framework (MOF) (Essam et al., 2024), layered double hydroxide (LDH) (Elkartehi et al., 2021), and graphene (Singh et al., 2023). Lowering the costs and environmental implications from wastes will be an important component to investigate for the widespread use of innovative adsorbents in wastewater technology. Despite most of the recent adsorbents having high adsorption capability in theoretical simulations, the number of recently produced materials, such as MXenes, remains restricted. As a result, large-scale MXene synthesis has emerged as an important research topic. The removal efficiency and the tunable properties of adsorbents used in water remediation increased after functionalization or modification. Subsequently, there are still numerous modified or functionalized adsorbents that are worth studying for future research. Biologically activated carbons are widely utilized in advanced wastewater remediation in nations like Japan, Holland, Switzerland, and America as well as industrial wastewater remediation (Jin et al., 2013). Separation nanotechnology and adsorption, which include the utilization of nano-sized metal oxides and graphene, are now the most commercially viable. There have been few investigations that conducted adsorption tests under actual circumstances, making metal oxides less useful as adsorbents. Furthermore, these Nano materials have demerits such as water leakage, difficult separation, easy aggregate formation, and the potential to harm the environment and human health, all of which prevent them from being effectively employed in full-scale wastewater treatment systems. The recent tendency is to produce adsorbents from waste, including hydrochars and biochars. Hydrochar/biochar adsorption is now being used in lab- and pilot-scale wastewater treatment due to its excellent adsorption impact on heavy metals and dyes removal. However, the quantitative characterization of biochar/hydrochar, disposal of biochar after adsorption to prevent environmental concerns, and the adsorption mechanism still need more investigations prior to upscaling (Chai et al., 2021). Moreover, different investigations have reported on the performance of clay minerals in adsorption experiments, and several effective elements like pH, temperature, adsorbent dose, duration period, and adsorbate concentration impact the clay mineral's adsorption capabilities. The adsorption characteristics of clay minerals revealed that montmorillonite and bentonite from the smectite were the most effective adsorbents among other clays. Smectite clay minerals are naturally nanoscale in size; thus, they surely offer several environmental protection benefits. Clay minerals are widely used as effective adsorbents; however, modification routes like thermal treatment, acid activation, organo clay, development of porous clay heterostructure, and pillared clays can enhance the adsorption capacity, selectivity, and recyclability of the clays (Barakan and Aghazadeh, 2021).

19.2.3 Advanced Oxidation Process

The advanced oxidation process (AOP) can be done via different mechanisms such as electrochemical, photolysis, ozone, Fenton, and sonolysis where hydroxyl radicals (OH) or sulfate radicals (SO_4^-), as powerful oxidizing agents, are produced in sufficient amounts during the AOP treatment of wastewater to either eliminate certain inorganic pollutants, traceable organic contaminants, or refractory organic matter, or to increase the biodegradability of the wastewater as a pretreatment followed by a biological treatment. The chemical and physical characteristics of the target pollutants, the operating circumstances, and the kind of AOP used all have a significant impact on treatment efficiency. It is important to remember that different processes might also take place during the AOP treatment and help to reduce the target pollutants, in addition to hydroxyl radical or sulfate radical-based oxidation (Deng and Zhao, 2015). Important factors in these processes include the pH and existence of particles and other ions in the wastewater solution, the quantity of catalysts and oxidation agents, the irradiation time, the light intensity, and the initial concentration of the target compounds. However, there are still unanswered questions regarding the generation of unknown by-products or intermediates that, in some cases, may be more dangerous than the parent molecules, and the comparatively high operating costs of these processes as a result of the use of pricey chemicals and increased energy consumption. Moreover, each of these methods is susceptible to non-target substances scavenging hydroxyl radicals. Doping TiO_2 with non-metallic elements including carbon, nitrogen, and sulfur can extend its absorption wavelength range and boost photocatalytic activity. Treatment results may also be enhanced by the sequential or simultaneous administration of different AOPs. When combined with TiO_2 photocatalysis, ultrasonic (US) irradiation may increase the quantity of hydroxyl radicals produced and hasten the mass transfer of reagents onto the TiO_2 surface (Saravanan et al., 2022). A few of these include keeping an eye on the pH, chemicals, and organic matter in water sources; eliminating heavy metals that dissolve in them; maintaining dosage and operating errors; lacking operational standards and bandwidth; and the high cost of setting up machinery. Since removing hydrogen peroxide from treated water is a major process and requires a large number of •OH molecules, improved solutions for these shortcomings should be developed together with development measures. It is crucial to use renewable energy sources because photochemical AOPs powered by sunlight provide a definite benefit for both simultaneous energy generation and water treatment. The area of AOPs may benefit greatly from breakthroughs in materials research, which lead to the discovery of novel materials with intriguing properties. Recent successful testing of graphene materials as activators in sulfate-radical AOPs, for instance, emphasizes that AOPs are an efficient route for wastewater treatment that various sectors must collaborate to face the challenges (Saravanan et al., 2022).

Process scaling is a key yet necessary phase in the development of a practical electrochemical system for the oxidation of organic compounds. The major elements involved in scaling up include plant size, efficiency, and economic feasibility. Organic pollutants can be successfully eliminated using anodic oxidation or electro-Fenton-based methods. However, there are still other obstacles to overcome before the industrial implementation. To begin, the cost of electrode materials should be decreased, allowing less important raw materials to be used more frequently. A second major feature that research must explore and enhance is the layout design of the electrochemical reactor, with a focus on cell architecture that promotes efficient mass transfer and increases the kinetic rate. These initiatives will increase the volume of treated wastewater, hence enhancing the competitiveness of such technology (Clematis et al., 2023).

19.2.4 Microbial Fuel Cells

Over the past 25 years, research interest in microbial fuel cells (MFCs) has increased due to growing worldwide demands for clean energy and water. MFCs have gone from being a laboratory breakthrough to prototype units and some commercial systems. Nevertheless, technology currently misses the appropriate design foundation that is present in current treatment procedures. It was reported that there are significant correlations between the rate and load of wastewater treatment, even in the face of significant changes in operational conditions, wastewater origin, environments/climates, performance, and architecture in terms of production. A framework for constructing MFCs that provide HRT and reactor size to remediate any type of wastewater at different scales was established using these relationships. This meta-analysis compares a wide range of research, standardizes the data, and makes generalizations regarding the design and performance of MFCs. Better, more consistent reporting might reinforce this framework, increase data accessibility, and eventually support additional studies and the commercialization of MFCs as an efficient WWTT (Bird et al., 2022).

19.2.5 Membrane-based Processes

The current research has been working on the development and application of efficient membrane materials, such as polyvinylidenefluoride-hexafluoropropylene and -tetrafluoroethylene. For the past decade, most efforts have focused on using new and creative ways to address the issue of membrane fouling in MBRs. The most recent investigations have focused on the circular economy, utilization of biological principles, nanomaterials and nanocomposites, and electrically based techniques for membrane fouling. These innovative membranes' fouling control approaches have shown remarkable efficacy. Carbon nanotubes (CNTs) and a bucky-paper membrane are also being investigated (Bera et al., 2022). Around 95% of the investigations were conducted on a laboratory scale. It was discovered that there are only eight large-scale and fourteen pilot-scale studies. The research on pilot-scale units treating industrial wastewater has shown that membrane processes were effective in effluent treatment, allowing water recycling within the sector. According to the discharge criteria, properties of the wastewater to be treated, membrane technologies, particularly hybrid systems, have been found acceptable. Further studies are needed on both water reuse possibilities and the improvement of the system's performance. Furthermore, cleaning procedures should be devised for fouling management, which is critical for operational conditions. Feasibility studies should be conducted to determine the most efficient membrane procedure in terms of cost considerations and treatment performance (Keskin et al., 2021).

19.2.6 Flocculation and Coagulation

Granular FeO(OH) was used on a pilot scale to remove vanadium from mining wastewater (Figure 19.1). Two pilot (X and Y) filters were installed in distinct streams. The vanadium contents in the influents were 6.46–99.1 mg/L and 0.443–2.33 mg/L for X and Y filters, respectively. The HRTs of filters X and Y were 51 and 127 days. Water quality tests found that vanadium was collected successfully in both filters. The XPS revealed the presence of oxidized vanadium (5+) in the FeO(OH), as well as an increase in carbon content caused by sorbed organic molecules. Moreover, ferric sulfate was used as a coagulant with an optimum dose of 0.35 g/L at the original pH of the influent. However, the needed coagulant quantity dropped when the influent pH was changed to 4.6–4.8 (Zhang et al., 2021).

Artificial neural network (ANN) optimization was combined with experimental investigations and commercial design studies to remediate wastewater. Three coagulants were found to have ideal conditions through batch coagulation/flocculation experiments: ferric chloride ($FeCl_3 \cdot 6H_2O$), poly aluminum chloride (PAC), and aluminum sulfate ($Al_2(SO_4)_3$). The goal was to remove as much total nitrogen (TN), total suspended solids, and COD as possible from slaughterhouse and meat processing wastewater (SMW). The results show the promising effectiveness of PAC, with eradication rates of 74.6%, 82.0%, and 89.1% for COD, TSS, and TN, respectively, being attained. These ideal results were obtained with the

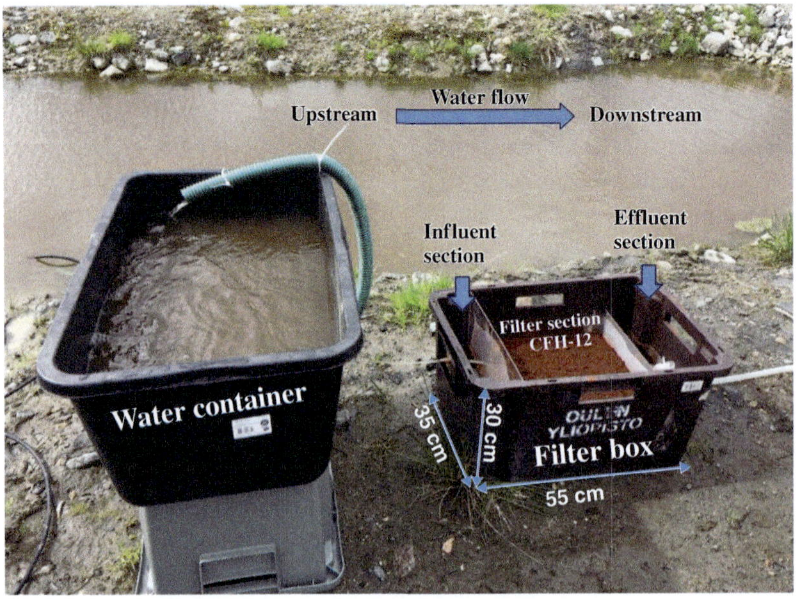

Figure 19.1 Experimental setup of filters X and Y. *Source:* Zhang et al. (2021)/with the permission of Elsevier.

following parameters: pH 7, 0.100 g/L of the dosage, 1 minute of quick mixing at 200 rpm, gradual mixing at 30 rpm for 20 minutes, and a settling period of 0.5 hours. Using PAC coagulant, an ANN model was constructed, trained, and validated in MATLAB to forecast the effectiveness of pollutant removal. It took into account factors including pH, settling time, fast/slow mixing conditions, and dose. Multiple metrics demonstrated the strong performance of the ANN model: total regression (R) for COD, TSS, and TN was 0.91, 0.95, and 0.87, respectively. The results of the linear regression show a high degree of matching between the experimental and projected values, confirming the ANN models' superior performance. The treatment system has concrete tanks for durability and is rated for SMW of 4.166 m^3/h. For effective wastewater treatment, it has a 1500-L clariflocculation unit (30 rpm) and a quick mixing tank (75 L) (Badawi et al., 2024).

19.2.7 Hybrid Systems

Hybrid systems have been investigated to enhance the overall efficiency of the treatment systems. For example, the production of bioelectricity and the reduction of COD from synthetic wastewater were assessed using a novel design of a pilot-scale built wetland MFC by Kumar et al. (2023). The system was made up of two aerobic chambers and nine separate constructed wetland (CW)-MFC units arranged in a rectangular tank reinforced with fiber (working volume: 135.5 L). In addition to the standard sediment-configured MFC, two MFCs integrated with terracotta-based separator electrode assemblies were added to each CW-MFC unit's anodic zone to improve the system's electrical performance. The system's CW-MFC unit eliminated COD from wastewater (around 97.56%) while operating in fed-batch mode. The auxiliary MFCs were able to reach the current densities and power of 0.2296 A/m^2 and 0.0585 W/m^2, respectively. The pilot-scale system removed COD from 4.1 m^3 of wastewater in the continuous up-flow mode, with a removal efficiency of 82.8 $\pm$ 1.9%. The current density and greatest power that the auxiliary MFCs were able to achieve were 0.2833 A/m^2 and 0.0414 W/m^2, respectively (Kumar et al., 2023). The CW was also integrated with adsorption (bark and biochar). When compared to the CW filled with typical material sand, the removal of 11 of the 12 investigated MPs was enhanced when bark-biochar was used as the support matrix. The inclusion of bark-biochar resulted in the most enhanced elimination of irbesartan, carbamazepine, benzotriazole, and hydrochlorothiazide. The Micropollutants (MPs) removal were unchanged when the bark biochar CW bed was aerated. Additionally, during the 10-month operation, the bark-biochar enhanced the elimination of total nitrogen (TN), but no improvement was shown in the removal of total organic carbon or total phosphorus. Bark-biochar, which will function as an MP adsorbing matrix, can be used in place of sand in a batch-operated CW to significantly improve MP removal (Lei et al., 2023). The CWs are also investigated for the potential and effectiveness of the removal of industrial pollutants. To undertake the proposed treatability investigation, four pilot-scale modified free water surface (FWS) built wetland systems were erected inside an industrial park. About 170 mg of COD, 0.08 g of biochemical oxygen demand (BOD), 0.09 g of suspended solids (SS), and 0.032 g of NH_3–N are present in the typical influent. Four distinct wetland plant species were assessed in the plant-selection study: floating plants (*I. aquatica* (*Ipomoea aquatica*) and *P. stratiotes* (*Pistia stratiotes* L.)) and emergent plants (*T. orientalis* (*Typha orientalis* Presl.) and *P. communis* (*Phragmites communis* L.). However, only *P. communis*, an emergent plant, was able to live and proliferate when fed 0.4 m^3 of raw wastewater per day. Therefore, *P. communis* was employed in the next therapy investigation. The treatment research looked at two distinct control parameters: media (small gravels with a diameter of one centimeter and ceramic bioballs) and HRT of 3, 5, and 7 days. The system that used vesicle ceramic bioballs as the medium and an HRT of 5 days had the best and most acceptable pollutant removal efficiency (Chen et al., 2006).

A cutting-edge biotechnology of the 21st century, aerobic granular sludge (AGS) was integrated with activated carbon (Nancharaiah et al., 2023). For more than four years, the granular sequencing batch reactors (gSBRs) were run in a tropical climate (T≈303.15 K) to assess the impact of GAC on granular stability, biological nutrient removal (BNR), and granulation at the pilot scale. Granule development was seen in less than three months. Within six months, gSBRs without GAC particles and those with GAC particles showed mixed liquor suspended solids (MLSS) values of 4000 and 8000 mg/L, respectively. The granules' SVI5 was 22 mL/g, and their average size was 1.2 mm. The primary method of eliminating ammonium in the gSBR without GAC was nitrate production. However, because of the washout of nitrite-oxidizing bacteria in the presence of GAC, ammonium was eliminated by nitrification via nitrite. Also, phosphorus removal was significantly greater. Following three months, the removal efficiency of phosphorus was 75% and 15%, respectively, with and without GAC particles. The bacterial community was moderated, and the population of organisms that accumulate polyphosphate was enhanced by the addition of GAC. This suggests that investigating the prototype (Figure 19.2) for a sufficient time is also an important parameter in the prototype's plants (Nancharaiah et al., 2023).

The optimization and development of a novel hybrid photocatalytic nanofiltration reactor (PCNFR) prototype achieves synergistic merits of the size exclusion

Figure 19.2 The pilot-scale SBRs (2 m^3) utilized for AGS production and the remediation of actual household wastewater in tropical climates are depicted schematically in diagram (a) and in photo (b). In order to treat sewage, pilot-scale granular SBRs were used both with and without granular activated carbon particles. *Source*: Nancharaiah et al. (2023)/with permission of Elsevier.

capacity and photocatalytic degradation activity of micropollutants. These two characteristics are now associated with several membrane monoliths. After determining the titania materials' physicochemical and photoinduced decontamination capabilities in powder form, the resulting titania-modified membrane monoliths' structural and morphological characterization was completed. With promising removal efficiency against pesticides like acetamiprid and thiabendazole, the PCNFR can work at low pressures and high recovery of water. It can also recycle 15 m^3/d of actual agro-wastewater (Theodorakopoulos et al., 2023).

19.3 Technology Requirement for Scaling up WWTPs

The treatment efficiency and the economic perspective should be taken into account in order to reduce issues that arise during scaling up. These include area requirements, energy requirements, system designs, environmental effects, and operation and maintenance. Maintaining the system's performance during the scale-up process requires careful attention to keeping the operational parameters consistent. There are some technical requirements for the pilot-scale wastewater treatment: optimizing the operating conditions and design of the pilot-scale plant.

Due to resource scarcity, existing industrial plants are shifting away from waste treatment, like biomass and wastewater treatment, and toward resource recovery. Wastewater and activated sludge (AS) may be converted into biofuels, organic acids, manure, insecticides, and other high value bioproducts. This will aid in the shift from a linear economy to a circular economy while also contributing to long-term growth. However, extracting materials from wastewater and AS to manufacture value-added goods is relatively expensive when compared to traditional treatment approaches. Furthermore, the majority of antioxidant technologies are still in the laboratory and have not yet reached commercial scale. To encourage resource recovery technology innovation, the different techniques of treating AS and wastewater to develop nutrients, energy, and biofuels should be further explored, such as and not limited to chemical stabilization, biochemistry, and thermochemistry. The limitations of AS and wastewater treatment procedures are predicted based on biochemical properties, environmental, and economic variables. Biofuels produced from the third generation feedstocks, such as wastewater, achieved the circular economy and sustainable development principles. Microalgal biomass is utilized to make bioethanol, biohydrogen, biooils, biogas, biodiesel, charcoal, biofertilizers, biopesticides, and bioplastics. Modern regulations and technologies can help to drive a circular economy that depends on biological materials (Awasthi et al., 2023). There are innovative technologies for resource recovery from AS and wastewater, like biofuel production, environmental impact reduction, and nutrients (NPK) recovery. Wastewater technology lacks the implementation of the 10Rs of sustainability: R1: Refuse, R2: Rethink, R3: Reduce, R4: Reuse, R5: Repair, R6: Refurbish, R7: Remanufacture, R8: Repurpose, R9: Recycle, R10: Recover. Although the recovery systems have greater running and capital expenses than traditional treatment systems, R1, R9, and R10 are applied to some extent in industrial wastewater treatment. To operate smarter WWTPs, harmful industrial processes should be rejected, and greener wastewater technologies should be proposed (R1). Restructuring the WWTPs and optimizing the design of the processes inside the treatment facility will enhance the principle of Rethink (R2). The third principle of sustainability (Reduce) can be obtained by minimizing the consumption of natural resources before and during the wastewater treatment process. Reusing the treated wastewater for different purposes, such as irrigation and industrial processes, will promote R4 (Reuse) and R5 (Repair), respectively. Refurbishing (R6) and Remanufacturing (R7) the current WWTPs with the highest technical specification like internet of things (IoTs), and the integration with other processes have a strong contribution to sustainable development of wastewater treatment. Utilizing the same materials/processes to treat different kinds of wastewaters falls under the umbrella of Repurpose (R8). Processing or regeneration of materials to be used many times in wastewater treatment, along with recycling of wastewater to be used in the same industrial plant, will strengthen R9 (Recycle). Optimizing processes to recover valuable materials as well as energy is in agreement with R10 (Recover). However, in order to make such systems economically feasible, policymakers need to develop laws that assure fair market prices for these items while also encouraging technological advancements. In conclusion, technical advancements in resource recovery from AS and wastewater represent a significant step toward more effective waste management and recycling (Awasthi et al., 2023). Hence, the combination of the current wastewater treatment processes with recycling technologies is important.

Combining IoT with water quality prediction and management of wastewater can transform existing processes and face pressing issues, given the world's desire for sustainable systems of clean water. These processes include modeling water quality parameters, membrane filtration, adsorption, monitoring water quality indices, chlorination, detecting river level, and monitoring or automating effluent water in aquaculture units. However, machine learning (ML) models and artificial intelligence (AI) have been applied in various domains. A promising solution is the development of an IoT-based automated water quality monitoring system that employs machine learning techniques and cloud computing to efficiently handle the current challenges. The IoTs has been used to simulate, optimize, and automate a variety of tasks, such as water/wastewater treatment processes, monitoring and controlling natural systems, and water-related agriculture activities such as hydroponics and aquaponics (Alprol et al., 2024).

19.4 Parameters Affecting the Scale-up of Treatment Technology

There are different parameters affecting the scaling-up of the WWTP. These factors are varied according to the type of technology. However, the main controlling factors are listed below.

19.4.1 Hydraulic Retention Time

HRT is a significant aspect in determining the effectiveness of the treatment process. Hydraulic loading is another key factor in calculating HRT, impacting reactor size. It directly influences performance efficiency by regulating hydraulic loads. To achieve optimal removal efficiency, each system must have an optimized HRT value. It is shown that raw wastewater with low strength typically needs a low HRT of 3–12 hours to achieve acceptable effluent characteristics. Wastewater with high strength or wastewater with unique components may necessitate significant HRTs to achieve the required sewage quality. In conclusion, HRT is entirely dependent on the research objectives, operational parameters, experimental design, influent wastewater, and effluent quality requirements (Waqas et al., 2023a).

19.4.2 Solid Retention Time

Bacteria should be aggregated or flocculated to allow for effective gravity settling in the clarifier. Bio-flocculation is caused in part by the development of extracellular polymers, which bind bacteria together. In activated sludge processes (ASPs), biomass needs a minimum solid retention time (SRT) of 3–4 days to form a sticky slime layer, while too long SRTs result in poorly settleable flocs for a variety of reasons, including filamentous bacterial development. SRT was limited to 4–5, 27, 30, and 35 days. In addition to ensuring adequate floc settleability, the upper bound's primary objective was to minimize sludge mineralization, which affects the sludge calorific value. It was assumed that one kilogram of dry sludge has an average calorific value of 15 MJ (unaffected by organic matter concentration) (Hreiz et al., 2015).

19.4.3 Extracellular Polymeric Substance

The extracellular polymeric substances (EPSs) improve microbial cell adherence to solid surfaces by modifying the surface characteristics of the cells. The EPS are typically made up of proteins and carbs. Other components discovered in EPS include uronic acids, inorganic components, lipids, and nucleic acids in various matrices. EPS synthesis is essential for the attachment of microorganisms to the surfaces of solids. The excessive EPS synthesis may result in viscous thickening, limiting organic diffusion, and delaying full breakdown. High biomass on the carrier medium may lead to increased EPS generation. The intricate link between the C/N ratio and EPSs must be investigated to assess the system efficiency (Waqas et al., 2023a).

19.4.4 Organic Loading Rate

The OLR calculates the quantity of organic materials supplied to the system per unit time. The capacity of microorganisms in the biofilm to break down organic matter has a considerable effect on the wastewater treatment system's function. At very high OLR, the microbes cannot keep up with the nutrients and oxygen demand, causing the buildup of organic waste and a loss in treatment effectiveness; and at too low OLR, microbes cannot digest all the organic materials in the raw wastewater, causing low removal efficiency and ineffective treatment (Waqas et al., 2023b).

19.4.5 Dissolved Oxygen

The presence of O_2 is critical for the aerobic microorganisms' activities involved in nitrification, organic materials oxidation, in addition to any other biological reactions. Aerobic bacteria require sufficient dissolved oxygen (DO) to maintain their metabolic activities. Adequate oxygen levels allow for effective conversion of NH_3 to nitrate during nitrification and oxidation of organic contaminants. High DO concentrations boost metabolic rates and microbial activity, which improves therapy efficacy. Inadequate DO levels inhibit the oxidation of organic materials, causing reduced removal efficiency. Also, low DO levels cause the accumulation of residual organic molecules, inefficient breakdown of organic contaminants, and limit treatment efficacy. Keeping an adequate DO level prevents the accumulation of refractory chemicals and enables efficient organic materials' removal. (Waqas et al., 2023a)

19.4.6 Biofilm and Floc

Biofilm and flocs are crucial components of the wastewater treatment process, both helping to remove pollutants and organics. Flocs and biofilms are unique microbial compositions. The biofilm is a coating of microorganisms that adheres to the media surface in the bioreactor. It ensures a large surface area for microbial growth and may remove a variety of pollutants. This biofilm is often made up of several bacteria, fungi, and protozoa that collaborate to degrade and eliminate nutrients and organics. It is extremely successful in removing pollutants from wastewater since it ensures a safe haven for microbes. Moreover, it protects bacteria against harmful organisms, as well as varying effluent quality. Furthermore, it encourages the development of nitrifying bacteria that are crucial in eliminating nitrogen from wastewater via nitrification (Waqas et al., 2023a).

19.4.7 Temperature

Seasonal fluctuations and temperature can have a considerable influence on microbial activity and biological processes, which affect system stability and treatment efficacy. As the temperature rises, so do metabolic rates and microbial activity, which improve the elimination of organic matter and nutrients. High temperature encourages the development of thermophilic bacteria that break down specific chemicals more effectively. Temperature affects the pace of biological processes in wastewater remediation. The rate of nitrification/denitrification and other biological processes strongly rely on temperature. Warmer temperatures cause less oxygen solubility, resulting in oxygen limits. Inadequate oxygen delivery impairs the efficacy of aerobic processes, including nitrification and organic matter oxidation (Waqas et al., 2023a).

19.4.8 Other Parameters

The availability and location of land, climate, and nearness to urban areas are also important factors in the scaling-up of anaerobic lagoons (Mittal, 2006). For membrane-based processes, membrane configuration is one of the components in the reverse osmosis (RO) desalination system that should be considered while designing the process to achieve both optimal economy and performance. The number of elements per pressure vessel and the array's structure should be taken into account throughout the design process. The RO unit configuration plays in the overall system efficiency. The advantages of having more components per vessel translate into lower investment costs, which in turn translate into the need for fewer vessels to achieve the same membrane area. A single-stage unit requires less power since its pressure loss is just around 1 bar as opposed to 3.4 bars for a two-stage device. Moreover, there are other criteria affecting scaling up such as the reactor design criteria, energy cost, feed water characteristics, product quality requirements (Wenten et al., 2016), operating costs, and maintenance (Rashid et al., 2021).

Phototrophic microorganisms thrive primarily on light, inorganic water, inorganic nutrients, and carbon dioxide under appropriate growth circumstances (temperature and pH). All of these factors are inextricably linked, complicating photobioreactor design. Photobioreactor scalability requires careful consideration of a variety of characteristics, including light-dark frequencies, light intensity, light dispersion, environment (pH, temperature, and nutrients), and hydrodynamics. The presence of light for each cell may be the most essential component in photobioreactor scaling up. Moreover, robust microalgae must be picked to avoid failure during scale-up due to contamination. Additionally, microalgae growth kinetics dependent on light should serve as the foundation for model-based scale-up (Benner et al., 2022).

19.5 Stepwise Usage of Multiple Technologies for Field Implementation

Treatment of wastewater has primarily been undertaken at laboratory scales with capacities under 1 L. Efforts to improve adsorbents/catalysts and operating parameters have resulted in significant advances. This section highlights recent advancements in using various technologies for wastewater treatment, indicating their maturity in recent years. However, the strategy is threefold: (i) the requirement for scale-up via an integrated experimental and modeling method, (ii) combining traditional treatment systems as part of a larger WWTP, and (iii) integration of IoT into WWTTs. For example, the removal of algae electrolytically from water was applied in continuous and batch units to identify operational parameters influencing algal removal effectiveness and to produce design correlations that may be relevant for scaling up and operating at full-scale. The hybrid system combined electroflotation and electroflocculation by employing a cathode of Ti (inactive) alloy to generate gaseous microbubbles required for flotation and an anode of polyvalent aluminum alloy for flocculant development. The batch system tests validated prior research showing increased electrical input power. To handle substantial temperature and pH rises, it may be necessary to balance electrolysis time and electrical power input. Although liquid circulation happens naturally during electrolysis, additional mixing may be useful with low electrical input power. This spontaneous mixing increases with increased electrical input power; therefore, external mixing may be less important in this situation. Increasing mixing reduced efficiency by dispersing microbubbles of entrapped gas inside algae clusters or destroying the flocs. Batch system experiments have revealed that electroflotation alone cannot completely remove algae. The recovery of algae was low (40–50%), because of the lower trapping of gas microbubbles while algae are scattered as individual cells instead of aggregated algae clusters. The continuous electrolytic recovery of algae took into account optimum operation conditions to maximize the efficiency of algal removal while releasing as little surplus aluminum as possible. A charge dosage of 250 coulombs was found to be most effective in removing algae while releasing minimal aluminum. Because the optimal charge dosage may be utilized to connect to the electrolysis duration or operating current, it may be a valuable operating and scale-up factor of a continuous system (Alfafara et al., 2002). Also, membrane-based technologies have become a major player in the water business. Membrane fouling by the disposal of concentrated brine rejects, as well as chemical agents, is a significant drawback. Using electrochemical advanced oxidation processes (EAOPs) to break down fouling compounds before membrane filtration or as a posttreatment for concentrates is a promising approach (Garcia-Rodriguez et al., 2022).

Moreover, anaerobically digested sludge (TOC 0.5235, COD 2.62, and TSS 3.13 g/dl) was managed using an integrated system that included electrochemical peroxidation (ECP) with Fe electrodes, followed by an electro-Fenton (EF) with a carbon brush cathode and a boron-doped diamond anode. Using AOP and EC during ECP for 2 hours at optimum conditions resulted in 85.6% and 89.3% of TSS and COD removal, respectively. This reduced the mass of solid sludge discharged from the process and increased sludge dewaterability. After the solid sludge was separated from the effluent, the ECP-treated wastewater was submitted to further EF treatment (this residue is rich in fertilizing nutrients). The EF effectively mineralized the residual organic matter under ideal circumstances (0.025 A/cm^2 and pH 3), to 0.0163 g/L and 0.0246 g/L of TOC and COD after a duration time of 8 hours. This two-stage electrochemical technique has proven to be an effective strategy for dealing with extremely contaminated and viscous sludge (Olvera-Vargas et al., 2019).

19.6 Case Studies: From Lab to Scaled-up System

Scaling up wastewater treatment systems is a difficult operation, and researchers are still failing to produce satisfactory results when treating real wastewater on a field size. There are numerous case studies focused on scaling up lab-scale systems. For example, the use of ferric coagulation, AS at lab scale, Fenton oxidation at lab and full scales was assessed to treat olive mill effluent. Initially, coagulation at the real effluent pH was done, resulting in a COD reduction up to 38% at lab scale. Also, this technique produced some dissolved iron, which can be employed as a catalyst in the subsequent Fenton's peroxidation phase. At the lab scale, the Fenton technique removed 75% of the COD. Furthermore, this technique improved biodegradability, with an initial BOD5/COD average ratio of 0.03 growing to 0.37 after the operation. This integrated system was then attempted at full scale, and after additional optimization, Fenton oxidation and coagulation were able to remove 62% and 43% of COD, respectively. It was able to optimize an industrial WWTP using the intriguing results obtained at lab scale. The use of Fenton's peroxidation after ferric coagulation resulted in up to 90% efficiency and a final COD of 0.2 gO_2/L. The resultant effluent can be released into municipal sewers or recycled for re-use following in situ biological treatment (Amaral-Silva et al., 2017). Another case study discussed a horizontal subsurface flow constructed wetland (HSFCW) system to manage wastewater from the glass production sector. To reduce the danger of the treatment strategy, a prototype consisting of an HSFCW and a settling tank was evaluated for 120 days. The results of the pilot research were utilized to develop a CW system at large scale, which treated 10 m^3/d. Overall, the design recorded high removal rates of COD (90%), TP (>90%), TN (>90%), BOD5 (90%), and TSS (99%). The tested system's high efficiency allowed for the recycling of the treated wastewater in the glass manufacturing process, lowering the glass industry's freshwater usage (Gholipour et al., 2020).

19.7 Summary

Numerous investigations have been carried out at the lab scale to treat different kinds of wastewaters. This chapter focuses on the factors affecting the scale-up of WWTTs, such as solid retention time, biofilm and floc, extracellular polymeric substances, organic loading rate, temperature, hydraulic retention time, dissolved oxygen, and other parameters. However, the efficiencies of these processes are decreased at the upscaling owing to different reasons, including the complexity of the real wastewater, the initial and operating cost, and the presence of other controlling parameters at a large scale. Hence, scaling-up studies should be continued to optimize the operating conditions, minimize the cost, investigate the integration of two or more technologies in one hybrid system, and implement the 10Rs of sustainability in wastewater treatment.

Review Questions

1. What are the merits of applying 10Rs in wastewater treatment?
2. What are the criteria for selecting two or three wastewater treatment processes to be involved in the hybrid systems?
3. Suggest a hybrid system to treat different types of wastewaters.
4. Is scaling up an easy process?
5. What is the best reactor design for the AS process?
6. How do you manage technical issues in hybrid wastewater treatment systems?
7. How do you keep up with the latest technologies in wastewater treatment at a large scale?
8. Discuss the policies and legislative laws that promote the scaling up of new wastewater treatment systems.

9 Compare decentralized treatment units and centralized (large-scale) wastewater treatment systems.
10 Suggest an innovative sewage treatment process.

Acknowledgments

They said, "Exalted are You; we have no knowledge except what You have taught us. Indeed, it is You who is the Knowing, the Wise."

References

Alfafara, C.G., Nakano, K., Nomura, N. et al. (2002). Operating and scale-up factors for the electrolytic removal of algae from eutrophied lakewater. *Journal of Chemical Technology & Biotechnology: International Research in Process, Environmental & Clean Technology*, **77**, 871–876.

Alprol, A.E., Mansour, A.T., Ibrahim, M.E.E.-D. et al. (2024). Artificial intelligence technologies revolutionizing wastewater treatment: current trends and future prospective. *Water*, **16**, 314.

Amaral-Silva, N., Martins, R.C., Nunes, P. et al. (2017). From a lab test to industrial application: scale-up of Fenton process for real olive mill wastewater treatment. *Journal of Chemical Technology and Biotechnology*, **92**, 1336–1344.

Amer, A.M., El-Dek, S.I., Farghali, A.A. et al. (2024). Management of ibuprofen in wastewater using electrospun nanofibers developed from PET and PS wastes. *Chemosphere*, **359**, 142313. https://doi.org/10.1016/j.chemosphere.2024.142313

Anantharaj, C., Veeraraghavan, A., and Ashok Kumar, N. (2020). Effects of pH and temperature on biological hydrogen production using mixed wastewater by dark fermentation in a continuous stirred tank reactor. *Research Journal of Chemistry and Environment*, **24**, 5.

Askari, S.S., Giri, B.S., Basheer, F. et al. (2024). Enhancing sequencing batch reactors for efficient wastewater treatment across diverse applications: a comprehensive review. *Environmental Research*, **260**, 119656.

Awasthi, M.K., Ganeshan, P., Gohil, N. et al. (2023). Advanced approaches for resource recovery from wastewater and AS: a review. *Bioresource Technology*, **384**, 129250. https://doi.org/10.1016/j.biortech.2023.129250

Badawi, A.K., Hassan, R., Farouk, M. et al. (2024). Optimizing the coagulation/flocculation process for the treatment of slaughterhouse and meat processing wastewater: experimental studies and pilot-scale proposal. *International Journal of Environmental Science and Technology*, **21**, 1–16.

Barakan, S. and Aghazadeh, V. (2021). The advantages of clay mineral modification methods for enhancing adsorption efficiency in wastewater treatment: a review. *Environmental Science and Pollution Research*, **28**, 2572–2599. https://doi.org/10.1007/s11356-020-10985-9

Benner, P., Meier, L., Pfeffer, A. et al. (2022). Lab-scale photobioreactor systems: principles, applications, and scalability. *Bioprocess and Biosystems Engineering*, **45**, 791–813. https://doi.org/10.1007/s00449-022-02711-1

Bera, S.P., Godhaniya, M., and Kothari, C. (2022). Emerging and advanced membrane technology for wastewater treatment: a review. *Journal of Basic Microbiology*, **62**, 245–259.

Bird, H., Heidrich, E.S., Leicester, D.D. et al. (2022). Pilot-scale Microbial Fuel Cells (MFCs): a meta-analysis study to inform full-scale design principles for optimum wastewater treatment. *Journal of Cleaner Production*, **346**, 131227.

Carneiro, R.B., Gomes, G.M., Sabatini, C.A. et al. (2023). Enhancing organic micropollutants removal in wastewater with an innovative two-stage anaerobic fixed-film bioreactor: role of acidogenic and methanogenic steps. *ACS ES&T Engineering*, **4**, 354–364.

Chai, W.S., Cheun, J.Y., Kumar, P.S. et al. (2021). A review on conventional and novel materials towards heavy metal adsorption in wastewater treatment application. *Journal of Cleaner Production*, **296**, 126589. https://doi.org/10.1016/j.jclepro.2021.126589

Chan, Y.J., Chong, M.F., Law, C.L. et al. (2009). A review on anaerobic–aerobic treatment of industrial and municipal wastewater. *Chemical Engineering Journal*, **155**, 1–18.

Chen, T.Y., Kao, C.M., Yeh, T.Y. et al. (2006). Application of a constructed wetland for industrial wastewater treatment: a pilot-scale study. *Chemosphere*, **64**, 497–502.

Clematis, D., Delucchi, M., and Panizza, M. (2023). Electrochemical technologies for wastewater treatment at pilot plant scale. *Current Opinion in Electrochemistry*, **37**, 101172.

Deng, Y. and Zhao, R. (2015). Advanced Oxidation Processes (AOPs) in wastewater treatment. *Current Pollution Reports*, **1**, 167–176. https://doi.org/10.1007/s40726-015-0015-z

Duan, J., Kitamura, K., Tsukamoto, H. et al. (2024). Enhanced granulation of AS in an airlift reactor for organic carbon removal and ammonia retention from industrial fermentation wastewater: a comparative study. *Water Research*, **251**, 121091.

Ealias, A.M., Meda, G., and Tanzil, K. (2024). Recent progress in sustainable treatment technologies for the removal of emerging contaminants from wastewater: a review on occurrence, global status and impact on biota. *Reviews of Environmental Contamination and Toxicology*, **262**, 16. https://doi.org/10.1007/s44169-024-00067-z

Elkartehi, M.E., Mahmoud, R., Shehata, N. et al. (2021). LDH nanocubes synthesized with zeolite templates and their high performance as adsorbents. *Nanomaterials*, **11**, 3315.

Essam, A., Eldek, S.I., and Shehata, N. (2024). Management of caffeine in wastewater using MOF and perovskite materials: optimization, kinetics, and adsorption isotherm modelling. *Journal of Environmental Health Science and Engineering*, **22**, 345–360.

Fadali, O.A., Nassar, M.M., Ibrahim, I.E. et al. (2014). Effect of lag-phase in the transition from oxic to anoxic conditions on the performance of the sequencing batch reactor. *International Journal of Innovation and Applied Studies*, **7**, 393.

Garcia-Rodriguez, O., Mousset, E., Olvera-Vargas, H. et al. (2022). Electrochemical treatment of highly concentrated wastewater: a review of experimental and modeling approaches from lab-to full-scale. *Critical Reviews in Environmental Science and Technology*, **52**, 240–309.

Gholipour, A., Zahabi, H., and Stefanakis, A.I. (2020). A novel pilot and full-scale constructed wetland study for glass industry wastewater treatment. *Chemosphere*, **247**, 125966.

Hreiz, R., Latifi, M.A., and Roche, N. (2015). Optimal design and operation of AS processes: state-of-the-art. *Chemical Engineering Journal*, **281**, 900–920.

Jin, P., Jin, X., Wang, X. et al. (2013). Biological activated carbon treatment process for advanced water and wastewater treatment. In: *Biomass Now-Cultivation and Utilization*, 153–192. IntechOpen.

Keskin, B., Ersahin, M.E., Ozgun, H. et al. (2021). Pilot and full-scale applications of membrane processes for textile wastewater treatment: a critical *review*. *Journal of Water Process Engineering*, **42**, 102172.

Kumar, V.K., Manangath, S.P., and Gajalakshmi, S. (2023). Innovative pilot-scale constructed wetland-microbial fuel cell system for enhanced wastewater treatment and bioelectricity production. *Chemical Engineering Journal*, **460**, 141686.

Lei, Y., Wagner, T., Rijnaarts, H. et al. (2023). The removal of micropollutants from treated effluent by batch-operated pilot-scale constructed wetlands. *Water Research*, **230**, 119494.

Mainardis, M., Buttazzoni, M., and Goi, D. (2020). Up-flow anaerobic sludge blanket (UASB) technology for energy recovery: a review on state-of-the-art and recent technological advances. *Bioengineering*, **7**, 43.

Mittal, G.S. (2006). Treatment of wastewater from abattoirs before land application—a review. *Bioresource Technology*, **97**, 1119–1135.

Mumtaz, F., Li, B., Al Shehhi, M.R. et al. (2024). Treatment of phenolic-wastewater by hybrid technologies: a review. *Journal of Water Process Engineering*, **57**, 104695. https://doi.org/10.1016/j.jwpe.2023.104695

Najafpour, G.D., Hashemiyeh, B.A., Asadi, M. et al. (2008). Biological treatment of dairy wastewater in an upflow anaerobic sludge-fixed film bioreactor. *American Eurasian Journal of Agricultural and Environmental Science*, **4**, 251–257.

Nancharaiah, Y.V., Sarvajith, M., and Mohan, T.V.K. (2023). Pilot-scale aerobic granular sludge reactors with granular activated carbon for effective nitrogen and phosphorus removal from domestic wastewater. *Science of the Total Environment*, **894**, 164822. https://doi.org/10.1016/j.scitotenv.2023.164822

Nelson, M.J., Nakhla, G., and Zhu, J. (2017). Fluidized-bed bioreactor applications for biological wastewater treatment: a review of research and developments. *Engineering*, **3**, 330–342.

Olvera-Vargas, H., Zheng, X., Garcia-Rodriguez, O. et al. (2019). Sequential "electrochemical peroxidation–Electro-Fenton" process for anaerobic sludge treatment. *Water Research*, **154**, 277–286.

Rashid, R., Shafiq, I., Akhter, P. et al. (2021). A state-of-the-art review on wastewater treatment techniques: the effectiveness of adsorption method. *Environmental Science and Pollution Research*, **28**, 9050–9066. https://doi.org/10.1007/s11356-021-12395-x

Saravanan, A., Deivayanai, V.C., Kumar, P.S. et al. (2022). A detailed review on advanced oxidation process in treatment of wastewater: mechanism, challenges and future outlook. *Chemosphere*, **308**, 136524. https://doi.org/10.1016/j.chemosphere.2022.136524

Shehata, N., Sayed, E.T., Abdelkareem, M.A. et al. (2021). Bio-based adsorbents in water/wastewater treatment. In: *Encyclopedia of Smart Materials*, 830–838. Elsevier.

Shekho, M.S. and Hassan, N.E. (2024). A review on techniques for the cleaning of wastewater. *GSC Advanced Research and Reviews*, **18**, 118–128.

Sikosana, M.L., Sikhwivhilu, K., Moutloali, R. et al. (2019). Municipal wastewater treatment technologies: a review. *Procedia Manufacturing*, **35**, 1018–1024.

Singh, S., Naik, T.S.S.K., Shehata, N. et al. (2023). Novel insights into graphene oxide-based adsorbents for remediation of hazardous pollutants from aqueous solutions: a comprehensive review. *Journal of Molecular Liquids*, **369**, 120821. https://doi.org/10.1016/j.molliq.2022.120821

Soo, A., Kim, J., and Shon, H.K. (2024). Technologies for the wastewater circular economy – a review. *Desalination and Water Treatment*, **317**, 100205. https://doi.org/10.1016/j.dwt.2024.100205

Theodorakopoulos, G.V., Arfanis, M.K., Sánchez Pérez, J.A. et al. (2023). Novel pilot-scale photocatalytic nanofiltration reactor for agricultural wastewater treatment. *Membranes (Basel)*, **13**, 202.

Tyagi, N., Sharma, G., Ashraf, W. et al. (2024). Review on recent advancements of 2D MXene-based nanostructures on different possible approaches for wastewater treatment. *Chemical Physics Letters*, **845**, 141320.

Waqas, S., Harun, N.Y., Sambudi, N.S. et al. (2023a). Effect of operating parameters on the performance of integrated fixed-film AS for wastewater treatment. *Membranes (Basel)*, **13**, 704. https://doi.org/10.3390/membranes13080704

Waqas, S., Harun, N.Y., Sambudi, N.S. et al. (2023b). A review of rotating biological contactors for wastewater treatment. *Water*, **15**, 1913.

Wenten, I.G., Khoiruddin, K., Aryanti, P.T.P. et al. (2016). Scale-up strategies for membrane-based desalination processes: a review. *Journal of Membrane Science and Research*, **2**, 42–58.

Zhang, M., Lee, E., Vonghia, E. et al. (2020). Introduction to aerobic membrane bioreactors: current status and recent developments. *Current Developments in Biotechnology and Bioengineering*, 1–23. Elsevier. https://doi.org/10.1016/B978-0-12-819809-4.00001-2

Zhang, R., Walder, I., and Leiviskä, T. (2021). Pilot-scale field study for vanadium removal from mining-influenced waters using an iron-based sorbent. *Journal of Hazardous Materials*, **416**, 125961.

20

Membrane Technology Enhanced by Nanostructures for Wastewater Circular Economy

Alaa El Din Mahmoud[1,2], Nourhan S. Sultan[3], Nikolay G. Chechenin[4], Mayar M. Osman[5], E.M. El-Maghraby[5], and E. M. Elsehly[5]**

[1]Environmental Sciences Department, Faculty of Science, Alexandria University, Alexandria, Egypt
[2]Green Technology Group, Faculty of Science, Alexandria University, Alexandria, Egypt
[3]Biotechnology and Biomolecular Chemistry Department, Faculty of Science, Cairo University, Giza, Egypt
[4]Skobeltsyn Institute of Nuclear Physics, Lomonosov Moscow State University, Moscow, Russia
[5]Physics Department, Faculty of Science, Damanhour University, Damanhour, Egypt

*Corresponding authors: alaa-mahmoud@alexu.edu.eg; elsahli@sci.dmu.edu.eg

20.1 Introduction

The scarcity of clean water presents a significant threat to people's health, the long-term viability of the planet, and biodiversity (Gleick and Cooley, 2021). Recent research indicates that around four billion individuals, which accounts for two-thirds of the world's population, experience severe water scarcity for at least one month each year (Mekonnen and Hoekstra, 2016). The primary factors contributing to dwindling water resources include globalization, population growth, altered weather conditions, shifts in consumer habits, and irrigation expansion (Qin et al., 2019). Just 2.5% of the world's water supply is freshwater, with most being salty. Furthermore, approximately 2.5% of the water is solidified in the Arctic and Antarctic. Approximately 30% is found in far underground water sources, which are challenged to reach. Consequently, only 0.01% of the world's freshwater resources are readily attainable. Industrial facilities pollute water bodies, and fertilizers/pesticides are used extensively for agricultural purposes and exploration of hydrocarbons (Le and Nunes, 2016; Kathi and El Din Mahmoud, 2024; Chaukura et al., 2022).

Wastewater to Resource Recovery: Applying the Circular Economy Toward Sustainable Development, First Edition. Edited by Arvind Kumar Mungray et al. Published 2026 by John Wiley & Sons, Inc.

Wastewater treatment facilitates the maintenance of environmental purity. Nevertheless, the criteria for sustainable development goals for wastewater treatment standards rely on cost-effective, uncomplicated, and environmentally friendly technologies (Obaideen et al., 2022; Mahmoud, 2020a). The absence of wastewater treatment technology renders the 2030 standards unattainable. Therefore, assessing and enhancing wastewater treatment technology is crucial to fulfilling green environmental development goals (Mahmoud and Krasucka, 2023).

Membrane-based methods are becoming more prevalent in wastewater treatment due to their effectiveness, affordability, and technical feasibility. Microfiltration, ultrafiltration, nanofiltration, reverse osmosis (RO), and forward osmosis (FO) are conventional membrane processes that use transmembrane pressure to separate solutes with varying molecular weights through a membrane. Membrane technologies are advanced and noninvasive procedures utilized in what is known as the Global Recovery Process in the second, third, and fourth phases of the downstream processing process (Galanakis, 2012). These properties determine whether a membrane is isotropic or anisotropic. Isotropic membranes have a homogeneous composition and a homogeneous physical structure. Microporous materials display higher permeation fluxes compared to nonporous (dense) materials, which exhibit lower rates of permeation. Microfiltration membranes are commonly constructed with isotropic microporous membranes. Alternatively, anisotropic membranes contain discrete layers, each with a different topology and composition, throughout the membrane region. These membranes are made up of a small, selective layer supported by a broader, highly permeable layer (Mahmoud and Mostafa, 2023; Jarrar et al., 2024). Pressure-based separation techniques have been the subject of extensive study in sewage treatment, desalination, pharmaceutics, biotechnology, food, and chemicals. Membrane technologies have benefits such as operational simplicity, minimal energy usage, excellent stability, environmental friendliness, flexibility, controllability, and versatility. Integrating various membrane activities can enhance performance, reduce plant size, optimize energy consumption, and minimize the environmental footprint (Charcosset, 2016). Since the 1960s, acetate membranes have been used to desalinate water. Subsequently, various membrane-based techniques have advanced, including microfiltration, ultrafiltration, nanofiltration, RO, and FO (Matsuura, 2002) (Figure 20.1). Understanding the features of membranes has provided insights into the water purification process. Table 20.1 shows the pros and cons of different membrane technologies for treating wastewater.

Using nanotechnology to clean water, the advancement of nanotechnology has created countless opportunities for water purification. In environmental mitigation, metals at the nanoscale (iron, silver, gold, and titanium) and their oxidants

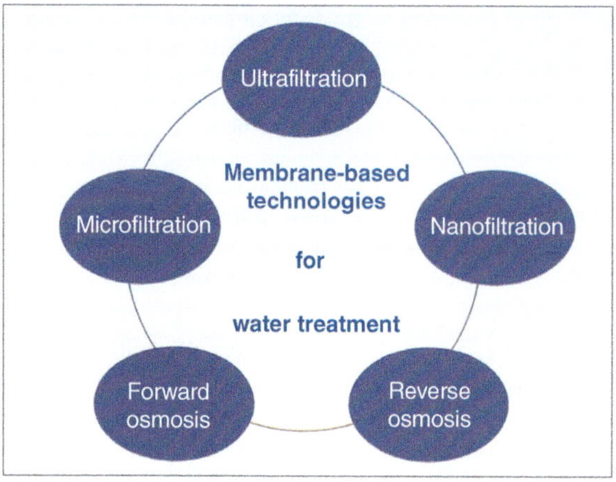

Figure 20.1 Different types of water treatment membranes.

have been frequently employed. It has been demonstrated that biological pollutants, including bacteria, viruses, and fungi, may be effectively removed by silver nanoparticles (Mahmoud, 2020b). Redox processes involving micropollutants have been altered by titanium nanoparticles (Mahmoud, 2022). Microbial, odorous, and phenolic-recalcitrant molecules may all be broken down into innocuous species by the photocatalyst nano-TiO_2 (Canela and Jardim, 2008; Nasrabadi and Foroutan, 2011). Increased surface porosity in the majority of nanomaterials prevents macrovoid formation and enhances salt rejection (Goh et al., 2013). The degradation of several resistant colors, halogenated chemicals, and the removal of heavy metals using bacteria that cause infection have all been made possible by nanomaterials. Gold and iron nanoparticles are particularly well-suited for eliminating inorganic heavy metals from surface and waste waters (Khin et al., 2012). In addition to being efficient at dissolving a variety of pollutants, nanomaterials are also active in the bimetallic coupling of metal oxides and other metals, which enhances catalysis for pollution in a complementary way. Composite membranes may be made with a variety of nanomaterials.

This improves the capacity to retain salt and reduces desalination expenses, land area, and energy. For instance, a thin film RO membrane is created by combining zeolite nanoparticles with a polymer matrix (Fathizadeh et al., 2011; Jeong et al., 2007). It improves salt retention capacity to 99.7% and water transport (Jadav and Singh, 2009). RO polymer matrices were doped into silica nanoparticles to desalinate water (Jadav and Singh, 2009). It enhanced the transport characteristics, pore sizes, and polymeric networks. Because of their exceptional adsorption capabilities, graphene and CNTs were utilized in adsorption-based desalination (Mahmoud et al., 2018; Mahmoud et al., 2020).

Table 20.1 Advantages and drawbacks of treatment methods for cleaning produced water.

Membrane technologies	Advantages	Disadvantages
Microfiltration	• Lowest cost and easiest membrane to use. • Removal of a manageable quantity of large particles of contaminants, such as bacteria and algae.	• Ineffective at the removal of tiny, emulsified oil droplets. • Absence of total dissolved solids clearance.
Ultrafiltration	• High flux. • Removal of turbidity and pigment. • Eliminates larger molecules, such as viruses and proteins.	• Insufficient elimination of all organic carbon. • Slight removal of the total amount of dissolved solids. Operates at higher pressure than microfiltration.
Nanofiltration	• Better removal of total dissolved solids and smaller molecules. • Effective for desalination.	• Oil nanodroplets may escape from the membrane. • Reduced flow rates compared to ultrafiltration and microfiltration. • More costly than other membrane varieties.
Reverse osmosis (RO)	• Total oil elimination. • High elimination of both total dissolved solids and total organic carbon. • Suitable for the production of drinking water. • Effectiveness at high levels.	• Operating costs are high. • Excessive energy consumption. • High operating pressure. • This is most prone to irreversible fouling and scaling of all membrane technologies.
Forward osmosis (FO)	• Permits regulating contaminants to be treated. • Consumption of energy is low. • Compared to RO, lower operating pressure. • Suitable for desalination and wastewater treatment.	• Water is transferred to another polluted stream. • Intense concentration polarization. • Needs a draw solution.

Nanomaterials are useful for desalinating water; however, they are not without restrictions. The following disadvantages must be considered: high pressure requirements, fouling, precipitation of pollutants, pore blockage, poor inflow, sluggish reaction, creation of hazardous intermediates, generation of newly produced ion particles, and aggregation during storage. Major issues also include low odds of reusability and unknown dangers to ecosystems (Khin et al., 2012). Before commercialization can occur, a number of issues such as stability, scaling up, and inadequate data quality on long-term operation must be handled (Goh et al., 2013). Thus, the most critical technology for reclaiming and purifying water is a water treatment system that is dependable, secure, sensitive, and reasonably priced.

In this chapter, we explore each type of membrane technology within the circular economy framework for wastewater management. In addition, we provide a comprehensive understanding of integrating nanotechnology into membranes for sustainable sewage disposal.

20.2 Types of Membranes for Wastewater Treatment

20.2.1 Microfiltration

Wastewater originates from industrial sources, agriculture, and water usage at home. In many countries, untreated wastewater results in costly issues. However, several countries release wastewater into exposed reservoirs that come into direct contact with the environment. Regretfully, larger suspended particles fall near the reservoir bottom. In addition to total dissolved solids, total suspended solids also contribute to the presence of pollutants that must be eliminated (Campos et al., 2002). Numerous filtration techniques are needed to minimize solid particles in wastewater. The microfiltration technique (Figure 20.2) employs a separator during the initial and final phases of the bioreactor. Filters using membranes are constructed from carbonate, polycarbonate, polypropylene, and polyethene materials. The preparatory step includes prefiltration through filter

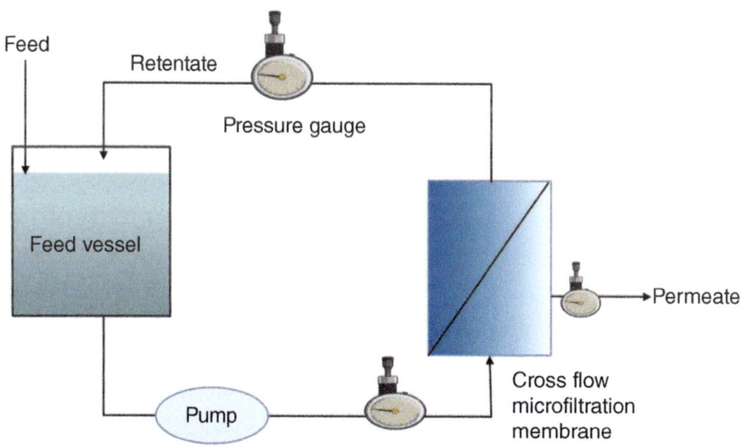

Figure 20.2 Diagrammatic representation of the microfiltration process.

cloths, gelation, flocculation, adsorption, and relaxation (Islam et al., 2022). Membrane microfiltration can significantly reduce particulate matter in the 0.1–10 μm range. Therefore, these membranes are designed to remove all dispersed particles, microorganisms, sediments, and contaminants from the effluent system. Nevertheless, the challenge persists because ions and particles smaller than 10 μm continue to move and stay within the treatment facility (Anis et al., 2019). The species encompasses ionic salts, metallic elements, organics, colloidal fractions, viruses, and coliforms. Additionally, the waste produced by this technique, such as by-products of pretreatment or discharge, must be processed appropriately or safely removed (Hakami et al., 2020).

20.2.2 Ultrafiltration

Ultrafiltration (Figure 20.3) is undoubtedly effective for removing particles not trapped within microfilters. Microfiltration differs from this process in that it retains a wider range of particle sizes and preserves membrane function by applying more osmotic pressure (Lin et al., 2016). The filter also inhibits the movement of mineral particles, pathogens, and organic contaminants. Regrettably, the utilization of this method is dependent on the presence of greasy wastes (Jiang et al., 2018). Unfortunately, a mixture of oil and water is considered an organic pollutant that requires an economically efficient treatment method. The simplicity and efficacy of the ultrafiltration method make it a preferred method for treating oily wastewater (Siagian et al., 2021). Nevertheless, it is imperative to tackle these setbacks. The effectiveness of this technique is hindered by the buildup of drops of oil, which cause fouling and clogging of membrane filters. Several aspects influence ultrafiltration membrane filters,

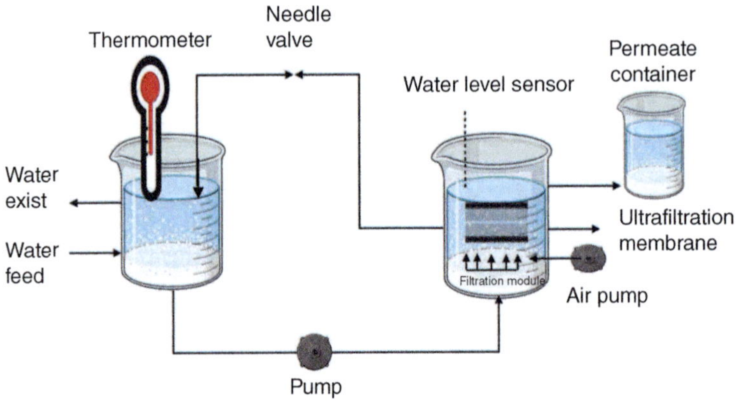

Figure 20.3 Schematic representation of ultrafiltration procedure.

including filtration characteristics, operational conditions, and optimization procedures (Chew et al., 2018). Furthermore, ions and solids that have been dissolved and reduced to nanoscale particles continue to move and persist within the processing facility. These species contain hazardous ion contaminants and other chemical pollutants. To improve technological efficiency, it is imperative to choose substances with a high attraction to water. This will modify the membrane and facilitate ultrafiltration performance (Goh et al., 2015). To minimize fouling and achieve high liquid permeation, it is necessary to use materials with high affinity. Furthermore, optimal pH, temperature, salinity, and oil content levels play a crucial role in maximizing the effectiveness of this technique (Al Aani et al., 2020).

20.2.3 Nanofiltration

The process of nanofiltration (Figure 20.4) is a membrane-based technology positioned between RO and ultrafiltration. The procedure requires pressures ranging from 150 to 500 psi, less than those required for RO. Nanofiltration membranes have been utilized to treat water and wastewater to address their distinctive properties, including a strong affinity for divalent and multivalent ions while allowing the passage of single ions and small particles (<100 Da) (Ahmed et al., 2022). The pressure-based membrane separation process known as nanofiltration is making significant strides in the domains of treating wastewater and water purification because of its exceptional charge-dependent repulsion characteristic and a high percentage of penetration. They selectively remove pigments and hazardous ions such as chromium, lead, tin, carbohydrates, and certain salts through filtration.

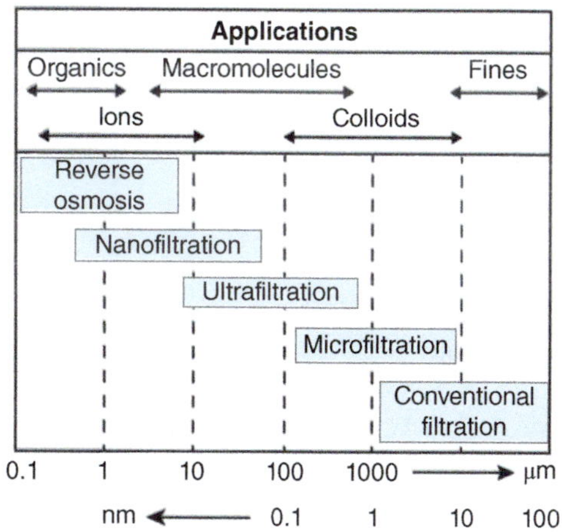

Figure 20.4 Approaches for membrane-based filtration that take into account the types and effective sizes of particles that the membrane usually removes.

20.2.4 Reverse Osmosis

Many treatment techniques have been developed to produce safe drinking water. Nevertheless, conventional thermal desalination methods provide several advantages, such as ease of use (Abdelaziz et al., 2021). A fixed surface charge is present in several NF membranes, which allows for the specific binding of different pollutants within the fluid. As a result, in terms of the mechanism of separation, compared to RO membranes, NF membranes are one step ahead. Figure 20.5 shows the water softening procedure utilizing the NF approach. Since the NF membranes' pores are incredibly tiny—usually ranging from 1 to 5 nm—virtually all solutes are effectively rejected by the membrane. The hard water's monovalent ions can flow through the membranes due to their surface charge, but the multivalent ions are retained. For uncharged species in NF, rejection mechanisms primarily involve size exclusion; however, for ionic species, size exclusion as well as electrostatic contact are responsible for rejection (Choi et al., 2002; Verliefde et al., 2009).

The limitations associated with these technologies include their limited or minimal capabilities (Peng et al., 2021). Therefore, they are unsuitable for extensive use. Membrane-based desalination methods like RO are known for their high capacity. The RO process encounters some obstacles, including a significant amount of energy (Greenlee et al., 2009), and, similar to other technologies based on membranes, a chemical wash is needed to minimize membrane fouling (Matshetshe et al., 2021). This method is the prevailing method for freshwater processing globally due to its economic advantages. Several measures must be taken to optimize RO process cost-efficiency, recycle energy, decrease fuel consumption, and improve total energy efficiency.

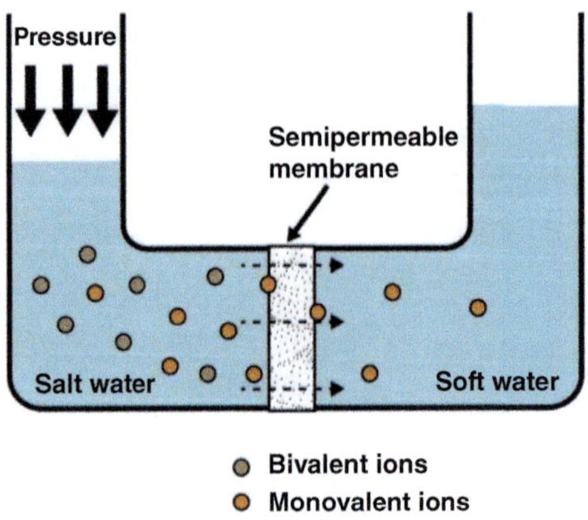

Figure 20.5 Schematic illustration of the nanofiltration method of softening water.

20.2.5 Forward Osmosis

A semipermeable membrane is used in FO to isolate water from dissolved materials (Obaid et al., 2020). FO membranes are used to separate feed solutions from high-salt concentration solutions. Due to osmotic pressure, purified water will move over the membrane from the feed solution to the opposite side (Zhao et al., 2011). FO may employ several water sources, including saltwater, water from rivers, lake water, and sewage, as inputs to provide drinkable water. Draw solutions can contain sugars, polyelectrolytes, and salts as solutes (Zhao et al., 2012). Unlike in RO, the feed solution is not subjected to extra pressure in this procedure. A preliminary treatment step must be performed before separating pure water from the draw solution. During this stage, the solute is subjected to ultrafiltration, nanofiltration, and thermal breakdown. Once the FO procedure has been completed, it is commonly combined with another type of treatment to create a combined treatment approach (Emadzadeh et al., 2019). FO was combined with membrane distillation to regulate the oil/water separator effluent. The system removed over 93% of organic and nonorganic substances and oil-containing impurities. The quality of the treated water is high, making it appropriate for commercial reuse. FO had high concentration levels, ranging from 77% to 84%. This resulted in a small volume of waste needing disposal. Hybrid systems' long-term viability and usability depend on their lack of chemicals or water (Ansari et al., 2017). Moreover, there was no evidence of membrane wetness or microorganisms. Several technical difficulties were encountered, including ion polarization and decreased FO flow due to $CaSiO_3$ colloidal scaling on the support layer. $CaSO_4$ and NaCl crystals on the active layer decreased FO flow. Organic nanopollutants and vaporized fatty acids can also pass through the filtrate (Saqib Nawaz et al., 2022).

As of today, there does not appear to be a miraculous solution to the structural problems associated with the purification of water and sewage. Therefore, a mix of multiple membrane technologies, particularly forward and reverse osmosis, is imperative to efficiently address water as well as sewage disposal. The rate at which we adopt FO relies on economic analysis supporting an affordable and straightforward membrane design (Vinardell et al., 2020).

20.3 Nanostructure-based Membrane Technology

Significant advancements in the subject of filtration of water have been made in recent years with regard to the use of nanostructured materials, including zeolites, dendrimers, metal/oxide nanoparticles, and carbon nanotubes (CNTs).

20.3.1 Carbon-based Nanomaterials

To produce membranes of the future generation with great flux and selectivity, and minimal hydrophobic, hydrophilic, and biofouling issues, membranologists must take note of the impressive qualifications that CNTs have to offer. Any membrane process has an intrinsic restriction on the trade-off between flow and specificity, which the CNTs have the ability to overcome.

Since CNTs are very mechanically strong, easy to prepare, and have excellent rejection ability, these are some of the most studied nanomaterials for membrane applications and construction. The use of hollow CNTs in membranes has been reported to exhibit a high percentage of contaminant solvent permeability and rejection (Noy et al., 2007). Only water is able to flow through the CNTs' tiny pore width, which is normally between 1 and 10 nm, whereas biological and chemical pollutants are prevented. Membranes based on one of the advantages of CNTs is their flexibility, such as polymeric membranes and sturdy like ceramic membranes. Furthermore, water percolates through CNTs quite quickly (Elsehly et al., 2016; Wan et al., 2009).

A membrane having a pore diameter of approximately 1 nm, made of free-standing CNTs, was recently described by Karan et al. (2012). The researchers assert that the rejection rates of organic dyes are raised by their CNT-based membrane by nearly three orders of magnitude when compared to membranes that are accessible commercially. Similarly, Srivastava et al. (2004) demonstrated the effective elimination of bacterial pollutants such as the waterborne poliovirus and *Escherichia coli* by using CNTs to remove a number of large hydrocarbons. In addition to their capacity for the exclusion of sizes, CNTs have demonstrated a potent adsorption of many pollutants, both biological and chemical, found within water. Parham and colleagues produced a CNT/ceramic composite filter, which they showed to be highly efficient at eliminating heavy metal ions from water (nearly 100%) and yeast (98%) (Parham et al., 2013).

For material scientists, chemists, and engineers, the inner hollow chamber of CNTs presents significant opportunities for nanoscale R&D. The incredibly effective passage of molecules of water via these very thin molecular pipelines is a unique phenomenon caused by the extraordinarily high aspect ratios of hydrophobic graphitic walls that are molecularly smooth and have nanoscale inner diameters in CNTs. Compared to other pores of similar size, water molecules travel across orders of magnitude nanotube pores faster. The transport and thermodynamic characteristics of confined water are significantly different from those of bulk water due to the smaller diameter of CNTs, which is in the order of nanometers. When it came to water, the focus of scientific inquiry was first on how it behaved in brief constricted tubes containing a single-wall nanotube (SWNT) that was submerged in bulk water and measured 0.8 nm in diameter and

1.35 nm in length (Hummer et al., 2001). Water molecules have been observed to periodically travel via bursts of molecules connected by hydrogen bonds along the nanotube axis. Nonetheless, it has been demonstrated that water molecules may organize into one-dimensional structures that depend on the diameter of CNTs with lower diameters. Through functionalization, CNTs may be made to work better and exhibit enhanced separation characteristics (also known as "gate-keeper supervised" separation) without compromising speed. Furthermore, adding organic moieties improves the nanotube anchoring in host materials, which improves composite reinforcement. The limited research that is now accessible attributes the antimicrobial activity of silver nanotubes (SNTs) to bacteria that are both Gram-positive and Gram-negative. The resulting harm is ascribed to oxidative stress or physical contact that compromises the integrity of the cell membrane (Kang et al., 2008; Narayan et al., 2005). Therefore, it may be possible to prevent microbial adhesion and the development of biofouling on surfaces by using CNTs. Apart from preventing biofouling, CNTs' surface as well as tip can be suitably adapted to address both hydrophilic and hydrophobic fouling.

While many investigations and applications of developed/functionalized CNTs have been conducted in the field of water purification concerning arsenic, fluoride, heavy metals, and hazardous organics, it is imperative to remember that the maximum utility of CNTs can only be realized by CNTs being implanted in the host matrix (by creating a membrane-based structure that preserves the nanotube symmetry) and then making the tubes functional to incorporate the capacity for particular division. Lines of CNTs arranged in an array were initially integrated onto a polymer sheet in 2004 (Hinds et al., 2004) to build an organized structure of nanoporous membrane. Consequently, years have been spent by a team of researchers (Holt et al., 2006) examining the flow of fluid through nanoscale structures. They discovered that they were carrying 1000 times the volume of water than was anticipated. Functionalized CNT tip membranes (Majumder et al., 2005) showed the capacity to control the movement of molecules via the cores of CNTs for possible uses in terms of chemical sensing and separations. The researchers established an aligned CNTs' microgeometry for water filtration applications (Dasgupta et al., 2008) and provided a methodology for developing an integrated membrane system (Kar et al., 2008). A highly focused elimination of impurities based on the functionalization of the CNT tip may lead to physicochemical interactions between species and the functional group covering the tip, which can incorporate the necessary physicochemical features into the membrane surface. Apart from both high selectivity and flux, the main benefit of CNT immobilization in a host matrix for applications involving separation is that it reduces the potential health risk posed by the toxicity of CNTs in their growing state when they are used as sorbents used in applications for purifying water, where the final product may become polluted with CNTs (Ali et al., 2019).

20.3.2 Fundamental CNT Membranes Technology

When functionalized CNTs, i.e. such CNTs which are chemically modified or "functionalized" by attaching molecules or groups to their surface (f-CNTs), are used as membranes for desalination or water treatment, the membrane selectivity against permeability trade-off—which has proven difficult for membranes of the present generation—is mitigated and CNT materials can be used safely and repeatedly. Table 20.2 compares CNT and conventional membranes. There is no need for pretreatment or posttreatment when using CNT membranes for a range of wastewater treatment or desalination applications. Three main f-CNT membrane types have been created or are being researched. These membrane types are mixed matrix membranes (MMMs), horizontally aligned CNT (HACNT) membranes, and vertically aligned CNT (VACNT) membranes.

20.3.3 Vertically Aligned CNT Membranes

Using CNTs to extract water pollutants is a promising method of removing contaminants from water (Yang et al., 2013). Utilizing CNT fillers in conjunction with polystyrene film, Hinds et al. (2004) produced a multi-walled VACNT membrane with inner pores of 6 nm. To fabricate membranes made of VACNTs, CNTs are typically arranged perpendicularly across the support layer, which can be composed of different types of materials (Das et al., 2014). A stacked CNT layer must be fabricated on a substrate using chemical vapor deposition to create the appropriate membrane architecture. The cavities between tubes can be covered with polymers to reduce the loss of water (Lee et al., 2016). Subsequently, a cleaning procedure is carried out to remove excess material from the membrane's topography and allow the CNT edges to open, creating precisely defined nanoscale pores (Goh et al., 2016). Thus, membranes are effective in eliminating a variety of pollutants from water (Das et al., 2014). It is challenging to develop VACNT membranes for this purpose because the CNTs must be aligned over a wide enough area for extensive sewage processing. An innovative method for MWCNT alignment in a membrane made of polystyrene under a varying voltage of 2000 V cm^{-1} was created by Wu et al. (2014). This procedure yields a type of membrane identified as MWCNTs/PS aligned by electrocasting a nanocomposite membrane. The results of the investigation showed that increasing the frequency range of the electric field from 1 to 100 Hz greatly enhanced the permeability of the VACNT membrane and the dispersion of MWCNTs. Lately, a new kind of membrane of VACNTs, known as an outside-the-creation-of-wall-CNT membrane, has been developed. This membrane utilizes the spaces amongst CNT fillers to eliminate impurities rather than the void channel inside CNT (Lee et al., 2015).

Table 20.2 CNT-derived membrane contrasted with traditional membranes.

	CNT-derived membrane	RO	Nanofiltration	Ultrafiltration	Microfiltration
Description	It is a polymer composite oriented perpendicular or an impermeable filler matrix with an open-tip single hollow structure	A technique that selectively pushes solvent against the osmotic pressure differential using transmembrane pressure	A process that separates particles less than 2 nm and dissolved macromolecules	A technique for removing solutes larger than 100 nm from a solution	A procedure that extracts solutes larger than 100 nm from the solvent
Applications of water treatment	Pollutant removal from complex matrices, desalination, water, and wastewater treatment	Water purification, desalination, reuse, and producing deionized water	Removal of natural organic compounds, ions, and softening	Virus particle rejection	Rejects germs, protozoa, and suspended particles
Materials	Ceramics, polymers, and CNTs	Organic polymers	Porous polymers	Cellulose, acrylic, polysulfone, and other materials	Polysulfone, polypropylene, polyurethane, and others
Types and sizes of pores	Micro-, meso-, or macropores: around 0.8–100 nm	Not porous	Micropores less than 2 nm	Mesopores are 2–50 nm in size	Macropores, 50–500 nm
Thickness	Varied with type	Approximately 0.1–0.2 micrometers	Approximately 0.05 micrometers	Between 150 and 300 μm	Between 50 and 100 μm
Water permeability	Approximately 7×10^{-7} mPa^{-1} s^{-1}	Approximately 3×10^{-12} mPa^{-1} s^{-1}	Approximately 40×10^{-12} mPa^{-1} s^{-1}	Approximately 0.5×10^{-10} mPa^{-1} s^{-1}	------------

(Continued)

Table 20.2 (Continued)

	CNT-derived membrane	RO	Nanofiltration	Ultrafiltration	Microfiltration
Working pressure	varied depending on the application type	between 30 and 60 bar	between 20 and 40 bar	between 1 and 10 bar	less than 1.0 bar
Characteristics	• High durability • Low energy consumption • Excellent performance • Resistance to adverse environmental conditions • Resistance to fouling • Economical	• High energy usage • Excellent performance • Reduced ability to withstand extreme weather conditions • Limited robustness • Suffering from fouling • Not as economical	• High energy usage • Excellent performance • Reduced resilience to adverse environmental conditions • Limited robustness • Fouling vulnerable • Not as cost-effective	• Moderate performance, moderate energy consumption, less resilience to unfavorable weather circumstances, and inadequate durability • Fouling susceptible • Less cost-effective	• Moderate energy consumption, poor performance, less ability to withstand unfavorable environmental conditions, low durability, fouling susceptibility, and less expensive

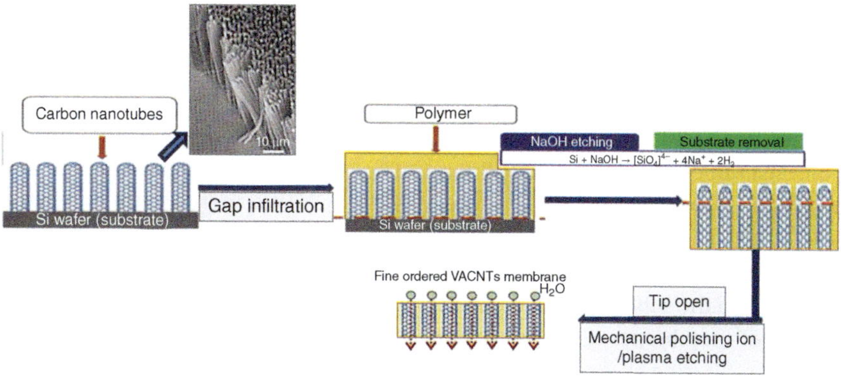

Figure 20.6 VACNT membrane fabrication process. *Source:* Yuan et al. (2012)/IOP Publishing.

Figure 20.6 illustrates a membrane VACNT and the microscale composition of a synthetic VACNT arrangement. The fabrication of VACNT membranes involves three steps: Initially, chemical vapor deposition is used to create the VACNT array (Barreiro et al., 2006; Yang et al., 2011; Yuan et al., 2012). CNTs have length, diameter, alignment, and other properties that are influenced by temperature, operating pressure, metallic catalyst selection, hydrocarbon content, reaction duration, and substrate type (Prasek et al., 2011; Szabó et al., 2010). Heating causes ethylene (C_2H_4) to react with ferrocene (Fe $(C_5H_5)_2$) in the role of a catalyst precursor. Within the furnace, CNTs develop upon silicon chips covered with aluminum. Here, aluminum serves as a model for the development of the CNTs. The pore width of CNTs can be adjusted by varying the catalyst precursor feed rate or by coating the substrate with an alternative catalyst (for example, Cu, Pd, Al, Ni, and Pt, among others). Second, for the purpose of preserving the composition of the membrane without damaging its assortment, fillers of low viscosity, such as epoxy (Baek et al., 2014), polystyrene (Elsehly, 2020), and parylene (Kim et al., 2014), are infused into the substrate's VACNT array. The fabrication process for VACNT membranes is shown in Figure 20.6 (Lee et al., 2016). Vacuum pressure is used to fill in the spaces between the VACNT array, allowing the water molecule to flow via the CNT channels. NaOH etching is used to remove the Si substrate following gap penetration. Ultimately, ion etching, plasma etching, mechanical polishing, and/or microtoming are used to open the tip of the CNT. However, because the VACNT membrane is hydrophobic, it takes a lot of pressure to get water molecules into the nanochannel. In Hinds' group, the tip of the VACNT arrays was opened by an H_2O plasma-enhanced oxidation procedure, which attached a carboxylic group to the CNT tip, making the CNT entry hydrophilic (Elsehly, 2020).

20.3.4 CNT Membranes with Mixed Matrixes

In this instance, CNTs are integrated within the matrix of polymers rather than helping them to form membrane structures (like the membranes of HACNTs and VACNTs). The goal of CNT mixed-matrix membranes is to improve the separation capabilities of polymeric membranes. The physicochemical characteristics of polymeric membranes may be changed by adding CNTs as inorganic fillers, which have a major impact on permeability, rejection efficiency, and fouling (Elsehly et al., 2017). The pore's vertical structure is asymmetrical (range of micrometers of holes within the sublayer, and nanometer range of pores on top). When compared to VACNT membranes, the process of creating CNT mixed-matrix NF/RO membranes for filtering is very simple. The process consists of two steps: (i) creating a layer of polyamide acting as a filter and (ii) preparing the membrane MF/UF acting as a support layer. Phase inversion is used to create a UF membrane support layer in the first stage. The phase inversion stage regulates its thickness, surface porosity, and pore size, which impact the permeate flow of NF/RO membranes. The creation of a thin-film selective PA layer using polymerization across interfaces is the second phase.

A less than 1 nm polyamide layer is deposited on a porous support layer formed by the reaction of trimesolychloride and m-phenylene diamine (Paul and Jons, 2016; Yip et al., 2010). Low permeate flow is produced by the PA layer despite its strong selectivity to NaCl. Furthermore, free chlorine, Cl_2, has the ability to degrade PA's amide group. In order to increase the permeable flow, a number of research projects concentrated on adding modified CNTs to the layer of support (Song et al., 2015; Zhao et al., 2017; Paul and Jons, 2016). More precisely, CNTs are added to the hexane/organic TMC mixture (in the PA selective layer) or the aqueous MPD solution (the support layer) to improve durability, chemical resistance, and permeability (Baroña et al., 2013; Zheng et al., 2017). CNTs can also be inserted between the PSf support layer and the thin-film layer of selective PA. A CNT interlayer can increase the efficiency of adsorbing PA to polymerize it and retaining the solution of amino monomers, according to Wu et al. (2016).

20.3.5 CNTs with a Horizontal Alignment (Bucky Paper)

The randomly placed CNT structures produce HACNT membranes (Dumée et al., 2011). It is manufactured using (i) vacuum filtering of the CNT dispersion in ethanol for an appropriate membrane structure, and (ii) CNT functionalization for uniform CNT dispersion. Consistent oven drying is employed to distribute the CNTs on the filter, remove moisture and solvents from the films, and detach the substrate's films (Endo et al., 2005; Kim et al., 2006; Muramatsu et al., 2005). The HACNT membrane's thickness and diameter are regulated by the vacuum infiltration device's scale. The layout and production process define its porosity and pore size. It is between a few nanometers and micrometers in size. HACNT membranes can adsorb large amounts of material and are microporous because of their array structure. Nevertheless, the application of HACNT membranes to the membrane process driven by high pressure may be hampered by their inferior mechanical strength when in contrast to membranes composed of CNTs.

20.3.6 Metal Oxides

One more reasonably priced choice is creating metal oxides in NF membranes. Furthermore, in the presence of light, the majority of metal oxides exhibit photocatalytic activity, which aids in the degradation of several organic and inorganic pollutants, including a number of dangerous bacteria found in water, turning the membranes into reactive materials as opposed to merely physical barriers. Humic acid and total organic carbon (TOC) were filtered out of wastewater using a titanium dioxide membrane (TiO_2) lattice of nanowires, which was manufactured by Leckie and colleagues (Zhang et al., 2008).

Under ultraviolet light irradiation, the authors claimed nearly 100% removal of humic acid and a clearance rate of over 90% of TOC through the combination of the filtering and photocatalytic properties of the TiO_2-based membrane. When TiO_2 and γ-alumina were combined, the membranes of NF also demonstrated a noteworthy rate of organic dye retention (Romanos et al., 2012). Zirconium dioxide (ZrO_2) nanoparticles were used in conjunction with TiO_2 particles by Van Gestel et al. (2006) to improve the TiO_2-based membranes' resistance to corrosion. They showed that the membranes were very stable throughout a broad pH range of values (from 1 to 13). In a recent work, Alsalhy et al. (2013) showed that adding zinc oxide (ZnO) nanoparticles to poly(phenyl sulfone) membranes increased the water flow from 76 to 107 ($Lm^{-2}\,h^{-1}\,bar^{-1}$). The inclusion of ZnO membranes with nanoparticles in them has boosted their hydrophilicity, which is primarily responsible for the improvement in flux density. Still, the solute retention membrane activity both with and without ZnO nanoparticles was almost the same, according to the investigators' observations. Additionally, it has been shown that incorporating silica (SiO_2) nanoparticles into polymeric NF membranes improves the membranes' hydrophilicity, permeability, and thermal stability (Jin et al., 2012).

20.3.7 Zeolites

Zeolites are aluminosilicate crystals—materials that have micropores and have pore diameters that vary from sub-nanoscale to nanometer (Bertolini et al., 2022). Because the aluminosilicate crystals are inert, zeolites have exceptional thermal and chemical stability, making them appropriate for use in the processes of RO and NF. It has been observed that zeolite-filled polydimethyl-siloxane (PDMS) NF membranes are effective NF membranes in terms of thermal stability, minimizing PDMS network swelling, and retaining numerous organic pollutants (Gevers et al., 2005, 2006). Zeolites have also been employed in the separation of saturated and unsaturated hydrocarbons (Nikolakis et al., 2001) and new metal complexes (Turlan et al., 2001).

20.4 Membrane Technology's Application in Wastewater Treatment

Biological and mechanical separation operations, such as RO, ultrafiltration, nanofiltration, and microfiltration, require membranes made of organic and inorganic materials, most commonly polyethylene, polypropylene, and polyethylene terephthalate (Aliyu et al., 2018). Silica, metals, zeolites, and ceramics are some of the materials used to make nonorganic membranes. Due to their stability in both chemical and thermal aspects, they are usually employed in industrial applications such as microfiltration, ultrafiltration, and hydrogenation (Mallada and Menéndez, 2008). Different driving forces govern how media move across the membranes. Membrane processes can be classified as equilibrium-based, not dependent on equilibrium, driven by pressure, or not driven by pressure (Jhaveri and Murthy, 2016). Figure 20.7 summarizes several of these methods based on their motivating factors. Each of these membrane approaches is covered separately in the following text.

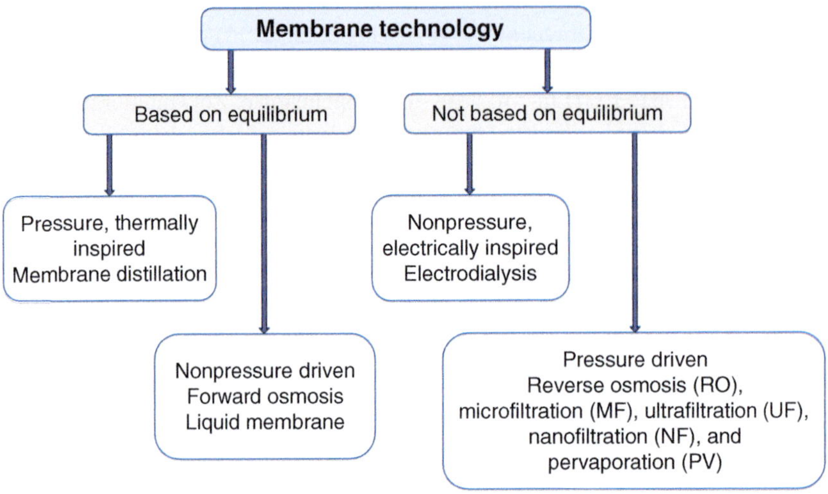

Figure 20.7 Diagrammatic illustration of a few membrane functions.

20.4.1 Membrane Mechanisms Driven by Pressure

The most popular membrane processes for treating wastewater, from the beginning to the end of treatment, are propelled by pressure. Hydraulic pressure is necessary for these procedures to produce separation. These procedures come in four primary forms: MF, UF, NF, and RO. Aside from their pressure needs, these processes differ mostly in their membrane pore diameters (Chollom, 2014). A description of the key elements of these procedures is given in Table 20.3. Pressure-driven membrane processes, such as RO, are renowned for their ability to separate tiny particles, microorganisms, and monovalent particles 99.5% of the time (Muro et al., 2012; Hankins and Singh, 2016).

RO has led the way in water reclamation for years by treating wastewater and removing salt from seawater. RO produces a hydrostatic pressure powerful enough

Table 20.3 A few characteristics of membranes driven by pressure.

Membrane process	Molecular weight cut off (kilo Dalton)	Diameters retained (μm)	Needed pressure (bar)	Type of membrane	Permeability on average (L/m^2 h bar)	Availed solutes
MF	100:500	10^{-1}:10	1:3	Symmetric, asymmetric, or porous	500	Micro-particles, organics, grease, oil, fats, and bacteria
UF	20:150	10^{-3}:1	2:5	Tiny pores and asymmetry	150	oils, Proteins, sugar, colors, organic materials, and microplastics
NF	2:20	10^{-3}:10^{-2}	5:15	Tight, asymmetric, thin-film composite with pores	10:20	Sulfates, lactose, sucrose, divalent ions, pigments, and salt
RO	0.2:2	10^{-4}:10^{-3}	15:75	Oriented, fibrous composite (semi-porous)	5:10	All contaminants, even monovalent ions

to overcome the feed's natural osmotic pressure. This stands against the natural process of osmosis. Under pressure, the membrane absorbs water particles. Once these molecules have diffused through the membrane material, they desorb from the membrane and are collected on the permeate side of the membrane (Greenlee et al., 2009).

Several membrane processes driven by pressure combinations have been used in various wastewater treatment contexts. They sometimes act as a pretreatment for further unit operations. In one experiment, Nataraj et al. (2006) treated distillery effluent with NF and RO in combination, and successfully eliminated 98% of the pollutants. In a different application, reactive dye printing wastewater was treated using UF and RO together in a small-scale system. The RO permeate was suitable for removal and repurposing, but the UF permeate still failed to meet the discharge restrictions. Reactive dye, sodium alginate, urea, and oxidizing agents were among the contaminants that were effectively eliminated (Šostar-Turk et al., 2005). Table 20.4 displays some more applications of these pressure-driven membrane technologies.

As demonstrated in most of the abovementioned applications, different types of membranes generally act as preparatory procedures preceding RO. This will improve the retention of steady flow and lessen fouling of the RO membrane. Additionally, this serves as a multi-stage treatment to eliminate pollutants from effluent (Bartels et al., 2005; Bergman et al., 2003). Undoubtedly, pressurized membrane techniques have rendered wastewater-to-water recycling a viable possibility. But because of the pressure, the problem with energy requirements still exists.

20.4.2 Electrodialysis

In order to remove dissolved ions from water, electrodialysis and electrodialysis reversal use electricity and ion-permeable membranes. An electric potential ion-permeable membrane is used in these procedures (Campione et al., 2018). Two different kinds of ion exchange membranes are used in electrodialysis. One is cation-rejecting and permeable to anions, whereas the other is anion-rejecting and permeable to cations. Additionally, there are two ways of solving the problem: feed concentrates and dilutes. Ions from the dilute move through oppositely charged membranes into the concentrate when an electric current flows through the system. The positively charged anion-exchange membrane (AEM) then holds onto the cations. In a similar vein, the cation-exchange membrane (CEM) holds on to the anions. As a result, the concentrate stream becomes rich in ions while the feed stream becomes depleted of ions (Mahmoud and Mostafa, 2023).

Table 20.4 A few pressure-driven membrane technologies used in the treatment of wastewater.

Membrane Processes	Wastewater reclaimed	Outcomes	Reference
UF	Factory producing vegetable oil	TSS (100%), TOC (87%), PO_4^{3-} (85%), Cl^- (40%), COD (91%).	Mohammadi and Esmaeelifar (2004)
MF–RO	Urban polluted water	Drugs and pesticides were eliminated to the discharge limit.	Rodriguez-Mozaz et al. (2015).
MF	Municipal wastewater (phosphorus removal and disinfection)	Removing contaminants to below the limit of detection.	Dittrich et al. (1996)
MF	Emulsified synthetic oily wastewater	95% of organic pollutants are eliminated.	Wang et al. (2009)
NF–RO	Leachate dumped	95% water recovery.	Rautenbach et al. (2000)
UF	Effluent from poultry slaughterhouses	Removal of lipids (99%), suspended materials (98%), and COD and BOD (>94%)	Yordanov (2010)
NF	Fabric	Color (100%), salinity (30%), and COD (57%).	Ellouze et al. (2012)
UF–RO	Metal finishing industry	90–99% removal of different Contaminants.	Petrinic et al. (2015)
UF–RO	Oily wastewater	100% oil and grease, 98% TOC, 98% COD, 95% TDS, and 100% turbidity.	Salahi et al. (2011)
UF–NF/RO	Phenolic waste from the paper industry	95.5% COD and 94.9% phenol.	Sun et al. (2015)

TSS: total suspended solids, TOC: total organic carbon, TDS: total dissolved solids, COD: chemical oxygen demand.

20.5 Role of Membrane Technology in Resource Recovery from Wastewater

Chemical substances are separated in membrane separation procedures by a variation in transport rate across the membrane interface. The driving power, mobility, and concentration of each component within the interface all affect this transfer rate. The primary determinants of an effective chemical component separation are solute molecular size, membrane morphology, and chemical affinity.

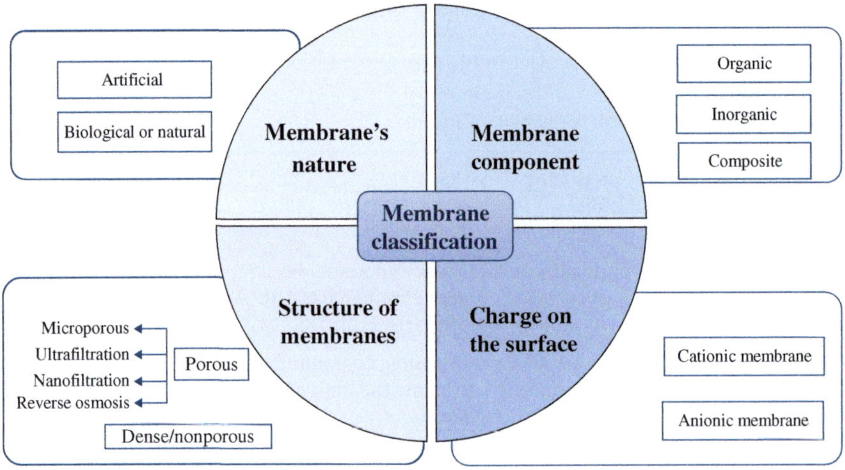

Figure 20.8 Diagrammatic representation of many membrane classifications utilized in wastewater treatment.

Membrane kinds and modules affect how well they separate (Sonune and Ghate, 2004). As seen in Figure 20.8, types of membranes include anisotropic and isotropic, organic or inorganic, porous or nonporous, or composite membranes. The membrane's nature, structure, material, and surface charge are taken into consideration while classifying it (Bera et al., 2022).

Membrane technology works on the basis of selectively allowing some substances to flow through the membrane while obstructing the passage of others. It usually takes an outside motivator to make this procedure easier. Driving forces such as pressure differentials and concentration differences categorize membrane processes in wastewater. Ion exchange membranes facilitate solute partitioning, diffusion, and electrochemical reactions (De Gisi and Notarnicola, 2017). In addition to the different, distinct membrane techniques, new hybrid and integrated technologies have also been created recently. Synthetic organic polymers are used to manufacture all membranes, whether they are noncellulosic or cellulosic, that are used in municipal water treatment. Numerous materials, including PVDF, cellulose acetate, poly-sulfone, polyethersulfone, and polypropylene, are used to make MF and UF membranes. Because of their varied characteristics, the many materials that went into making the membrane have unique pH, surface charges, and hydrophobicities. The properties of the membrane material have a significant impact on the membrane's overall usefulness. Membrane-based separation approaches involve the selection of a suitable procedure with the right membrane, size, shape, and driving force in order to achieve desired separation. Pressure, thermal, electrical, and concentration processes are the several categories of membrane separation procedures (Obotey Ezugbe and Rathilal, 2020). The following classification is compiled in Table 20.5.

Table 20.5 Membrane separation technique classification.

Procedure for membrane separation	Kind of membrane	Pore size	Motivating force	Principle of separation (the process of division)	Application	References
Process of membrane division driven by pressure						
Microfiltration	Porous	0.05–10 μm	Variation in pressure (0.1 : 2 bar)	Sieving	Food, beverage, and pharmaceutical industries	Radjenovic et al. (2008)
Ultrafiltration	Porous	1 : 100 nm	Pressure difference (1–10 bar)	Sieving	Textile, food, dairy, pharmaceutical, and water treatment industries	Gitis and Hankins (2018)
Nanofiltration	-----------	0.1–10 nm	Pressure difference (10–25 bar)	Solution-diffusion	Wastewater treatment and brackish water desalination	Bolong et al. (2009)
Reverse osmosis	-----------	<2 nm	Pressure difference (15–80 bar)	Solution-diffusion	Concentration of milk and juice, desalination of brackish and seawater	Mulligan et al. (2001)
Process of separation driven by concentration						
Pervaporation	Nonporous	-----	Difference in vapor pressure (0.001 : 1 bar)	Solution-diffusion	Recovery of hydrogen and helium	Takht Ravanchi et al. (2009)
Gas separation	Porous/ nonporous	<1 μm	Disparity in concentration; difference in partial pressure	Solution/diffusion (nonporous membranes); Knudsen flow (porous membranes)	Elimination of organic matter from water	Abedini and Nezhadmoghadam (2010)

(Continued)

Table 20.5 (Continued)

Procedure for membrane separation	Kind of membrane	Pore size	Motivating force	Principle of separation (the process of division)	Application	References
Process of membrane separation powered by electricity						
Electrodialysis	Nonporous	-----	Difference in electrical potential	Exclusion mechanism based on Donnan	Seawater desalination, separation of amino aids	Ghaly et al. (2013)
Temperature-driven separation process						
Membrane distillation	--------------	0.2:1 μm	Difference in vapor pressure	Equilibrium between vapor and liquid	Desalination of seawater and the semiconductor industry	Ismail et al. (2018)

20.6 Techno-economic Analysis

Wastewater treatment is attracting significant attention due to concerns regarding untreated wastewater disposal in nearby water bodies. TEA, or techno-economic analysis, is a powerful method for assessing products' technical and economic performance. It involves utilizing mathematical modeling to estimate capital cost, operating cost, and income (Murthy, 2022). The TEA tool helps choose the most cost-efficient approach to technology. It also enables the R&D team to assess and enhance the performance of a newly developed process or improve an existing one. The wastewater treatment system modeling uses Intelligen SuperPro and Aspen Plus (Mahmud et al., 2021).

20.6.1 Techno-economic Analysis's Technical Aspect

A model is created that outlines the steps involved in the TEA.

20.6.1.1 Step 1: Process Design

This stage encompasses the entire process from beginning to end (Kumar et al., 2020) (Figure 20.9). After the process is completed, the flow of material determines the balance of material during the process. This balance provides information on the amount of feed that will be converted into the selected final product.

20.6.1.2 Step 2: Cost Calculation

Estimating process fees involves determining the total capital expenditures associated with the process, which can be classified into immediate and indirect charges (Llamas et al., 2021). Direct costs include site acquisition, sourcing supplies, labor charges, and machinery purchases. Indirect charges involve transportation, insurance, construction overhead, and contingency charges. There are two methods for computing indirect costs: determining how much each direct cost component contributes and averaging them. Another method calculates the proportion of total direct spending (Tavakkoli et al., 2017).

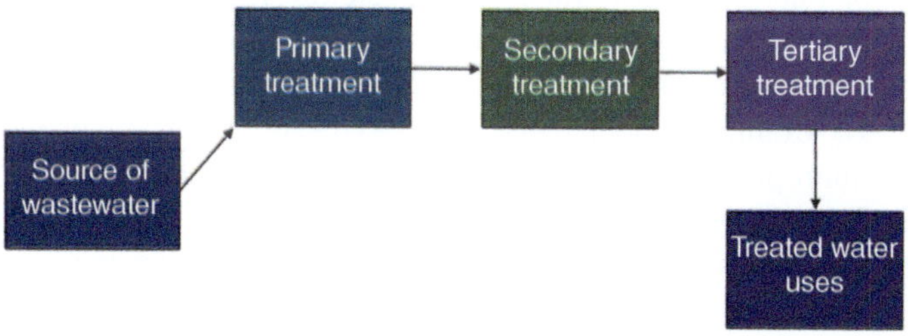

Figure 20.9 An overview of the basic process diagrams employed in the techno-economic evaluation of sewage treatment.

20.6.1.3 Step 3: Operational Costs

Operational costs are estimated using raw materials, electricity, employees, machinery replacement, and maintenance (Al-Anzi et al., 2020). Labor and transport expenses associated with collecting feedstock are also included (Kumar et al., 2020). Nevertheless, when it comes to wastewater treatment, whether it is brackish or seawater, the expenses associated with feeding can be regarded as negligible since the effluent acts as an ingredient (Llamas et al., 2021).

20.6.1.4 Step 4: Developing and Validating the Model

Model testing is essential in process design when simulations are examined through experimentation or pilot plant operations. If the outcomes align with the prediction, the plan might be suitable for future use (Dong et al., 2017). Sensitivity analysis is employed to evaluate input fluctuations' influence on a process's output, particularly in treating wastewater constituents, such as chloride, biological oxygen demand, and total dissolved solids (Cluff et al., 2014).

20.6.2 A Technical and Economic Assessment of Effluent Processing Methods

Wastewater treatment expense depends on various aspects, including geographical location, treatment methodology, waste disposal methods, and target use. RO systems have higher operating costs than freshwater systems (Edalat and Hoek, 2020). However, by combining direct contact membrane distillation with heat, treatment costs can be reduced by a considerable factor of 7.7 (Tavakkoli et al., 2017). Nevertheless, a thorough investigation is required to ascertain membrane methods' economic and technological viability in treating wastewater.

20.7 Prospects for the Future

The management of water and wastewater is being significantly enhanced via membrane technology. It has several uses and is thought to be an extremely helpful wastewater treatment technique. To create new, more effective membrane materials, for example, copolymers like PVDF–HFP (polyvinylidene fluoride and hexafluoropropylene) and PVDF–TFE (polyvinylidene fluoride–tetrafluoroethylene), new research is now being conducted. To address the problem of membrane fouling in MBRs, the majority of efforts over the past 10 years have focused on the application of cutting-edge techniques (Bera et al., 2022). NMs, CE, biological concepts, and electrically based methods have been the focus of the majority of contemporary research on membrane fouling management. Research has shown that these new methods for managing membrane fouling are quite effective (Obotey Ezugbe and Rathilal, 2020). The application of bucky-paper membranes and CNT are also being experimented with. Subsequent research and investigation are necessary before implementing these for large-scale MBRs. More than one solution is also needed to regulate membrane contamination (Bera et al., 2022). In the field of membrane technologies, particularly biomembrane technology, membrane fouling remains a major issue that needs to be addressed in the upcoming years (Ismail et al., 2018).

It is imperative to develop and improve alternative membrane filters in the future to reduce reliance on chemical and biological engineering. Activated carbon chambers with varying particle sizes and green energy technology can treat fouled membranes and reuse them. Furthermore, membrane components must be cost-effective and designed for optimal reuse and recycling (Yusuf et al., 2020). Additional investigation is warranted to explore the integration of machine learning, simulation, 3D, and 4D technologies with membrane-based systems.

A simultaneous TEA and life cycle assessment (LCA) of wastewater treatment methods is recommended to evaluate upcoming treatment technologies' technical, economic, and environmental feasibility in the coming years. More extensive investigations are required to assess treatment procedures' effectiveness and their ecological consequences. This can include establishing protocols to evaluate the effectiveness of treatment and environmental impacts. It can also use sophisticated analytical methods to characterize produced water and treatment by-products. Effective wastewater management system development necessitates collaboration between industry and academics to provide a link between scientific study and actual implementation.

20.8 Challenges and Opportunities

20.8.1 Critical Assessment of Developments

Fouling occurs with membrane-enabled water treatment systems. The creation of innovative CNT membranes that can overcome the challenges posed by inorganic, organic, and biological fouling has been a constant focus of membrane scientists. Reducing fouling and microbial adherence on CNT membranes can be achieved through the appropriate chemical group functionalization of CNT surfaces. However, because of the compromise between reduced water permeability and high removal efficiency caused by aquatic materials that either deposit inside membrane holes or are adsorbed on membrane surfaces, CNT membranes continue to have fouling problems.

20.8.2 Knowledge Gaps

The following scientific issues must be resolved in order to create long-lasting, anti-fouling CNT membranes:

1. How can CNTs' toxicity increase to inactivate biological pollutants in biofouling management without compromising their properties?
2. On the interior and exterior surfaces of CNT membranes, what kind(s) of interaction(s) predominate in the foulant attachment?
3. What function does a nanocomposite composition perform in fouling formation?

20.8.3 Future Research Directions

Membranes with antifouling qualities can be created by tailoring f-CNTs using various techniques, like attaching different metal oxides to CNT surfaces or adding hydrophilic groups. Furthermore, creating small-diameter CNTs and constructing f-CNT membranes with high conductivity combined with disturbances to electrical pulses may result in effective ways to manage biofouling. The membrane community needs to conduct additional studies to determine the potential uses for f-CNT membranes for fouling reduction given their qualities. Furthermore, there is a lot of promise in using fresh and innovative f-CNT desalination membranes for water treatment. Therefore, for CNT membranes to be successful in the future in uses for environmental remediation, including water treatment and desalination, there is an urgent need to develop f-CNT membranes that display increased resistance to fouling (Ali et al., 2019).

20.9 Conclusion

This chapter examines the present developments in membrane-based systems for treating wastewater and water, including applications. It emphasizes the significance of these technologies in contributing to accomplishing sustainable goals and their involvement in the circular economy. This study presents primary findings as follows. The most common membrane-related procedures include FO, RO, ultrafiltration, nanofiltration, and microfiltration. Integrating more than two membrane processes is advantageous for overcoming each process's limitations, reducing energy consumption, and improving overall performance efficiency. The growing fascination with these methods stems from their ease of use, minimal environmental footprint, and exceptional selectivity.

Utilizing standard membrane techniques in order to treat wastewater resulting from the production of agro-food by-products has been confirmed during the past several years. But the most promising use of membrane technology in the coming years will be the restoration of high-value added components through the valuation of wastewater, approaches that have been acknowledged as emerging technologies. The effectiveness of membrane mechanisms is typically dependent on multiple previously mentioned factors (pore size, parameters of operation, kind of membrane, structure of membrane, and so on). However, the benefits of membrane processes—such as superior selectivity and productivity, and the lack of further stages—surmount the drawbacks of other conventional techniques. The creation of unified membrane systems in recent times may represent a revolutionary strategy for by-product fractionation with the goal of achieving the "Universal Recovery Process." These iterative designs can address a variety of R&D problems, including: (i) recovering high-value chemicals; (ii) disposing of byproducts; and (iii) reducing contamination of the environment and water. Fractionation may be used to produce water (water for recycling), depending on the system that is in place and the kind of wastewater. Do they qualify as environmentally friendly processes? It has been verified at the lab scale. Thus, the final response pertains to the food processing industries through large-scale application. Lastly, procedure enhancement, cost cutting, low-quality energy, biofilm, selectivity, and pretreatment management need to be the main areas of concentration for attempts to enhance membrane technology's environmental performance.

Current TEA research has shown that combining multiple processing techniques is extremely beneficial and cost-effective for treating wastewater. Recent LCA studies suggest that including renewable and co-energy production in the treatment process improves sustainability. Future research should focus on conducting concurrent TEA and LCA studies to optimize process performance. This will contribute to sewage processing techniques' economic and environmental sustainability.

Review Questions

1. What are the different membrane-based technologies for water/wastewater treatment?
2. How do you characterize various membranes based on their pore size or operating pressures?
3. What is a forward osmosis membrane and its mechanism? How is it different from RO?
4. What are the limitations and advantages of RO and FO membranes?
5. What are the properties that can be changed by using nanoparticles? Explain how.
6. What are the different driving forces governing various membranes for water/wastewater treatment?
7. What are the various challenges of membranes in resource recovery from wastewater?

8 What are the various opportunities in membrane development for resource recovery from wastewater?
9 How are TEA and LCA considered good tools for applications of membranes?
10 How can CNTs improve various properties of membranes?

References

Abdelaziz, G.B., Algazzar, A.M., El-Said, E.M.S. et al. (2021). Performance enhancement of tubular solar still using nano-enhanced energy storage material integrated with V-corrugated aluminum basin, wick, and nanofluid. *Journal of Energy Storage*, **41**, 102933.

Abedini, R. and Nezhadmoghadam, A. (2010). Application of membrane in gas separation processes: its suitability and mechanisms. *Petroleum and Coal*, **52**, 69–80.

Ahmed, S.F., Mehejabin, F., Momtahin, A. et al. (2022). Strategies to improve membrane performance in wastewater treatment. *Chemosphere*, **306**, 135527.

Al-Anzi, B., Al-Hammadi, S., Yoon, J. et al. (2020). Techno-economic analysis of multi-stage ion concentration polarization with recirculation for treatment of oil produced water. *Journal of Environmental Management*, **269**, 110788.

Al Aani, S., Mustafa, T.N., and Hilal, N. (2020). Ultrafiltration membranes for wastewater and water process engineering: a comprehensive statistical review over the past decade. *Journal of Water Process Engineering*, **35**, 101241.

Ali, S., Rehman, S.A.U., Luan, H.-Y., et al. (2019). Challenges and opportunities in functional carbon nanotubes for membrane-based water treatment and desalination. *Science of the Total Environment*, **646**, 1126–1139.

Aliyu, U.M., Rathilal, S., and Isa, Y.M. (2018). Membrane desalination technologies in water treatment: a review. *Water Practice and Technology*, **13**, 738–752.

Alsalhy, Q.F., Ali, J.M., Abbas, A.A. et al. (2013). Enhancement of poly(phenyl sulfone) membranes with ZnO nanoparticles. *Desalination and Water Treatment*, **51**, 6070–6081.

Anis, S.F., Hashaikeh, R., and Hilal, N. (2019). Microfiltration membrane processes: a review of research trends over the past decade. *Journal of Water Process Engineering*, **32**, 100941.

Ansari, A.J., Hai, F.I., Price, W.E. et al. (2017). Forward osmosis as a platform for resource recovery from municipal wastewater - A critical assessment of the literature. *Journal of Membrane Science*, **529**, 195–206.

Baek, Y., Kim, C., Seo, D.K. et al. (2014). High performance and antifouling vertically aligned carbon nanotube membrane for water purification. *Journal of Membrane Science*, **460**, 171–177.

Baroña, G.N.B., Lim, J., Choi, M. et al. (2013). Interfacial polymerization of polyamide-aluminosilicate SWNT nanocomposite membranes for reverse osmosis. *Desalination*, **325**, 138–147.

Barreiro, A., Hampel, S., Rümmeli, M.H. et al. (2006). Thermal decomposition of ferrocene as a method for production of single-walled carbon nanotubes without additional carbon sources. *The Journal of Physical Chemistry B*, **110**, 20973–20977.

Bartels, C.R., Wilf, M., Andes, K. et al. (2005). Design considerations for wastewater treatment by reverse osmosis. *Water Science and Technology*, **51**, 473–482.

Bera, S.P., Godhaniya, M., and Kothari, C. (2022). Emerging and advanced membrane technology for wastewater treatment: a review. *Journal of Basic Microbiology*, **62**, 245–259.

Bergman, R.A., Joffe, D., Adams, N. et al. (2003). Gwinnett county water reclamation with 50 MGD ultrafiltration–proof testing and design. In: *Proceedings of the AWWA Membrane Technology Conference, American Water Works Association, Atlanta, GA, USA*, 2–5. American Water Works Association (AWWA).

Bertolini, T.C.R., Fungaro, D.A., and Mahmoud, A.E.D. (2022). The influence of separately and combined bentonite and kaolinite as binders for pelletization of NaA zeolite from coal fly ash. *Cerâmica*, **68**, 375–384.

Bolong, N., Ismail, A.F., Salim, M.R. et al. (2009). A review of the effects of emerging contaminants in wastewater and options for their removal. *Desalination*, **239**, 229–246.

Campione, A., Gurreri, L., Ciofalo, M. et al. (2018). Electrodialysis for water desalination: a critical assessment of recent developments on process fundamentals, models and applications. *Desalination*, **434**, 121–160.

Campos, J.C., Borges, R.M.H., Oliveira Filho, A.M. et al. (2002). Oilfield wastewater treatment by combined microfiltration and biological processes. *Water Research*, **36**, 95–104.

Canela, M.C. and Jardim, W.F. (2008). Identification and photocatalytic destruction of malodorous compounds in sewage. *Environmental Technology*, **29**, 673–679.

Charcosset, C. (2016). Ultrafiltration, microfiltration, nanofiltration and reverse osmosis in integrated membrane processes. In: *Integrated Membrane Systems and Processes*, 1–22. Wiley Online Library.

Chaukura, N., Muzawazi, E.S., Katengeza, G. et al. (2022). Chapter 20 - Remediation technologies for contaminated soil systems. In: *Emerging Contaminants in the Terrestrial-Aquatic-Atmosphere Continuum* (ed. W. Gwenzi), Elsevier.

Chew, C.M., Aroua, M.K., and Hussain, M.A. (2018). Advanced process control for ultrafiltration membrane water treatment system. *Journal of Cleaner Production*, **179**, 63–80.

Choi, J.-H., Dockko, S., Fukushi, K. et al. (2002). A novel application of a submerged nanofiltration membrane bioreactor (NF MBR) for wastewater treatment. *Desalination*, **146**, 413–420.

Chollom, M.N. (2014). *Treatment and reuse of reactive dye effluent from textile industry using membrane technology*. PhD Thesis, Durban University of Technology, South Africa.

Cluff, M.A., Hartsock, A., Macrae, J.D. et al. (2014). Temporal changes in microbial ecology and geochemistry in produced water from hydraulically fractured Marcellus shale gas wells. *Environmental Science & Technology*, **48**, 6508–6517.

Das, R., Ali, M.E., Hamid, S.B.A. et al. (2014). Carbon nanotube membranes for water purification: a bright future in water desalination. *Desalination*, **336**, 97–109.

Dasgupta, K., Kar, S., Venugopalan, R. et al. (2008). Self-standing geometry of aligned carbon nanotubes with high surface area. *Materials Letters*, **62**, 1989–1992.

De Gisi, S. and Notarnicola, M. (2017). Industrial wastewater treatment. In: *Encyclopedia of Sustainable Technologies*, 23–42. Elsevier.

Dittrich, J., Gnirss, R., Peter-Fröhlich, A. et al. (1996). Microfiltration of municipal wastewater for disinfection and advanced phosphorus removal. *Water Science and Technology*, **34**, 125–131.

Dong, X., Trembly, J., and Bayless, D. (2017). Techno-economic analysis of hydraulic fracking flowback and produced water treatment in supercritical water reactor. *Energy*, **133**, 777–783.

Dumée, L., Germain, V., Sears, K. et al. (2011). Enhanced durability and hydrophobicity of carbon nanotube bucky paper membranes in membrane distillation. *Journal of Membrane Science*, **376**, 241–246.

Edalat, A. and Hoek, E.M.V. (2020). Techno-economic analysis of RO desalination of produced water for beneficial reuse in California. *Water*, **12**, 1850.

Ellouze, E., Tahri, N., and Amar, R.B. (2012). Enhancement of textile wastewater treatment process using Nanofiltration. *Desalination*, **286**, 16–23.

Elsehly, E.M. (2020). Enhanced removal of Ni (II) from aqueous solutions by effective acid functionalization of carbon nanotube based filters. *Egyptian Journal of Chemistry*, **63**, 3861–3871.

Elsehly, E.M., Chechenin, N.G., Makunin, A.V. et al. (2017). Functionalized carbon nanotubes based filters for chromium removal from aqueous solutions. *Water Science and Technology*, **75**, 1564–1571.

Elsehly, E.M., Chechenin, N.G., Makunin, A.V. et al. (2016). Characterization of functionalized multiwalled carbon nanotubes and application as an effective filter for heavy metal removal from aqueous solutions. *Chinese Journal of Chemical Engineering*, **24**, 1695–1702.

Emadzadeh, D., Matsuura, T., Ghanbari, M. et al. (2019). Hybrid forward osmosis/ultrafiltration membrane bag for water purification. *Desalination*, **468**, 114071.

Endo, M., Muramatsu, H., Hayashi, T. et al. (2005). 'Buckypaper' from coaxial nanotubes. *Nature*, **433**, 476–476.

Fathizadeh, M., Aroujalian, A., and Raisi, A. (2011). Effect of added NaX nano-zeolite into polyamide as a top thin layer of membrane on water flux and salt rejection in a reverse osmosis process. *Journal of Membrane Science*, **375**, 88–95.

Galanakis, C.M. (2012). Recovery of high added-value components from food wastes: conventional, emerging technologies and commercialized applications. *Trends in Food Science & Technology*, **26**, 68–87.

Gevers, L.E.M., Vankelecom, I.F.J., and Jacobs, P.A. (2005). Zeolite filled polydimethylsiloxane (PDMS) as an improved membrane for solvent-resistant nanofiltration (SRNF). *Chemical Communications*, 2500–2502. https://doi.org/10.1039/B500401B

Gevers, L.E.M., Vankelecom, I.F.J., and Jacobs, P.A. (2006). Solvent-resistant nanofiltration with filled polydimethylsiloxane (PDMS) membranes. *Journal of Membrane Science*, **278**, 199–204.

Ghaly, A.E., Ananthashankar, R., Alhattab, M.T. et al. (2013). Production, characterization and treatment of textile effluents: a critical review. *Journal of Chemical Engineering Process Technology*, **5**, 1–18.

Gitis, V. and Hankins, N. (2018). Water treatment chemicals: trends and challenges. *Journal of Water Process Engineering*, **25**, 34–38.

Gleick, P.H. and Cooley, H. (2021). Freshwater scarcity. *Annual Review of Environment and Resources*, **46**, 319–348.

Goh, K., Karahan, H.E., Wei, L. et al. (2016). Carbon nanomaterials for advancing separation membranes: a strategic perspective. *Carbon*, **109**, 694–710.

Goh, P.S., Ismail, A.F., and Ng, B.C. (2013). Carbon nanotubes for desalination: performance evaluation and current hurdles. *Desalination*, **308**, 2–14.

Goh, P.S., Ng, B.C., Lau, W.J. et al. (2015). Inorganic nanomaterials in polymeric ultrafiltration membranes for water treatment. *Separation & Purification Reviews*, **44**, 216–249.

Greenlee, L.F., Lawler, D.F., Freeman, B.D. et al. (2009). Reverse osmosis desalination: water sources, technology, and today's challenges. *Water Research*, **43**, 2317–2348.

Hakami, M.W., Alkhudhiri, A., Al-batty, S. et al. (2020). Ceramic microfiltration membranes in wastewater treatment: filtration behavior, fouling and prevention. *Membranes*, **10**, 248.

Hankins, N. and Singh, R. (2016). *Emerging Membrane Technology for Sustainable Water Treatment*. Elsevier.

Hinds, B.J., Chopra, N., Rantell, T. et al. (2004). Aligned multiwalled carbon nanotube membranes. *Science*, **303**, 62–65.

Holt, J.K., Park, H.G., Wang, Y. et al. (2006). Fast mass transport through sub-2-nanometer carbon nanotubes. *Science*, **312**, 1034–1037.

Hummer, G., Rasaiah, J.C., and Noworyta, J.P. (2001). Water conduction through the hydrophobic channel of a carbon nanotube. *Nature*, **414**, 188–190.

Islam, A., Raghupathy, B.P.C., Sivakumaran, M.V. et al. (2022). Ceramic membrane for water filtration: addressing the various concerns at once. *Chemical Engineering Journal*, **446**, 137386.

Ismail, F., Khulbe, K.C., and Matsuura, T. (2018). *Reverse Osmosis*. Elsevier.

Jadav, G.L. and Singh, P.S. (2009). Synthesis of novel silica-polyamide nanocomposite membrane with enhanced properties. *Journal of Membrane Science*, **328**, 257–267.

Jarrar, R., Abbas, M.K.G., and Al-Ejji, M. (2024). Environmental remediation and the efficacy of ceramic membranes in wastewater treatment—a review. *Emergent Materials*, **7**, 1295–1327.

Jeong, B.-H., Hoek, E.M.V., Yan, Y. et al. (2007). Interfacial polymerization of thin film nanocomposites: a new concept for reverse osmosis membranes. *Journal of Membrane Science*, **294**, 1–7.

Jhaveri, J.H. and Murthy, Z.V.P. (2016). A comprehensive review on anti-fouling nanocomposite membranes for pressure driven membrane separation processes. *Desalination*, **379**, 137–154.

Jiang, M., Ye, K., Deng, J. et al. (2018). Conventional ultrafiltration as effective strategy for dye/salt fractionation in textile wastewater treatment. *Environmental Science & Technology*, **52**, 10698–10708.

Jin, L., Shi, W., Yu, S. et al. (2012). Preparation and characterization of a novel PA-SiO2 nanofiltration membrane for raw water treatment. *Desalination*, **298**, 34–41.

Kang, S., Herzberg, M., Rodrigues, D.F. et al. (2008). Antibacterial effects of carbon nanotubes: size does matter! *Langmuir*, **24**, 6409–6413.

Kar, S., Bindal, R.C., Prabhakar, S. et al. (2008). Potential of carbon nanotubes in water purification: an approach towards the development of an integrated membrane system. *International Journal of Nuclear Desalination*, **3**, 143–150.

Karan, S., Samitsu, S., Peng, X. et al. (2012). Ultrafast viscous permeation of organic solvents through diamond-like carbon nanosheets. *Science*, **335**, 444–447.

Kathi, S. and El Din Mahmoud, A. (2024). Trends in effective removal of emerging contaminants from wastewater: a comprehensive review. *Desalination and Water Treatment*, **317**, 100258.

Khin, M.M., Nair, A.S., Babu, V.J. et al. (2012). A review on nanomaterials for environmental remediation. *Energy & Environmental Science*, **5**, 8075–8109.

Kim, S., Fornasiero, F., Park, H.G. et al. (2014). Fabrication of flexible, aligned carbon nanotube/polymer composite membranes by in-situ polymerization. *Journal of Membrane Science*, **460**, 91–98.

Kim, Y.A., Muramatsu, H., Hayashi, T. et al. (2006). Fabrication of high-purity, double-walled carbon nanotube buckypaper. *Chemical Vapor Deposition*, **12**, 327–330.

Kumar, A., Saini, K., and Bhaskar, T. (2020). Hydochar and biochar: production, physicochemical properties and techno-economic analysis. *Bioresource Technology*, **310**, 123442.

Le, N.L. and Nunes, S.P. (2016). Materials and membrane technologies for water and energy sustainability. *Sustainable Materials and Technologies*, **7**, 1–28.

Lee, B., Baek, Y., Lee, M. et al. (2015). A carbon nanotube wall membrane for water treatment. *Nature Communications*, **6**, 7109.

Lee, J., Jeong, S., and Liu, Z. (2016). Progress and challenges of carbon nanotube membrane in water treatment. *Critical Reviews in Environmental Science and Technology*, **46**, 999–1046.

Lin, J., Ye, W., Baltaru, M.-C. et al. (2016). Tight ultrafiltration membranes for enhanced separation of dyes and Na2SO4 during textile wastewater treatment. *Journal of Membrane Science*, **514**, 217–228.

Llamas, B., Suárez-rodríguez, M.C., González-lópez, C.V. et al. (2021). Techno-economic analysis of microalgae related processes for CO2 bio-fixation. *Algal Research*, **57**, 102339.

Mahmoud, A.E. and Mostafa, E. (2023). Nanofiltration membranes for the removal of heavy metals from aqueous solutions: preparations and applications. *Membranes*, **13**, 789.

Mahmoud, A.E.D. (2020a). Graphene-based nanomaterials for the removal of organic pollutants: insights into linear versus nonlinear mathematical models. *Journal of Environmental Management*, **270**, 110911.

Mahmoud, A.E.D. (2020b). Nanomaterials: green synthesis for water applications. *Handbook of Nanomaterials and Nanocomposites for Energy and Environmental Applications*, 1–21. Springer International Publishing.

Mahmoud, A.E.D. (2022). Recent advances of TiO2 nanocomposites for photocatalytic degradation of water contaminants and rechargeable sodium ion batteries. In: *Advances in Nanocomposite Materials for Environmental and Energy Harvesting Applications* (eds. A.E. Shalan, A.S. Hamdy Makhlouf, and S. Lanceros-Méndez), Cham: Springer International Publishing.

Mahmoud, A.E.D., Franke, M., Stelter, M. et al. (2020). Mechanochemical versus chemical routes for graphitic precursors and their performance in micropollutants removal in water. *Powder Technology*, **366**, 629–640.

Mahmoud, A.E.D. and Krasucka, P. (2023). Global water challenges and sustainability. In: *Artificial Intelligence and Modeling for Water Sustainability*. CRC Press.

Mahmoud, A.E.D., Stolle, A., and Stelter, M. (2018). Sustainable synthesis of high-surface-area graphite oxide via dry ball milling. *ACS Sustainable Chemistry & Engineering*, **6**, 6358–6369.

Mahmud, R., Moni, S.M., High, K. et al. (2021). Integration of techno-economic analysis and life cycle assessment for sustainable process design – a review. *Journal of Cleaner Production*, **317**, 128247.

Majumder, M., Chopra, N., and Hinds, B.J. (2005). Effect of tip functionalization on transport through vertically oriented carbon nanotube membranes. *Journal of the American Chemical Society*, **127**, 9062–9070.

Mallada, R. and Menéndez, M. (2008). *Inorganic Membranes: Synthesis, Characterization and Applications*. Elsevier.

Matshetshe, K., Sikhwivhilu, K., Ndlovu, G. et al. (2021). Antifouling and antibacterial β-cyclodextrin decorated graphene oxide/polyamide thin-film nanocomposite reverse osmosis membranes for desalination applications. *Separation and Purification Technology*, **278**, 119594.

Matsuura, T. (2002). Chapter 30 Recent progresses in membrane separation science and technologies: a review. *Comprehensive Analytical Chemistry*, **37**, 983–1004. Elsevier.

Mekonnen, M. and Hoekstra, A. (2016). Four billion people facing severe water scarcity. *Science Advances*, **2**, e1500323–e1500323.

Mohammadi, T. and Esmaeelifar, A. (2004). Wastewater treatment using ultrafiltration at a vegetable oil factor. *Desalination*, **166**, 329–337.

Mulligan, C.N., Yong, R.N., and Gibbs, B.F. (2001). Surfactant-enhanced remediation of contaminated soil: a review. *Engineering Geology*, **60**, 371–380.

Muramatsu, H., Hayashi, T., Kim, Y.A. et al. (2005). Pore structure and oxidation stability of double-walled carbon nanotube-derived bucky paper. *Chemical Physics Letters*, **414**, 444–448.

Muro, C., Riera, F., and Díaz-Nava, C. (2012). Membrane separation process in wastewater treatment of food industry. In: *Food Industrial Processes: Methods and Equipment*. BoD–Books on Demand.

Murthy, G.S. (2022). Chapter Two - Techno-economic assessment. In: *Biomass, Biofuels, Biochemicals* (eds. G.S. Murthy, E. Gnansounou, and S.K. Khanal et al.). Elsevier.

Narayan, R.J., Berry, C.J., and Brigmon, R.L. (2005). Structural and biological properties of carbon nanotube composite films. *Materials Science and Engineering: B*, **123**, 123–129.

Nasrabadi, A.T. and Foroutan, M. (2011). Ion-separation and water-purification using single-walled carbon nanotube electrodes. *Desalination*, **277**, 236–243.

Nataraj, S.K., Hosamani, K.M., and Aminabhavi, T.M. (2006). Distillery wastewater treatment by the membrane-based nanofiltration and reverse osmosis processes. *Water Research*, **40**, 2349–2356.

Nikolakis, V., Xomeritakis, G., Abibi, A. et al. (2001). Growth of a faujasite-type zeolite membrane and its application in the separation of saturated/unsaturated hydrocarbon mixtures. *Journal of Membrane Science*, **184**, 209–219.

Noy, A., Park, H.G., Fornasiero, F. et al. (2007). Nanofluidics in carbon nanotubes. *Nano Today*, **2**, 22–29.

Obaid, M., Abdelkareem, M.A., Kook, S. et al. (2020). Breakthroughs in the fabrication of electrospun-nanofiber-supported thin film composite/nanocomposite membranes for the forward osmosis process: a review. *Critical Reviews in Environmental Science and Technology*, **50**, 1727–1795.

Obaideen, K., Shehata, N., Sayed, E.T.et al. (2022). The role of wastewater treatment in achieving sustainable development goals (SDGs) and sustainability guideline. *Energy Nexus*, **7**, 100112.

Obotey Ezugbe, E. and Rathilal, S. (2020). Membrane technologies in wastewater treatment: a review. *Membranes*, **10**, 89.

Parham, H., Bates, S., Xia, Y. et al. (2013). A highly efficient and versatile carbon nanotube/ceramic composite filter. *Carbon*, **54**, 215–223.

Paul, M. and Jons, S.D. (2016). Chemistry and fabrication of polymeric nanofiltration membranes: a review. *Polymer*, **103**, 417–456.

Peng, G., Sharshir, S.W., Wang, Y. et al. (2021). Potential and challenges of improving solar still by micro/nano-particles and porous materials - a review. *Journal of Cleaner Production*, **311**, 127432.

Petrinic, I., Korenak, J., Povodnik, D. et al. (2015). A feasibility study of ultrafiltration/reverse osmosis (UF/RO)-based wastewater treatment and reuse in the metal finishing industry. *Journal of Cleaner Production*, **101**, 292–300.

Prasek, J., Drbohlavova, J., Chomoucka, J. et al. (2011). Methods for carbon nanotubes synthesis—review. *Journal of Materials Chemistry*, **21**, 15872–15884.

Qin, Y., Mueller, N.D., Siebert, S.et al. (2019). Flexibility and intensity of global water use. *Nature Sustainability*, **2**, 515–523.

Radjenovic, J., Matošić, M., Mijatović, I. et al. (2008). Membrane Bioreactor (MBR) as an advanced wastewater treatment technology. In: *Handbook of Environmental Chemistry, Volume 5, Water Pollution*, 5. Springer.

Rautenbach, R., Linn, T., and Eilers, L. (2000). Treatment of severely contaminated waste water by a combination of RO, high-pressure RO and NF — potential and limits of the process. *Journal of Membrane Science*, **174**, 231–241.

Rodriguez-Mozaz, S., Ricart, M., Köck-Schulmeyer, M. et al. (2015). Pharmaceuticals and pesticides in reclaimed water: efficiency assessment of a microfiltration–reverse osmosis (MF–RO) pilot plant. *Journal of Hazardous Materials*, **282**, 165–173.

Romanos, G.E., Athanasekou, C.P., Katsaros, F.K. et al. (2012). Double-side active TiO_2-modified nanofiltration membranes in continuous flow photocatalytic reactors for effective water purification. *Journal of Hazardous Materials*, **211–212**, 304–316.

Salahi, A., Badrnezhad, R., Abbasi, M. et al. (2011). Oily wastewater treatment using a hybrid UF/RO system. *Desalination and Water Treatment*, **28**, 75–82.

Saqib Nawaz, M., Alamoudi, T., Soukane, S. et al. (2022). Performance and implications of forward osmosis-membrane distillation hybrid system for simultaneous treatment of different real produced water streams. *Chemical Engineering Journal*, **450**, 138479.

Siagian, U.W.R., Khoiruddin, K., Wardani, A.K. et al. (2021). High-performance ultrafiltration membrane: recent progress and its application for wastewater treatment. *Current Pollution Reports*, **7**, 448–462.

Song, X., Wang, L., Tang, C.Y. et al. (2015). Fabrication of carbon nanotubes incorporated double-skinned thin film nanocomposite membranes for enhanced separation performance and antifouling capability in forward osmosis process. *Desalination*, **369**, 1–9.

Sonune, A. and Ghate, R. (2004). Developments in wastewater treatment methods. *Desalination*, **167**, 55–63.

Šostar-Turk, S., Simonič, M., and Petrinić, I. (2005). Wastewater treatment after reactive printing. *Dyes and Pigments*, **64**, 147–152.

Srivastava, A., Srivastava, O.N., Talapatra, S. et al. (2004). Carbon nanotube filters. *Nature Materials*, **3**, 610–614.

Sun, X., Wang, C., Li, Y. et al. (2015). Treatment of phenolic wastewater by combined UF and NF/RO processes. *Desalination*, **355**, 68–74.

Szabó, A., Perri, C., Csató, A. et al. (2010). Synthesis methods of carbon nanotubes and related materials. *Materials*, **3**, 3092–3140.

Takht Ravanchi, M., Kaghazchi, T., and Kargari, A. (2009). Application of membrane separation processes in petrochemical industry: a review. *Desalination*, **235**, 199–244.

Tavakkoli, S., Lokare, O.R., Vidic, R.D. et al. (2017). A techno-economic assessment of membrane distillation for treatment of Marcellus shale produced water. *Desalination*, **416**, 24–34.

Turlan, D., Urriolabeitia, E.P., Navarro, R. et al. (2001). Separation of Pd complexes from a homogeneous solution using zeolite membranes. *Chemical Communications*, **24**, 2608–2609. Royal Society of Chemistry.

Van Gestel, T., Kruidhof, H., Blank, D.H.A. et al. (2006). ZrO_2 and TiO_2 membranes for nanofiltration and pervaporation: Part 1. Preparation and characterization of a corrosion-resistant ZrO2 nanofiltration membrane with a MWCO<300. *Journal of Membrane Science*, **284**, 128–136.

Verliefde, A.R.D., Cornelissen, E.R., Heijman, S.G.J. et al. (2009). Construction and validation of a full-scale model for rejection of organic micropollutants by NF membranes. *Journal of Membrane Science*, **339**, 10–20.

Vinardell, S., Astals, S., Mata-Alvarez, J. et al. (2020). Techno-economic analysis of combining forward osmosis-reverse osmosis and anaerobic membrane bioreactor technologies for municipal wastewater treatment and water production. *Bioresource Technology*, **297**, 122395.

Wan, R., Lu, H., Li, J. et al. (2009). Concerted orientation induced unidirectional water transport through nanochannels. *Physical Chemistry Chemical Physics*, **11**, 9898–9902.

Wang, Y., Chen, X., Zhang, J. et al. (2009). Investigation of microfiltration for treatment of emulsified oily wastewater from the processing of petroleum products. *Desalination*, **249**, 1223–1227.

Wu, B., Li, X., An, D. et al. (2014). Electro-casting aligned MWCNTs/polystyrene composite membranes for enhanced gas separation performance. *Journal of Membrane Science*, **462**, 62–68.

Wu, M.-B., Lv, Y., Yang, H.-C. et al. (2016). Thin film composite membranes combining carbon nanotube intermediate layer and microfiltration support for high nanofiltration performances. *Journal of Membrane Science*, **515**, 238–244.

Yang, H.Y., Han, Z.J., Yu, S.F. et al. (2013). Carbon nanotube membranes with ultrahigh specific adsorption capacity for water desalination and purification. *Nature Communications*, **4**, 2220.

Yang, X., Yuan, L., Peterson, V.K. et al. (2011). Open-ended aligned carbon nanotube arrays produced using CO2-assisted floating-ferrocene chemical vapor deposition. *The Journal of Physical Chemistry C*, **115**, 14093–14097.

Yip, N.Y., Tiraferri, A., Phillip, W.A. et al. (2010). High performance thin-film composite forward osmosis membrane. *Environmental Science & Technology*, **44**, 3812–3818.

Yordanov, D. (2010). Preliminary study of the efficiency of ultrafiltration treatment of poultry slaughterhouse wastewater. *Bulgarian Journal of Agricultural Science*, **16**, 700–704.

Yuan, D., Lin, W., Guo, R. et al. (2012). The fabrication of vertically aligned and periodically distributed carbon nanotube bundles and periodically porous carbon nanotube films through a combination of laser interference ablation and metal-catalyzed chemical vapor deposition. *Nanotechnology*, **23**, 215303.

Yusuf, A., Sodiq, A., Giwa, A. et al. (2020). A review of emerging trends in membrane science and technology for sustainable water treatment. *Journal of Cleaner Production*, **266**, 121867.

Zhang, X., Du, A.J., Lee, P. et al. (2008). TiO_2 nanowire membrane for concurrent filtration and photocatalytic oxidation of humic acid in water. *Journal of Membrane Science*, **313**, 44–51.

Zhao, S., Zou, L., and Mulcahy, D. (2011). Effects of membrane orientation on process performance in forward osmosis applications. *Journal of Membrane Science*, **382**, 308–315.

Zhao, S., Zou, L., Tang, C.Y. et al. (2012). Recent developments in forward osmosis: opportunities and challenges. *Journal of Membrane Science*, **396**, 1–21.

Zhao, X., Li, J., and Liu, C. (2017). A novel TFC-type FO membrane with inserted sublayer of carbon nanotube networks exhibiting the improved separation performance. *Desalination*, **413**, 176–183.

Zheng, J., Li, M., Yu, K. et al. (2017). Sulfonated multiwall carbon nanotubes assisted thin-film nanocomposite membrane with enhanced water flux and anti-fouling property. *Journal of Membrane Science*, **524**, 344–353.

21

Economic Benefits of Source Separation of Wastewater

*Anushree R. Gahlowt[1], Ambika Arkatkar[1], and Anand Kishore Kola[2]**

[1]*Department of Biotechnology, Veer Narmad South Gujarat University, Surat, Gujarat, India*
[2]*Department of Chemical Engineering, National Institute of Technology Warangal, Warangal, Telangana, India*

*Corresponding author: kola@nitw.ac.in

21.1 Introduction

Wastewater treatment involves many systems, each with feasibility and economic compatibility. The development and acceptance of these systems help design goals that further contribute to different sectors of the country. One of the key factors that evaluates the standards to be incorporated in the action plan includes revenue generation as opposed to investments in the project or the systems, along with the robustness of the strategies involved.

The first man-made drains were found in the Mesopotamian Empire in Iraq (4000–2500 BC). The drainage system underwent a major expansion around 5000 years ago during the Bronze Age. In the Indus Valley, the resident Harappan civilization and the Minoans on the island of Crete (c. 3200–1100 BC) created effective sewerage systems. Sewerage systems that first appeared in North America and Europe about 150 years ago were on par with these ancient systems (de Feo et al., 2014).

Over the years, the idea of wastewater treatment gained approval, and centralized wastewater treatment systems emerged. These traditional wastewater systems often mix various streams of wastewater before treatment and are thus source-to-sink systems. There are some management systems classified in this approach, mainly stormwater management that prevents discharge retardation and works

Wastewater to Resource Recovery: Applying the Circular Economy Toward Sustainable Development, First Edition. Edited by Arvind Kumar Mungray et al. Published 2026 by John Wiley & Sons, Inc.

for rainwater treatment. Stormwater is the rainwater runoff and melted snow water on the roof that runoff. The basic idea is to prevent water runoff to drains by collecting it in stormwater detention tanks (SWDTs) (D'ambrosio et al., 2021; Orth, 2007). Other systems include centralized sewage treatment plants (STPs) where the combined sewage from the sewer network is treated. The combined sewage includes domestic wastewater, stormwater, and field and industrial discharge. The effluent is treated at various levels, focusing on mechano-biological treatment. Following treatment, the wastewater is released into the bodies of water (Orth, 2007). The process is complex and increases the consumption of resources, thus increasing operational costs and posing a lack of efficient resource recovery. India has developed a network of STPs in cities, but the growing demand is putting excess pressure on the existing centralized treatment facility. This has created a scenario of inadequate treatment of wastewater (Figure 21.1).

Source separation of wastewater can help in optimizing the segregated waste and deriving components that generate revenue. The decentralized approach introduces a theme that favors a sustainable approach, thus closing the loop for a circular economy. The greywater can be separated efficiently from the source of origin as compared to the blackwater (Ghaitidak and Yadav, 2013; Larsen et al., 2015). Therefore, newer systems can derive a favorable cost-based analysis based on recovery and reuse efficiency.

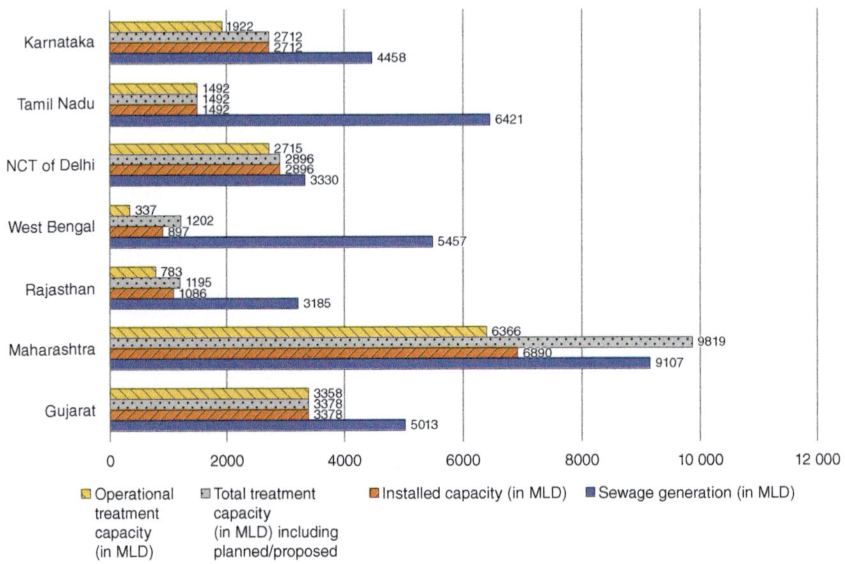

Figure 21.1 The graph representing the sewage generation and treatment facilities of states having metropolitan cities in India.

In this chapter, we will study the notion of source separation of wastewater by challenging its applications and limitations, and segueing to some case studies, followed by a perspective that evaluates the economics behind it.

21.2 Why Source Separation Is Important for the Treatment of Wastewater?

Source separation of wastewater deals with segregation at the source. The scale of source separation will depend on its application at the individual or community level. The decentralization of procurement of wastewater enables treatment of the segregated components accordingly to optimize more accuracy (Figure 21.2). A loop is created where the byproducts obtained after consumption are recycled to procure a potable sense. The key objective projects the idea of longer availability of products and raw materials in the economy, while the generated waste should be treated as secondary raw material. (CPCB, 2024) This derails the notion of a linear-driven resource economy in favor of a circular economy. The circular economy aims at recycling products to replenish the raw materials and waiving off the excess burden on exhaustible natural reserves.

The gap between the treatment capacity and sewage generation can be studied in the report published by CPCB in 2020. The report provides a detailed inventory of the STPs in India. The following graph (Figure 21.1) shows the gap between available treatment capacities and actual sewage generated (CPCB, 2021).

Central Public Health and Environmental Engineering Organisation (CPHEEO) has prescribed standards for reuse of treated wastewater for a myriad of purposes like horticulture, irrigation, non-contact impoundments, and washing (CPCB, 2024). Approximately 60% of India's population is dependent on on-site sanitation systems (OSS) and the rest depends on off-site systems (Ministry of Housing and Urban Affairs (MoHUA), 2021). The centralized treatment systems do not recover the raw materials efficiently. The decentralized systems segregate the components of wastewater by facilitating an efficient recycling process. The optimization of these systems provides newer alternatives and techniques to recycle the resources.

Source separation of wastewater yields resources that can be utilized again. This includes nutrient recovery, energy production, and processing of potable and non-potable water (based on requirements). In accordance with the National Building Code of India (2016), residential complexes would employ sullage (grey water) and treated sewage (black water) for gardening, fire safety, and toilet flushing (Goyal and Kumar, 2022).

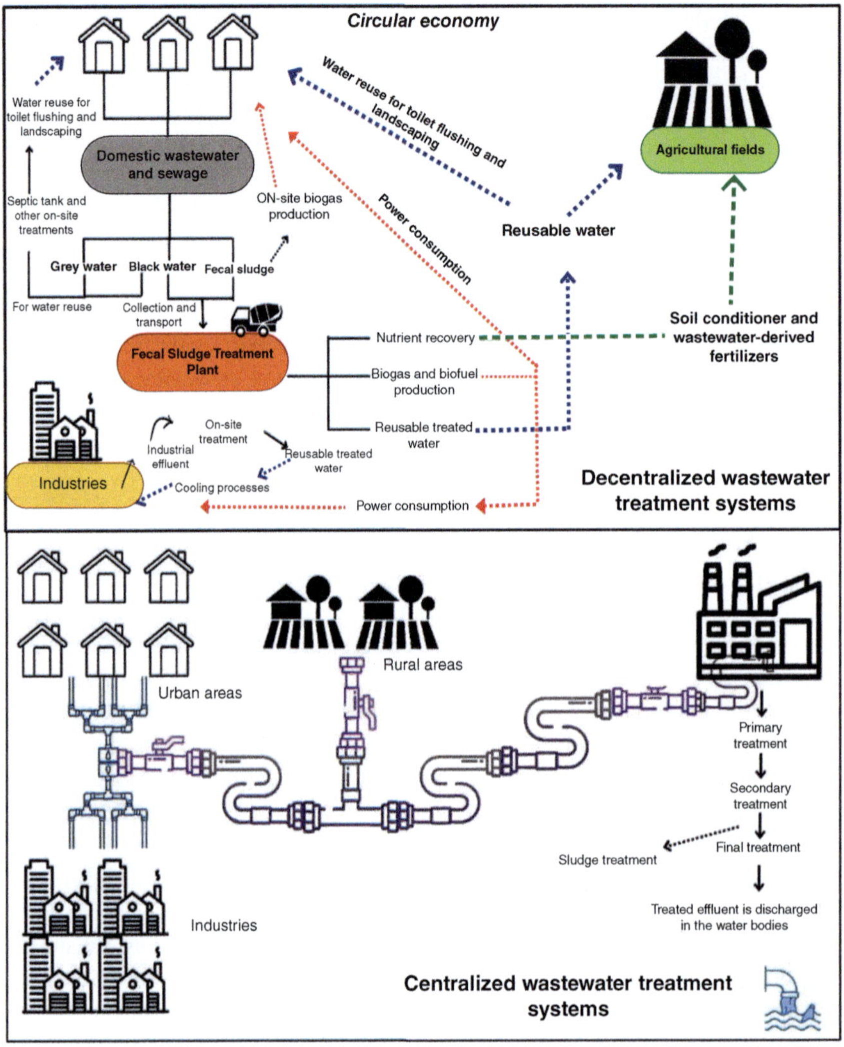

Figure 21.2 Comparison of decentralized and centralized wastewater treatment systems.

21.2.1 Nutrient Recovery

Excessive erosion, contamination, and degradation of natural resources have depleted nutrients from the soil over the years. The macronutrient deficiency of nitrogen, phosphorus, and potassium, as well as other nutrients like zinc, has affected crop yield over the years (Larsen et al., 2015; McKinsey, 2013). Therefore, the use of fertilizers is opted for a better yield. The fertilizer in untreated wastewater often leads to eutrophication, thereby affecting the ecosystem of the water body by lowering the dissolved oxygen levels as a result of increased vegetative growth. The wastewater is treated for a nutrient recovery process to extract nutrients like nitrogen, phosphorus, and potassium. This refills the inventory of required nutrients, which can be commercialized as fertilizers while combating water pollution. The use of recovered nutrients guides toward a sustainable approach (de Boer et al., 2018).

Although the nutrients are recovered, their plant availability is significant. The "plant availability" refers to the uptake of nutrients by the plants. The higher availability indicates easier uptake of nutrients by the plants. Struvite ($NH_4MgPO_4{\cdot}6H_2O$) and polyphosphates can be recovered from the wastewater, and they have more plant availability as opposed to mineral fertilizers. Therefore, the focus can be guided toward recovery instead of phosphate elimination. Phosphate elimination is the aftermath of dealing with eutrophic water bodies. However, nutrient recovery acts as a preventive measure and transforms the obtained nutrients into secondary raw materials (De-Bashan and Bashan, 2004; Günther et al., 2018; Satyaprakash et al., 2017).

In water recovery operations, membrane techniques like reverse osmosis (RO) are employed. These membrane processes produce RO brine. RO brine is produced by filtering a filtrate with low salts and rejected concentrates through an RO membrane under high pressure. It is laborious to dispose of RO brine that is obtained from thermal power plants. Brine handling is significant when considering the environmental risks and high management costs. The incorporation of water recovery from the brine can guide toward a circular economy approach. Up to 80% of water could be recovered from RO brine streams for textile mills in Turkey. The expected annual cost savings for salt and water, respectively, would have been \$37 000 and \$176 256 for a full-scale operation in the textile mill (Partal et al., 2022). The brine has a mineral content 5–10 times higher than the raw reclaimed water, and it contains 10–20% of the incoming water supply. Although flash evaporation can be applied, it is costly for the thermal power plant. Alternatively, the RO brine can play a role as a phosphate precipitant for its recovery from the source-separated urine. The results were effective and cost-efficient (Alhazmy, 2014; Tian et al., 2016).

21.2.2 Energy Savings

A circularly managed Wastewater Treatment Plant (WWTP) aims to recycle the by-products, and the biogas produced could be utilized to satisfy the internal energy demands of the wastewater treatment plant (Shanmugam et al., 2022).

An Excel-based model was developed that focused on the literature data to observe and compare energy and water balance, recovery of nutrients, chemical usage, effluent quality, and land area requirement. The model was designed for the European system, focusing mainly on the Netherlands. Four sanitation concepts were studied:

1. Centralized system
2. Centralized system with source-separation of urine
3. Source-separation system with segregation of black water, kitchen refuse, and grey water
4. Source-separation system with segregation of urine, feces, kitchen refuse, and grey water.

The concept of centralized sanitation had the highest primary energy consumption of 914 MJ/cap/year. When taking indirect energy gain into account, the centralized system with urine source separation decreased energy consumption from 687 MJ/cap/year to 501 MJ/cap/year. The indirect energy gain was attributed to water saving, reuse, and recovery of nutrients. The third sanitation concept, a source-separation system that separated black water, kitchen waste, and grey water, had a primary power usage of 767 MJ/cap/year, and the power usage with an indirect energy gain was observed to be 522 MJ/cap/year. The source-separation-based sanitation concept for urine separation listed the power consumption at 567 MJ/cap/year with the gravity urine-diverting toilet (UDT), and 555 MJ/cap/year with the vacuum urine-diverting toilet UDT. While including the indirect energy gain, the gravity UDTs attained a lower power consumption reading of 208 MJ/cap/year, while the vacuum UDTs attained 190 MJ/cap/year. The comparison derived while assessing these models aided in understanding the feasibility of these systems based on power consumption (Tervahauta et al., 2013).

21.2.3 Biofuel

Fossil fuels are exhaustible natural resources. The exploitative utilization of fossil fuels for industrial and urban usage has burdened the natural reserves. The rate of consumption far exceeds the rate of formation. Exploring alternatives to fossil fuels is recommended to prevent the jeopardy of the sectors heavily dependent on energy consumption. Biofuels and biodiesels are obtained from plants, algae, animal and agricultural waste. The cultivation of these crops pilots a sustainable approach and waives some burden off of natural fossil fuel reserves. The aviation and automotive industries may soon find usage for the biofuels produced by integrated microalgae and jatropha biorefineries. Fuels hydroprocessed from algae and jatropha oils will be tested at the US Air Force Research Laboratory (Giwa et al., 2018).

Native to tropical climates, *Jatropha curcas L.* is a perennial bush that bears oil and can withstand droughts. The seeds of Jatropha are used to extract the biodiesel. Algal alternatives are also recommended and introduced in the biodiesel research. In India, the study supported the idea of algae-based biodiesel as an alternative or to be supplemented as an additional source to jatropha biodiesel for the country's biodiesel system. While analyzing the life cycle assessment (LCA) impact results, the reduction in GHG emissions for both biodiesels was 36–40% (Ajayebi et al., 2013). Utilization of wastewater can save some nutrient costs. This salvages the idea of using treated sewage effluents (TSE) to irrigate the biodiesel crops. The study conducted on the Jatropha plantation yield in the Southern Moroccan region irrigated the crops with TSE. Jatropha is salt-sensitive, and recurrent leaching and increased salinity reduced the crop yield. Regulation and maintenance of salt concentration were advisable (Rajaona et al., 2012).

The biomass production cost for algae is high, but the industrial potential and associated benefits are more sought-after when driving a comparison with other biomass crops. Most research in microalgal biofuel production often faces a common factor of hindrance, which is the cost involved in microalgal cultivation. The cost of growth nutrients increases the cultivation costs. However, cultivation of these microalgae in wastewater saves the nutrient cost and reduces the operation cost of municipal wastewater treatment by 32–35%. The algal alternatives fare well when treated with TSE and also cover the nutrient costs (Ashokkumar et al., 2019).

Dunaliella, a microalga, when cultivated on dairy wastewater (as a culture medium), was observed to have more biomass content than when grown on artificial seawater medium (Mansouri and Eskandari Godar, 2024). This microalgae can be employed in biorefineries and used for its biofuel capacity (Kalra et al., 2022; Mansouri and Eskandari Godar, 2024). Similarly, the algal species *Synechocystis* could be employed as a biomass material for integrated biorefineries and the production of liquid biofuels at the commercial level. The study used municipal wastewater (brown-grey wastewater) collected from the waste treatment plant (WTP) in Chennai, India (Ashokkumar et al., 2019).

21.2.4 Urine-derived Products

The source-separated urine can be utilized to recover products for its nutrient value as a fertilizer. Field-scale experiments in Finland used urine as a fertilizer for growing barley. The barley yield obtained is equivalent to the yield obtained by the crop grown with mineral fertilizer. The experiments conducted in Sweden in 1996–1998 also showed that urine used as barley fertilizer was as efficient as commercial fertilizer (Johansson, 2000; Viskari et al., 2018). However, an average of 2–10% loss of nitrogen from the field to the atmosphere is observed when the urine fertilizer is used. This loss can amount to up to 30–40% (Johansson, 2000).

The novel fertilizer-producing urinals aid in nutrient recovery. The scaling potential of urine is being explored, and calcium phosphate, which is formed in the collection tank of the urinal, can be incorporated into producing bricks. The process is induced by microbes, where human urine is treated with calcium hydroxide that offers a urea-rich solution for microbe-induced calcium carbonate precipitation (MICP). This method is adopted in the formation of bio-bricks. 31 L of urine is required to produce a single bio-brick with a compressive strength of 2.7 MPa (Lambert and Randall, 2019). A study compared the brick-compression strength of bio-bricks, wherein the compression strength increased up to 70% through the addition of natural and synthetic fibers. These bio-bricks can be employed in infrastructures like pavement tiles (Goyal and Kumar, 2022; Li et al., 2020).

21.3 Challenges of Source Separation Systems and Solutions

The approach of decentralization deals with unburdening the sink resource. Many technologies, projects, and policies are designed to attain a pan-level feasibility status for the populace. The task, in itself, becomes arduous to achieve since it tends to strike an equilibrium across the factors involved altogether.

21.3.1 Lack of Monitoring Systems and Construction Spaces

Efficient hygiene practices are compromised when there is an inadequacy of regular monitoring of community-based sanitation systems. Consequently, the natural resources are contaminated, ultimately encouraging the spread of diseases. Diarrhea is the second most prevalent cause of death in children under the age of five (Wardlaw et al., 2010). The on-site systems (OSS) often face difficulty in functioning. Due to the absence of soakways, the untreated wastewater is directly discharged into open drains. Manual collection of sludge instead of using de-sludging equipment exposes the operator to sludge (Shi et al., 2010). Deficit of adequate treatment facilities that deal with collection, transportation, and further treatment of waste creates a gap between quality and quantity of generated and treatment of wastewater. The cost of OSS may be lower than sewer networks, but household investments incur more costs (Dodane et al., 2012). The unavailability of space in densely populated cities hinders the construction of OSS (Wankhade, 2015). Therefore, policy frameworks shall enumerate and address such constraints.

21.3.2 Inconsistent Regulations on Waste-derived Fertilizers

The Waste Framework Directive's End-Of-Waste (EoW) criteria (Articles 6(1) and (2)) specify whether designated product regulation can be applied to waste-recovered materials rather than waste-related treatment procedures. When waste-derived items are employed for a specified purpose and there is a market or demand for them, then they cease to be considered waste. The product must comply with existing legislation and product standards. The product must not have any detrimental effects on the environment or human health (Hukari et al., 2016).

The phosphate recovery processes vary from region to region, with their respective regulations. For instance, the existing legislation in the Netherlands leverages a higher ground for conventional fertilizers over waste-derived products. This stems from the variable composition and production processes (Barquet et al., 2020; de Boer et al., 2018). This monopoly affects the interests of the stakeholders. The stakeholders play a significant role in commercializing the products and availing them to the masses. Therefore, the regulations must be flexible, and incentives may be provided to allow the waste-derived products to compete in the market.

21.3.3 Presence of Pharmaceuticals, Heavy Metals, and Other Compounds

Conventional STPs fail to practice specific measures for the removal of pharmaceuticals. The presence of pharmaceuticals in the discharged effluent and the reusable effluent poses a threat to the feasibility of recovered water and other wastewater-derived products. Removal of these pharmaceuticals prevents the unwanted distribution of these compounds to farmlands and food chains. Some conventional STPs opt for biodegradation as the removal pathway for antibiotics. The sorption of antibiotics exhibits a marginal difference, and nevertheless, they are discharged (Chen et al., 2022).

In general, there is no evidence that conventional treatment systems remove antibiotics more effectively than decentralized treatment systems. However, some drugs or pharmaceuticals can be removed efficiently when treated under decentralized systems. A study was conducted that showed the presence of analgesic paracetamol and antidiabetic metformin at high concentrations in source-separated black water, in comparison to conventional wastewater. The pharmaceuticals were concentrated in loads in blackwater, and the subsequent anaerobic and aerobic degradation allowed for better removal. The removal efficiency of metoprolol, trimethoprim, naproxen, and hydrochlorothiazide was shown to be higher in the examined source-separated sanitation system than in the typical STP (Butkovskyi et al., 2015). The source-separated effluents can also contain pharmaceuticals and heavy metals. Direct use of such effluent is not advisable.

The accumulation of oil and grease from the greywater irrigation may decrease the capillary rise by 60% (Travis et al., 2008). Under conditions with higher levels of lead (Pb) and cadmium (Cd), the co-presence of these elements in the treated wastewater increases Cd uptake by plants like star grass by a 2.6-fold rate. This leads to the accumulation of these metals in the irrigated crops (Madyiwa et al., 2004). The photocatalysis of anti-cancer drugs present in laboratory-grade water can be considered as a solution. This would in turn allow for more exploration in designing recovery procedures (Janssens et al., 2020).

21.3.4 Concerns Toward Source-separated Urine

Source-separated urine establishes itself as a fertilizer due to its evidently higher nutrient content. However, the hydrolysis of urea causes ammonia volatilization. Stabilization of urine requires an urge to act to evaluate the potential fertilizer and take note of nutrient loss due to hydrolysis (Lv et al., 2020). Although domestic wastewater is managed efficiently, the hydrolysis reaction raises the potential for odor emission. The initial release of carbon dioxide and ammonia, along with the formation of volatile fatty acids in high volume, contributed to odor issues (Zhang et al., 2013). The precipitation of substance in the pipe increases the chances of clogging and blockage, which is observed in the case of waterless urinal setups with urine valves and sealant liquid cartridges (Yan et al., 2021).

The source-separated fresh urine can be stabilized by the heat-activated treatment of peroxydisulfate (PDS). The "stabilized urine" maintains its stability and resists secondary contamination by continuous and slow pH decrease due to PDS decomposition during the period of storage. A nitrogen loss of less than 8% was detected after 22 days of storage due to a reduction in ammonia volatilization (Lv et al., 2020). Dehydration of urine (from urine-diverting dry toilet, UDDT) in an alkaline medium concentrates the nutrients in the urine. The dry-end product obtained after liquid elimination can be easily transported and considered safe to be used as a fertilizer after four days of storage. The concerns related to precipitation in the UDDTs need more research and projects to find apt solutions to this major hindrance and aid in driving focus toward large-scale implementation (Senecal et al., 2018; Yan et al., 2021).

21.3.5 Perception Toward Source Separation

A study was conducted that surveyed countries and evaluated the interest and acceptance rates among the respondents regarding the consumption of food grown with human urine. About 68% of the survey-takers were in favor of human urine recycling, 59% stated their willingness to consume urine-fertilized food, whereas 11% of respondents were in disfavor of the notion and they believed that urine may pose risks to human health. A total of 89% of the participants agreed to consume food that was fertilized with cow excreta (Simha et al., 2021). A study evaluated the attitudes and perceptions of the rural community toward the eco-toilet system (ETS) in Mulanay, Quezon, as the study area in the Philippines. The ETS appeared most appealing to the citizens if water conservation was assured. Potential ETS users who were educated about the nutritional reuse of human excreta and willing to collect it had more positive sentiments toward the ETS (Ignacio et al., 2018).

21.4 Case Studies

Case studies can be considered as models to study and understand field situations and evaluate the possible future of a given technology or approach.

21.4.1 Case Studies on Centralized Treatment of Wastewater Systems

21.4.1.1 Case 1–Reuse of Treated Sewage by Surat's Industries

The Gujarat Pollution Control Board (GPCB) states that Surat, a city in Gujarat state of India, witnesses a total sewage generation of 1100 MLD. Surat Municipal Corporation (SMC) utilizes its two STPs for recycling and reusing sewage. The treated sewage of 115 MLD after its tertiary treatment is supplied to about 220 industries in Sachin and Pandesara Industrial Estates (PIE). Out of the two STPs, the Bamroli STP works at a treatment capacity of 215 MLD. The STP has been recycling and reusing 75 MLD sewage in two streams (40 MLD and 35 MLD). The plant receives water from houses through the sewage network. PIE prescribed their water quality requirement with TDS at <500 mg/lit. The water after treatment from the 40 MLD unit was mixed with potable water of equal capacity to form blended water or industrial-grade water. This ensured the maintenance of TDS levels and a revenue resource for SMC (Ministry of Housing and Urban Affairs (MoHUA), 2021). The capital cost of the STP was INR 85.10 Crore for 40 MLD units, with the O&M cost/year of INR 18.64 crore. The 35 MLD unit commenced its operation in March 2020 with a capital cost of INR 104.00 crore and an operation and maintenance cost of INR 12.66 crore. The treatment units comprise primary treatment, secondary treatment, and tertiary treatment. UASB and Sequential Batch Reactor (SBR) +UV have been used for secondary treatment. Activated carbon filter (ACF) for tertiary treatment and other treatments such as ultra-filtration (UF) and RO have been employed for efficient treatment of sewage. The RO reject is treated with a Flash Mixer and Clariflocculator and finally discharged via an outlet to the creek of the Arabian Sea. The RO permeate had been distributed to member industries in the PIE and Sachin GIDC (Gujarat Industrial Development Corporation). The permeate distribution is facilitated via pipeline networks and storage in the collection tanks. Flow meters had been installed, and a booked quantity of water was distributed to individual industries. The revenue generated by October 2022 has been enumerated to INR 248.95 crore and INR 65.81 crore for 40 MLD and 35 MLD units, respectively. The treated water cost in 2014 was INR 18.20 per kL, which has been increased to INR 32 per kL (CPCB, 2024).

21.4.1.2 Case 2–Koyambedu Treatment Plant of Chennai

Chennai Metropolitan Water Supply and Sewerage Board (CMWSSB) commissioned two tertiary treatment reverse osmosis (TTRO) projects to supply water to industrial zones in Chennai. Earlier, in 1991, Kodungaiyur STP supplied secondary-treated effluent to chemical and fertilizer industries like Manolo Petro Chemicals (Ministry of Housing and Urban Affairs (MoHUA), 2021). The TTRO plant located at Koyambedu, Chennai, has a treatment capacity of 45 MLD. The capital cost of the TTRO was INR 396.50 crore, and the operation cost for 15 years is INR 197.5 crore. The treated water has been supplied to the SIPCOT industries and other small industries in Pillaipakkam and Vallam Vadagal from 18 December 2019 onward. The treatment stage includes rapid sand gravity filters and pre-chlorination procedures. Basket strainers and ultrafiltration systems had been allotted for secondary treatment, whereas cartridge filters, RO system, and ozonation had been employed for tertiary treatment. One of the key benefits of this project was the diversion of freshwater toward Chennai citizens instead of earlier supplies to industries. The rate/tariff supply of treated water to the industries is fixed at INR 65 per kL. The total revenue generated thus far is INR 19.67 crore (CPCB, 2024).

21.4.2 Case Studies on Decentralized Treatment of Wastewater

21.4.2.1 Case 3–Fecal Sludge Management (FSM) in Devanahalli, Bangalore

Devanahalli lies in the North of Bangalore. The town is located 10 km away from the Bangalore International Airport. The area has faced water deficits in turn, affecting the establishment of new projects and new offices. Sewer networks are absent, and on-site sanitation systems (OSS) are prevalent. The sludge was being emptied by vacuum trucks into the stormwater drains or surface water bodies. Lack of sanitation and erratic desludging of collected waste along with the reuse of fecal sludge without any treatment for agriculture raised concerns (NITI Aayog, 2021). Nearly 78% of toilets had a single pit, 10% with septic tanks, and 3% had twin pits. The remaining 6% discharged into the stormwater drains. The transport latrine needed to be desludged every 2–5 weeks (Rath et al., 2020). The fecal sludge treatment plant (FSTP) was built at a capital cost of INR 90 lakhs and an operational cost of INR 24 lakhs. The revenue was generated with a model comprising composting. The plant was set up after the discussions between the

Consortium for DEWATS Dissemination India (CDD) and Devanahalli's Town Municipal Corporation (TMC). In April 2019, the FSTP was handed over to the TMC. For the last six months, the FSTP (6 KLD) was utilized to its full capacity (Consortium for DEWATS Dissemination Society, 2020). The Devanahalli's TMC was keen on identifying ways to obtain pathogen and contaminant-free fecal sludge, which aligned with CDD's search for a site to demonstrate aspects of Fecal Sludge Management (FSM). The soil conditioner obtained was pathogen-free, and it contributed to 29% of revenue in FSSM. The FSTP performance was enhanced by the addition of an Anaerobic Stabilization Reactor (ABR) and drying beds (NITI Aayog, 2021). There were no mobile parts installed in the FSTP; the sludge moved between modules through the pressure that was exerted by generated biogas or via gravitational pull. Greenhouse Solar Drier Roofs (GHSD) were constructed over the existing unplanted sludge drying bed. This aided in the maintenance of a high temperature inside the unit (around 60 °C), thus inactivating the pathogens and drying the sludge. After 15–18 days, the sludge cakes were dry enough to be considered for composting. The soil conditioner efficacy was verified and supported by the farmers with results reporting a twofold increase in their crop yield. The revenue generated included sales brought in by the sale of co-compost, user fees from desludging service provision, tipping fees, sale of vegetables, advertisement and land lease, and registration of trucks from private sector operators. The revenue amounted to INR 13.20 lakh (Consortium for DEWATS Dissemination Society, 2020).

21.4.2.2 Case 4–DEWATS at Arvind Eye Hospital, Puducherry

Aravind Eye Hospital (AEH) is located in Puducherry and functions on a mission to provide good quality and affordable eye care. The architects of the project focused on a sustainable system. Three organizations were involved in this project: the Auroville Centre for Scientific Research from Auroville, the Bremen Overseas Research and Development Association, BORDA, and the Malanadu Society from Delhi. The DEWATS or the decentralized wastewater treatment system was commissioned in 2003. The plant was robust with affordable maintenance. The tropical climate of Puducherry favored the installation of a combination of a planted gravel filter (PGF) and an anaerobic baffled reactor (ABR). AEH generated around 307 m^3 of domestic wastewater per day. The treated domestic wastewater was recycled for irrigating the surrounding landscape and the vegetable and fruit garden, whereas some amount of it was used for toilet flushing. According to CDD, the polishing pond in the treatment system was used for polishing water and pathogen control. The only visible part of the system was the polishing pond and the PGF. *Canna indica* and other floating aquatic plants were grown in the filter bed. The estimated capital cost incurred was INR 1.1 crore (for a 25-year life), whereas the operation cost (inclusive of electricity cost, operator fees, sludge removal cost, and expenses related to filter bed cleaning) was estimated as INR 54 720 per month. Approximately 20–160 GWh of electricity can be estimated for savings. This system ultimately reduces the load on the municipal wastewater treatment and recycling system by decentralizing the treatment hierarchy and promoting the principles of circular economy. The system allowed savings of INR 44–50 million over the course of 25 years. DEWATS adopts simple treatment technology and are sustainable for regions that lack sewer coverage (Rath et al., 2020).

21.4.2.3 Case 5-Fecal Sludge Management in Leh, Ladakh

Leh, the union territory of Ladakh, is located in the hilly regions of India. The town is situated at a higher altitude and lacks a sewerage network. The flush toilets with soak pits have been prevalent, where the toilet waste from households is directly discharged into the groundwater. This untreated wastewater contaminates the potable resources and nitrogen and disease-causing pathogens like *Escherichia coli*. This poses a threat to the tourism potential of the town. By 2017, the Municipal Committee of Leh (MCL) and the Ladakh Autonomous Hill Development Council (LAHDC) began actively searching for solutions. The MCL invited BORDA and CDD for technological and commercial evaluation. The recommendations suggested the implementation of Fecal Sludge Management (FSM) services and considering a public–private partnership (PPP). Blue Water Company (BWC) was identified to manage the treatment plant and desludging operations. The contract period was five years and could be extended by two years. The land and the suction trucks were provided by MCL. The capital cost for the treatment plant (12 KLD) was INR 78 lakh, and the cost incurred for the trucks was INR 28 lakh. The O&M cost was enumerated as INR 3.1 lakh. BWC was paid 90% of the collected fee after the service delivery and the record submission.

Planted drying beds equipped with gravity-based aero-stabilization and polishing ponds were used. The fecal solids were dried on the drying beds for three to five years and then disposed of. The water obtained from the polishing pond was recycled by implementing its usage for landscaping. The experiment team used a separate off-board motor along with the transport truck's main suction motor to facilitate desludging. This increased the desludging distance from 30–50 feet to 400 feet. The motor aided in pulling sludge on inclined slopes (Rath et al., 2020).

21.5 Economic Analysis of Source Separation

Business models can be defined in this sector based on the scale of operation, service and resource recovery. The scale of operation may range from household-level, community-level to pan-level. The service provided by the business may include desludging waste, emptying and collecting waste, and transporting waste to treatment sites. The resource recovery includes the procurement of nutrients and energy. The ownership of the business may be public, private, or PPP (Rao et al., 2020). According to a study by the Council on Energy, Environment and Water (CEEW), the market value of the currently existing and available quota of treated wastewater has been estimated to be over INR 630 million, and the figures will substantially project an increase to INR 1.9 billion in revenue. According to the estimates, a revenue of INR 966 billion would have been observed to be generated if the available treated wastewater was used for landscaping or irrigation purposes in 2021 (Bassi et al., 2023). The price determination of treated wastewater (TWW) has been mentioned in the guidelines of respective states. For Gujarat and Haryana, the price is determined by considering the cost of capital, cost incurred during the operation status, or simply put, the operation and maintenance (O&M) costs of STPs, the distribution networks, and the existing tertiary treatment facilities (Bassi et al., 2023).

The production costs of up to 20 and 21% were recovered by the compost sale in Nilgiris and Devanahalli, the districts of Tamil Nadu and Karnataka, respectively in 2020. Better planning and improving the system and self-sufficiency of the system eliminate the need for direct financial support (Fendel et al., 2022). Therefore, revenue generation through resource recovery and cost recovery is important. According to a report on the National Inventory of STPs in 2021 prepared by CPCB, the urban centers of our country have a sewage generation of 72 368 MLD. Out of this, 20 235 MLD was utilized while the remaining was disposed of untreated (CPCB, 2024). The gap must be addressed to prevent the loss of secondary raw materials (Table 21.1). It can be seen from the economic analysis of the case studies that the capital cost of the case studies is the major one time investment. Once working the treatment facilities' O and M cost is minor expenses. The revenue generation is comparatively considerate.

The wastewater possesses untapped potential. The visionary approach to utilize the segregated components and enhance the recovery process while understanding the boundaries of resource consumption serves as a way to pave newer routes to a sustainable domain. In 2021, India had 11 622 million cubic meters (MCM) of treated wastewater available for reuse. Estimated savings of INR 50 million could have been generated if around 6000 metric tonnes (MT) of nutrients were recovered. The estimated figures based on treatment capacities and sewage generation suggest that this figure will expand to 15 288 MCM by 2025 and 35 178 MCM by 2050 (Bassi et al., 2023). In India, various ministries promote the reuse of TW. These include NITI Aayog, the Ministry of Water Resources, River Development and Ganga Rejuvenation, the Ministry of Housing and Urban Affairs (MoHUA), the Ministry of Power, and the Ministry of Environment, Forest and Climate Change (MoEFCC). Multiple stakeholder consultation on existing state policies and draft newer policies.

Table 21.1 Economic summary of case studies.

Case no.	Location	Type of treatment plant	Capacity of the plant	Functioning	Use of treated effluent	Capital costs (INR) Crore	O&M costs (INR) Crore	Revenue/savings (INR) Crore
1	Bamroli, Surat	STP – secondary & tertiary	40 MLD 35 MLD	Town-level treatment plant for TW reuse	Industrial-grade water	85.10 104.00	18.64 12.66	248.995 65.81
2	Koyambedu, Chennai	TTRO	45 MLD	Town-level treatment plant for TW reuse and sale	Industrial-grade water	396.5	197.5	19.67
3	Devanahalli, Bangalore	FSTP	6 KLD	Town-level treatment plant focused on NR and TW reuse	Soil conditioner	0.9	0.24	0.132
4	Aravind Eye Hospital, Puducherry	DEWATS	307 m^3 of domestic wastewater per day	Institutional-level treatment setup focused on TW reuse	Landscaping, irrigation of gardens, and toilet flushing	1.1	0.0054 720/ month	Savings 0.5 over 25 years
5	Leh, Ladakh	FSTP	12 KLD	Town-level treatment plant focused on desludging the OSS and TW reuse	landscaping	0.78 + 0.28	0.031	90% of collected fees paid to BWC

Abbreviation: STP: sewage treatment plant, TTRO: tertiary treatment reverse osmosis, FSTP: fecal sludge treatment plant, DEWATS: decentralized wastewater treatment system, MLD: megaliters per day, NR: nutrient recovery, KLD: kiloliters per day, TW: treated water, OSS: onsite sanitation system, BWC: Blue Water Company, O&M: operations and maintenance.

The potential of the source separation approach can be tapped to fortify the revenue generated through the processes and explore alternatives to existing products in the market at the commercial level. The wastewater can be segregated and utilized for nutrient recovery, biofuel and biogas production, and procuring potable and industrial-grade water for reusable purposes. The inventory for the reusable resources can be exploited further, and newer products can be introduced for consumers. This urges to evaluation of the economic value of the said products and ensures that it garners interest from the respected shareholders.

21.6 Summary

The approach toward decentralization is gaining momentum with various field experiments and project implementations. The feasibility of adopting the newer measures is weighed against the existing technologies and consequently designed to be made compatible with legislated regulations. Source separation aids in the decentralization of technologies by fragmenting the wastewater treatment process and ensuring higher rates of secondary resource recovery. The circular economy constructed in the process vouches for a sustainable approach and initiates the notion of building robust systems that support the economic sector of the nation. Implementation of source separation of wastewater can be a paragon when dealing with generated waste. The approach and the framework can be developed to utilize the wastewater and obtain feasible resources that will help in a circular economy. The case studies discussed in the sections exhibit the practical application of the concept and understanding of the economic feasibility to intrigue and empower existing and newer stakeholders in the process.

The role of stakeholders must be defined clearly to prevent confusion. The guidelines for the nutrient recovery and reuse projects must be defined. The variation in the tariff system between different states and the lack of information on legal actions in case of non-compliance with policies should be addressed. The institution's monitoring capacity must be trained to strengthen its functioning and to leverage increased coordination between key stakeholders and the local community. A blended finance model or viability gap funding to boost PPP. Institutional barriers, geopolitical factors, political barriers, and measures to safeguard public health can be considered as parameters to assess and define the role of stakeholders. This etches the ideals of sustainability and encourages to finding of equipotent alternatives to make the existing systems more robust while being compatible with the user interface of the procedure and enhancing its chances of implementation at the community level.

Review Questions

1. How do source-separation systems impact wastewater treatment costs and energy consumption?
2. What are the potential revenue streams generated by the source-separation system?
3. How do legislative regulations impact the source-separation system? Explain the role of stakeholders associated with it.
4. Evaluate the challenges encountered while dealing with source-separation systems.
5. How does the quotient of untreated sewage promote the economic benefits of source-separation ideology?
6. How does the presence of inorganic antibiotics in treated wastewater affect the efficiency of the procedures and the quality of the wastewater-derived product obtained?
7. Write a note on the potential of the wastewater-derived fertilizer market and the challenges faced on the commercial front.
8. The desludging process of collected waste is important in fecal sludge management. How do the FSTPs at Devanahalli and Leh deal with the emptying and transport of collected waste? What treatment units are installed that process the end products?
9. The treatment unit at Aravind Eye Hospital focuses on the system that recycles the water for reuse. Evaluate the key objectives that must be considered while designing such systems at the household level and community level. Is it feasible to replicate such systems for residential complexes?
10. How does the revenue generated from the business model aid in the sufficiency of source-separation systems?

References

Ajayebi, A., Gnansounou, E., and Kenthorai Raman, J. (2013). Comparative life cycle assessment of biodiesel from algae and jatropha: a case study of India. *Bioresource Technology*, **150**, 429–437. https://doi.org/10.1016/J.BIORTECH.2013.09.118

Alhazmy, M.M. (2014). Economic and thermal feasibility of multi stage flash desalination plant with brine-feed mixing and cooling. *Energy*, **76**, 1029–1035. https://doi.org/10.1016/j.energy.2014.09.022

Ashokkumar, V., Chen, W.H., Ngamcharussrivichai, C. et al. (2019). Potential of sustainable bioenergy production from Synechocystis sp. cultivated in wastewater at large scale – a low cost biorefinery approach. *Energy Conversion and Management*, **186**, 188–199. https://doi.org/10.1016/J.ENCONMAN.2019.02.056

Barquet, K., Järnberg, L., Rosemarin, A. et al. (2020). Identifying barriers and opportunities for a circular phosphorus economy in the Baltic Sea region. *Water Research*, **171**, 115433. https://doi.org/10.1016/j.watres.2019.115433

Bassi, N., Gupta, S., and Chaturvedi, K. (2023). *Reuse of Treated Wastewater in India: Market Potential and Recommendations for Strengthening Governance*. Council on Energy, Environment and Water. https://doi.org/10.13140/RG.2.2.29603.32803

Butkovskyi, A., Hernandez Leal, L., Rijnaarts, H.H.M. et al. (2015). Fate of pharmaceuticals in full-scale source separated sanitation system. *Water Research*, **85**, 384–392. https://doi.org/10.1016/J.WATRES.2015.08.045

Chen, C.X., Aris, A., Yong, E.L. et al. (2022). Evaluation of the occurrence of antibiotics at different treatment stages of decentralised and conventional sewage treatment plants. *International Journal of Environmental Science and Technology*, **19**(6), 5547–5562. https://doi.org/10.1007/s13762-021-03519-4

Consortium for DEWATS Dissemination Society. (2020). *Insights from Faecal Sludge Management in Devanahalli - Five years of operations*. Consortium for DEWATS Dissemination (CDD) Society.

CPCB. (2021). *National Inventory of Sewage Treatment Plants*. Central Pollution Control Board.

CPCB. (2024). *Guidelines for Reuse of Treated Sewage in Reference to Item of Circular Economy*. Central Pollution Control Board.

D'ambrosio, R., Longobardi, A., Balbo, A. et al. (2021). Hybrid approach for excess stormwater management: combining decentralized and centralized strategies for the enhancement of urban flooding resilience. *Water*, **13**(24), 3635. https://doi.org/10.3390/W13243635

de Boer, M.A., Romeo-Hall, A.G., Rooimans, T.M. et al. (2018). An assessment of the drivers and barriers for the deployment of urban phosphorus recovery technologies: a case study of the Netherlands. *Sustainability*, **10**(6), 1790. https://doi.org/10.3390/SU10061790

de Feo, G., Antoniou, G., Fardin, H.F. et al. (2014). The historical development of sewers worldwide. *Sustainability*, **6**(6), 3936–3974. https://doi.org/10.3390/SU6063936

De-Bashan, L.E. and Bashan, Y. (2004). Recent advances in removing phosphorus from wastewater and its future use as fertilizer (1997–2003). *Water Research*, **38**(19), 4222–4246. https://doi.org/10.1016/J.WATRES.2004.07.014

Dodane, P.H., Mbéguéré, M., Sow, O. et al. (2012). Capital and operating costs of full-scale fecal sludge management and wastewater treatment systems in Dakar, Senegal. *Environmental Science and Technology*, **46**(7), 3705–3711. https://doi.org/10.1021/es2045234

Fendel, V., Kranert, M., Maurer, C. et al. (2022). Stakeholder assessment on closing nutrient cycles through co-recycling of biodegradable household kitchen waste and black water between Rural and Urban areas in South India. *Recycling*, **7**(4), 49. https://doi.org/10.3390/RECYCLING7040049

Ghaitidak, D.M. and Yadav, K.D. (2013). Characteristics and treatment of greywater-a review. *Environmental Science and Pollution Research*, **20**(5), 2795–2809. https://doi.org/10.1007/s11356-013-1533-0

Giwa, A., Adeyemi, I., Dindi, A. et al. (2018). Techno-economic assessment of the sustainability of an integrated biorefinery from microalgae and Jatropha: a review and case study. *Renewable and Sustainable Energy Reviews*, **88**, 239–257. https://doi.org/10.1016/J.RSER.2018.02.032

Goyal, K. and Kumar, A. (2022). A comprehensive view of existing policy directives and future interventions for water reuse in India. *Water Policy*, **24**(7), 1195–1207. https://doi.org/10.2166/WP.2022.073

Günther, S., Grunert, M., and Müller, S. (2018). Overview of recent advances in phosphorus recovery for fertilizer production. *Engineering in Life Sciences*, **18**(7), 434–439. https://doi.org/10.1002/elsc.201700171

Hukari, S., Hermann, L., and Nättorp, A. (2016). From wastewater to fertilisers—technical overview and critical review of European legislation governing phosphorus recycling. *Science of The Total Environment*, **542**, 1127–1135. https://doi.org/10.1016/J.SCITOTENV.2015.09.064

Ignacio, J.J., Malenab, R.A., Pausta, C.M. et al. (2018). Perceptions and attitudes toward eco-toilet systems in rural areas: a case study in the Philippines. *Sustainability*, **10**(2), 521. https://doi.org/10.3390/SU10020521

Janssens, R., Cristóvão, B.M., Bronze, M.R. et al. (2020). Photocatalysis using UV-A and UV-C light sources for advanced oxidation of anti-cancer drugs spiked in laboratory-grade water and synthetic urine. *Industrial and Engineering Chemistry Research*, **59**(2), 647–653. https://doi.org/10.1021/acs.iecr.9b04608

Johansson, M. (2000). Urine separation – closing the nutrient cycle. Final report on the R&D project: source separated human urine – a future source of fertilizer for agriculture in the Stockholm region. In: *Stockholm Vatten, Stockholmshem & HSB National Federation*. https://sswm.info/sites/default/files/reference_attachments/JOHANSSON%202000%20Urine%20Separation%20-%20Closing%20the%20Nutrient%20Cycle_0.pdf

Kalra, R., Gaur, S., and Goel, M. (2022). Harnessing the potential of microalgal species Dunaliella: a biofuel and biocommodities perspective. *Algal Biotechnology: Integrated Algal Engineering for Bioenergy, Bioremediation, and Biomedical Applications*, 259–279. Elsevier. https://doi.org/10.1016/B978-0-323-90476-6.00008-X

Lambert, S.E. and Randall, D.G. (2019). Manufacturing bio-bricks using microbial induced calcium carbonate precipitation and human urine. *Water Research*, **160**, 158–166. https://doi.org/10.1016/J.WATRES.2019.05.069

Larsen, T.A., Udert, K.M., and Lienert, J. (2015). *Source Separation and Decentralization for Wastewater Management*. Iwa Publishing. https://doi.org/10.2166/9781780401072

Li, Y., Wen, K., Li, L. et al. (2020). Experimental investigation on compression resistance of bio-bricks. *Construction and Building Materials*, **265**, 120751. https://doi.org/10.1016/J.CONBUILDMAT.2020.120751

Lv, Y., Li, Z., Zhou, X. et al. (2020). Stabilization of source-separated urine by heat-activated peroxydisulfate. *Science of the Total Environment*, **749**, 142213. https://doi.org/10.1016/J.SCITOTENV.2020.142213

Madyiwa, S., Chimbari, M.J., and Schutte, F. (2004). Lead and cadmium interactions in *Cynodon nlemfuensis* and sandy soil subjected to treated wastewater application under greenhouse conditions. *Physics and Chemistry of the Earth, Parts A/B/C*, **29**(15–18), 1043–1048. https://doi.org/10.1016/J.PCE.2004.08.006

Mansouri, H. and Eskandari Godar, Z. (2024). Enhancement of biomass production and some biochemical compound of Dunaliella salina cultivated in dairy wastewater by plant growth regulators. *Journal of Applied Phycology*, **36**, 1–11. https://doi.org/10.1007/s10811-024-03354-3

McKinsey. (2013). *India as an Agriculture and High Value Food Powerhouse: A New Vision for 2030*. McKinsey & Company.

Ministry of Housing and Urban Affairs (MoHUA). (2021). *Circular Economy in Municipal Solid and Liquid Waste*. https://mohua.gov.in/pdf/627b8318adf18Circular-Economy-in-waste-management-FINAL.pdf

NITI Aayog. (2021). *Faecal Sludge and Septage Management in Urban Areas: Service and Business Models*.

Orth, H. (2007). Centralised versus decentralised wastewater systems? *Water Science and Technology*, **56**(5), 259–266. https://doi.org/10.2166/WST.2007.579

Partal, R., Basturk, I., Murat Hocaoglu, S. et al. (2022). Recovery of water and reusable salt solution from reverse osmosis brine in textile industry: a case study. *Water Resources and Industry*, **27**, 100174. https://doi.org/10.1016/J.WRI.2022.100174

Rajaona, A.M., Sutterer, N., and Asch, F. (2012). Potential of waste water use for jatropha cultivation in arid environments. *Agriculture*, **2**(4), 376–392. https://doi.org/10.3390/AGRICULTURE2040376

Rao, K.C., Velidandla, S., Scott, C.L. et al. (2020). Business models for fecal sludge management in India. *Resource Recovery and Reuse*, **2020**(18 Special Issue), 13–184. https://doi.org/10.5337/2020.209

Rath, M., Schellenberg, T., Rajan, P. et al. (2020). *Decentralized Wastewater and Fecal Sludge Management: Case Studies from India*. Asian Development Bank Institute.

Senecal, J., Nordin, A., Simha, P. et al. (2018). Hygiene aspect of treating human urine by alkaline dehydration. *Water Research*, **144**, 474–481. https://doi.org/10.1016/J.WATRES.2018.07.030

Shanmugam, K., Gadhamshetty, V., Tysklind, M. et al. (2022). A sustainable performance assessment framework for circular management of municipal wastewater treatment plants. *Journal of Cleaner Production*, **339**, 130657. https://doi.org/10.1016/J.JCLEPRO.2022.130657

Shi, L., Kone, D., Lumbao, L. et al. (2010). A rapid assessment of septage management in Asia: policies and practices in India, Indonesia, Malaysia, the Philippines, Sri Lanka, Thailand, and Vietnam. *AECOM International Development and the Swiss Institute of Aquatic Sciences and Technology, Switzerland*, 126. https://www.ircwash.org/sites/default/files/AECOM-2010-Rapid.pdf

Simha, P., Barton, M.A., Perez-Mercado, L.F. et al. (2021). Willingness among food consumers to recycle human urine as crop fertiliser: evidence from a multinational survey. *Science of the Total Environment*, **765**, 144438. https://doi.org/10.1016/J.SCITOTENV.2020.144438

Satyaprakash, M., Nikitha, T., Reddi, E.U.B. et al. (2017). Phosphorous and phosphate solubilising bacteria and their role in plant nutrition. *International Journal of Current Microbiology and Applied Sciences*, **6**(4), 2133–2144. https://doi.org/10.20546/IJCMAS.2017.604.251

Tervahauta, T., Hoang, T., Hernández, L. et al. (2013). Prospects of source-separation-based sanitation concepts: a model-based study. *Water (Switzerland)*, **5**(3), 1006–1035. https://doi.org/10.3390/w5031006

Tian, X., Wang, G., Guan, D. et al. (2016). Reverse osmosis brine for phosphorus recovery from source separated urine. *Chemosphere*, **165**, 202–210. https://doi.org/10.1016/J.CHEMOSPHERE.2016.09.037

Travis, M.J., Weisbrod, N., and Gross, A. (2008). Accumulation of oil and grease in soils irrigated with greywater and their potential role in soil water repellency. *Science of The Total Environment*, **394**(1), 68–74. https://doi.org/10.1016/J.SCITOTENV.2008.01.004

Viskari, E.L., Grobler, G., Karimäki, K. et al. (2018). Nitrogen recovery with source separation of human urine—preliminary results of its fertiliser potential and use in agriculture. *Frontiers in Sustainable Food Systems*, **2**, 358254. https://doi.org/10.3389/fsufs.2018.00032

Wankhade, K. (2015). Urban sanitation in India: key shifts in the national policy frame. *Environment and Urbanization*, **27**(2), 555–572. https://doi.org/10.1177/0956247814567058

Wardlaw, T., Salama, P., Brocklehurst, C. et al. (2010). Diarrhoea: why children are still dying and what can be done. *The Lancet*, **375**(9718), 870–872.

Yan, Z., Cheng, S., Zhang, J. et al. (2021). Precipitation in urine source separation systems: challenges for large-scale practical applications. *Resources, Conservation and Recycling*, **169**, 105479. https://doi.org/10.1016/J.RESCONREC.2021.105479

Zhang, J., Giannis, A., Chang, V.W.C. et al. (2013). Adaptation of urine source separation in tropical cities: process optimization and odor mitigation. *Journal of the Air & Waste Management Association*, **63**(4), 472–481. https://doi.org/10.1080/10962247.2013.763306

22

Integration of Different Technologies to Enhance the Efficiency of Wastewater Treatment

Nabila Shehata[1] *and Davidson Egirani*[2]*

[1]*Environmental Science and Industrial Development Department, Faculty of Postgraduate Studies for Advanced Sciences, Beni-Suef University, Beni-Suef, Egypt*
[2]*Faculty of Science, Niger Delta University, Wilberforce Island, Nigeria*

*Corresponding author: davidsonegirani@ndu.edu.ng; eenonidavidson@yahoo.com

22.1 Introduction

Many water communities have established water recycling systems to meet growing water demands driven by climate change, continued population growth, and water misuse. Treated water is typically used in industry, agriculture, and landscaping applications that do not require potable water. The increasing concentration of pollutants and the reuse of water are popular research topics. The presence of both established and new contaminants in water supplies, such as cosmetics, pharmaceuticals, endocrine-disrupting substances, heavy metals, natural organics, and dyes, is especially concerning. Ni, Cu, Cr, Cd, Zn, Pb, and Co are among the dangerous heavy metal anions and cations that can contaminate the biosphere. Groundwater and surface water have been found to contain dangerously high concentrations of dyes from the paper, textiles, and cosmetics industries. Discharge of inefficiently treated sewage into freshwater bodies has led to the accumulation of endocrine-disrupting substances and xenobiotics, causing ecological hazards even at low concentrations ($<1\ \mu g/L$) (Kim et al., 2022).

In wastewater treatment, various technologies have been identified as having some drawbacks that make them inadequate for complete wastewater treatment. An integrated strategy utilizing the amalgamation of these technologies is being

Wastewater to Resource Recovery: Applying the Circular Economy Toward Sustainable Development, First Edition. Edited by Arvind Kumar Mungray et al. Published 2026 by John Wiley & Sons, Inc.

implemented to enhance wastewater treatment effectiveness. Specifically, eliminating unknown and known contaminants from wastewater depends on the physicochemical properties of pollutants and the treatment methods used. Biological, chemical, and physical treatments, including anaerobic digestion, sonication, UV radiation, membrane filtration, ozonation, adsorption over granular or powdered activated carbon (AC), coagulation-flocculation-sedimentation-filtration devices, chlorination, etc., are among the frequently employed methods for eliminating contaminants (Kim et al., 2022).

Membrane-based technologies are quite effective for wastewater treatment and utilize semi-permeable membranes that separate contaminants from wastewater based on their size, charge, and molecular interactions. Commonly used membrane technologies include microfiltration, ultrafiltration, nanofiltration, reverse osmosis, forward osmosis, and membrane bioreactors. However, the fouling of the membrane, resulting in a reduction of water flux, high cost of maintenance and operation, are the limitations of these technologies. Hence, integration of membrane technologies with other treatment methods like advanced oxidation, biological process, hybrid membrane processes, photocatalysis, adsorption, electrocoagulation, etc., has been researched to counter membrane fouling and cost.

Combining membrane separation with other methods like photocatalysis can improve energy efficiency, prevent membrane fouling, and have a positive synergistic effect. This method works especially well for sorting mixtures of materials. Studies on the typical designs of hybrid photocatalytic membrane reactors (PMRs) and their possible uses in treating industrial wastewater were conducted. The permeate flow, membrane fouling, and permeate quality of several photocatalysis membrane reactor setups were compared. The many factors include light intensity, wavelength, photocatalyst, pH, temperature, flow rate, aeration, residence time, initial pollutant concentration, inorganic ions, reactor design and geometry, and the types of membranes affecting PMR performance were investigated. The best configuration for treating industrial wastewater was determined to be the suspension PMR configuration due to its high pollutant removal efficiency and favorable operating and design costs (Khader et al., 2023).

Water and wastewater resources may be effectively cleaned of impurities by integrating membrane filtration and adsorption. In recent years, the combination of these two separation methods has been shown to be a state-of-the-art approach to water treatment, offering improved removal efficiency and fewer drawbacks than each method alone. Membrane-adsorption bioreactors (MABRs) and low-pressure membrane (LPM)-adsorption coupling are two combined designs that have shown promise in creating freshwater that meets drinking water requirements. They must, however, optimize several factors, such as hydraulic retention time (HRT), adsorbent dosage, and particle size, design a regeneration cycle that is in sync with the primary process, and perform techno-economic and life-cycle analyses to get the best results with the least amount of energy and process cost before they can fully mature for industrial use. To make this process more environmentally viable, eco-friendly, and preferred over conventional treatment methods, it is also advised to design additional studies, such as techno-economic, life cycle, and operational variable optimization. It is difficult to optimize hybrid and integrated adsorption membrane processes using traditional techniques because of their intricacy and the involvement of numerous operating variables. Multivariate/multi-objective optimization strategies, however, present very promising methods for identifying and altering the ideal operating conditions (Davoodbeygi et al., 2023).

To sum up, hybrid/integrated processes can revolutionize the way wastewater treatment is done. The best option for conventional treatment routes can be chosen after a comprehensive plan that considers the distinct qualities of each hybrid/integrated layout, process optimization, integration feasibility, risk assessment, and cost estimation. Therefore, the primary goal of this chapter is to examine modern methods for treating wastewater, hybrid systems, multidisciplinary approaches, and improving resource recovery.

22.2 Usage of Modern Technologies for Wastewater Treatment

22.2.1 Nanotechnology

The majority of traditional technologies, including filtration, (AC) adsorption, and periodic chemical oxidation, are frequently expensive and time-consuming. Biodegradation usually takes a while, but it is inexpensive and environmentally friendly. The ability to quickly, efficiently, and economically remove hazardous substances from wastewater to a safe level is critical. In this context, nanotechnology has the potential to be very important. Research to create novel nanomaterials and nanocomposites with improved selectivity and capacity for certain pollutants is progressing rapidly. Due to their greater surface area, sequestering capacity, and reactivity, nanoparticles can be advantageous to use. A variety of nanomaterials, including zeolite, titanium oxide, carbon nanotubes (CNTs), metal-organic framework, layered double hydroxide, biopolymers, self-assembled monolayers on mesoporous supports, zero-valent iron nanoparticles, single enzyme and nanoscale semiconductor photocatalysts, nanometals and nanometal oxides with different functions, are currently being researched for remediation of surface water, groundwater, drinking water, and industrial wastewater (Theron et al., 2008).

Water and wastewater treatment technologies enabled by nanotechnology have demonstrated promise in lab studies, but there is significant variation in their commercialization readiness. Many barriers stand in the way of these products' future development and commercialization, including technological difficulties, cost effectiveness, and possible hazards to human health and the environment. Before nanotechnology can be widely accepted for wastewater treatment, two important research areas must be resolved. The first step is to assess how well various nanotechnologies treat actual natural water and wastewater; to verify sensing technologies based on nanomaterials, more realistic settings must be used for future research. Second, the effectiveness of these nanotechnologies in the long run is largely unknown because the majority of laboratory experiments have been completed quickly (Qu et al., 2013).

Despite the tremendous advantages of nanomaterials in wastewater treatment, there have also been worries expressed regarding the possible harm these particles could cause to both humans and the environment. Research indicates that the size, shape, structure, reactivity, and attractiveness of nanomaterials may make them dangerous. However, the lack of established protocols and tools makes it challenging to evaluate how nanomaterials affect the environment and human health. Additionally, there aren't any standard frameworks for risk management, risk assessment, or risk research at the moment. To guarantee that nanoparticles are as safe as possible while realizing their full potential, these procedures must be created and researched. Notwithstanding these gaps in knowledge, it is obvious that the development of novel nanomaterials, especially in water and wastewater treatment, will be essential to meet the growing demand for drinking water worldwide and guarantee a sufficient supply of high-quality water (Theron et al., 2008).

22.2.2 Activated Carbon

The AC is the material most commonly used in several fields due to its exceptional adsorption properties. A large surface area, a finely shaped internal pore structure, and a range of functional groups, including carboxyl, phenol, and carbonyl, are some of these attributes. The AC is usually produced using lignite, coal, or lignocellulosic material particles. This process is called pyrolysis and involves burning the material under controlled conditions at a temperature of around 700 °C and a limited supply of oxygen. The particles are "activated" when exposed to high temperatures (900 °C) and oxidizing gases such as CO_2 and/or steam, causing the carbon to form a highly porous network within the particles. According to Metcalf & Eddy Inc. (2003), this led to the formation of different pore sizes: (i) macropores, where pore sizes are larger than 25 nm; mesopores, which have pore sizes between 1 and 25 nm; and micropores, which have pores smaller than one nanometer. Optimizing the effective parameters is critically important when producing AC from various lignocellulosic biomass sources. Activator type, impregnation time and ratio, pyrolysis temperature and time, inert gas used (if a chemical activation is used), and functionalization/modification conditions are some of the factors that need to be considered. Greater surface area, uniform shape, and stability are the goals of these optimization efforts (Rasouli et al., 2023).

22.2.3 Hydrochar/Biochar

Recently, there has been a lot of interest in the creation of biochar (BC), a valuable substance made from the pyrolysis of biomass. BC is produced by hydrothermal carbonization, pyrolysis, gasification, and torrefaction, whereas BC-based nanocomposite materials are produced by hydrothermal and solgel ultrasonic processes. These BC-based nanocomposites' intrinsic qualities enable them to overcome the drawbacks of pure BC and photocatalysts. Strong adsorption and photocatalytic activity, quicker recovery times, and enhanced reusability were all displayed by the BC-based nanocomposites. As a result, BC and its nanocomposites have been suggested as a potential remedy for wastewater pollution through the use of both adsorption and photocatalytic degradation processes. These nanocomposite materials enable improved pollutant removal from industrial wastewater by combining the effects of adsorption and photocatalysis. Factors influencing the control of pollutants in wastewater through photocatalytic degradation and adsorption pathways should be examined to optimize efficacy (Rodriguez-Narvaez et al., 2023).

22.2.4 Molecularly Imprinted Polymers

These synthetic polymers have cavities for specific target molecules. To make them, a complex made of a template and monomers with functional groups that specifically interact with the template through covalent or noncovalent bonds is copolymerized using a copolymerization agent. A few holes that are the exact size, shape, and functional groups of the stencil molecule are left behind after the embossing stencil is removed. Molecularly imprinted polymers (MIPs) are appropriate materials for wastewater treatment because of their preset selectivity. Many of the MIP-based wastewater treatment methods that have been developed or are being developed are still in the experimental or pilot stages. Therefore, there is much space for development (Huang et al., 2015). First, it is still mostly unknown what basic mechanisms underlie the recognition and synthesis of MIP ligands because they have been relatively understudied (Nicholls et al., 2011). Before the polymerization, the stability and length of the monomer-template complex are essential for the effective development of MIPs. The monomer and the template engage in covalent or noncovalent bonding interactions, including electrostatic, hydrophobic, and hydrogen bonding. Second, conventional polymers are frequently made with organic solvents and have special swelling properties in aqueous solutions. MIPs are unable to precisely identify the target in the water because of their changed structure after swelling. Additionally, the competition of water molecules with the template weakens or eliminates the noncovalent bond between the functional monomer and the template. Nonetheless, there has been some advancement in the creation of MIPs that function well with water. Adding hydrophilic qualities to the polymer is a novel way to create MIPs that are compatible with water. Finally, it's critical to take into account the potential for residual template molecules to escape during the adsorption processes (Huang et al., 2015).

22.2.5 Microbial Fuel Cells

The world is currently seeing a trend toward urbanization and industrialization, which is driving up energy consumption. The production of energy depends heavily on fossil fuels, which raises environmental problems. A significant environmental issue brought on by population growth and industry expansion is wastewater. The potentially useful development of microbial fuel cells (MFCs) can help overcome these challenges. Wastewater's chemical energy components are converted into electrical energy by the microorganisms in MFCs. These fuel cells offer a consistent supply of green bioenergy throughout treatment. This strategy can reduce greenhouse gas emissions and water pollution because it is environmentally benign. There are several disadvantages to large-scale MFCs as well, including high capital expenses and low energy density. If MFCs are incorporated into the current treatment system, they may prove to be a viable alternative. Two strategies were highlighted to address the issue of large-scale power degradation in MFC systems: reactor expansion is one approach, but it has the drawback of increasing the distance between the anode and cathode. Consequently, this increases internal resistance and reduces the amount of electricity produced. On the other hand, it seems more practical to connect several MFC reactors to build large-scale systems. Its disadvantages include difficult operating problems and voltage return. MFCs' restricted ability to produce power is another drawback. Larger cells (more than 2 L) have a reduced power density, usually less than 30 (W/m^3), even if they display high power density in smaller dimensions (less than 50 mL) (Oh and Logan, 2007).

Enhancing MFCs' efficacy in remediating different contaminants has been the field's main emphasis. To remove pollutants, recent research has employed electrode materials that have undergone minor alterations, such as replacing carbon-based materials with nanomaterials. Modifications to reactor architecture, including flow patterns and electrode spacing, have been studied to enhance mass transfer and maximize MFC performance. Furthermore, the process of selecting appropriate microbial consortia, such as enriched or modified communities, has been investigated to improve the breakdown of certain pollutants. Mechanisms for integrating different remediation approaches, including chemical oxidation, precipitation, or adsorption, have been studied to increase overall treatment efficiency and synergistically remove pollutants (Kaya et al., 2023). Significant strides have been achieved in recent years to improve MFC efficiency through the development of novel microbial catalysts, optimization of electrode materials, power output enhancement, and commercialization. When it comes to increasing power output and accelerating electron transport kinetics, the use of nanomaterials like graphene and CNTs in electrode manufacturing has produced encouraging results. Research has also looked at how different electrode layouts and surfaces affect the stability and lifetime of MFCs. Furthermore, customized biofilm formation techniques have been developed to maximize the result of advances in our understanding of microbial communities and their interactions inside MFCs (Prashanthi, 2023).

22.3 Multidisciplinary Approach for Wastewater Treatment

Achieving sustainable water treatment requires a comprehensive view of wastewater treatment systems. In addition to cost, treatment efficiency, and energy components, factors such as recycling and social concerns should also be considered when evaluating a sustainable wastewater treatment plant or system and selecting the best system for a given scenario. Engineers must collaborate with social scientists, economists, biologists, health officials, and members of the public in interdisciplinary teams to solve this issue (Jenssen et al., 2007).

Through national and international research societies, scientists, wastewater treatment service providers, and related industrial communities seized the chance to develop a variety of technologies that could produce high-quality wastewater with lower energy consumption, lower capital costs, and lower operating costs—all while achieving a form of resource recovery funding. Many different recovered products have been produced at different stages of technological maturity, including biogas, struvite, heavy metals, syngas, and bioplastics. The majority of technical innovations occur at the university level. A select few prestigious universities are in a good position to work with businesses and obtain capital to advance technology. However, the standardization of laboratory setup and experimental planning that results from the globalization and standardization of financing procedures also makes it possible for research findings to be consistent across borders. Unconventional thinking is frequently ignored or unsupported by peers. Before allocating additional funds, it is crucial to thoroughly evaluate the advantages, accounting for the expenses, time, and effort involved in creating new technologies. Large water service providers and consultants are crucial for addressing market gaps because they possess the necessary cost models and benefit calculation expertise, but their innovation and research are frequently insufficient (Soares, 2020).

Water management has changed dramatically due to technological advances that have made it possible to install sensors and other devices in water systems to collect information about water quality, flow rates, and consumption trends. By analyzing this real-time data, it is possible to anticipate demand, find leaks, and improve distribution. Machine learning and artificial intelligence (AI) are increasingly being used to predict water demand trends, identify potential system problems, and improve wastewater treatment processes. Weather-related impacts on water availability and demand can be predicted using AI models, improving the ability of water systems to adapt to climatic unpredictability. Using satellites and drones, remote sensing technology provides detailed information about water resources, land use, and environmental factors that influence water cycles. This technique is essential for detecting changes in water availability, assessing aquatic environments, and monitoring large water basins (Olatunde et al., 2024).

22.4 Integration of Treatments to Enhance Resource Recovery

Wastewater contains substances, energy, nutrients, and water, all of which can have a very positive impact on resource recovery. It is estimated that phosphorus will run out in the next 50–100 years. Nevertheless, 22% of the world's phosphorus needs could be met through wastewater recovery using existing systems, like the Stickney Wastewater Treatment Plant's (WWTP) recovery technologies, which presently attain high recovery rates (85%). With chemical energy and biogas recovery rates of about 20% and 80%, respectively, wastewater also has a high energy value (Soo et al., 2024).

According to Taghvaie Nakhjiri and co-workers (2022), industrial wastewater contains several precious metals, including copper, silver, gold, and platinum. The low profitability of wastewater, the infrastructure invested in, the policies supporting the circular economy in treatment systems, and the lack of widespread adoption of advanced technologies are the main obstacles to the recovery of energy and nutrients from wastewater (Soo et al., 2024). Currently, recovering heat and nutrients from wastewater has the potential to meet 25% of market demand (Renfrew et al., 2022). A significant amount of water is used for irrigation and agriculture; methods such as hydroponics are being researched to reduce water consumption while recovering water and nutrients (Magwaza et al., 2020). Wastewater can be used as a resource to recover biofuels, biogases, charcoal, biofertilizers, biooils, and bioplastics. However, resource recovery is associated with higher costs than pure discharge (Awasthi et al., 2023). On the other hand, recovered goods have a market value and can be sold to raise money to cover recovery costs and perhaps even break even. Due to financial constraints, wastewater treatment plant operators are not encouraged to implement resource recovery.

Energy recovery can be done via methane gas produced during the anaerobic digestion technology for wastewater treatment. It may be captured and used as a source of thermal and electrical energy. The technique of methane-rich biogas has the benefits of sustainability and ease of production, allowing for effective energy recovery. Additionally, there is no need for a distribution network as all of the energy produced in the wastewater treatment system may be used locally, which lowers indirect greenhouse gas emissions. Furthermore, it is expected that the cost of producing power using biogas technology will be lower than that of other renewable energy sources and fossil fuels (Zhou et al., 2024).

In 2023, wastewater was treated using an up-flow anaerobic sludge blanket (UASB) reactor; results show that 2.6 $kWh/kgCOD_{rem}$ of electric energy and 260.9 $NL/kgCOD_{rem}$ of methane were recovered (NL: liters at standard conditions) (Pereira Silva et al., 2023). Biogas has a lower heating value than natural gas, though, because it is primarily made up of CH_4 (55–60%) and CO_2 (40–45%), with smaller amounts of N_2, H_2, H_2S, and water vapor. Therefore, strategies for increasing methane production are crucial for raising power conversion efficiency and promoting the use of biogas as a sustainable energy alternative to fossil fuels (Cheng et al., 2020).

Algal biorefinery routes reduce carbon footprints, conserve water, and enhance soil health, providing viable answers to environmental problems related to conventional farming. Sustainability is promoted along the whole agricultural value chain by creating a closed-loop system that combines the production of algal biomass with nutrient waste streams from agriculture. Products made from algae reduce the need for chemical inputs, supporting the objectives of regenerative agriculture to improve biodiversity and ecosystem services. These methods improve the organic matter content, nutrient enrichment, and soil structure; they are particularly helpful in resource-constrained, degraded soil locations. Algal genetic and metabolic engineering offers promising avenues for future study. The efficiency, scalability, and cost-effectiveness of algal biorefineries must be improved, adoption obstacles must be removed, and research and development must be supported to promote sustainable farming practices (Samoraj et al., 2024).

Microalgae can be used to produce numerous bioactive compounds, such as nitrogen and phenolic compounds, polysaccharides, carotenoids, lipids, lutein, peptides, different vitamins, eicosatetraenoic acid, astaxanthin, brassica sterol, phycocyanin, β-carotene, superoxide dismutase, α-tocopherol, and γ-aminobutyric acid, in addition to their applications in heavy metal uptake from wastewater. Algae culture can also be used to develop biofuels such as biodiesel, syngas, bio-char, bio-oil, bioethanol, bio-butanol, and bio-methanol (Olabi et al., 2023).

22.5 Hybrid Systems: An Overview of Their Achievements

Hybrid system processes arise by combining one or more wastewater treatment technologies with other unit processes to achieve performance greater than that of any single technology. Using each system's ability to compensate for the deficiencies of the other, the hybrid process produces more high-quality purified water. Due to the increased risk of membrane fouling, hybrid processes are increasingly being researched. To lessen methane fouling, for instance, extra unit processes like sedimentation, flocculation, coagulation, or other membrane processes are added as retreatment procedures. In response to more stringent discharge regulations and the intense search for alternate water sources, these techniques are being investigated. AC has been used in numerous applications to treat contaminated groundwater for human consumption in combination with low-pressure membranes (ultrafiltration, or UF, and microfiltration, or MF). Drinking water was rendered usable by the combination's capacity to eradicate harmful bacteria, dissolved organic residues, Giardia, Cryptosporidia, and particulate matter (Stoquart et al., 2012). Similarly, combined coagulation was used as a pretreatment for UF in China's highly muddy Taihu River, producing 10 000 m^3 of drinking water daily. The combination was able to maintain a specific flow of up to 190.2 m^3/m^2 h·bar by effectively reducing UF membrane fouling (Xiangli et al., 2008; Asheghmoalla and Mehrvar, 2024). Based on the above discussion, Figure 22.1 lists various technologies like biological, electrochemical, membrane-based, and thermal, and accordingly, the list of generated products through these technologies. Each technology has some drawbacks; therefore, integration and even hybrid technologies are suggested to overcome the limitations for better efficiencies and resources as products in a cost-effective manner. Certain questions should be answered before integration/hybrid systems (Davoodbeygi et al., 2023); (i) what the possible configurations are, (ii) their advantages or disadvantages, (iii) which one is suitable for a certain product/application, (iv) technical challenges and their strategies, (v) readiness for their full-scale application and commercialization, and so on.

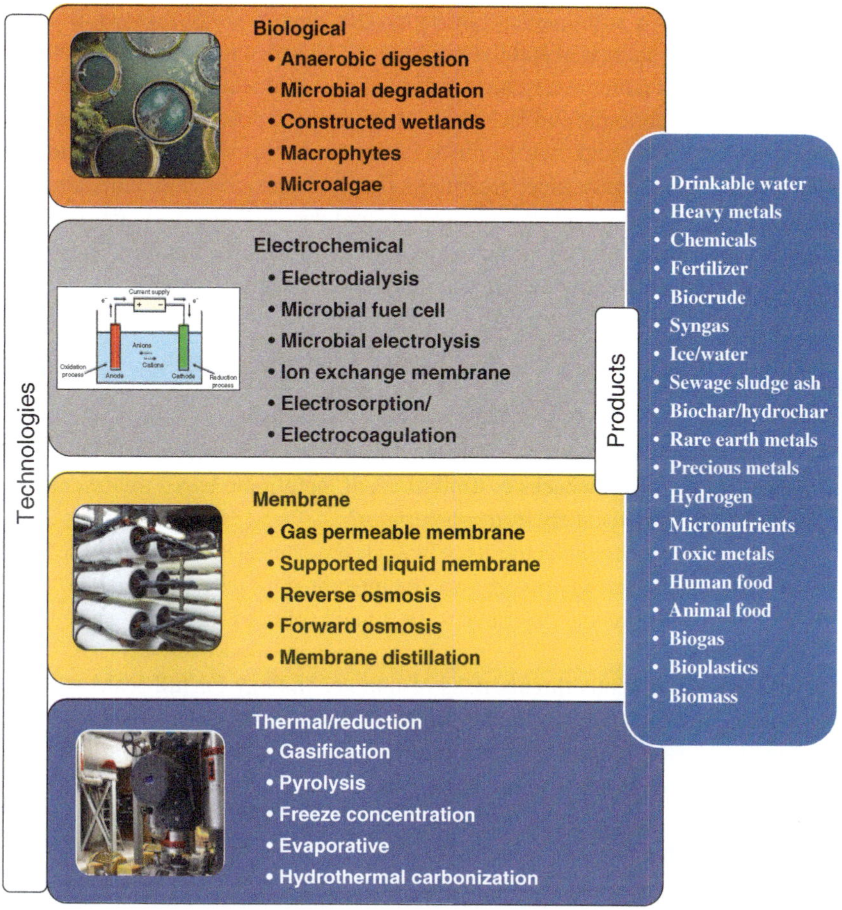

Figure 22.1 The thermal, electrochemical, biological, membrane, and chemical technologies that can be used to recover materials and energy, and the products that can be recovered include materials, nutrients, water, metals, and energy. *Source:* Sorin/Adobe Stock Photos; Electro Oxidation (EO/Anodic/Electrochemical Oxidation) – Aquacare Solution Enviro Engineers https://www.aquacaresee.co.in/aquacare-solution-enviro-engineers.html (Last accessed: 17 July 2025); Sodel Vladyslav/Adobe Stock Photos; Wastewater Digest.

22.6 Summary

The current book chapter examines the advantages and disadvantages of modern wastewater treatment technologies, including MIPs, AC, nanotechnology, biochar/hydrochar, and MFCs. The hybrid wastewater treatment systems show encouraging results in both technical and financial terms. In addition, these systems can ensure highly effective and sustainable wastewater treatment technologies while overcoming the disadvantages of standalone systems. Most of these systems are still in their infancy, and further research is needed to fully understand their development, their large-scale economic and technical feasibility, their potential risks to humans and the environment, and their commercialization. In this situation, a multidisciplinary approach to wastewater treatment should emphasize close collaboration between industrial and academic researchers and decision-making teams capable of developing current processes and technologies and securing funding. The aim is not only wastewater treatment, but also material and energy recovery.

Review Questions

1. Why is there a need for the integration of different technologies?
2. What can be the drawbacks or limitations of membrane-based treatment systems for product recovery from wastewater?
3. What are the merits and demerits of biological wastewater systems, and accordingly, suggest which technologies can be integrated for resource recovery?
4. Why is nanotechnology assumed to be a good solution for water/wastewater treatment?
5. What are the various threats/risks of nanotechnology for full-scale applications toward water/wastewater treatment?
6. How is activated carbon or hydrochar useful for the removal of various contaminants from water?
7. What are the various potentials of microbial fuel cells and limitations toward their scale-up or commercial applications?
8. Suggest remedial measures for microbial fuel cells toward the limitations during scale-up applications.
9. Why is it termed as a biorefinery? How do various algae remove contaminants from water/wastewater and can convert them into products?
10. For a successful integration process, what are the different aspects available should be answered?

References

Asheghmoalla, M. and Mehrvar, M. (2024). Integrated and hybrid processes for the treatment of actual wastewaters containing micropollutants: a review on recent advances. *Processes*, **12**, 339.

Awasthi, M.K., Ganeshan, P., Gohil, N. et al. (2023). Advanced approaches for resource recovery from wastewater and activated sludge: a review. *Bioresource Technology*, **384**, 129250. https://doi.org/10.1016/j.biortech.2023.129250

Cheng, P., Chen, D., Liu, W. et al. (2020). Auto-flocculation microalgae species Tribonema sp. and Synechocystis sp. with T-IPL pretreatment to improve swine wastewater nutrient removal. *Science of the Total Environment*, **725**, 138263. https://doi.org/10.1016/j.scitotenv.2020.138263

Davoodbeygi, Y., Askari, M., Salehi, E. et al. (2023). A review on hybrid membrane-adsorption systems for intensified water and wastewater treatment: process configurations, separation targets, and materials applied. *Journal of Environmental Management*, **335**, 117577. https://doi.org/10.1016/j.jenvman.2023.117577

Huang, D.-L., Wang, R.-Z., Liu, Y.-G. et al. (2015). Application of molecularly imprinted polymers in wastewater treatment: a review. *Environmental Science and Pollution Research*, **22**, 963–977.

Jenssen, P.D., Vråle, L., and Lindholm, O. (2007). Sustainable wastewater treatment. In: *Proceedings of International Conference on Natural Resources and Environmental Management and Environmental Safety and Health*. Kuching, Malaysia. https://sswm.info/sites/default/files/reference_attachments/JENSSEN%202007%20Sustainable%20Wastewater%20Treatment.pdf

Kaya, S.I., Ozcelikay, G., Cetinkaya, A. et al. (2023). 13—Emerging trends—Current developments in nanobioremediation via electrochemistry. In: *Nanoremediation* (eds. C.M. Hussain and N.N.B.T.-N. Nassar), 363–389. Elsevier. https://doi.org/10.1016/B978-0-12-823874-5.00002-4

Khader, E.H., Mohammed, T.J., Albayati, T.M. et al. (2023). Current trends for wastewater treatment technologies with typical configurations of photocatalytic membrane reactor hybrid systems: a review. *Chemical Engineering and Processing—Process Intensification*, **192**, 109503. https://doi.org/10.1016/j.cep.2023.109503

Kim, S., Nam, S.-N., Jang, A. et al. (2022). Review of adsorption–membrane hybrid systems for water and wastewater treatment. *Chemosphere*, **286**, 131916. https://doi.org/10.1016/j.chemosphere.2021.131916

Magwaza, S.T., Magwaza, L.S., Odindo, A.O. et al. (2020). Hydroponic technology as decentralised system for domestic wastewater treatment and vegetable production in urban agriculture: a review. *Science of the Total Environment*, **698**, 134154. https://doi.org/10.1016/j.scitotenv.2019.134154

Metcalf & Eddy, Inc. (2003). *Wastewater Engineering Treatment and Reuse*, 4e. McGraw-Hill

Nicholls, I.A., Andersson, H.S., Golker, K. et al. (2011). Rational design of biomimetic molecularly imprinted materials: theoretical and computational strategies for guiding nanoscale structured polymer development. *Analytical and Bioanalytical Chemistry*, **400**, 1771–1786.

Oh, S.-E. and Logan, B.E. (2007). Voltage reversal during microbial fuel cell stack operation. *Journal of Power Sources*, **167**, 11–17.

Olabi, A.G., Shehata, N., Sayed, E.T. et al. (2023). Role of microalgae in achieving sustainable development goals and circular economy. *Science of the Total Environment*, **854**, 158689. https://doi.org/10.1016/j.scitotenv.2022.158689

Olatunde, T.M., Adelani, F.A., and Sikhakhane, Z.Q. (2024). A review of smart water management systems from Africa and the United States. *Engineering Science & Technology Journal*, **5**, 1231–1242.

Pereira Silva, T., Guimarães de Oliveira, M., Marques Mourão, J.M. et al. (2023). Monte Carlo-based model for estimating methane generation potential and electric energy recovery in swine wastewater treated in UASB systems. *Journal of Water Process Engineering*, **51**, 103399. https://doi.org/10.1016/j.jwpe.2022.103399

Prashanthi, R. (2023). A review on microbial fuel cell and green energy. *Ionics (Kiel)*, **29**, 1667–1697.

Qu, X., Alvarez, P.J.J., and Li, Q. (2013). Applications of nanotechnology in water and wastewater treatment. *Water Research*, **47**, 3931–3946. https://doi.org/10.1016/j.watres.2012.09.058

Rasouli, K., Rasouli, J., Mohtaram, M.S. et al. (2023). Biomass-derived activated carbon nanocomposites for cleaner production: a review on aspects of photocatalytic pollutant degradation. *Journal of Cleaner Production*, **419**, 138181. https://doi.org/10.1016/j.jclepro.2023.138181

Renfrew, D., Vasilaki, V., McLeod, A. et al. (2022). Where is the greatest potential for resource recovery in wastewater treatment plants? *Water Research*, **220**, 118673. https://doi.org/10.1016/j.watres.2022.118673

Rodriguez-Narvaez, O.M., Nadarajah, K., Suarez-Toriello, V.A. et al. (2023). Engineered hydrochar production methodologies, key factors influencing agriculture wastewater treatment, and life cycle analysis: a critical review. *Journal of Water Process Engineering*, **56**, 104483. https://doi.org/10.1016/j.jwpe.2023.104483

Samoraj, M., Çalış, D., Trzaska, K. et al. (2024). Advancements in algal biorefineries for sustainable agriculture: biofuels, high-value products, and environmental solutions. *Biocatalysis and Agricultural Biotechnology*, **58**, 103224. https://doi.org/10.1016/j.bcab.2024.103224

Soares, A. (2020). Wastewater treatment in 2050: challenges ahead and future vision in a European context. *Environmental Science and Ecotechnology*, **2**, 100030. https://doi.org/10.1016/j.ese.2020.100030

Soo, A., Kim, J., and Shon, H.K. (2024). Technologies for the wastewater circular economy – A review. *Desalination and Water Treatment*, **317**, 100205. https://doi.org/10.1016/j.dwt.2024.100205

Stoquart, C., Servais, P., Bérubé, P.R. et al. (2012). Hybrid membrane processes using activated carbon treatment for drinking water: a review. *Journal of Membrane Science*, **411**, 1–12.

Taghvaie Nakhjiri, A., Sanaeepur, H., Ebadi Amooghin, A. et al. (2022). Recovery of precious metals from industrial wastewater towards resource recovery and environmental sustainability: a critical review. *Desalination*, **527**, 115510. https://doi.org/10.1016/j.desal.2021.115510

Theron, J., Walker, J.A., and Cloete, T.E. (2008). Nanotechnology and water treatment: applications and emerging opportunities. *Critical Reviews in Microbiology*, **34**, 43–69. https://doi.org/10.1080/10408410701710442

Xiangli, Q., Zhenjia, Z., Nongcun, W. et al. (2008). Coagulation pretreatment for a large-scale ultrafiltration process treating water from the Taihu River. *Desalination*, **230**, 305–313.

Zhou, L., Liang, M., Zhang, D. et al. (2024). Recent advances in swine wastewater treatment technologies for resource recovery: a comprehensive review. *Science of the Total Environment*, **924**, 171557. https://doi.org/10.1016/j.scitotenv.2024.171557

23

Utilization of Artificial Intelligence (AI) for Modernization of Wastewater Treatment and Resource Recovery

Priyanka Ray Choudhury[1], *Muskan Syed*[1], *Alakto Choudhury*[1], *Sheetal Kumari*[2, 3], *Abhishek Sengupta*[1], *Priyanka Narad*[4*], *Smriti Agarwal*[5], *and Manoj Chandra Garg*[2]

[1]*Amity Institute of Biotechnology, Amity University, Noida, Uttar Pradesh, India*
[2]*Amity Institute of Environmental Sciences, Amity University, Noida, Uttar Pradesh, India*
[3]*School of Eco-environment, Harbin Institute of Technology, Shenzhen, China*
[4]*Indian Council of Medical Research, Ansari Nagar, New Delhi, India*
[5]*Motilal Nehru National Institute of Technology Allahabad, Prayagraj, Uttar Pradesh, India*

*Corresponding author: priyanka.narad@gmail.com

23.1 Introduction

Water is a vital resource used in daily life. Due to contemporary lifestyles and rapid industrial growth, the amount of pollutants reaching water bodies is tremendous (Wang et al., 2023). The polluted water bodies, therefore, become rich in wastes, microplastics, microbial growth, chemicals, dyes, and suspended solids (Chahal et al., 2016). Depending on the level and extent of pollution, along with other factors such as the type of pollutant, physical and chemical composition, traditional wastewater treatment (WWT) systems relied on contamination removal via a combination of physical, chemical, and biological processes (Zaibel et al., 2022; Wang et al., 2023). As a result, there is often a certain degree of uncertainty and variation in the WWT system (Zhao et al., 2020). These fluctuations could affect effluent costs, quality, and the energy requirements for treatment (Zhang et al., 2023). This makes it imperative to adopt innovative technologies for the improvement of efficiency, reduction of expenses, and the energy requirements in WWT.

Wastewater to Resource Recovery: Applying the Circular Economy Toward Sustainable Development, First Edition. Edited by Arvind Kumar Mungray et al. Published 2026 by John Wiley & Sons, Inc.

Emerging technologies of the current decade include Artificial Intelligence (AI), which is a disruptive technology first introduced in the mid-1950s (Muthukrishnan et al., 2020) and since then has been adopted across various sectors such as healthcare, life sciences, environmental, finance, etc. as a solution for different challenges faced in these sectors owing to the complexities in the nature of the problems as well as the heterogeneous nature of data generated in these fields (Bajwa et al., 2021; Holzinger et al., 2023; Zamani et al., 2023; Bahoo et al., 2024). Lately, it has generated renewed interest due to the emergence of generative AI (Ooi et al., 2025). Consequently, AI solutions are also being rapidly adopted for WWT and have found applicability in resource recovery too (Wang et al., 2023). Since the process of WWT is complex and influenced by various elements, it can benefit from AI utilization to address the issues regarding uncertainties in pollutant composition, effluent quality, operational control, etc. (Zhao et al., 2020). Additionally, by addressing these issues, it can also help in lowering the costs of WWT system operation and management and increasing the sustainability of the WWT plant.

The main goal of this chapter is to explore the effects of the adoption of AI in WWT and resource recovery as a means to modernize WWT systems and also to identify the benefits of doing so. It briefly covers the current practices and technologies implemented in WWT, highlighting the challenges that could be mitigated via the utilization of AI. It also gives applications and examples of existing AI usage in WWT to exemplify its benefits and impact. Despite its benefits, AI suffers from challenges such as different pollutants, treatment costs, etc., which can severely impact the efficient use and implementation of this technology. These challenges are described serially toward the end, highlighting feasible solutions to enable rational AI adoption in WWT.

23.2 Traditional Processes Involved in WWT

Processes typically involved in WWT include different chemical, biological, and physical techniques. The selection of either of these methods relies on the composition of pollutants in the water body, costs associated with the treatment, and the feasibility of implementation methods (Nishat et al., 2023). Some of the commonly used methods are briefly discussed in the following text.

23.2.1 Physical Methods

These methods involve the physical separation of the pollutants from the water body and involve processes such as adsorption, membrane filtration, nano-filters, etc. The process of adsorption can be achieved by physical processes, i.e. physisorption, or by chemical, i.e. chemisorption process, where the former possesses weaker van der Waals forces resulting in a reversible process. Being an easy-to-use and cost-effective method, adsorption has been extensively utilized for the separation and removal of heavy metal toxicity and small amounts of harmful substances (Zhang et al., 2013; Song et al., 2015). However, it is also a nonselective method; it also heavily relies on different types of adsorbents according to the type of pollutants, and the efficiency of the existing adsorbent diminishes over time due to waste particles remaining on the adsorbent's surface, thus leading to increased regeneration costs (Chai et al., 2021).

Filtration of wastewater can be achieved by using membranes or nano-filters. Membrane technology is known to be a prevalent technique for removing contaminants from wastewater and relies on a barrier separating two phases via selective permeation (Obotey Ezugbe and Rathilal, 2020). The key drivers involved in membrane separation methods are usually a pressure gradient system or an externally applied electrical potential. Separation of contaminants via the membrane depends on the pore diameter and size of the contaminant to be separated. Based on the aforementioned membrane dependency, membrane filtration could be classified into different categories such as microfiltration, ultrafiltration, nanofiltration, and reverse osmosis (RO), with nanofiltration techniques being one of the more contemporary approaches in the filtration process (Nishat et al., 2023). While it removes the need for chemicals and is relatively easier to use, it is also one of the costliest processes for WWT due to its requirements of regular maintenance for optimum function, and it also possesses a high demand for energy to maintain the membrane flux (Visvanathan et al., 2000).

23.2.2 Chemical Methods

These methods employ oxidation processes for a range of catalytic applications. These processes are crucial for achieving desired chemical transformations, often enhancing reaction efficiency and selectivity. By facilitating reactions that might otherwise require more extreme conditions, chemical oxidation plays a vital role in various industrial and environmental contexts. These reactions include the usage of hydroxyl radicals, chlorine, photo-electrochemical oxidation, electrochemical oxidation, ozonation, and UV-assisted Fenton oxidation. These techniques are usually dependent on the pH of the wastewater and often require catalysts (Nishat et al., 2023). These processes are often used in conjunction with membrane-based filtration methods and are usually responsible for converting harmful chemical pollutants to less harmful substances via advanced oxidation reactions (Saleh et al., 2020).

23.2.3 Biological Methods

Biological methods employ biological metabolic processes to purify wastewater. These processes are generally considered to be eco-friendly since they do not generate secondary pollution and can be classified into biosorption, biodegradation, and bioaccumulation techniques based on their mechanism of function. Bioaccumulation describes the mechanism by which living beings accumulate harmful compounds through their metabolic activities. Biosorption is based on the presence of biological macromolecules like proteins and polysaccharides on the organism's surface, which have an affinity for and can bind with the harmful compounds in the environment. Lastly, biodegradation is the process in which toxic compounds are converted into harmless or less toxic compounds by the metabolic process of organisms (Song et al., 2023). A good example of such biological methods includes microbial fuel cells and microalgae, which are gaining popularity as a cost-effective technology for sustainable WWT (Bhat et al., 2021).

23.2.4 Challenges in Current WWT Processes

Current WWT systems face a number of social and technical challenges that need to be addressed. The adoption of new technologies can greatly assist in mitigating these challenges, some of which are mentioned in the following text.

23.2.4.1 Social Challenges

In developing countries, the amount of sewage produced tends to rise in tandem with urban population growth. If the current WWT systems are unable to manage the pace of urbanization, it can result in the accumulation of untreated wastewater. This situation poses significant environmental and public health risks, as inadequate sewage management could possibly lead to water source contamination, the spread of diseases, and an adverse impact on ecosystems. Ensuring the expansion and modernization of WWT infrastructure is essential for addressing these challenges and supporting sustainable urban development. Also, since the operational costs can be quite high, a huge number of nonoperational WWT plants have also been observed in such nations, raising concerns regarding sustainable and equitable WWT systems. Other factors, such as compliance, maintenance of existing functional WWT plants, and integration of the informal sector, are also some social issues that need to be resolved (Kumar et al., 2021; Vaidya et al., 2023).

23.2.4.2 Technical Challenges

Technical challenges mainly involved the variety of contaminants, which may be vastly different in different wastewater sources, the presence of nonchemical contaminants such as microplastics and nanoparticles, which cannot be removed easily, and pathogenic contaminants, which are difficult to kill or remove. These pollutants must be especially taken care of in the case of potable water implementation and require thorough treatment followed by disinfection processes such as UV treatment, chlorination, chemical oxidation, etc. (Villarín and Merel, 2020). Wastewater reuse and recovery is another field that is difficult to address.

The increasing awareness and demand for carbon-footprint reduction, adoption of circular economy, and sustainable practices in WWT technologies have incentivized processes like wastewater reuse and resource recovery. Despite the use of advanced WWT and monitoring to meet quality standards, direct potable reuse still involves direct human exposure, causing public concern. The primary issue is the presence of common wastewater contaminants, including therapeutic and drug waste, hygiene product wastes, and artificial sweeteners like acesulfame and sucralose, which are not entirely eliminated during the treatment process. Also, it can be quite challenging to detect treatment failures and prevent the consumption of such "treated" water (Malchi et al., 2014; Dias and Petit, 2015; Villarín and Merel, 2020).

23.2.4.3 Economic Challenges

As mentioned previously, the upkeep costs of WWT plants in developing nations undergoing rapid urbanization are a major challenge in effective WWT. To attain higher water quality, more sophisticated processes are required, which in turn elevate operational costs and energy requirements. Although electricity usage tends to decrease as the scale of treatment facilities grows, an increase in treatment level, driven by rising demand, results in higher energy consumption (Diaz-Elsayed et al., 2020). Regarding costs, operational and distribution systems showcase diseconomies of scale. Several factors typically influence distribution costs, with the most significant being the area's topography, the size and locality of the treatment facility, and the density of the population (Rezaei et al., 2019). As a result, depending on the scale, the costs involved in distribution can significantly add to the total expenses.

These issues lead to approximately 25% of the world population lacking access to safe potable water, around 50% without sufficient sanitation services, and roughly 30% without adequate hygiene facilities. AI has the potential to improve insights into supply chains, enhance catchment management and emergency response, optimize the design, operation, and maintenance of treatment plants and distribution networks, and better manage service availability and demand. In this way, AI can assist in addressing the challenges present in current WWT strategies (Richards et al., 2023). AI is increasingly employed to tackle the shortcomings in global water systems.

23.3 AI Technologies in WWT

"Industry 4.0" is the currently ongoing fourth industrial revolution, which seeks to enhance and automate industrial operations using digital technologies such as the Internet of Things (IoT), Big Data, and AI. In WWT, Industry 4.0 is tasked with utilizing these technologies to accelerate technological progress, support a shift toward sustainability, and drive digital transformation (Cairone et al., 2024). The growth

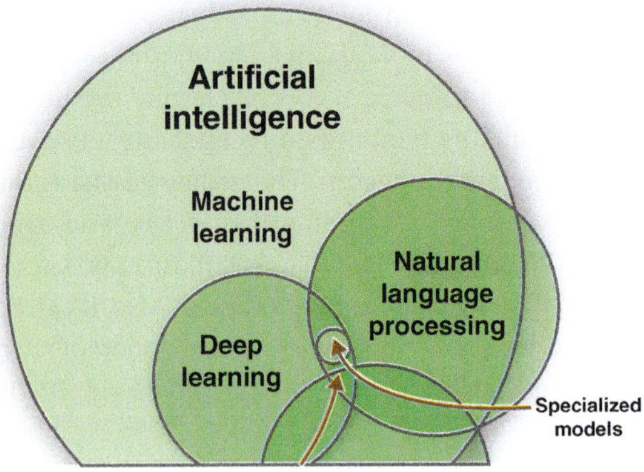

Figure 23.1 Types of artificial intelligence.

and advancement of AI have found applications in different fields, including but not limited to healthcare, natural language processing (NLP), fraud detection, computer vision, and autonomous driving, and industries such as agriculture, medicine, environmental protection, and scientific research. This is because its efficiency allows for the classification and regression analysis of large data sets, greatly benefiting industries and promoting development across all sectors. With the rise of big data and enhanced computation power, AI is getting increasingly popular (Wang et al., 2023) and is addressing gaps in water applications overlooked by traditional methods. By 2030, AI spending in the water industry is expected to make up nearly 10% of a $90 billion investment (Lowe et al., 2022).

AI is a broad discipline focused on developing and understanding computer systems that can execute tasks typically dependent on human intelligence and intervention. Its essence revolves around comprehending and emulating human cognitive functions. This endeavor has led to the development of several fundamental concepts that are shown in Figure 23.1.

23.3.1 Machine Learning

Machine Learning (ML) is a branch of AI focused on developing algorithms that can learn from data without requiring direct programming. The main approaches in ML include supervised learning, where models are trained using labeled data; unsupervised learning, which uncovers patterns in unlabeled data; and reinforcement learning, where agents refine their actions by interacting with an environment and receiving feedback (Tyagi and Chahal, 2020).

23.3.1.1 Supervised Learning

Supervised algorithms require human involvement to supply both the input data and the corresponding output labels. Additionally, during the training process, they require feedback on the accuracy of the model's predictions, helping to refine and improve the algorithm's performance (Saravanan and Sujatha, 2018).

23.3.1.2 Unsupervised Learning

These techniques fundamentally differ from supervised learning by not requiring training with labeled data. Instead, these methods analyze unlabeled data to uncover patterns, structures, and relationships within the dataset without preexisting knowledge of the outcomes. This capability to function without predefined instructions makes unsupervised learning especially valuable for exploratory data analysis and for identifying hidden patterns in large and complex datasets (Saravanan and Sujatha, 2018).

23.3.1.3 Reinforcement Learning

Reinforcement learning (RL) is an ML approach in which an agent improves decision-making by interacting with its environment, aiming to maximize cumulative rewards. The agent assesses different actions and updates its approach based on feedback, carefully balancing the need to explore new possibilities with the need to leverage known successful strategies. The core components of RL include the environment, agent, actions, states, and rewards (Szepesvári, 2022).

23.3.2 Deep Learning

Deep Learning (DL) is a specialized field within ML involving the employment of deep neural networks containing multiple layers to identify complex patterns and features in extensive datasets. These networks autonomously learn hierarchical features from raw input data, enabling advanced modeling capabilities across varied domains such as image visualization, speech recognition, NLP, and beyond, as mentioned earlier. The depth of the network architecture, coupled with extensive computational resources, facilitates the extraction and abstraction of complex data features, making DL a cornerstone of contemporary AI research (LeCun et al., 2015). It encompasses techniques like artificial neural networks (ANNs), convolutional neural networks (CNNs), long short-term memory (LSTM) networks, recurrent neural networks (RNNs), adaptive neuro-fuzzy inference systems (ANFISs), fuzzy inference systems (FISs), and transformer models (Lowe et al., 2022; Wang et al., 2023; Alprol et al., 2024).

Integration of Industry 4.0 technologies, particularly AI along with its subsets like ML and DL, presents transformative potential for WWT. These advanced methodologies enable efficient analysis of large, complex datasets, allowing for improved monitoring, prediction, and optimization of WWT processes. Automating and optimizing these processes with AI-driven insights improves operational efficiency while promoting sustainable water management practices.

As AI investment in the water industry grows, its applications in WWT are poised to revolutionize the sector, facilitating a greener transition and propelling digital transformation. Consequently, it holds immense potential in modernizing WWT processes.

23.4 Applications of AI Technologies in WWT

The WWT process is mainly concerned with processes that include monitoring water quality, optimizing existing processes, fault detection, and designing processes (Figure 23.2). In recent years, AI models have become increasingly prominent in wastewater management, where they are utilized to forecast and enhance the WWT process (Wang et al., 2023).

23.4.1 Monitoring and Control

Historically, monitoring and evaluating surface water quality have depended on labor-intensive, costly, and time-consuming studies and manual sampling techniques (Rana et al., 2023). A key benefit of AI in WWT is its ability to process and analyze vast amounts of data from multiple sensors and monitoring devices in real time. AI algorithms can constantly evaluate factors like flow rates, turbidity, chemical concentrations, dissolved oxygen (DO), temperature, and pH. This capability enables AI systems to detect anomalies, predict potential issues, and provide proactive recommendations, thereby preventing problems before they arise (Rana et al., 2023). The ability to anticipate and address

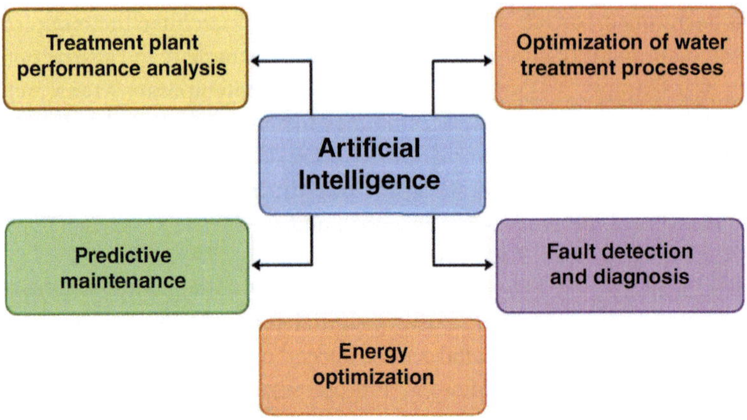

Figure 23.2 Application of AI in WWT.

issues in real time is vital for maintaining the stability and efficiency of WWT operations. It is also capable of dynamically modifying the dosage of disinfectants, flocculants, and coagulants in accordance with the properties and quality of the incoming wastewater. By ensuring that chemical consumption is minimized, this minimizes the unwanted effects that excessive chemical discharge has on the environment while also reducing operating expenses (Essahlaoui et al., 2024). AI can greatly decrease energy use in aeration processes by adjusting blower operations based on DO levels (Zhu et al., 2018). When assessing influent and effluent water, essential quality parameters—including biological oxygen demand (BOD), chemical oxygen demand (COD), pH levels, DO, flow rate, temperature, and pollutant concentration—can be readily measured. These factors serve as inputs for AI models to analyze and determine water quality (Lowe et al., 2022). These parameters are also employed in the assessment of the effectiveness of WWT or the performance of WWT plants (Wang et al., 2023). The monitoring of the aforementioned parameters can help in preventing poor-quality water output. Apart from monitoring water quality, AI can also enable real-time monitoring of WWT processes to further enhance and optimize the treatment workflow, thus saving time, resources, and costs in the process. Continuous and real-time water quality monitoring allows agencies to be alerted to any fluctuations in assessment parameters. The gathered data supports informed decision-making for effective water management and the protection of aquatic ecosystems (Park et al., 2022; Kamyab et al., 2023).

Currently, various different AI-based WWT models have been developed for monitoring water quality and its management. For example, ML-based models have been created to assess and predict the water quality index for different rivers such as Ganga and Yamuna Rivers in India, Wailuku River in Hawaii, Kelantan, Muda, Kinta and Johor River in Malaysia, Chao Phraya River in Thailand, Link and Klamath Rivers in the United States and Tabriz River in Iran (Alizadeh et al., 2018; Elkiran et al., 2018, 2019; Najah Ahmed et al., 2019; Abba et al., 2020; Kisi et al., 2020; Nourani et al., 2021; Saboe et al., 2021; Nur Adli Zakaria et al., 2021). Another example includes the potential for real-time adjustments by predicting the quality of effluents using the PCA-NARX model. This model, developed by Yang et al., demonstrated that their dynamic model performed more optimally than the static ANN models in modeling effluent quality (Yang et al., 2022).

23.4.2 By-product Management

Most WWT processes result in by-product formation, which, in higher concentrations, could pose as pollutants themselves. An example of such a process is that of chlorination, which is commonly used as a disinfectant to address microbial contaminants in water. Excess chlorine content in water can be carcinogenic and toxic to humans (Richardson, 2003). Several AI models have been developed to tackle this challenge. For example, Godo-Pla et al. designed an advanced control system that leverages AI techniques, including fuzzy inference systems (FIS), to optimize the combined dosage of sodium hypochlorite and chlorine dioxide in the primary disinfection process (Godo-Pla et al., 2021). In another study, Mahato and Gupta applied AI-based methods to predict trihalomethanes in drinking water supplies. They performed seasonal variation assessment and correlation analysis to identify patterns in contaminant presence under different conditions and developed different AI models, out of which, based on performance metrics, the ANN model was found to be the best predictor of trihalomethanes in treated water (Mahato and Gupta, 2022). These are some of the few examples of AI in wastewater by-product management; however, there have been multiple studies dedicated to doing the same (Cordero et al., 2021; Xu et al., 2022; Peleato, 2022; Okoji et al., 2022).

23.4.3 Process Design and Optimization

The WWT process often requires complex design processes and operating conditions. AI models excel in simplifying these complexities by effectively establishing relationships between various input and output variables. It also has the potential to significantly increase effluent quality, save costs, and enhance efficiency in WWT systems. AI-driven process optimization constantly fine-tunes different treatment phases through the use of sophisticated algorithms and ML techniques (Wang et al., 2023). AI's potential in process design and optimization can be exemplified by the study by Nayak et al., where they developed an optimized process for enhancing microalgal biomass production utilizing domestic wastewater and coal-fired flue gas, employing an artificial ANN and genetic algorithm (GA) for optimization (Nayak et al., 2018). GA and ANN-based models have also been useful in predicting optimal operational conditions for the treatment of dye wastewater. Picos-Benítez et al. utilized these models to predict and optimize the treatment of sulfate wastewaters with Bromophenol blue dye using an electro-oxidation process (Picos-Benítez et al., 2020). Other applications involve assessing agricultural waste for effective biosorption of textile wastewater, improving cleaner production techniques in industrial WWT, predicting water quality, and optimizing process parameters (Li et al., 2021; Zhu et al., 2022; Aghilesh et al., 2023).

23.4.4 Improving Water Accessibility

AI can help address social challenges related to fair access to clean water by optimizing water distribution. By analyzing data from water resources, consumption patterns, and other sources, AI provides insights that enable more informed decisions for efficient and equitable management and allocation of water resources (Kamyab et al., 2023). It can also support infrastructure maintenance by identifying imperfections and other inefficiencies in water systems, thereby reducing wastewater and optimizing water usage patterns. In a different study, Salloom et al. improved neural network models for multi-step time-series forecasting by integrating a method based on the proportional-integral-derivative (PID) control technique. This method adjusted the predicted values at each step to better align with actual values, improving water demand predictions. The PID-based method notably increased prediction accuracy while maintaining system simplicity. Furthermore, it was tested for hourly energy consumption forecasting, proving effective across various types of periodic time-series data (Salloom et al., 2022). Moreover, as previously mentioned, technological advancements, such as IoT (Internet of Things) and Big Data analytics in conjunction with AI models, can also aid in assessing water demands accurately, thus contributing to efficient resource management (Fu et al., 2023; Raghavendar et al., 2023).

23.4.5 Simulating Contaminant Removal

AI facilitates novel applications in WWT strategies by allowing the simulation of various adsorbents and evaluating their efficacy in the removal of specific contaminants. For example, Bisaria et al. employed ANNs and ANFIS models to simulate the adsorption of chlorpyrifos (CPS) using *Trapa bispinosa* peel. Their research found that the peel was an efficient and sustainable adsorbent, offering a low equilibrium time along with high adsorption capacity (Bisaria et al., 2024). In another study, Nnaji et al. made use of RSM, ANN, and ANFIS models to estimate the removal efficiencies of COD and CTSS from textile wastewater using complex salt–*Luffa cylindrica* seed extract (CS-LCSE) as a coagulant. From their results, they concluded that the ANFIS model provided the most optimal performance (Nnaji et al., 2022). Another significant example involves forecasting the efficiency of removing COD from domestic wastewater with the aid of Fe/Cu nanoparticles under different situations of operation. In this study, Mahmoud et al. developed an ANN model and found that Fe/Cu nanoparticles were highly efficient in removing COD (Mahmoud and Mahmoud, 2021).

23.4.6 Improving Process Performance

AI can also assist in predicting the performance of a proposed WWT system, thereby preemptively aiding in the better design of WWT plants. Commonly used performance predictions for processes typically encompass pollutant removal efficiency, performance of the WWT plant, optimal process conditions, and effluent quality (Wang et al., 2023). For example, Nourani et al. evaluated the performance of the Nicosia WWT plant to forecast its effectiveness using both individual AI models and ensemble approaches. They discovered that the ANFIS model outperformed other single AI models, while the neural network ensemble model offered the most accurate predictions among the ensemble methods (Nourani et al., 2021). Furthermore, AI models have been applied in wastewater pretreatment processes to enhance both the accuracy of pretreatment and the adaptive control precision of the system (Chan and Nai-He, 2020; Şenol, 2021).

23.4.7 Fault Detection

One crucial use of AI in WWT systems is fault detection, which greatly improves the effectiveness and dependability of predictive maintenance plans. In order to detect early failures, abnormalities in the process, and operational inefficiencies, this advanced method makes use of complex computational algorithms. This allows for proactive interventions and improves system performance. A lot of fault detection techniques are built on the fundamentals of SPC. AI is used to adapt and improve methods like exponentially weighted moving average (EWMA) control charts and Cumulative Sum control charts, which are used in the identification of small perturbations in process parameters that might signal impending defects (Mercorelli, 2024).

Various statistical and learning methods for automatic fault detection in wastewater systems have shown promise, but the nonlinear dynamics and complex variable interactions require even more advanced methods with greater learning capabilities. Mamandipoor et al. utilized DL methods such as LSTM to detect faults in WWT plants; their study demonstrated that DL models could effectively identify and predict faults, improving the reliability and efficacy of WWT processes (Mamandipoor et al., 2020). Chang et al. also detected a fault in the WWT process with the aid of a broad slow feature neural network (BSFNN) having an incremental learning capability. Due to the leveraging of slow feature windows for precise monitoring while maintaining low time overhead through pseudoinverse strategies and their ability to adapt in real-time with flexible node adjustments, they were able to prove that their proposed method for fault detection in WWT plants outperforms other existing models (Chang et al., 2024).

23.5 AI for Resource Recovery in WWT

Traditional WWT processes focus on waste and pollutant removal from water; however, the increased levels of nutritionally rich waste (previously deemed contaminant) in wastewater provide a unique opportunity to be harvested for other commercial uses. For instance, wastewater rich in contaminants such as nitrogen, potassium, ammonium, phosphorus, etc. can be repurposed for use in agriculture (Villarín and Merel, 2020; Al-Hazmi et al., 2023). There are other varied ways in which WWT and recovery processes can be employed, such as energy recovery, nutrient recovery, mineral recovery, and water reuse.

23.5.1 Nutrient Recovery

The recovery of nutrients from WWT holds immense potential for applications in other sectors. Nutrients can be recovered from wastewater through methods like urine source separation, toilet composting, or fertigation (using nutrient-rich recovered water), and then locally reused for plant fertilization (Anand and Apul, 2014). Urine waste, although only making up 1% of the total volume of municipal wastewater, is a major source of nutrients for agriculture. It contains approximately 80% of the nitrogen (N), 50% of the phosphorus (P), and 70% of the potassium (K) present in the wastewater (Wilsenach et al., 2003). Biosolids, such as sludge, can also be a good source for nutrient recovery (Awasthi et al., 2023) and can contain around 30–40% of phosphorus and 20–30% of nitrogen in aerobic WWT. Techniques such as precipitation or enhanced phosphorus removal can enrich the sludge to contain approximately 95% phosphorus, thus making it suitable for land applications in agriculture (Cornel et al., 2011). Transportation costs and efforts for sludge can be made easier via dewatering to reduce the sludge volume. A good example of biosolids nutrient recovery for the agricultural sector is the Beaver Dam WWT plant, where, during the warm months, biosolids from secondary digesters (containing 2% solids) are applied to neighboring farms as fertilizer. Another example is the Watsonville Area Water Recycling Project in California, which supports water as well as nutrient recovery through fertigation in a tertiary treatment facility. This system retains up to 27.4 mg/L of total nitrogen, which is subsequently utilized for agricultural purposes (Diaz-Elsayed et al., 2019). Over 50 technologies are available for recovering phosphorus from wastewater and its by-products including sewage sludge, sludge ash, and digester supernatant. These methods include techniques such as wet chemical processes, precipitation, crystallization, and thermochemical processes (Egle et al., 2015). Also, there are various commercial crystallization and precipitation processes that have been developed such as FIX-Phos, Berliner Verfahren, Phosnix, Ostara, Kurita, CSIR, DHV Crystalactor, PRISA, and PhoStrip (Mehta et al., 2015). Opportunities for nutrient recovery across all scales are offered by fertigation, with trends largely following those of water reclamation systems. Economies of scale are shown in the treatment of wastewater, while distribution is affected by scale-related diseconomies (Diaz-Elsayed et al., 2020). For instance, Cornejo et al. found that electricity usage in household drip irrigation having nutrient recovery (1.4 kWh/m^3) was three to seven times more than the energy needed for water distribution at community and city scales. As a result, the specific embodied energy and carbon footprint per cubic meter of treated water were significantly higher at the household level compared to larger-scale systems (Cornejo et al., 2016).

AI models have also been reportedly developed for predicting nutrient concentration in effluent for nutrient recovery strategies. For instance, Inbar et al. developed an ML model for the estimation of phosphorus in wastewater. Their XGBoost model was found to have the highest prediction accuracy of 87% and precision of 85% and helped them to enhance the WWT plants monitoring as well as mitigate the discharge of low-quality effluent (Inbar and Avisar, 2024).

23.5.2 Energy Recovery

Given that WWT plants consume significant amounts of energy, recovering some of that energy can be both cost-effective and sustainable. Thermal energy can be recovered when greywater is separated from the wastewater flow. Although this is not yet widely practiced, cold climates offer the best conditions for thermal energy recovery from wastewater, as the process depends on temperature differences between the wastewater and its environment. Examples of thermal energy recovery technologies, such as NEXheater and EcoDrain, can capture energy from household wastewater and help reduce the energy required for household heating (Diaz-Elsayed et al., 2019). Along with nutrient recovery, biosolids can be used in energy recovery as well. Anaerobic digestion is the most commonly employed method for producing biogas, which is then utilized for energy. Another popular method is the incineration of biosolids, which can also reduce the waste volume and destroy pathogens. Because biogas comprises carbon dioxide, methane, water vapor, and trace compounds, it is recommended to pretreat the raw gas through methods such as drying and hydrogen sulfide removal. Following this, further purification using techniques like pressure swing adsorption, water or chemical scrubbing, cryogenic separation, or membrane separation allows the biogas to be utilized as a natural gas supplement (Stillwell et al., 2010; Yadav et al., 2024). As the scale of operations of WWT plants increases, so does their potential for energy recovery (Diaz-Elsayed et al., 2020). It has been noted that, on a larger scale, it is possible to achieve energy self-sufficiency and generate net positive energy. For instance, the Marselisborg WWT plant in Denmark, after optimizing its treatment processes and increasing biogas production, resulted in the plant generating over 150% of its energy needs (Diaz-Elsayed et al., 2019).

While energy recovery via AI implementation has been less explored and implemented, AI has been implemented for energy optimization and reduction processes. For example, Filipe et al. minimized electrical energy consumption by pumps in WWT plants using data-driven approaches and AI.

They used uncertainty forecasts for wastewater intake rates and sensor data available through the Supervisory Control and Data Acquisition system to propose a method that blends statistical analysis with deep reinforcement learning. When applied in a 90-day real-world case study, their results showed a 16.7% reduction in electrical energy consumption and a 97% decrease in the number of alarms (tank level exceeding 7.2 m), compared to the previous system with a fixed set-point level (Filipe et al., 2019). In another study, Zhu et al. identified that a vast majority of WWT plants often consume excess energy for aeration to meet the BOD of the water. In an attempt to balance and minimize the energy demand at the Metropolitan Water Reclamation District of Greater Chicago (MWRDGC) Calumet, they implemented an ANN taking 95% upper confidence limit (UCL) of multiple linear regression (MLR) to overestimate high carbonaceous five-day BOD concentrations to minimize the risk of insufficient treatment (Zhu et al., 2018). These studies demonstrate how AI offers significant potential for minimizing energy consumption and enhancing energy recovery in WWT, leading to more sustainable and efficient operations, and how their implementation can modernize WWT.

23.5.3 Metal Recovery

Wastewater produced by the mining industry contains significant amounts of metal contaminants, as large quantities of water are used in mining and subsequently released into the environment. These metal contaminants often exceed the discharge limits set by the World Health Organization, making it necessary to filter or remove them (Matebese et al., 2024). Various physical and chemical methods exist for recovering metals from leachate in municipal landfills, including adsorption, membrane filtration, chemical precipitation, ion exchange, and ion flotation (Krishnan et al., 2021). The chemical methods for metal recovery include cementation, chemical precipitation, ion exchange, and ion flotation methods. Chemical precipitation is the most cost-effective method for metal recovery, involving the removal of soluble metal ions by altering the solution to form insoluble metal complexes. This process typically involves adjusting the pH to 9–10 to convert dissolved metal ions into solid metal hydroxides (Chiu et al., 2019). Cementation is a precipitation process that extracts metals from waste streams by reacting a metal-containing solution with a specific metal powder or scrap, resulting in the formation of a metal sponge. For efficient cementation, a significant difference in electromotive force between the metals is necessary. Valuable metals like gold, platinum, palladium, rhodium, copper, and zinc, along with stop baths, stabilizers, and solution tailings, are recovered from waste solutions using this method (Kazakova et al., 2020; Jeon et al., 2020). Ion exchange methods are also commonly employed for metal recovery. This process is a reversible one in which an ion from a wastewater solution exchanges places with a similarly charged ion bound to a fixed solid particle (Krishnan et al., 2021). Ion flotation is an increasingly popular method for removing heavy metal ions from wastewater. This technique is favored for its simplicity, low energy consumption, fast processing time, minimal space requirements, versatility, small sludge production, low residual ion concentration, and cost-effectiveness. The process involves the addition of surfactants or collectors carrying opposite charges to the target ions, leading to the formation of a surfactant complex. The ions are then extracted by introducing gas bubbles into the solution, which bind to the complexes and float them to the surface for removal (Chang et al., 2019; Taseidifar et al., 2019). Physical methods for metal recovery from wastewater include physical separation or filtration by membranes or adsorption. Frequently used adsorbents include biomaterials, industrial solid waste, zeolites, clay minerals, and activated carbons (Arora, 2019).

Recent progress in metal recovery from wastewater highlights electrochemical technologies like coupled redox fuel cells (CRFCs), photoelectrochemical fuel cells (PFCs), and microbial fuel cells (MFCs). Over the past decade, these technologies have seen notable scientific and technical advancements. They are now capable of directly converting chemical energy in waste into electricity while simultaneously recovering metal resources. This dual functionality makes them highly promising for applications in energy generation, resource recovery, and WWT (Chen et al., 2023). However, the application of AI in metal recovery is still primarily focused on understanding the uncertain, nonlinear patterns of metal removal across various treatment techniques (Bhagat et al., 2020).

23.5.4 Water Reuse

The reuse of non-potable water is the most common form of water reuse at different scales. On a smaller scale, it typically involves applications such as irrigation and toilet flushing. Additionally, methods like using wetlands, membrane bioreactor (MBR)-based treatments, and activated sludge systems are used to facilitate water reuse. Projects aiming for indirect potable reuse of water, on the other hand, incorporate an environmental buffer before the water is subjected to further treatment at a WWT plant for potable use (Diaz-Elsayed et al., 2019). A standard treatment process for indirect potable reuse (IPR) projects typically involves secondary treatment, followed by ultra/microfiltration, reverse osmosis (RO), and UV light disinfection. RO plays a crucial role in water reuse strategies due to its effectiveness in removing a wide range of microbial contaminants, including bacteria, viruses, heavy metals, nitrates, and pesticides (Warsinger et al., 2018). Various groundwater recharge methods also involve treating wastewater through passive, natural systems like plant beds and constructed wetlands. With proper water treatment, the treated water can be suitable for direct potable use as well. A notable example is that of the New Goreangab Water Reclamation Plant (NGWRP) located in Windhoek, Namibia, which could cater to approximately 25% of the potable water demand for Windhoek in the year 2007 (Lahnsteiner and Lempert, 2007).

Integrating AI models with traditional methods and IoT architecture can improve the intelligent design systems for WWT and reuse (Alprol et al., 2024). For instance, in a study led by JC Chen et al., neural network-based models were utilized for predicting nitrogen content in treated wastewater for wastewater reuse and groundwater recharge (Chen et al., 2003).

23.6 Challenges and Future Perspectives

Despite their advantages and successes in WWT, AI models also have limitations that should not be overlooked. One of the primary issues is the issue of the "black box." While AI models can capture complex nonlinear relationships, they, however, lack insight into underlying mechanisms, therefore serving as black-box tools. While they show promise in WWT applications, integrating them with traditional approaches could enhance understanding and reduce data requirements (Hassija et al., 2024). It is also heavily reliant on data, and based on the type and volume of data, any bias in data collection and reporting could heavily influence the model and can give very different outputs, especially if different features from the data are used in the training process (Whang et al., 2023). This brings up another challenge related to the absence of standardized methods for determining the most effective network architectures, which often depend on the specific problem. Trial and error tends to be the main approach, but it frequently leads to overfitting or underfitting (Wang et al., 2023). Issues regarding data management and reproducibility are also prevalent, especially since, as previously noted, AI models need large volumes of data, and poor data quality can lead to inaccurate predictions. Therefore, it is essential to have systems in place that gather precise data. While many treatment plants and natural systems are equipped with sensors to comply with regulatory standards, these sensors often rely on long-term averages, which may not be backed by detailed historical data. A primary issue is the presence of missing values, which can arise from poor connections, sensor malfunctions, or weakened signal strength (Mamandipoor et al., 2020). To address this, storing detailed data for future use in training AI and ML systems is recommended. Transfer learning can help by applying knowledge from pretrained models to new applications with limited data, improving efficiency and results without starting from scratch. Also, lack of transparency due to non-reporting of code or avoiding making source code and data open hinders the fair comparison of various AI model performances (Lowe et al., 2022).

The integration of AI presents numerous opportunities to enhance WWT, with ongoing technological advancements increasing its potential to tackle complex WWT challenges. With the adoption and increasing implementation of AI in WWT strategies, it has become possible to modernize the process and improve upon existing WWT systems and mitigate some of the challenges faced in traditional WWT processes. In the future, AI applications in WWT could leverage high-resolution spatial and temporal data from remote sensing technologies like satellites and drones. Integrating these datasets with advanced AI algorithms will provide more accurate insights for targeted and timely interventions. Also, IoT devices could continuously collect real-time data on water quality and conditions, which could then be analyzed by AI to deliver quick insights, enabling rapid responses and proactive management (Kamyab et al., 2023). Other cutting-edge technologies, such as technologies like digital twins, drones, big data analytics, cloud computing, blockchain, VR/AR, and robotics, also showcase significant potential to enhance automation and advanced process control within WWT plants and might see more integration with AI in WWT processes (Cairone et al., 2024).

23.7 Conclusion

The incorporation of AI into WWT and resource recovery processes represents a major advancement in modernizing this essential sector. AI has the potential to greatly enhance efficiency, cut operational costs, and promote sustainability within WWT systems. Its applications, ranging from real-time monitoring and control to resource recovery and water reuse, highlight its transformative potential in managing wastewater and reclaiming valuable resources. However, the successful implementation of AI in existing systems will require overcoming challenges related to data quality, system integration, and cost. Looking ahead, AI is poised to play a key role in advancing WWT technologies, but addressing these challenges will be essential to unlocking its full potential.

This chapter highlights the transformative potential of AI in advancing and enhancing WWT applications and resource recovery. While traditional WWT methods have been effective, they are often faced with significant issues, such as high costs, maintenance challenges, and the need to address diverse pollutants. The chapter also highlights how the implementation of AI in WWT systems has facilitated rapid modernization by presenting innovative solutions to these challenges by offering substantial improvements in both efficiency and effectiveness across various domains. AI has revolutionized traditional WWT practices through enhanced real-time monitoring and control, optimized management of by-products, and refined process design and performance.

AI technologies enable sophisticated real-time analysis of water quality parameters and fault detection, facilitating proactive management and reducing operational costs. They also improve by-product management, such as minimizing excess chlorine, and optimize process design to enhance overall treatment efficiency. Furthermore, AI's capacity to simulate contaminant removal and predict system performance exemplifies its potential to drive significant advancements in WWT. Moreover, AI has also enabled improved water accessibility and efficient resource distribution, addressing both technical and social aspects of water management. AI applications in nutrient recovery, such as predicting phosphorus concentrations and optimizing recovery techniques, demonstrate their ability to enhance agricultural outputs and sustainability. Energy recovery through AI also shows promise, with advancements in reducing energy consumption and improving biogas production. Furthermore, AI's role in metal recovery and water reuse highlights its potential to refine processes and support efficient resource management. As AI technologies continue to advance, they hold the promise of further innovations, paving the way for more sustainable and effective WWT solutions.

Looking ahead, the ongoing development and application of AI in wastewater management will be vital in tackling global issues such as water scarcity, environmental conservation, and sustainable resource use. Although challenges like data quality concerns and the need for skilled personnel persist, the advantages of integrating AI far surpass these hurdles. To ensure the successful evolution of WWT plants into smart, efficient, and sustainable resource recovery hubs, it is essential for researchers, industry experts, and policymakers to work together to drive further advancements in AI applications within this field.

Review Questions

1. What are the processes involved in traditional wastewater treatment systems?
2. What are the various challenges involved in traditional wastewater treatment systems?
3. Why is there a need to integrate AI in wastewater treatment systems?
4. What is AI and its various components/types? Explain each in brief.
5. What are the different sectors where AI can be implemented in wastewater treatment systems?
6. What are the various resources available in wastewater?
7. How is nutrient recovery done from wastewater, and give the quantification of N, P, and K in wastewater?
8. How can AI assist in energy recovery processes in wastewater treatment systems?

9 Give some examples of intelligent systems for wastewater treatment and reuse.

10 What are the various challenges available in the implementation of AI in resource recovery from wastewater treatment plants and, accordingly, their probable solutions?

References

Abba, S.I., Linh, N.T.T., Abdullahi, J. et al. (2020). Hybrid machine learning ensemble techniques for modeling dissolved oxygen concentration. *IEEE Access*, **8**, 157218–157237. https://doi.org/10.1109/ACCESS.2020.3017743

Aghilesh, K., Kumar, A., Agarwal, S. et al. (2023). Use of artificial intelligence for optimizing biosorption of textile wastewater using agricultural waste. *Environmental Technology*, **44**(1), 22–34. https://doi.org/10.1080/09593330.2021.1961874

Al-Hazmi, H.E., Mohammadi, A., Hejna, A. et al. (2023). Wastewater reuse in agriculture: prospects and challenges. *Environmental Research*, **236**(Pt 1), 116711. https://doi.org/10.1016/j.envres.2023.116711

Alizadeh, M.J., Kavianpour, M.R., Danesh, M. et al. (2018). Effect of river flow on the quality of estuarine and coastal waters using machine learning models. *Engineering Applications of Computational Fluid Mechanics*, **12**(1), 810–823. https://doi.org/10.1080/19942060.2018.1528480

Alprol, A.E., Mansour, A.T., Ibrahim, M.E.E-D. et al. (2024). Artificial intelligence technologies revolutionizing wastewater treatment: current trends and future prospective. *Water*, **16**(2), 314. https://doi.org/10.3390/w16020314

Anand, C.K. and Apul, D.S. (2014). Composting toilets as a sustainable alternative to urban sanitation—a review. *Waste Management*, **34**(2), 329–343. https://doi.org/10.1016/j.wasman.2013.10.006

Arora, R. (2019). Adsorption of heavy metals–A review. *Materials Today Proceedings*, **18**, 4745–4750. https://doi.org/10.1016/j.matpr.2019.07.462

Awasthi, M.K., Ganeshan, P., Gohil, N. et al. (2023). Advanced approaches for resource recovery from wastewater and activated sludge: a review. *Bioresource Technology*, **384**, 129250. https://doi.org/10.1016/j.biortech.2023.129250

Bahoo, S., Cucculelli, M., Goga, X. et al. (2024). Artificial intelligence in finance: a comprehensive review through bibliometric and content analysis. *SN Business & Economics*, **4**(2), 23. https://doi.org/10.1007/s43546-023-00618-x

Bajwa. J., Munir. U., Nori. A. et al. (2021) Artificial intelligence in healthcare: transforming the practice of medicine. *Future Healthcare Journal*, **8**(2), e188–e194. https://doi.org/10.7861/fhj.2021-0095

Bhagat, S.K., Tung, T.M., and Yaseen, Z.M. (2020). Development of artificial intelligence for modeling wastewater heavy metal removal: state of the art, application assessment and possible future research. *Journal of Cleaner Production*, **250**, 119473. https://doi.org/10.1016/j.jclepro.2019.119473

Bhat, S.A., Cui, G., Li, W. et al. (2021). Chapter 11 - Challenges and opportunities associated with wastewater treatment systems. In: *Current Developments in Biotechnology and Bioengineering* (eds. S. Kumar, R. Kumar, and A. Pandey), 259–283. Elsevier.

Bisaria, K., Singh, R., Gupta, M. et al. (2024). Novel acoustic-activated alkali-functionalized Trapa bispinosa peel biochar for green immobilization of chlorpyrifos from wastewater: artificial intelligence modelling and experimental validation. *Biomass Conversion and Biorefinery*, **14**(6), 7763–7782. https://doi.org/10.1007/s13399-022-02898-z

Cairone, S., Hasan, S.W., Choo, K-H. et al. (2024). Revolutionizing wastewater treatment toward circular economy and carbon neutrality goals: pioneering sustainable and efficient solutions for automation and advanced process control with smart and cutting-edge technologies. *Journal of Water Process Engineering*, **63**, 105486. https://doi.org/10.1016/j.jwpe.2024.105486

Chahal, C., van den Akker, B., Young, F. et al. (2016). Chapter two - Pathogen and particle associations in wastewater: significance and implications for treatment and disinfection processes. In: *Advances in Applied Microbiology* (eds. S. Sariaslani and G. Michael Gadd), 63–119. Academic Press.

Chai, W.S., Cheun, J.Y., Kumar, P.S. et al. (2021). A review on conventional and novel materials towards heavy metal adsorption in wastewater treatment application. *Journal of Cleaner Production*, **296**, 126589. https://doi.org/10.1016/j.jclepro.2021.126589

Chan, H. and Nai-He, Y. (2020). A pretreatment method of wastewater based on artificial intelligence and fuzzy neural network system. *Journal of Intelligent & Fuzzy Systems*, **39**(2), 1711–1720. https://doi.org/10.3233/JIFS-179945

Chang, L., Cao, Y., Fan, G. et al. (2019). A review of the applications of ion floatation: wastewater treatment, mineral beneficiation and hydrometallurgy. *RSC Advances*, **9**(35), 20226–20239. https://doi.org/10.1039/c9ra02905b

Chang, P., Xu, Y., Meng, F. et al. (2024). Fault detection in wastewater treatment process using broad slow feature neural network with incremental learning ability. *IEEE Transactions on Industrial Informatics*, **20**(3), 4540–4549. https://doi.org/10.1109/TII.2023.3324971

Chen, J.C., Chang, N.B., and Shieh, W.K. (2003). Assessing wastewater reclamation potential by neural network model. *Engineering Applications of Artificial Intelligence*, **16**(2), 149–157. https://doi.org/10.1016/S0952-1976(03)00056-3

Chen, Q-Y., Lu, R-C., Zhu, Y-C. et al. (2023). Metal recovery and electricity generation from wastewater treatment: the state of the art. *Processes*, **11**(1), 88. https://doi.org/10.3390/pr11010088

Chiu, K-L., Shen, Y-H., Chen, Y-H. et al. (2019). Recovery of valuable metals from spent lithium ion batteries (LIBs) using physical pretreatment and a hydrometallurgy process. *Advances in Materials*, **8**(1), 12–20. https://doi.org/10.11648/j.am.20190801.12

Cordero, J.A., He, K., Janya, K. et al. (2021). Predicting formation of haloacetic acids by chlorination of organic compounds using machine-learning-assisted quantitative structure-activity relationships. *Journal of Hazardous Materials*, **408**, 124466. https://doi.org/10.1016/j.jhazmat.2020.124466

Cornejo, P.K., Zhang, Q., and Mihelcic, J.R. (2016). How does scale of implementation impact the environmental sustainability of wastewater treatment integrated with resource recovery? *Environmental Science & Technology*, **50**(13), 6680–6689. https://doi.org/10.1021/acs.est.5b05055

Cornel, P., Meda, A., and Bieker, S. (2011). 4.12 - Wastewater as a source of energy, nutrients, and service water. In: *Treatise on Water Science* (ed. P. Wilderer), 337–375. Oxford: Elsevier.

Dias, E.M. and Petit, C. (2015). Towards the use of metal–organic frameworks for water reuse: a review of the recent advances in the field of organic pollutants removal and degradation and the next steps in the field. *Journal of Materials Chemistry A*, **3**(45), 22484–22506. https://doi.org/10.1039/C5TA05440K

Diaz-Elsayed, N., Rezaei, N., Guo, T. et al. (2019). Wastewater-based resource recovery technologies across scale: a review. *Resources, Conservation and Recycling*, **145**, 94–112. https://doi.org/10.1016/j.resconrec.2018.12.035

Diaz-Elsayed N., Rezaei N., Ndiaye A. et al. (2020). Trends in the environmental and economic sustainability of wastewater-based resource recovery: a review. *Journal of Cleaner Production*, **265**, 121598. https://doi.org/10.1016/j.jclepro.2020.121598

Egle, L., Rechberger, H., and Zessner, M. (2015). Overview and description of technologies for recovering phosphorus from municipal wastewater. *Resources, Conservation and Recycling*, **105**, 325–346. https://doi.org/10.1016/j.resconrec.2015.09.016

Elkiran, G., Nourani, V., and Abba, S.I. (2019). Multi-step ahead modelling of river water quality parameters using ensemble artificial intelligence-based approach. *Journal of Hydrology*, **577**, 123962. https://doi.org/10.1016/j.jhydrol.2019.123962

Elkiran, G., Nourani, V., Abba, S.I. et al. (2018). Artificial intelligence-based approaches for multi-station modelling of dissolve oxygen in river. *Global Journal of Environmental Science and Management*, **4**(4), 439–450. https://doi.org/10.22034/gjesm.2018.04.005

Essahlaoui, F., Elhajrat, N., Halimi, M. et al. (2024). Intelligent real-time monitoring system for wastewater management using artificial neural network. In: *Sustainable and Green Technologies for Water and Environmental Management* (eds. M. Azrour, J. Mabrouki, and A. Guezzaz), 15–24. Cham: Springer Nature Switzerland.

Filipe, J., Bessa, R.J., Reis, M. et al. (2019). Data-driven predictive energy optimization in a wastewater pumping station. *Applied Energy*, **252**, 113423. https://doi.org/10.1016/j.apenergy.2019.113423

Fu, X., Wu, R., Qi, H. et al. (2023). Long-term trends in surface water quality of China's seven major basins based on water quality identification index and big data analysis. *Environmental Impact Assessment Review*, **100**, 107090. https://doi.org/10.1016/j.eiar.2023.107090

Godo-Pla, L., Rodríguez, J.J., Suquet, J. et al. (2021). Control of primary disinfection in a drinking water treatment plant based on a fuzzy inference system. *Process Safety and Environmental Protection*, **145**, 63–70. https://doi.org/10.1016/j.psep.2020.07.037

Hassija, V., Chamola, V., Mahapatra, A. et al. (2024). Interpreting black-box models: a review on explainable artificial intelligence. *Cognitive Computation*, **16**(1), 45–74. https://doi.org/10.1007/s12559-023-10179-8

Holzinger, A., Keiblinger, K., Holub, P. et al. (2023). AI for life: trends in artificial intelligence for biotechnology. *New Biotechnology*, **74**, 16–24. https://doi.org/10.1016/j.nbt.2023.02.001

Inbar, O. and Avisar, D. (2024). Enhancing wastewater treatment through artificial intelligence: a comprehensive study on nutrient removal and effluent quality prediction. *Journal of Water Process Engineering*, **61**, 105212. https://doi.org/10.1016/j.jwpe.2024.105212

Jeon, S., Tabelin, C.B., Park, I. et al. (2020). Ammonium thiosulfate extraction of gold from printed circuit boards (PCBs) of end-of-life mobile phones and its recovery from pregnant leach solution by cementation. *Hydrometallurgy*, **191**, 105214. https://doi.org/10.1016/j.hydromet.2019.105214

Kamyab, H., Khademi, T., Chelliapan, S. et al. (2023). The latest innovative avenues for the utilization of artificial Intelligence and big data analytics in water resource management. *Results in Engineering*, **20**, 101566. https://doi.org/10.1016/j.rineng.2023.101566

Kazakova, N., Lucheva, B., and Iliev, P. (2020). A study on the cementation process of non-ferrous metals from a brine leaching solution. *Journal of Chemical Technology and Metallurgy*, **55**(1), 223–227

Kisi, O., Alizamir, M., and Docheshmeh Gorgij, A. (2020). Dissolved oxygen prediction using a new ensemble method. *Environmental Science and Pollution Research*, **27**(9), 9589–9603. https://doi.org/10.1007/s11356-019-07574-w

Krishnan, S., Zulkapli, N.S., Kamyab, H. et al. (2021) Current technologies for recovery of metals from industrial wastes: an overview. *Environmental Technology & Innovation*, **22**, 101525. https://doi.org/10.1016/j.eti.2021.101525

Kumar, S., Kumar R., and Pandey, A. (2021). Chapter 1 - Solid waste and wastewater management: a social and global perspective. In: *Current Developments in Biotechnology and Bioengineering* (eds. S. Kumar, R. Kumar, and A. Pandey), 1–22. Elsevier.

Lahnsteiner. J. and Lempert. G. (2007). Water management in Windhoek, Namibia. *Water Science & Technology*, **55**(1–2), 441–448. https://doi.org/10.2166/wst.2007.022

LeCun. Y., Bengio. Y., and Hinton. G. (2015). Deep learning. *Nature*, **521**(7553), 436–444. https://doi.org/10.1038/nature14539

Li, X., Yi, X., Liu, Z. et al. (2021). Application of novel hybrid deep leaning model for cleaner production in a paper industrial wastewater treatment system. *Journal of Cleaner Production*, **294**, 126343. https://doi.org/10.1016/j.jclepro.2021.126343

Lowe, M., Qin, R., and Mao, X. (2022). A review on machine learning, artificial intelligence, and smart technology in water treatment and monitoring. *Water*, **14**(9), 1384. https://doi.org/10.3390/w14091384

Mahato, J.K. and Gupta, S.K. (2022). Exploring applicability of artificial intelligence and multivariate linear regression model for prediction of trihalomethanes in drinking water. *International Journal of Environmental Science and Technology*, **19**(6), 5275–5288. https://doi.org/10.1007/s13762-021-03392-1

Mahmoud, M.S. and Mahmoud, A.S. (2021). Wastewater treatment using nano bimetallic iron/copper, adsorption isotherm, kinetic studies, and artificial intelligence neural networks. *Emergent Materials*, **4**(5), 1455–1463. https://doi.org/10.1007/s42247-021-00253-y

Malchi, T., Maor, Y., Tadmor, G. et al. (2014). Irrigation of root vegetables with treated wastewater: evaluating uptake of pharmaceuticals and the associated human health risks. *Environmental Science & Technology*, **48**(16), 9325–9333. https://doi.org/10.1021/es5017894

Mamandipoor, B., Majd, M., Sheikhalishahi, S. et al. (2020). Monitoring and detecting faults in wastewater treatment plants using deep learning. *Environmental Monitoring and Assessment*, **192**(2), 148. https://doi.org/10.1007/s10661-020-8064-1

Matebese, F., Mosai, A.K., Tutu, H. et al. (2024). Mining wastewater treatment technologies and resource recovery techniques: a review. *Heliyon*, **10**(3), e24730. https://doi.org/10.1016/j.heliyon.2024.e24730

Mehta, C.M., Khunjar, W.O., Nguyen, V. et al. (2015). Technologies to recover nutrients from waste streams: a critical review. *Critical Reviews in Environmental Science and Technology*, **45**(4), 385–427. https://doi.org/10.1080/10643389.2013.866621

Mercorelli, P. (2024). Recent advances in intelligent algorithms for fault detection and diagnosis. *Sensors*, **24**(8), 2656. https://doi.org/10.3390/s24082656

Muthukrishnan, N., Maleki, F., Ovens, K. et al. (2020). Brief history of artificial intelligence. *Neuroimaging Clinics of North America*, **30**(4), 393–399. https://doi.org/10.1016/j.nic.2020.07.004

Najah Ahmed, A., Binti Othman, F., Abdulmohsin Afan, H. et al. (2019). Machine learning methods for better water quality prediction. *Journal of Hydrology*, **578**, 124084. https://doi.org/10.1016/j.jhydrol.2019.124084

Nayak, M., Dhanarajan, G., Dineshkumar, R. et al. (2018). Artificial intelligence driven process optimization for cleaner production of biomass with co-valorization of wastewater and flue gas in an algal biorefinery. *Journal of Cleaner Production*, **201**, 1092–1100. https://doi.org/10.1016/j.jclepro.2018.08.048

Nishat, A., Yusuf, M., Qadir, A. et al. (2023). Wastewater treatment: a short assessment on available techniques. *Alexandria Engineering Journal*, **76**, 505–516. https://doi.org/10.1016/j.aej.2023.06.054

Nnaji, P.C., Anadebe, V.C., Onukwuli, O.D. et al. (2022). Multifactor optimization for treatment of textile wastewater using complex salt–Luffa cylindrica seed extract (CS-LCSE) as coagulant: response surface methodology (RSM) and artificial intelligence algorithm (ANN–ANFIS). *Chemical Papers*, **76**(4), 2125–2144. https://doi.org/10.1007/s11696-021-01971-7

Nourani, V., Asghari, P., and Sharghi, E. (2021). Artificial intelligence based ensemble modeling of wastewater treatment plant using jittered data. *Journal of Cleaner Production*, **291**, 125772. https://doi.org/10.1016/j.jclepro.2020.125772

Nur Adli Zakaria, M., Abdul Malek, M., Zolkepli, M. et al. (2021). Application of artificial intelligence algorithms for hourly river level forecast: a case study of Muda River, Malaysia. *Alexandria Engineering Journal*, **60**(4), 4015–4028. https://doi.org/10.1016/j.aej.2021.02.046

Obotey Ezugbe, E. and Rathilal, S. (2020). Membrane technologies in wastewater treatment: a review. *Membranes*, **10**(5), 89. https://doi.org/10.3390/membranes10050089

Okoji, C.N., Okoji, A.I., Ibrahim, M.S. et al. (2022). Comparative analysis of adaptive neuro-fuzzy inference system (ANFIS) and RSRM models to predict DBP (trihalomethanes) levels in the water treatment plant. *Arabian Journal of Chemistry*, **15**(6), 103794. https://doi.org/10.1016/j.arabjc.2022.103794

Ooi, K-B., Tan, GW-H., Al-Emran, M. et al. (2025). The potential of generative artificial intelligence across disciplines: perspectives and future directions. *Journal of Computer Information Systems*, **65**(1),76–107. https://doi.org/10.1080/08874417.2023.2261010

Park, J., Lee, W.H., Kim, K.T. et al. (2022). Interpretation of ensemble learning to predict water quality using explainable artificial intelligence. *Science of the Total Environment*, **832**, 155070. https://doi.org/10.1016/j.scitotenv.2022.155070

Peleato, N.M. (2022). Application of convolutional neural networks for prediction of disinfection by-products. *Scientific Reports*, **12**(1), 612. https://doi.org/10.1038/s41598-021-03881-w

Picos-Benítez, A.R., Martínez-Vargas, B.L., Duron-Torres, S.M. et al. (2020). The use of artificial intelligence models in the prediction of optimum operational conditions for the treatment of dye wastewaters with similar structural characteristics. *Process Safety and Environmental Protection*, **143**, 36–44. https://doi.org/10.1016/j.psep.2020.06.020

Raghavendar, K., Batra, I., and Malik, A. (2023). A robust resource allocation model for optimizing data skew and consumption rate in cloud-based IoT environments. *Decision Analytics Journal*, **7**, 100200. https://doi.org/10.1016/j.dajour.2023.100200

Rana, R., Kalia, A., Boora, A. et al. (2023) Artificial intelligence for surface water quality evaluation, monitoring and assessment. *Water*, **15**(22), 3919. https://doi.org/10.3390/w15223919

Rezaei, N., Sierra-Altamiranda, A., Diaz-Elsayed, N. et al. (2019). A multi-objective optimization model for decision support in water reclamation system planning. *Journal of Cleaner Production*, **240**, 118227. https://doi.org/10.1016/j.jclepro.2019.118227

Richards, C.E., Tzachor, A., Avin, S. et al. (2023). Rewards, risks and responsible deployment of artificial intelligence in water systems. *Nature Water*, **1**(5), 422–432. https://doi.org/10.1038/s44221-023-00069-6

Richardson, S.D. (2003). Disinfection by-products and other emerging contaminants in drinking water. *TrAC Trends in Analytical Chemistry*, **22**(10), 666–684. https://doi.org/10.1016/S0165-9936(03)01003-3

Saboe, D., Ghasemi, H., Gao, M.M. et al. (2021). Real-time monitoring and prediction of water quality parameters and algae concentrations using microbial potentiometric sensor signals and machine learning tools. *Science of the Total Environment*, **764**, 142876. https://doi.org/10.1016/j.scitotenv.2020.142876

Saleh, I.A., Zouari, N., and Al-Ghouti, M.A. (2020). Removal of pesticides from water and wastewater: chemical, physical and biological treatment approaches. *Environmental Technology & Innovation*, **19**, 101026. https://doi.org/10.1016/j.eti.2020.101026

Salloom, T., Kaynak, O., Yu, X. et al. (2022). Proportional integral derivative booster for neural networks-based time-series prediction: case of water demand prediction. *Engineering Applications of Artificial Intelligence*, **108**, 104570. https://doi.org/10.1016/j.engappai.2021.104570

Saravanan. R. and Sujatha. P. (2018). A state of art techniques on machine learning algorithms: a perspective of supervised learning approaches in data classification. In: *2018 Second International Conference on Intelligent Computing and Control Systems (ICICCS)*, 945–949. IEEE.

Şenol, H. (2021). Methane yield prediction of ultrasonic pretreated sewage sludge by means of an artificial neural network. *Energy*, **215**, 119173. https://doi.org/10.1016/j.energy.2020.119173

Song, H.J., You, S., Jia, X.H. et al. (2015). MoS2 nanosheets decorated with magnetic Fe3O4 nanoparticles and their ultrafast adsorption for wastewater treatment. *Ceramics International*, **41**(10, Part A), 13896–13902. https://doi.org/10.1016/j.ceramint.2015.08.023

Song, Y., Wang, L., Qiang, X. et al. (2023). An overview of biological mechanisms and strategies for treating wastewater from printing and dyeing processes. *Journal of Water Process Engineering*, **55**, 104242. https://doi.org/10.1016/j.jwpe.2023.104242

Stillwell, A.S., Hoppock, D.C., and Webber, M.E. (2010). Energy recovery from wastewater treatment plants in the United States: a case study of the energy-water nexus. *Sustainability*, **2**(4), 945–962. https://doi.org/10.3390/su2040945

Szepesvári, C. (2022). *Algorithms for Reinforcement Learning*. Springer Nature.

Taseidifar, M., Ziaee, M., Pashley, R.M. et al. (2019). Ion flotation removal of a range of contaminant ions from drinking water. *Journal of Environmental Chemical Engineering*, **7**(4), 103263. https://doi.org/10.1016/j.jece.2019.103263

Tyagi, A.K. and Chahal, P. (2020). Artificial intelligence and machine learning algorithms. In: *Challenges and Applications for Implementing Machine Learning in Computer Vision*, 188–219. IGI Global.

Vaidya, R., Verma, K., Kumar, M. et al. (2023). Assessing wastewater management challenges in developing countries: a case study of India, current status and future scope. *Environment, Development and Sustainability*, **26**, 19369–19396. https://doi.org/10.1007/s10668-023-03540-2

Villarín, M.C. and Merel, S. (2020). Paradigm shifts and current challenges in wastewater management. *Journal of Hazardous Materials*, **390**, 122139. https://doi.org/10.1016/j.jhazmat.2020.122139

Visvanathan, C., Aim, R.B., and Parameshwaran, K. (2000). Membrane separation bioreactors for wastewater treatment. *Critical Reviews in Environmental Science and Technology*, **30**(1), 1–48. https://doi.org/10.1080/10643380091184165

Wang, Y., Cheng, Y., Liu, H. et al. (2023). A review on applications of artificial intelligence in wastewater treatment. *Sustainability*, **15**(18), 13557. https://doi.org/10.3390/su151813557

Warsinger, D.M., Chakraborty, S., Tow, E.W. et al. (2018). A review of polymeric membranes and processes for potable water reuse. *Progress in Polymer Science*, **81**, 209–237. https://doi.org/10.1016/j.progpolymsci.2018.01.004

Whang, S.E., Roh, Y., Song, H. et al. (2023). Data collection and quality challenges in deep learning: a data-centric AI perspective. *The VLDB Journal*, **32**(4), 791–813. https://doi.org/10.1007/s00778-022-00775-9

Wilsenach, J.A., Maurer, M., Larsen, T.A. et al. (2003). From waste treatment to integrated resource management. *Water Science Technology*, **48**(1), 1–9. https://doi.org/10.2166/wst.2003.0002

Xu, Z., Shen, J., Qu, Y. et al. (2022). Using simple and easy water quality parameters to predict trihalomethane occurrence in tap water. *Chemosphere*, **286**, 131586. https://doi.org/10.1016/j.chemosphere.2021.131586

Yadav, P., Singh, R.P., Singh, S.K. et al. (2024). Chapter Fifteen—Biogas production using waste water: methodologies and applications. In: *Advances in Chemical Pollution, Environmental Management and Protection* (eds. A. Kumar, M. Bilal, and LFR. Ferreira), 347–361. Elsevier.

Yang, Y., Kim, K-R., Kou, R. et al. (2022). Prediction of effluent quality in a wastewater treatment plant by dynamic neural network modeling. *Process Safety and Environmental Protection*, **158**, 515–524. https://doi.org/10.1016/j.psep.2021.12.034

Zaibel, I., Arnon, S., and Zilberg, D. (2022). Treated municipal wastewater as a water source for sustainable aquaculture: a review. *Reviews in Aquaculture*, **14**(1), 362–377. https://doi.org/10.1111/raq.12602

Zamani, E.D., Smyth, C., Gupta. S. et al. (2023). Artificial intelligence and big data analytics for supply chain resilience: a systematic literature review. *Annals of Operations Research*, **327**(2), 605–632. https://doi.org/10.1007/s10479-022-04983-y

Zhang, S., Jin, Y., Chen, W. et al. (2023). Artificial intelligence in wastewater treatment: a data-driven analysis of status and trends. *Chemosphere*, **336**, 139163. https://doi.org/10.1016/j.chemosphere.2023.139163

Zhang, X., Zhang, P., Wu, Z. et al. (2013). Adsorption of methylene blue onto humic acid-coated Fe3O4 nanoparticles. *Colloids and Surfaces a: Physicochemical and Engineering Aspects*, **435**, 85–90. https://doi.org/10.1016/j.colsurfa.2012.12.056

Zhao, L., Dai, T., Qiao, Z. et al. (2020). Application of artificial intelligence to wastewater treatment: a bibliometric analysis and systematic review of technology, economy, management, and wastewater reuse. *Process Safety and Environmental Protection*, **133**, 169–182. https://doi.org/10.1016/j.psep.2019.11.014

Zhu, J., Jiang, Z., and Feng, L. (2022). Improved neural network with least square support vector machine for wastewater treatment process. *Chemosphere*, **308**, 136116. https://doi.org/10.1016/j.chemosphere.2022.136116

Zhu, J-J., Kang, L., and Anderson, P.R. (2018). Predicting influent biochemical oxygen demand: balancing energy demand and risk management. *Water Research*, **128**, 304–313. https://doi.org/10.1016/j.watres.2017.10.053

24

Sustainability and Interdependency of Multi-step Scale Up of Wastewater Treatment Technologies

Ambika Arkatkar[1], Dipak A. Jadhav[2], Mohammed Hussein[3], Hai Yen Nguyen[3,4], Tasnim Eisa[3,4], Ahmed Bahaa[3,4], Arvind Kumar Mungray[5], and Kyu-Jung Chae[3,4]**

[1]Department of Biotechnology, Veer Narmad South Gujarat University (VNSGU), Surat, Gujarat, India
[2]School of Civil and Environmental Sciences, Faculty of Science and Technology, JSPM University Pune, Maharashtra, India
[3]Department of Environmental Engineering, College of Ocean Science and Engineering, Korea Maritime and Ocean University, Busan, Republic of Korea
[4]Interdisciplinary Major of Ocean Renewable Energy Engineering, Korea Maritime and Ocean University, Busan, Republic of Korea
[5]Department of Chemical Engineering, S. V. National Institute of Technology, Surat, Gujarat, India

*Corresponding author: ckjdream@kmou.ac.kr

24.1 Introduction

The treatment of any waste is a combination of many physical, chemical, and biological processes. The generation of wastewater is an ever-continuing process and will surely increase with the increasing demand. Though the demand increases, it is not going to increase the resources of potable water. This limitation has been fairly recognized by many activists, social reformers, and scientists all over the world. Many technologies and strategies are applied for the reuse and recycling of wastewater. The wastewater that is treated in any central facility comprises different types of chemicals (organic and inorganic) and physical (of different particle sizes) entities. As the waste is a mixture of components, the treatment strategies are also planned accordingly. The chronological order of the treatment technology depends on the ease of the system. For example, while treating industrial wastewater, we cannot choose sieve filtration as the first step,

Wastewater to Resource Recovery: Applying the Circular Economy Toward Sustainable Development, First Edition. Edited by Arvind Kumar Mungray et al. Published 2026 by John Wiley & Sons, Inc.

whereas sieve filtration is very much required as an initial step to start municipal wastewater treatment.

A single technology cannot be sufficient to treat wastewater. At present, the conventional system also uses a combination of different processes. The variations in the treatment methods are broadly classified in the form of physical, chemical, and biological methods. Each method demands different structural, operational, and maintenance requirements. The output of each method also differs with respect to the required quality as well as quantity of wastewater. Thus, forecasting or probability testing of any sewage treatment plant (STP) system becomes necessary for the smooth execution of the treatment. It is beneficial to check the environmental impact, energy utilization, and financial implications of the STP during its planning and also during its working phase. For example, a centralized STP demands energy right from the collection of the wastewater (the collection of the wastewater is a machine-driven process that pumps the water to the STP, whereas a decentralized system may or may not require any intense collection step. This small example explains how a simple point can affect the choice of technology and the decision of any stakeholder.

The sustainability of multi-step systems can be evaluated and optimized as and when required. This can be a stepwise process or can be only applicable to a single process, which can affect the overall output of the STP. The change and development of innovative technology always guide the administration to take sustainably beneficial steps. The changing environmental requirements for better conservation of natural resources have made it compulsory to innovate and replace the old technologies. The integration of a series of technologies is also considered a part of such innovation. In this integration process, one technology may assist or safeguard the other technology, for example, in the filtration process, the treated water is passed through an ultrafiltration unit only when high-grade potable water is required as an output of the treatment facility. The ultrafiltration membrane has a very small pore size and is prone to clogging if a direct flow of treated water is passed through it. In such a condition, the process will be slow, the retention of water will be high, and the lifespan of the membrane will be reduced. To avoid all such problems, the same filtration technique is used, but through a microfiltration membrane, which has a higher pore size compared to the ultrafiltration membrane. The effluent of microfiltration is then passed through the ultrafiltration process. Thus, one process is safeguarding the next step of the treatment facility. This change will have both economic and environmental benefits.

This chapter will discuss the existing interdependency of treatment methods and the importance of monitoring the STP for achieving sustainable treatment. The technology of simulation and beforehand assessment was not available in earlier days. In those days, the optimization, implementation, and construction of any new technology that was to be included in the STP were admissible based on good laboratory results. The shift of technology from lab to field many times goes through setbacks because of changes in the magnitude of usage, frequent changes in the operational conditions, etc. The availability of computerized and AI-based analysis has provided a good solution to these problems. The programs are available to check the most suitable combination of a multi-step system that can deliver the desired quality and quantity of treated water. The environmental regulatory requirements can be added as a possible input in these systems. The decision-making thus becomes easy and is supported by a factsheet.

24.2 Multi-step Wastewater Treatment System

The basic principle of separation is always applied in the multi-step treatment facility. The separation can be between the solid and liquid phases of the wastewater and between treatment processes, like aerobic and anaerobic processes. A conventional treatment setup is divided into primary, secondary, and tertiary treatment facilities. The screens and grit chamber are used for the removal of solid and physically separable waste from the raw wastewater as a first step in any STP (Silva, 2023). The sedimentation principle is applied before and after the biological treatment to achieve the separation of heavy and light materials in a wastewater stream. Coagulation and flocculation methods based on chemical interactions are also used to treat the sewage. The biological treatment may consist of aerobic and anaerobic sections, which may lead to the formation of different by-products like methane and treated water with low chemical oxygen demand (COD). For example, the common practice is to divert the effluent of the grit chamber to a primary sedimentation tank. This flow of wastewater indicates that the two physical processes are linked to each other. The grit chamber works as a pretreatment for the wastewater that further enters the primary sedimentation tank. Separation of solid waste in the grit chamber prior to sedimentation saves time as well as safeguards the quality of the sludge. The same workflow is followed in the next step of STP. The effluent from sedimentation enters for biological treatment. The wastewater that is devoid of sludge is easy to degrade as it only has suspended solids in it. These are physical and biological processes, but the treatment is completed after chemical treatment of the wastewater. Flocculation and coagulation require the addition of chemicals that can further assist in completing the treatment. The parameters for assessment of treated wastewater should be monitored on a regular basis and this includes monitoring of COD, biological oxygen demand (BOD), temperature, pH, total dissolved solid (TDS), total nitrogen (TN), total phosphorous (TP), etc. (Shamshad and Rehman, 2024; Sun et al., 2019).

Apart from the above-mentioned wastewater parameters, the economic evaluation of a multi-step treatment facility used for wastewater treatment was initially assessed on the basis of the quality of treated water and energy demand. This quality of wastewater can be detected at the last step of the STP. The energy requirement of these plants is high and is mostly required for operation and maintenance purposes. Energy is derived from electrical sources; it shows a positive correlation with the quality of wastewater. The treatment facilities are being experimented with a change in the energy source, such as the use of renewable energy options, solar, wind, and hydropower. Methane gas produced in the STP is also utilized to generate electricity for the operation of the facility.

Now, instead of focusing on the last step for analyzing the sustainability of STP, researchers propose taking into consideration a comprehensive model for the assessment of such STP. The impact of each step should be assessed, and optimization is recommended for each step. The impact can be seen on different levels, such as social, environmental (including all three spheres, soil, water, and air), economic, etc. (Zhang and Li, 2025).

24.3 Technologies Involved in Treatment Facility

The technologies, methods, and processes that are commonly applied in the STP are coagulation, adsorption, filtration, activated sludge, biological filters, membrane bed reactors, stabilization ponds, and anaerobic digestion, and advancement in these methods can be done like ion exchange technique, ionization, ozonation, Fenton reaction, etc. (Fung and Wibowo, 2013). These technologies are connected in a chain to achieve effective treatment of wastewater. In case it is decided to introduce a new technology in a working STP or apply a modification to the existing technology, an overall stepwise analysis is required to understand the impact of this change. As technological advancement is always considered to be beneficial for improving the quality of treated wastewater, such analysis focuses on the economic sustainability of the STP due to the proposed change.

The study conducted in Portugal took nearly 11 different companies into consideration for economic assessment. The parameters such as specific energy consumption and energy operating cost per population were calculated, and their data were collected from the Sabi database. It was noted that an algae-based solution was economical. It reduces the operation cost per inhabitant as compared to the conventional activated sludge and other methods. The operation cost included the expenses required for electricity, chemicals, sludge disposal, labor, etc., but in this study, only the energy cost was taken into consideration. The reduction of carbon footprint, which is a prominent component of sustainable goals, is also taken into consideration while shifting the technology from methane production to algal biomass production (Santos et al., 2022). Similarly, anaerobic digestor (ADs) are commonly used in the resource recovery process of methane, hydrogen, compost, volatile fatty acids, etc. This technology has now been associated with other technologies such as rotating bed filters, electrodialysis, microalgal treatment, adsorption, etc. The integration of these technologies helps in better energy and nutrient recovery. It has been reported to support a good amount of revenue generation (Soo et al., 2024).

Though a particular technology is very effective in the removal of a targeted pollutant, it may happen that during scale-up at the field level, some limitations will be noticed. For example, a technology like activated sludge process is considered cost-effective, but it might not be applicable for industrial wastewater. A popular method used in STP may develop limitations over the given time period, like the trickling filter, which is most used in STP but may face an issue of not being effective at high organic load. Technology like a membrane bioreactor will face the issue of membrane maintenance and fouling. A simple process like adsorption is only suitable for small amounts of wastewater treatment and nanofiltration, which can clear the turbidity of water up to 90% but often has to deal with the issue of membrane fouling that leads to high rejection. Constructed wetlands (CWs) are very easy to operate and an economically viable treatment option but they are applicable only for domestic wastewater and may face the problem of clogging (Obaideen et al., 2022). Thus, the choice of technology is not always dependent on the efficacy of the method but mostly on the field conditions. Once implemented, the method may be revised or improvised as per the needs of the field condition and the timeline of maintenance. The wear and tear of the machine that is applied for the execution of the method will always be considered for the economic analysis of the operational plant.

Energy generation from wastewater treatment is not applicable to all types of wastewater treatment plants. Instead of energy in the form of methane, the researchers have started to look for other resource recovery options. The treated water at present is also considered a resource. This water can be subjected to reuse, thus contributing to a circular economy and a zero-discharge treatment system. Another by-product after treatment of wastewater can be metals, nutrients, electricity, etc. (Shamshad and Rehman, 2024). The new technologies like wetland and bioelectrochemical system (BES) can contribute toward resource recovery from the wastewater.

The CWs are considered to be introduced in the existing STP treatment flow, as they require low maintenance but the treatment time is a little longer, and the space requirement is also large. Thus, this technology is more suitable for a small-scale treatment facility. It is mostly used for the on-site treatment of wastewater. It is a part of decentralized treatment facilities (Abdel-shafy and El-khateeb, 2013; Vymazal, 2011; Wang et al., 2017). Such treatment systems are termed as decentralized wastewater treatment system (DEWATS). DEWATS treats a small amount of wastewater on or near the site and also supports source separation of wastewater. These are generally targeted for the reuse of treated wastewater. The water can be used for agricultural purposes, washing, gardening, etc. Many small townships and educational institutes have taken the initiative to implement DEWATS as a sustainable treatment system. These small, planned facilities can again use different treatment methods in a stepwise manner. Processes like separation, settlement, CW, BES, etc., are used in DEWATS (Gutterer et al., 2009).

BES technology also contributes to resource recovery. These setups are capable of treating different types of wastewater and can generate different types of products ranging from electricity, nitrogen, phosphorus, fertilizer, etc. BES has been successful with the pilot-scale design. The recovery of nutrients like nitrogen and phosphorus is possible through on-site treatment of human urine using BES technology. These nutrients can be directly used as plant fertilizer. At the pilot as well as lab scale, the system has been tested for different modifications and optimizations for better efficacy. The economic viability, reproducibility, and infrastructure availability are a few parameters that still need optimization for commercial application of BES (Arkatkar et al., 2022; Yu et al., 2021).

The modification or advancement in the already existing technology is also considered by the researchers. The membrane technology is the costliest of all the technologies due to energy requirements, membrane clogging that reduces the lifespan of the membrane, and the cost of the membrane. Technologies like forward osmosis (FO), applicable to membranes, are being experimented. FO membranes require less energy, are less susceptible to fouling, and thus have better shelf life compared to reverse osmosis (RO) membranes (Pardeshi and Mungray, 2014; Patel et al., 2022).

24.4 Interdependency of Multi-step System

The systems discussed so far are based on different principles but achieve the collective goal of efficient treatment of wastewater. The outlet of one system becomes the inlet of the next system. Thus, they are connected in a channelized manner with respect to the need for the treatment facility. The choice of technology depends on the requirements of STP. A complete treatment facility is a sequential unit. The water flow is determined to move from one treatment method to another. The treatment course is designed accordingly. For example, the physical separation of waste material is done as a first process on the wastewater as it enters the STP. This step cannot be shifted after filtration; such a change will surely block the filters and will have operational and economic hazards to the treatment facility.

The above-mentioned channelized flow of wastewater is adopted in view of supporting the treatment methods. Sometimes innovations are applied using the available infrastructure in the STP. In the case study of an effluent treatment plant (ETP) operational in paper and pulp treatment, the probable usage of oxygen delignification is adopted. This technology helps to remove the lignin from pulp. It is considered an efficient option for a chlorine-based bleaching method. It reduces pollution due to the use of chemical reagents and is thus environmentally friendly. This technology, along with pulp washers, also reduces the overall water consumption in the paper and pulp mill. Thus, the amount of water and chemicals used depends on factors like the choice of technology, the quality of wastewater to be treated, the efficacy of the process, etc. (Dagar et al., 2022).

The integration of two technologies must be analyzed for possible energy usage and practical acceptability. It has been observed that biological treatment methods such as aerobic digestion and membrane bioreactors require a high amount of energy. The operational and maintenance cost of such STP is high only because of power consumption. It is often suggested to derive this power from a renewable energy resource. In one such case study at Ibn Tofail University of Kenitra, Morocco, it was documented that the STP used for the treatment of wastewater generated within the university campus requires 1061.76 kWh/day of energy to run its MBR unit. The power supply was shifted to a windmill and a photovoltaic system. This was helpful to cover 72% of the energy requirement of STP. The shifting saved around 106.76 euros per day for the STP (Bellar et al., 2023).

24.5 Sustainable Approaches for Scale-up Technology

Most of the individual systems target improving the performance efficiencies, where mutual impacts of communities, operational parameters, and seasonality on the treatment process have been under-evaluated. Meanwhile, combining the number of units with a multi-modular system promotes equal flow distribution and high treatment efficiency (Jadhav et al., 2024a, b, c). Scalable approaches in this direction may require integration of innovative solutions, sustainability, and adaptability across different scales (Jadhav et al., 2021a, b). Various physico-chemico-biological systems (membrane bioreactors, advanced oxidation processes) can be coupled with nature-driven processes (CWs, algal bioreactors), promoting the various decentralized units (Yadav et al., 2021). It will promote resource recovery options through an advanced treatment process along with circular economic incentives. However, complexity increases with the number of units coupled together during upscaling (Jadhav et al., 2022). Such complexity may arise due to the maintenance of uniform substrate conditions, backwashing, biofilm formation, longevity, and operational stability.

Additionally, Reymond et al. (2020) reported that the development of appropriate governance arrangements and policies can lead to the failure of scalable treatment plants. The successful implementation of field application can be assessed on the ground, and the analysis of governance aspects should be monitored regularly.

For regular monitoring and efficient governance, smart monitoring with automation can be linked with online platforms. The DEWATS, also termed as small-scale sanitation unit, was successfully monitored and reported under a joint project by Eawag's department under Sanitation, Water and Solid Waste for Development (Sandec Switzerland), Indian Institute of Technology (IIT) Madras (India), Bremen Overseas Research and Development Association (BORDA, Germany), CDD Society (India), Environment and Public Organization (ENPHO, Nepal), and other partners (Figure 24.1). These treatment units can be structurally designed as per the demand and available amenities. They can be planned indoors or outdoors, on-site or off-site, above or below the ground, nature-based or mechanized, etc. The Consultant Capacity Development (ConCaD) project, supported by the Bill & Melinda Gates Foundation, has developed a network with training institutes to provide an open source of training materials on citywide inclusive urban sanitation (CWIS) services (https://www.eawag.ch/en/department/sandec/projects/sesp/4s-small-scale-sanitation-scaling-up/).

The major consumption of energy for the operation of any STP is in the form of electrical energy. The financial implication is thus considered in the form of electrical energy consumption (Suransh et al., 2023). The energy can be utilized for pumping, aeration, machine operation, disinfection, etc. (Venkatesh and Brattebø, 2011). For the sustainability of the STP plant, the researchers have already experimented with the use of solar, wind, and power derived from renewable energy resources (Close et al., 2006; Jadhav et al., 2024a).

The concept of a sustainable system cannot be evaluated once and relied on forever; it requires a continuous monitoring approach. The manual monitoring

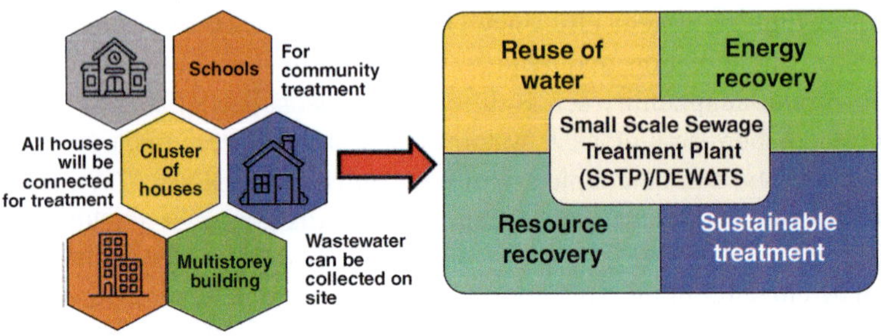

Figure 24.1 On-site treatment advantages of small-scale sewage treatment plant (SSTP)/DEWATS.

and time-consuming sample analysis are common practices in all STPs, but it is a cumbersome job. The data generated are then arranged to understand the performance of STP. This is done using a computer system. Now, researchers are putting efforts into reducing the manual testing and time required for performance analysis. The algorithms like fuzzy algorithms, neural networks, global optimal control, model predictive control, hybrid nonlinear model predictive control (NMPC), genetic algorithms, etc. are used to monitor process parameters like aeration, flow of wastewater, nutrient removal, dissolved oxygen (DO), and others that can help in different ways: sometimes it saves energy about 10–40%; the algorithm monitoring removes more than 95% of organic matter and maintains DO level (Sun et al., 2019).

The usage of Supervisory Control and Data Acquisition (SCADA) system for real-time data input in the monitoring system has been in use for a very long period. These systems always assist in the running of STP for timely monitoring and maintenance. The model helps maintain the operating speed of the pumping station, which saved around 17% of energy. The aeration system used in the biological treatment step of the STP was monitored using the SCADA system, which transfers the information to the online cloud server, which was used to check water quality and power meters. The information helped to save energy up to 20%. The artificial intelligence model that predicts the influent and effluent quality has good potential. These models were able to predict the possible changes two weeks ahead of the actual changes and thus contributed to the saving of 11% energy cost and 16% material cost (Cairone et al., 2024).

Online system monitoring is very well achieved by using AI-generated data; similarly, the sequence of methods to be applicable to sustainable treatment can also be done using computer modeling. These models help to make decisions about the method based on different parameters that impact economic, environmental, or even social requirements. The multi-criteria decision analysis (MCDA), a weighted sum model, was used to do a sustainable analysis of 13 wastewater treatment plants in Iraq. It considered the environmental, social, and technical factors related to the treatment plants. The techniques such as conventional treatment, oxidation ditches, aeration lagoons, and membrane bioreactors (MBR) were implemented in the treatment plants. It was noted that MBR technology had higher sustainability. Such model analysis must be used to monitor the environmental sustainability of the operating STPs (Omran et al., 2021). The modification in technology can be considered for a sustainable output. The MBR bioreactor considers both aeration and membrane filtration methods. The cleaning process of the membrane is a tedious part, and aeration requires energy. The modification, like the integration of photocatalytic membrane reactor, can be taken into consideration. It is an energy-saving integration but with some limitations (Obaideen et al., 2022). The idea of vermifiltration has also been experimented with, where earthworms are used to decompose the organic matter, and further, the same material can be used as fertilizer. Vermifiltration is reported to have successfully reduced the COD, BOD, and TSS of the wastewater (Arora and Kazmi, 2015).

To understand the interdependency and development of the sustainable system, researchers are focusing on collecting data to develop a Knowledge-Based Design System (KBDS). The database gives overall information regarding the various processes and flowsheets for wastewater treatment. The output can be in the form of compatibility between two processes, specification about the efficacy of the system to remove pollutants, environmental impacts analysis, structural requirements, etc. (Fung and Wibowo, 2013). These systems can be used to check the combination of multistep processes to be applicable even before the proposal of an STP. The process simulators are also used for such data analysis. These can further give ranking to the various combinations as per the requirements. Process selection also depends on other structural details of the STP, including the construction site, capital investment, operation cost, the toxicant removal rate, and the equipment age (Chen et al., 2018). All these monitoring systems particularly collect real-time data, streamlining and supporting processes from establishment to monitoring and identifying the problems associated with such a system. Such advancement will be beneficial for fostering coordination between the governmental agencies and enabling learning through big data analysis.

24.6 Summary

It can be noted that the usage of technology can be a primary concern after the quality of treated water is fixed. The quality of water and other resources to be recovered can be estimated by the knowledge of the components of wastewater that are planned to undergo treatment. The quantity of wastewater helps to decide the scale of the treatment facility. Once both quality and quantity are fixed, the planning should be done for the stepwise treatment methodology. The treatment methodology can be tested in the laboratory, but at most care should be taken while introducing the technology flow in the field. It is a common observation that the technologies will be interdependent. The facilities will many times assist the functionality of the next treatment step. The modification in one step can have a huge impact on the output of the treatment facility. Energy consumption should be low, and the overall quality of treated water must be within the limits of the environmental regulatory guidelines. To assess the possibilities of the outcome, various AI tools can be used prior to the proposal of a treatment facility.

Review Questions

The following review questions encourage analytical thinking and a deeper understanding of wastewater treatment processes, particularly targeting sustainable development.

1. Why is a combination of physical, chemical, and biological processes necessary for wastewater treatment?
2. How does the quality and quantity of wastewater influence the choice of treatment technology?
3. What are the limitations of conventional centralized wastewater treatment facilities, and how can they be addressed?
4. Why is a single technology insufficient for treating wastewater effectively?
5. How does the concept of interdependency play a role in multistep wastewater treatment systems?
6. Why must the sequence of treatment technologies be carefully planned for different types of wastewaters?
7. What are the major challenges in implementing a multistep wastewater treatment system?
8. How does resource recovery integrate into wastewater treatment technologies?
9. What factors determine whether a wastewater treatment facility should adopt a centralized or decentralized approach?
10. In what ways can wastewater treatment strategies contribute to sustainability and water conservation efforts globally?

References

Abdel-shafy, H.I. and El-khateeb, M.A. (2013). Integration of septic tank and constructed wetland for the treatment of wastewater in Egypt. *Desalination and Water Treatment*, **51**, 3539–3546. https://doi.org/10.1080/19443994.2012.749585

Arkatkar, A., Patel, A., Mungray, A.A. et al. (2022). Chapter 1 – fundamentals of bio-electrochemical systems for wastewater treatment: challenges and opportunities for resource recovery. In: *Novel Approaches Towards Wastewater Treatment and Resource Recovery Technologies* (eds. A. Mungray, A. Mungray, S. Sonawane et al.), 3–22. Elsevier. https://doi.org/10.1016/B978-0-323-90627-2.00004-6

Arora, S. and Kazmi, A.A. (2015). The effect of seasonal temperature on pathogen removal efficacy of vermifilter for wastewater treatment. *Water Research*, **74**, 88–99. https://doi.org/10.1016/j.watres.2015.02.001

Bellar, D., Choukai, O., Tahaikt, M. et al. (2023). A case study on the environmental and economic impact of photovoltaic systems in wastewater treatment plants. *Open Physics*, **21**(1), 20230158. https://doi.org/10.1515/phys-2023-0158

Cairone, S., Hasan, S.W., Choo, K.H. et al. (2024). Revolutionizing wastewater treatment toward circular economy and carbon neutrality goals: pioneering sustainable and efficient solutions for automation and advanced process control with smart and cutting-edge technologies. *Journal of Water Process Engineering*, **63**, 105486. https://doi.org/10.1016/j.jwpe.2024.105486

Chen, X., Xu, Z., Yao, L. et al. (2018). Processing technology selection for municipal sewage treatment based on a multi-objective decision model under uncertainty. *International Journal of Environmental Research and Public Health*, **15**(3), 448. https://doi.org/10.3390/ijerph15030448

Close, J., Ip, J., and Lam, K.H. (2006). Water recycling with PV-powered UV-LED disinfection. *Renewable Energy*, **31**(11), 1657–1664. https://doi.org/10.1016/j.renene.2005.08.034

Dagar, S., Singh, S.K., and Gupta, M.K. (2022). Economics of advanced technologies for wastewater treatment: evidence from pulp and paper industry. *Frontiers in Environmental Science*, **10**, 1–16. https://doi.org/10.3389/fenvs.2022.960639

Fung, K.Y. and Wibowo, C. (2013). Design of industrial wastewater treatment plants: a multi-faceted problem. *Current Opinion in Chemical Engineering*, **2**(4), 455–460. https://doi.org/10.1016/j.coche.2013.09.003

Gutterer, B., Sasse, L., Panzerbieter, T. et al. (2009). *Decentralised Wastewater Treatment Systems (DEWATS) and Sanitation in Developing Countries a Practical Guide* (eds. A. Ulrich, S. Reuter, and B. Gutterer), (Vol. 49, Issue 0). Water, Engineering and Development Center (WEDC), Loughborough University.

Jadhav, D.A, Chendake, A.D, Pandit, S. et al. (2024a). Is the bioelectrochemical system ready for industrial commercialization? Key players and market analysis for industrial utility. *Journal of Power Sources*, **629**, 236001.

Jadhav, D.A., Chendake, A.D., Vinayak, V. et al. (2022). Scale-up of the bioelectrochemical system: strategic perspectives and normalization of performance indices. *Bioresource Technology*, **363**, 127935.

Jadhav, D.A., Gunaseelan, K., Le, G.T.H. et al. (2024b). Sustainable vision toward development of microbial electrosynthesis for diverse resource recovery: circular economy. *Journal of Environmental Chemical Engineering*, **12**, 114027.

Jadhav, D.A., Mungray, A.K., Arkatkar, A. et al. (2021a). Recent advancement in scaling-up applications of microbial fuel cells: from reality to practicability. *Sustainable Energy Technologies and Assessments*, **45**, 101226.

Jadhav, D.A., Pandit, S., Sonawane, J.M. et al. (2021b). Effect of the membrane biofouling on the microbial electrochemical cell's performance and mitigation strategies. *Bioresource Technology Reports*, **15**, 100822.

Jadhav, D.A., Yu, Z., Hussein, M. et al. (2024c). Paradigm shift in nutrient-energy-water centered sustainable wastewater treatment system through synergy of bioelectrochemical system and anaerobic digestion. *Bioresource Technology*, **396**, 130404.

Obaideen, K., Shehata, N., Sayed, E.T. et al. (2022). The role of wastewater treatment in achieving sustainable development goals (SDGs) and sustainability guideline. *Energy Nexus*, **7**, 100112. https://doi.org/10.1016/j.nexus.2022.100112

Omran, I.I., Al-Saati, N.H., Al-Saati, H.H. et al. (2021). Sustainability assessment of wastewater treatment techniques in urban areas of Iraq using multi-criteria decision analysis (MCDA). *Water Practice and Technology*, **16**(2), 648–660. https://doi.org/10.2166/wpt.2021.013

Pardeshi, P. and Mungray, A. (2014). High flux layer by layer polyelectrolyte FO membrane: toward enhanced performance for osmotic microbial fuel cell. *International Journal of Polymeric Materials and Polymeric Biomaterials*, **63**(12), 595–601. https://doi.org/10.1080/00914037.2013.854232

Patel, A., Arkatkar, A., Chhatbar, M. et al. (2022). Advances in membrane technologies for industrial effluents for resource recovery. In: *Novel Approaches Towards Wastewater Treatment and Resource Recovery Technologies*, 167–184. Elsevier. https://doi.org/10.1016/B978-0-323-90627-2.00026-5

Reymond, P., Chandragiri, R., and Ulrich, L. (2020). Governance arrangements for the scaling up of small-scale wastewater treatment and reuse systems–lessons from India. *Frontiers in Environmental Science*, **8**, 72. https://doi.org/10.3389/fenvs.2020.00072

Santos, E., Albuquerque, A., Lisboa, I. et al. (2022). Economic assessment of energy consumption in wastewater treatment plants: applicability of alternative nature-based technologies in Portugal. *Water (Switzerland)*, **14**(13), 2042. https://doi.org/10.3390/w14132042

Shamshad, J. and Rehman, R.U. (2024). Innovative approaches to sustainable wastewater treatment: a comprehensive exploration of conventional and emerging technologies. *Environmental Science: Advances*, 189–222. Royal Society of Chemistry. https://doi.org/10.1039/d4va00136b

Silva, J.A. (2023). Wastewater treatment and reuse for sustainable water resources management: a systematic literature review. *Sustainability (Switzerland)*, **15**(14), 10940. https://doi.org/10.3390/su151410940

Soo, A., Kim, J., and Shon, H.K. (2024). Technologies for the wastewater circular economy – a review. *Desalination and Water Treatment*, **317**, 100205. https://doi.org/10.1016/j.dwt.2024.100205

Sun, Y., Lu, M., Sun, Y. et al. (2019). Application and evaluation of energy conservation technologies in wastewater treatment plants. *Applied Sciences*, **9**, 4501.

Suransh, J., Jadhav, D.A., Nguyen, D.D. et al. (2023). Scalable architecture of low-cost household microbial fuel cell for domestic wastewater treatment and simultaneous energy recovery. *Science of The Total Environment*, **857**, 159671. https://doi.org/10.1016/j.scitotenv.2022.159671

Venkatesh, G. and Brattebø, H. (2011). Energy consumption, costs and environmental impacts for urban water cycle services: case study of Oslo (Norway). *Energy*, **36**(2), 792–800. https://doi.org/10.1016/j.energy.2010.12.040

Vymazal, J. (2011). Constructed wetlands for wastewater treatment: five decades of experience. *Environmental Science & Technology*, **45**(1), 61–69. https://doi.org/10.1021/es101403q

Wang, Y., Zhao, Y., Xu, L. et al. (2017). Constructed wetland integrated microbial fuel cell system: looking back, moving forward. *Water Science and Technology*, **76**(2), 471–477. https://doi.org/10.2166/wst.2017.190

Yadav, A., Jadhav, D., Ghangrekar, M. et al., (2021). Effectiveness of constructed wetland integrated with microbial fuel cell for domestic wastewater treatment and to facilitate power generation. *Environmental Science and Pollution Research*, **29**, 51117–51129. https://doi.org/10.1007/s11356-021-17517-z

Yu, J., Park, Y., Widyaningsih, E. et al. (2021). Microbial fuel cells: devices for real wastewater treatment, rather than electricity production. *Science of the Total Environment*, **775**, 145904. https://doi.org/10.1016/j.scitotenv.2021.145904

Zhang, F. and Li, H. (2025). Comprehensive impact assessment of multi-stage wastewater treatment processes on the industrial environment. *Discover Applied Sciences*, **7**(4), 244. https://doi.org/10.1007/s42452-025-06666-2

25

Government Strategies to Economically Support the Cleaning of Water Bodies in India

Abdul Gani[1], *Athar Hussain*[1], *and Shray Pathak*[2]

[1]*Department of Civil Engineering, Netaji Subhas University of Technology, New Delhi, India*
[2]*Department of Civil Engineering, Indian Institute of Technology Ropar, Roopnagar, Punjab, India*

25.1 Introduction to Water Pollution and Its Impact on India's Water Bodies

Indian rivers, lakes, and wetlands are essential components within the country's natural, cultural, and economic fabric. These bodies of water provide vital supplies for drinking, agriculture, industry, and transportation, as well as habitat for a diverse range of species (Gani et al., 2024a). However, many of these bodies of water have become severely contaminated over time as a result of fast industrialization, urbanization, agricultural runoff, untreated sewage, and religious rituals (Gani et al., 2025a). The Indian government has recognized how urgent it is to clean these waterways in order to protect ecosystems, maintain public health, and promote sustainable economic growth. This chapter examines the economic policies and tactics implemented by the Indian government to promote the revitalization and cleansing of water bodies, looking at both regulatory frameworks and financial incentives.

25.1.1 Overview of Water Body Pollution in India

In India, water pollution has grown to be a serious environmental and public health issue, impacting around 70% of the surface water. Along with innumerable lakes and reservoirs, major rivers like the Ganges, Yamuna, and Brahmaputra are heavily contaminated. The sources of the pollution include plastic trash,

Wastewater to Resource Recovery: Applying the Circular Economy Toward Sustainable Development, First Edition. Edited by Arvind Kumar Mungray et al. Published 2026 by John Wiley & Sons, Inc.

industrial effluents, agricultural pesticides, and untreated urban sewage. This pollution has severe consequences: it reduces the drinking water supply, degrades ecosystems and agriculture, and incurs major health expenses. These repercussions have spurred the Indian government to put policies in place that financially assist water body cleansing, aquatic habitat restoration, and sustainable water use.

An ongoing water crisis may result from the failure to implement appropriate solutions, according to a report by the Government of India's think tank, the National Institution for Transforming India (NITI Aayog), which stated that 600 million people in India face extreme to high water stress (Jadeja et al., 2014; Shukla et al., 2018). With a decadal growth rate of 17%, the growing urban population is a major factor in both water pollution and demand due to discharges of industrial and domestic wastewater (Chai et al., 2021; Pathak et al., 2021). A large number of people from rural areas have moved to urban areas, increasing the demand for utility services and overusing groundwater (Gani et al., 2025b). In addition to population growth, industrialization causes an annual increase in water demand of roughly 4%, which leads to a significant rise in the volume of industrial wastewater (Makut et al., 2019). Water management in India is further complicated by the unequal distribution of natural water resources brought on by different regions, climate change, erratic monsoons, trade, and commerce (Gani et al., 2024c).

Insufficient freshwater has increased demand for unconventional water sources globally. India is currently ranked thirteenth on the list of water-stressed countries, with 21 major cities experiencing water depletion and other cities experiencing water supply curtailment (Vimalkumar et al., 2018). With 1.3 billion inhabitants, this second-most populous nation generates 15–135 L of wastewater per person per day, making it extremely difficult to create affordable wastewater treatment systems that satisfy current discharge regulations (Ciardelli and Ranieri, 2001; Panagopoulos and Haralambous, 2020). The management of water requires immediate attention due to water quality issues like the removal of newly discovered pollutants and the safe reuse of greywater (Gani et al., 2024b).

Wastewater treatments often employ a multi-stage process that includes primary, secondary, tertiary, and advanced treatments are used to gradually remove solids, organic matter, nutrients, and bacteria. For the removal of coarse solids and settleable particles, primary treatment includes screening, grit chambers, and sedimentation. Secondary Treatment oxidizes and breaks down colloidal and dissolved organic compounds through microbial activities. Activated sludge processes (ASPs), sequencing batch reactors, oxidation ponds, trickling filters, and moving bed biofilm reactors are the commonly used techniques used in the secondary treatment. Tertiary treatments are used when stringent effluent criteria need to be fulfilled. These includes filtering, disinfection, and nutrient removal. Advanced technologies are used for resource recovery and the removal of refractory contaminants including pesticides, heavy metals, and medications.

Around the world, innovative technologies that support wastewater treatment have been developed as a result of new concepts. In India, the trickling filter and ASP are common biological approaches used in wastewater treatment. Continuous initiatives are being made to solve the process's energy consumption, which is a significant contribution to the process cost, in the large-scale ASP treatment of wastewaters. Membrane technologies, in addition to ASPs, are frequently used in wastewater treatment. The membrane-based approach has been shown to be successful in the treatment and supply of water for home and industrial use, as well as in the food and beverage, pharmaceutical, metallurgical, and medical industries, thanks to the diverse options that membranes offer (Rahman et al., 2021). Together with the current posttreatment procedure, a cost-effective sequential biofiltration system was suggested since it could remove polychlorinated biphenyl and treat nutrients (Nayak and Hassan, 2021). In the secondary and tertiary phases of wastewater treatment, microalgae-based technologies have been employed to effectively remove heavy metals and address eutrophication brought on by the presence of inorganics in wastewaters (Debaere and Kapral, 2021).

Significant nonpoint sources of wastewater contamination include urban runoff, agricultural runoff, and livestock runoff, while point sources of wastewater pollution include discharges from municipal and industrial wastewater treatment facilities. A variety of wastewaters from sources, some of which are not taken into consideration in the design of the wastewater treatment process, contribute to the contamination of natural water resources. Agricultural runoffs, atmospheric depositions, and urban and rural runoffs are some of the major causes of water pollution that are not yet covered by the wastewater management plan.

India's water bodies are in a severe state of physical, chemical, and ecological degradation. Approximately 70% of surface water resources nationwide are deemed non-potable, primarily as a result of untreated municipal sewage discharge, uncontrolled industrial effluents, and diffuse agricultural runoff that is high in pesticides and fertilizers. As per the Central Pollution Control Board (CPCB, 2022a), 311 river segments continue to fall under the "priority polluted" category due to elevated levels of coliform and biochemical oxygen demand, which surpass the recommended criteria.

25.1.2 Impact on Public Health, Environment, and Economy

Human health is seriously impacted by unsafe water. The UNESCO 2021 World Water Development Report estimates that diarrhea brought on by contaminated drinking water, poor sanitation, and poor hand hygiene kills 829 000 people annually, including almost 300 000 children under the age of five, accounting for 5.3% of all deaths in this age group. According to a study comparing bottled water, purified water, and tap water, tap water was a major cause of gastrointestinal illnesses

(Payment et al., 1997). Diseases including cholera, trachoma, schistosomiasis, and helminthiasis are also more common where there is a lack of water and sanitation facilities. Cholera and polluted water are clearly related, according to data from research conducted in underdeveloped nations. Cholera can be prevented by treating and storing household water (Gundry et al., 2004; Gani et al., 2024e).

One of the main causes of childhood illnesses is water contamination. In 2016, 940 000 children died globally as a result of air, water, and soil pollution; two-thirds of these deaths happened in countries with low and moderate incomes, and the bulk of these deaths occurred in children under the age of five (Landrigan et al., 2018). In developing countries, the severity of industrial organic water pollution is strongly connected with infant and child mortality, and industrial water pollution is a major cause of infant and child mortality (Jorgenson, 2009; Gani et al., 2025c). Furthermore, arsenic in drinking water poses a potential cancer risk. According to Vladeva et al. (2000), children who consume water contaminated with nitrate may develop goiter.

25.1.3 Importance of Clean Water Bodies for Ecosystems and Communities

Clean water sources are essential for preserving natural equilibrium as well as for human health and economic stability. In addition to supporting a variety of ecosystems, healthy rivers, lakes, and reservoirs give populations access to vital resources. Clean water bodies are essential for both environmental integrity and community well-being. According to UNEP (2016), they preserve aquatic biodiversity, offer homes for fish, birds, and microbes, and control sediment movement and nutrient cycling to keep the ecosystem in balance. By reducing floods and replenishing groundwater, rivers, lakes, and wetlands serve as natural buffers that increase resilience to climate change (Millennium Ecosystem Assessment, 2005). Access to clean water improves human health by lowering the prevalence of water-borne diseases including cholera, typhoid, and hepatitis, which continue to be major public health issues in many developing nations, including India (WHO, 2022). According to the CPCB (2022a), clean water bodies are economically essential for hydropower, tourism, fisheries, agriculture, and other sources of income, but dirty water significantly raises treatment costs and lowers output. This section discusses why clean water bodies are essential to both natural environments and human societies.

25.1.3.1 Ecological Health and Biodiversity

Water bodies are essential for the survival of many plant and animal species. A balanced ecosystem that can support life at several levels of the food chain is created by clean water, which promotes the growth and well-being of both plants and animals (Pandey and Pande, 2023). The natural balance is upset when water bodies are contaminated, which results in:

1. **Loss of Biodiversity:** Pollutants cause habitat disruption and lower water oxygen levels, which can cause sensitive species, such as some fish and amphibians, to go extinct. The decline in biodiversity has a negative influence on ecosystems that rely on these organisms for equilibrium.
2. **Destruction of Wetlands:** Wetlands are vital for flood management, water filtration, and aquatic species breeding grounds. They frequently develop around bodies of water. These fragile ecosystems are harmed by pollution, which reduces their capacity to act as natural habitats and buffers.
3. **Threats to Migratory Species:** Clean water bodies serve as both food sources and migratory rest stops for numerous bird species. Wider ecosystems are disrupted when fish, insects, and other aquatic creatures that are essential to maintaining these migratory patterns disappear from polluted water bodies.

25.1.3.2 Human Health and Community Well-being

Human health depends on having access to clean water since contaminated water can spread illness, contain dangerous chemicals, and create conditions that are conducive to the growth of waterborne pathogens (Gani et al., 2024d). Communities that depend on contaminated water sources are at risk for many health issues, such as:

1. **Waterborne Diseases:** Water bodies that are contaminated frequently have bacteria, viruses, and parasites that can lead to cholera, typhoid, and hepatitis, among other illnesses. In underdeveloped and rural areas with little access to healthcare, these diseases are particularly hazardous.
2. **Chemical Contamination:** Severe health problems, including cancers, developmental disorders, and neurological damage, can result from industrial pollution, agricultural chemicals, and heavy metals in water bodies.
3. **Livelihood Impact on Fisherfolk and Farmers:** Communities that rely on fishing and agriculture are directly impacted when water bodies are polluted. Fish die-offs or contamination can harm fisherfolk's income, while polluted water used for irrigation affects crop quality and yield, further endangering food security.

25.1.3.3 Climate Regulation and Resilience

In order to control the climate and increase resilience to environmental challenges, healthy water bodies are crucial.

1. **Natural carbon sequestration:** Oceans, rivers, and wetlands absorb carbon dioxide, helping mitigate the effects of climate change by acting as carbon sinks.

2. **Flood and drought management:** Water bodies help regulate groundwater levels, manage runoff, and control flooding during monsoon seasons. Clean and well-maintained water systems reduce the risk of catastrophic flooding and provide a buffer during periods of drought.
3. **Temperature moderation:** Water bodies contribute to maintaining local climatic conditions and reducing temperature changes. They help in lowering the intensity of heatwaves and sustaining ecosystems that depend on steady temperatures by slowly absorbing and releasing heat.
4. Clean water bodies are critical to ecosystem health, human well-being, and economic prosperity. In addition to facilitating sustainable livelihoods, they promote cultural heritage, healthy drinking water, and biodiversity. Governments, communities, and businesses must collaborate to conserve and restore these crucial resources because of the varied significance of clean water.

25.2 National and State-level Initiatives for Water Conservation

Numerous policies and initiatives have been launched by the Indian government to combat water pollution and improve the condition of water bodies. While encouraging sustainable management techniques for upcoming generations, these initiatives seek to purify currently contaminated water bodies. The restoration of these vital water resources will require financial and economic support, which is intended to be provided by a number of important policies and programs.

25.2.1 National and State-level Policies

India is facing a serious water problem, with surface water bodies being contaminated, groundwater being depleted, and water shortages in many areas. To address these problems, the Indian government has set several national and state-level programs designed to preserve and manage the nation's water resources sustainably (Mohamed et al., 2024). These initiatives include water conservation programs, pollution control, groundwater replenishment, and public awareness campaigns.

25.2.1.1 National-level Initiatives

1. **Namami Gange Programme**
   The Namami Gange Programme, which was introduced in 2014 by the Ministry of Jal Shakti with a budget exceeding ₹20 000 crores, is among the most ambitious national projects. The program aims to rejuvenate and clean the Ganges River by tackling pollution from solid waste, industrial effluents,

and household sewage. It additionally promotes riverside development and establishes sustainable livelihoods for the communities that surround the river. It consists of several activities, including building sewage treatment facilities, cleaning the surface of rivers, planting trees, raising public awareness, and maintaining rural sanitation. The program's financial components include state government involvement, international support from institutions such as the World Bank and public–private partnerships (Rai et al., 2024). Despite its primary focus on pollution reduction, it also contains important water conservation initiatives that affect other Ganga-connected water bodies. Under Namami Gange, important economic strategies are:

A. **Subsidies for sewage treatment plants (STPs):** The government offers financial support for the building and operation of STPs to minimize pollution from untreated sewage, with funding allocated for both initial construction and regular maintenance.
B. **Sewage treatment infrastructure**: To prevent untreated waste from entering the river, more than ₹20 000 crore has been allocated for the construction and renovation of STPs in cities and towns along the Ganges. It also involves the installation of contemporary sewage treatment technologies.
C. **Industrial effluent control**: Strict regulations control the industrial discharge into rivers, and industries must treat their effluents before releasing them. Industry receives financial assistance for the establishment of wastewater treatment plants.
D. **Riverfront development and public awareness**: To guarantee sustainable use and tourism, it also entails the development and attractiveness of riverfronts.
E. **Waste management**: The program encourages the building of solid waste management facilities to stop trash from being thrown into the river.
F. **Afforestation and biodiversity conservation:** By enhancing biodiversity, lowering erosion, and improving soil quality, planting trees along the banks of the Ganga indirectly conserves water.

2. **National River Conservation Plan**

The National River Conservation Plan (NRCP) was introduced in 1995 to enhance the water quality in rivers across India. In particular, this plan addresses solid waste management and sewage treatment in large cities that are situated alongside rivers. States receive funding to build solid waste management systems, sewage treatment facilities, and riverfront development (Cullet et al., 2010).

The Ganga Action Plan (GAP) and Yamuna Action Plan (YAP) are two of the most prominent initiatives within the NRCP that aim to purify the

Ganges and Yamuna rivers, respectively. These initiatives are funded by the central government, with additional funds coming from the individual state governments.

3. **Swachh Bharat Mission**

   The main goal of the Swachh Bharat Mission, which was started in 2014, is to enhance waste management and sanitation at the local level. The Swachh Bharat Mission indirectly improves water quality by reducing the amount of untreated sewage that is deposited into rivers and lakes. The Swachh Bharat Mission's financial aid for the building of restrooms and sewage treatment facilities has a good ripple effect on the surrounding water bodies' health. Building sanitary facilities nationwide, especially in rural areas, is one of the Swachh Bharat Mission's main goals to reduce water pollution (Mahto, 2024). This program addresses the issue of open defecation, which is a major cause of water contamination.

   The Swachh Bharat Mission's primary economic initiatives include:

   A. **Subsidies for sanitation infrastructure:** Financial incentives are provided for households, particularly those in rural regions, to build toilets, which lowers the amount of human waste contamination in neighboring water bodies.

   B. **Financial incentives for waste management companies:** Companies that deal with solid trash are given tax exemptions and other incentives by the government, which helps to conserve water bodies indirectly by reducing the amount of rubbish that enters rivers and lakes.

4. **Atal Mission for rejuvenation and urban transformation**

   The Atal Mission for Rejuvenation and Urban Transformation was established in 2015 with the primary goal of enhancing urban infrastructure. Providing clean water and managing wastewater are two of its main objectives. The plan encourages the construction of stormwater drainage systems, sewage and septage management systems, and urban water body cleanup. Atal Mission for Rejuvenation and Urban Transformation has played a significant role in enhancing local urban bodies' financial ability to carry out water quality management initiatives.

5. **Jal Shakti Abhiyan**

   The Ministry of Jal Shakti launched the Jal Shakti Abhiyan (JSA) in 2019 as a national initiative to save water. The JSA promotes water conservation, with a particular emphasis on raising groundwater levels and avoiding surface water pollution in India. The campaign highlights the importance of local government, nongovernmental organizations, and citizens in addressing water-stressed districts. This project aims to turn water conservation into a mass movement by involving local communities and supporting rainwater gathering, watershed management, and sustainable irrigation practices.

25.2.1.2 State-level Initiatives

To clean their waterways, several state governments have also put in place regional initiatives. These programs are frequently designed to meet the unique requirements of the area and its waterways. Depending on local water concerns, topography, and climate, each Indian state has created its water conservation initiatives.

1. **Mukhya Mantri Jal Swavlamban Abhiyan (Rajasthan)**
   To inspire rural areas to become self-sufficient in water, the Mukhya Mantri Jal Swavlamban Abhiyan was started in Rajasthan, a state with limited water supplies. The main goals of this project are groundwater conservation, water waste reduction, and the construction of water harvesting structures (Kumar et al., 2024). The key actions of Mukhya Mantri Jal Swavlamban Abhiyan are:
   A. Construction of water harvesting structures
   B. Tree plantation
   C. Public awareness and community participation
2. **Jalyukt Shivar Abhiyan (Maharashtra)**
   The government of Maharashtra launched the Jalyukt Shivar Abhiyan effort to save water and eliminate drought in villages. This program was started in 2014 with an emphasis on groundwater recharging and rainwater collection in areas that are vulnerable to drought. The main tasks for this project included:
   A. Desilting of water bodies
   B. Micro-irrigation and efficient water use
   C. Check dam construction
3. **Sujalam Sufalam Jal Abhiyan (Gujarat)**
   The Gujarat government launched the Sujalam Sufalam Jal Abhiyan, a water conservation initiative. It seeks to increase groundwater levels, improve water storage, and desilt lakes and rivers through cooperative efforts. The program's key components consist of:
   A. Deepening and widening of rivers and lakes
   B. Groundwater recharge structures
   C. Public participation
4. **Neeru–Chettu Programme (Andhra Pradesh)**
   Andhra Pradesh started the Neeru–Chettu, or "Water-Tree," water conservation program to alleviate the state's water shortage. To increase water availability and enhance environmental quality, this initiative combines afforestation with water conservation. Essential tasks of the Neeru–Chettu Programme consist of:
   A. Afforestation
   B. Recharging groundwater
   C. Rainwater harvesting and storage

5. **Mission Kakatiya (Telangana)**
   A major initiative of the Telangana government, Mission Kakatiya aims to rejuvenate historic tanks and ponds to enhance agricultural and water storage. In honor of the Kakatiya dynasty, which constructed a system of tanks throughout the area in the past, the initiative aims to renovate these tanks for contemporary water conservation. Mission Kakatiya's salient features include:
   A. Tank restoration and desilting
   B. Groundwater recharge
   C. Community engagement

   To handle India's various water concerns, water conservation programs at both national and state levels are essential. A broad framework for conservation is established by national initiatives like Jal Shakti Abhiyan and Atal Bhujal Yojana, but state-specific initiatives are crucial for tackling regional water problems and community demands. To secure India's water future, these projects collectively highlight the significance of cooperation, sustainable practices, and public involvement.

25.3 Economic Strategies and Financial Support Mechanisms for Water Body Cleaning

Water body cleaning is a long-term, financially demanding undertaking. Several economic strategies and financial support mechanisms have been put in place by the Indian government to stimulate both public and private investment in water body restoration in order to fund the various cleaning programs.

25.3.1 Government Budgets and Allocations

The Indian government has allocated substantial funds to the preservation and cleansing of the nation's lakes, rivers, and groundwater supplies. Various national and state-level initiatives to reduce pollution, treat sewage, restore ecosystems, and adopt sustainable water management techniques are funded with these funds. Funding from the government has come from state revenues, national budgets, and foreign financial aid from institutions such as the World Bank. The main financial resources and budgetary allotments for water body cleanup in India are summarized in this section.

25.3.2 National-level Budgets and Allocations

The central government has started a number of large-scale programs aimed at reducing pollution and conserving water, particularly for large rivers and urban waterways. Important financial allocations at the national level consist of:

25.3.2.1 Namami Gange Programme

The Namami Gange Programme was first given a ₹20000 crore budget, which was to be spent over 5 years. The operations covered by this fund include river surface cleaning, rural sanitation, and sewage treatment facilities. To expand the program's operations, an extra ₹600 crore was allotted in 2020. This money came from the federal government as well as outside sources like the Japan International Cooperation Agency and the World Bank.

25.3.2.2 Atal Bhujal Yojana

Launched in 2020 with a total budget of ₹6000 crore, the Atal Bhujal Yojana focuses on conserving and managing groundwater in water-stressed states such as Maharashtra, Gujarat, Rajasthan, Uttar Pradesh, Karnataka, Madhya Pradesh, and Haryana. Funding for the project is provided in partnership with the World Bank, which provides half of the overall budget. Budget of Atal Bhujal Yojana areas consists of:

1. Community-led groundwater management
2. Incentives for groundwater recharge initiatives
3. Infrastructure development

25.3.2.3 National River Conservation Plan

The NRCP includes a specific budget for enhancing the quality of the water in India's rivers, including the Ganga. Through the installation of sewage treatment plants, lowering of industrial discharge, and the promotion of solid waste management, NRCP focuses on pollution abatement initiatives in different river basins. Highlights of the NRCP budget include:

1. Sewage infrastructure development
2. Pollution control initiatives
3. Solid waste management projects

25.3.3 State-level Budgets and Allocations

With specialized programs focused on rainwater collection, sustainable irrigation, and water body cleansing, Indian states have also allocated funds to solve regional water conservation issues. Prominent budgetary allocations at the state level consist of:

25.3.3.1 Maharashtra's Jalyukt Shivar Abhiyan

The Jalyukt Shivar Abhiyan, which was introduced in 2014 in Maharashtra, was given a budget of ₹7000 crore to address water scarcity by constructing infrastructure for water storage and rainfall harvesting. The objectives of budget allocation are:

1. Desilting and tank restoration
2. Water harvesting structures
3. Subsidies for micro-irrigation

25.3.3.2 Gujarat's Sujalam Sufalam Jal Abhiyan

By allocating a specific budget, Gujarat's Sujalam Sufalam Jal Abhiyan seeks to enhance water storage and construct water harvesting facilities. This program, which is funded by the government with hundreds of crores per year, focuses on boosting groundwater recharge, desilting ponds, and deepening lakes and rivers. The components of the budget include:

1. Deepening of water bodies
2. Community water management programs

25.3.3.3 Telangana's Mission Kakatiya

The goal of Telangana's Mission Kakatiya is to revitalize and restore the state's historic ponds and tanks. This project intends to restore more than 45 000 tanks in Telangana, with an initial investment of about ₹20 000 crore. Mission Kakatiya's budgetary allotments include:

1. Tank restoration
2. Infrastructure for water harvesting
3. Public engagement initiatives

25.3.3.4 Rajasthan's Mukhya Mantri Jal Swavlamban Abhiyan

The Mukhya Mantri Jal Swavlamban Abhiyan in Rajasthan aims to improve groundwater recharge and construct water-saving infrastructure in order to make the state water self-sustainable. This ₹3600 crore program emphasizes water conservation methods in both rural and urban areas. Some important budgetary allocations are:

1. Construction of water harvesting structures
2. Tree plantation and soil conservation
3. Capacity-building programs

25.3.4 International Aids

Apart from the national and state budgets, international organizations provide financial assistance to India for initiatives aimed at improving water quality and conserving water. The main contributors are:

1. **World Bank:** The World Bank has contributed approximately ₹3000 crore for pollution control and water resource management through initiatives like the Namami Gange and Atal Bhujal Yojana.
2. **Japan International Cooperation Agency:** The Namami Gange Programme is supported by the Japan International Cooperation Agency, which has given financial and technical support for the construction of sewage treatment plants (Rwelamila and Abdul-Aziz, 2020).

3. **Asian Development Bank:** The Asian Development Bank finances water conservation initiatives, particularly in rural regions, with a focus on irrigation modernization, sustainable agriculture, and infrastructure development for rainwater gathering.

25.3.5 Public–Private Partnerships

Public–private partnerships have become a more popular way for the government to raise funds and hire technical experts to clean up waterways. The establishment and upkeep of waste management systems, sewage treatment plants, and riverfront beautification initiatives are frequently encouraged for entrepreneurs through government subsidies and tax breaks. In exchange, they receive long-term agreements and a portion of the money made from ecotourism or municipal water bills.

25.3.6 Environmentally Sustainable Financing

In order to support projects that address water pollution, specifically from industrial effluents and agricultural runoff, the India Climate and Clean Energy Finance initiative was launched in 2018. Both domestic and foreign sources of funding contribute to these initiatives through low-interest loans. The government intends to offer ecologically friendly financing solutions for extensive water body restoration projects by utilizing green bonds and clean energy investments.

25.4 Regulatory Frameworks and Enforcement

In India, pollution management and water body conservation are governed by a strong regulatory framework. In order to avoid, control, and decrease pollution in rivers, lakes, and other bodies of water, this framework is made up of environmental laws, regulations, regulatory bodies, and judicial interventions. However, even with well-established policies, enforcing them is still a difficult task. This section explores the main regulatory frameworks, organizations, and enforcement strategies used in India to protect and purify water bodies.

25.4.1 Major Environmental Laws Governing Water Body Protection

The legal framework for water conservation in India consists of a number of historic laws and rules, the majority of which assign accountability for preventing and controlling water pollution to both municipalities and industries.

25.4.1.1 The Water Act, 1974

The first comprehensive law in India that was especially designed to address water pollution was the Water Act of 1974. It creates the legislative framework for controlling wastewater, preventing pollution, and regulating water quality (Gupta, 2020). Important clauses in the Water Act include:

1. **Pollution control boards:** To monitor and regulate the levels of pollution in water bodies, the act established Central and State Pollution Control Boards.
2. **Effluent standards:** It requires enterprises to adhere to certain effluent emission guidelines and punishes infractions with fines and penalties.
3. **Licensing for discharge:** To ensure that wastewater is cleansed to meet prescribed criteria, industries must acquire approval from the appropriate pollution control boards before releasing effluents into any body of water.

25.4.1.2 The Environment Act, 1986

The central government is given extensive authority to control and safeguard the environment by the Environment Act of 1986. This Act was passed to give India's environmental protection a cohesive structure after the Bhopal gas tragedy (Bhat, 2024). Important clauses in the Environment Act include:

1. **Power to set standards:** To control industrial pollution, the Environment Act gives the government the authority to establish national standards for effluent discharge and water quality.
2. **Pollution control orders:** Municipalities and enterprises that violate environmental regulations may be subject to penalties or closure orders from authorities.
3. **Hazardous substance regulation:** Hazardous waste management is covered under the Environment Act and is crucial to keeping hazardous substances out of rivers.

25.4.1.3 The National Green Tribunal Act, 2010

The National Green Tribunal (NGT) Act established the NGT, a specialized judicial entity responsible for resolving environmental disputes and implementing environmental regulations (Bhat, 2024). Important roles of the NGT include:

1. **Environmental dispute resolution:** The NGT resolves disputes connected to environmental protection, water pollution, and natural resource management, and frequently orders industries or municipal authorities to clean polluted water bodies.
2. **Enforcement orders:** The NGT has the authority to enforce environmental laws, impose fines, and issue orders for remediation.

3. **Public interest litigation:** The NGT accepts complaints from citizens and nongovernmental organizations, allowing the public to be involved in the implementation of legislation pertaining to pollution control and water conservation.

25.4.1.4 The Coastal Regulation Zone Notification, 2019

To avoid contamination of coastal and marine water bodies, activities along India's coastline are governed by the Coastal Regulation Zone (CRZ) Notification. This notification, issued under the Environment Act, intends to conserve coastal ecosystems by preventing pollution from industrial, residential, and tourism activities. Important CRZ regulations include:

1. Regulation of coastal development
2. Protected areas
3. Water quality standards

25.4.2 Regulatory Bodies and Their Roles

India's water conservation and pollution control framework is governed by a variety of government entities, each responsible for certain areas of enforcement and regulation.

25.4.2.1 Central Pollution Control Board

The Central Pollution Control Board (CPCB) is India's principal pollution control authority. The Ministry of Environment, Forests, and Climate Change oversees the CPCB, which was established under the Water Act (Bhat, 2024). It works in tandem with State Pollution Control Boards (SPCBs) to monitor the quality of water throughout the nation. The CPCB serves as a governing body in responsible for standard-setting, coordination, and development of policies. It establishes national norms for effluent discharge, waste management policies, and ambient air and water quality standards. It also monitors environmental quality through programs like the National Water Quality Monitoring Programme and the National Air Quality Monitoring Programme (CPCB, 2022a). Further, the CPCB conducts pollution control technology research and development, offers states technical assistance, and facilitates the execution of national initiatives like Namami Gange and the National Clean Air Programme (MoEFCC, 2019). The CPCB's functions include:

1. Water quality monitoring
2. Setting pollution standards
3. Compliance and inspections

25.4.2.2 State Pollution Control Boards

At the state level, pollution control measures are implemented by the SPCBs. They monitor and regulate water pollution within their territories by enforcing both federal statutes and state-specific rules. SPCBs serve as the state and union territory enforcement branches. They are responsible for monitoring adhering to CPCB standards, providing industries authorization to construct and operate within the terms of the Water and Air Acts, and taking legal or administrative action against those who violate the law (MoEFCC, 2020). Additionally, they draft state-specific action plans to reduce localized air and water pollution, rejuvenate contaminated river segments, and regulate industrial emissions. Additionally, through resolving complaints about pollution and implementing awareness campaigns, SPCBs are essential to public participation (CPCB, 2022b). SPCBs modify and implement these guidelines in accordance with local environmental conditions, whilst CPCB maintains consistency and coordination throughout India. The main functions of SPCBs are as follows:

1. Industrial regulation
2. Enforcement and penalties
3. Public awareness and education

25.4.2.3 National Mission for Clean Ganga

The Namami Gange Programme, which aims to clean and conserve the Ganga River, is managed by the National Mission for Clean Ganga (NMCG), a specialized organization. Under the Ministry of Jal Shakti, NMCG works with communities, NGOs, and state governments. Functions of the NMCG include:

1. Project implementation
2. Coordination and monitoring
3. Funding and resources

25.4.2.4 Ministry of Jal Shakti

The Ministry of Jal Shakti is the central government entity in charge of managing water resources, conserving rivers, and implementing water conservation measures. It was formed in 2019 when the Department of Water Resources and the Ministry of Drinking Water and Sanitation merged. The Ministry of Jal Shakti's primary functions include:

1. Policy formulation
2. River basin management
3. Inter-state water disputes

The legal framework for cleaning water bodies in India consists of a variety of environmental legislation, regulatory agencies, and court rulings. This structure has worked well for setting guidelines and implementing penalties.

25.5 Challenges in Economic Support and Implementation

Although the Indian government has set aside an extensive amount of money to clean water bodies, there are still several obstacles that prevent the funds from being used effectively:

1. **Coordination challenges**: State governments, the Ministry of Water Resources, the Ministry of Environment, and the Ministry of Urban Development are all involved in the multi-sectoral problem of water pollution. It is frequently difficult for various institutions to coordinate effectively.
2. **Weak enforcement of regulations**: Despite funding and legislation, anti-pollution measures are frequently not enforced effectively. Cleaning initiatives may be hampered or delayed at the local level by corruption, ignorance, and poor infrastructure.
3. **Sustainability issues**: Numerous initiatives to restore water bodies mainly depend on short-term funding, along with external investments. It will take ongoing investment and creative local management model development to ensure these initiatives' long-term viability.
4. **High capital costs and maintenance:** Particularly in rural locations, the substantial initial investment and continuous upkeep needed for infrastructure such as STPs can put a strain on municipal budgets.
5. **Compliance and enforcement issues:** Inconsistent enforcement of pollution regulations is partially caused by a lack of administrative resources.
6. **Lack of public awareness and participation:** Despite the success of awareness programs, it is still difficult to keep the public consistently involved in water conservation initiatives.
7. **Coordination between agencies:** Several ministries and agencies frequently operate in divisions, which results in inefficiencies and redundant tasks.

25.6 Conclusion: A Pathway to Sustainable Development

In India, cleaning up waterways is a costly and intricate process that calls for a large financial investment, creative business plans, and cooperation from a variety of parties. The Indian government has taken major action to address the problem of water pollution through programs including Namami Gange, the NRCP, and the Atal Mission for Urban Transformation. Nonetheless, there are still challenges with obtaining consistent funding, guaranteeing successful execution, and encouraging public–private cooperation. For India's attempts to clean water bodies to result in long-term outcomes, continuous investment, stronger regulation, and community engagement will be essential. To ensure a sustainable water future for its people and restore the health of its water bodies, India must take a comprehensive, multifaceted strategy.

Review Questions

1. What is the current scenario of water bodies in India?
2. What are the various technologies utilized in wastewater treatment, in general, and why?
3. What is the role/importance of clean water bodies in ecosystem and communities?
4. What are the various national-level programs and their approaches for controlling pollution?
5. What are the state-level programs and their approach for controlling pollution?
6. What are the international aids and their quantum for improving water quality and conserving water?
7. What are the various regulations toward pollution management and water body conservation?
8. What is the scope of CPCB and state-level pollution boards?
9. What are the various challenges toward implementing governmental initiatives in pollution management and water body conservation?
10. In spite of investing a huge amount of money every year and having a lot of regulatory framework, still level of pollution in water bodies is increasing. State the reasons and probable solutions.

References

Bhat, S. (2024). *Environmental Law and Policy in India*. Taylor & Francis.

Chai, W.S., Tan, W.G., Munawaroh, H.S.H. et al. (2021). Multifaceted roles of microalgae in the application of wastewater biotreatment: a review. *Environmental Pollution*, **269**, 116236.

Ciardelli, G. and Ranieri, N. (2001). The treatment and reuse of wastewater in the textile industry by means of ozonation and electroflocculation. *Water Research*, **35**(2), 567–572.

CPCB. (2022a). Water Quality Status of India 2022. Central Pollution Control Board Ministry of Environment, Forest and Climate Change, Government of India.

CPCB. (2022b). *Annual report 2021–22*. Central Pollution Control Board Ministry of Environment, Forest and Climate Change, Government of India. Retrieved from https://cpcb.nic.in/uploads/annualreports/Annual_Report_2021-22.pdf

Cullet, P., Gualtieri, A.G., Madhav, R. et al. (2010). *Water Law for the Twenty-First Century*. London: Routledge.

Debaere, P. and Kapral, A. (2021). The potential of the private sector in combating water scarcity: the economics. *Water Security*, **13**, 100090.

Gani, A., Hussain, A., Pathak, S. et al. (2024a). An empirical investigation on the elimination of heavy metals using bioremediation method for selected plant species. *Physics and Chemistry of the Earth, Parts A/B/C*, **134**, 103568.

Gani, A., Pathak, S., Hussain, A. et al. (2024b). Emerging pollutant in surface water bodies: a review on monitoring, analysis, mitigation measures and removal technologies of micro-plastics. *Environmental Geochemistry and Health*, **46**(7), 214. https://doi.org/10.1007/s10653-024-01992-7

Gani, A., Hussain, A., Pathak, S. et al. (2024c). Analysing heavy metal contamination in groundwater in the vicinity of Mumbai's landfill sites: an in-depth study. *Topics in Catalysis*, **67**(15), 1009–1023.

Gani, A., Pathak, S., and Hussain, A. (2024d). Comprehensive analysis of modified heavy metal pollution index and health risk assessment in the Yamuna river of Delhi, India: crucial study for environmental health management. *Journal of Hazardous, Toxic, and Radioactive Waste*, **28**(4), 04024021. https://doi.org/10.1061/JHTRBP.HZENG-1374

Gani, A., Singh, M., Pathak, S. et al. (2024e). Recent advancements in chlorine applications for water quality control. In: *Drinking Water Disinfection By-products: Sources, Fate and Remediation*, 35–58. Cham: Springer Nature Switzerland.

Gani, A., Pathak, S., Hussain, A. et al. (2025a). Water quality index assessment of river Ganga at Haridwar stretch using multivariate statistical technique. *Molecular Biotechnology*, **67**(8), 3130–3153. https://doi.org/10.1007/s12033-023-00864-2

Gani, A., Singh, M., Pathak, S. et al. (2025b). Groundwater quality index development using the ANN model of Delhi metropolitan city, India. *Environmental Science and Pollution Research*, 32(12), 7269–7284. https://doi.org/10.1007/s11356-023-31584-4

Gani, A., Hussain, A., Pathak, S. et al. (2025c). Impact of pollutants on groundwater quality and health risk assessment of quaternary aquifers in Northern India. *Journal of Hazardous, Toxic, and Radioactive Waste*, **29**(1), 04024039.

Gundry, S., Wright, J., and Conroy, R. (2004). A systematic review of the health outcomes related to household water quality in developing countries. *Journal of Water and Health*, **2**(1), 1–13. https://doi.org/10.2166/wh.2004.0001

Gupta, A. (2020). *Heavy Metal and Metalloid Contamination of Surface and Underground Water: Environmental, Policy and Ethical Issues*. CRC Press.

Jadeja, N.B., More, R.P., Purohit, H.J. et al. (2014). Metagenomic analysis of oxygenases from activated sludge. *Bioresource Technology*, **165**, 250–256.

Jorgenson, A.K. (2009). Foreign direct investment and the environment, the mitigating influence of institutional and civil society factors, and relationships between industrial pollution and human health. *Organization & Environment*, **22**(2), 135–157. https://doi.org/10.1177/1086026609338163

Kumar, R., Mishra, A., and Goyal, M.K. (2024). Water neutrality: concept, challenges, policies, and recommendations. *Groundwater for Sustainable Development*, **26**, 101306.

Landrigan, P.J., Fuller, R., Fisher, S. et al. (2018). Pollution and children's health. *Science of the Total Environment*, **650**(Pt 2), 2389–2394. https://doi.org/10.1016/j.scitotenv.2018.09.375

Mahto, M.K. (2024). Challenges and opportunities for water resources in India: an exploratory story of resilience. In: *Integrated Management of Water Resources in India: A Computational Approach: Optimizing for Sustainability and Planning*, 59–80. Springer.

Makut, B.B., Das, D., and Goswami, G. (2019). Production of microbial biomass feedstock via co-cultivation of microalgae-bacteria consortium coupled with effective wastewater treatment: a sustainable approach. *Algal Research*, **37**, 228–239.

Millennium ecosystem assessment. (2005). *Ecosystems and human well-being* (Vol. 5, p. 563). Washington, DC: Island press.

MoEFCC. (2019). *National Clean Air Programme*. Ministry of Environment, Forest and Climate Change Government of India.

MoEFCC. (2020). *National Clean Air Programme*. Ministry of Environment, Forest and Climate Change Government of India.

Mohamed, R.R., Manickam, N., Paulraj, A. et al. (2024). An investigation of coastal water quality and SEMBV disease in Penaeus monodon in coastal Southern India. In: *Integrated Management of Water Resources in India: A Computational Approach: Optimizing for Sustainability and Planning*, 159–171. Cham: Springer Nature Switzerland.

Nayak, M.A. and Hassan, W.U. (2021). A synthesis of drought prediction research over India. *Water Security*, **13**, 100092.

Panagopoulos, A. and Haralambous, K.J. (2020). Minimal Liquid Discharge (MLD) and Zero Liquid Discharge (ZLD) strategies for wastewater management and resource recovery–analysis, challenges and prospects. *Journal of Environmental Chemical Engineering*, **8**(5), 104418.

Pandey, P.K. and Pande, A. (2023). *Aquatic Environment Management*. CRC Press.

Pathak, S., Gupta, S., and Ojha, C.S.P. (2021). Assessment of groundwater vulnerability to contamination with ASSIGN index: a case study in Haridwar, Uttarakhand, India. *Journal of Hazardous, Toxic, and Radioactive Waste*, **25**(2), 04020081.

Payment, P., Siemiatycki, J., Richardson, L. et al. (1997). A prospective epidemiological study of gastrointestinal health effects due to the consumption of drinking water. *International Journal of Environmental Health Research*, **7**(1), 5–31. https://doi.org/10.1080/09603129773977

Rahman, A., Mondal, N.C., and Fauzia, F. (2021). Arsenic enrichment and its natural background in groundwater at the proximity of active floodplains of Ganga River, Northern India. *Chemosphere*, **265**, 129096.

Rai, P., Kumar, J.Y., and Sen, A. (2024). Ganga, GAP, and lockdown: potential threats to the biodiversity of the river. *Journal of Water and Climate Change*, **15**, jwc2024322.

Rwelamila, P.D. and Abdul-Aziz, A.R. (eds.) (2020). *Improving the Performance of Construction Industries for Developing Countries: Programmes, Initiatives, Achievements and Challenges*. Routledge.

Shukla, A.K., Ojha, C.S.P., Mijic, A. et al. (2018). Population growth, land use and land cover transformations, and water quality nexus in the Upper Ganga River basin. *Hydrology and Earth System Sciences*, **22**(9), 4745–4770.

UNEP. (2016). A Snapshot of the World's Water Quality: Towards a Global Assessment. United Nations Environment Programme.

Vimalkumar, K., Arun, E., Krishna-Kumar, S. et al. (2018). Occurrence of triclocarban and benzotriazole ultraviolet stabilizers in water, sediment, and fish from Indian rivers. *Science of the Total Environment*, **625**, 1351–1360.

Vladeva, S., Gatseva, P., and Gopina, G. (2000). Comparative analysis of results from studies of goitre in children from Bulgarian villages with nitrate pollution of drinking water in 1995 and 1998. *Central European Journal of Public Health*, **8**(3), 179.

World Health Organization (WHO). (2022). *Drinking-water*. Retrieved from https://www.who.int/news-room/fact-sheets/detail/drinking-water

Index

Note: Page numbers in *italics* and **bold** refers to figures and tables respectively.

a

absorption fields 320
AC. *see* activated carbon (AC)
ACF. *see* activated carbon filter (ACF)
acid mine drainage (AMD) 280–281
acidic methanogenesis 283
Act on the Promotion of Saving and Recycling of Resources (2007) 196
activated carbon (AC) 433, 518, 520–521
activated carbon filter (ACF) 505
activated sludge (AS) 441
activated sludge process (ASP) 2, 5, 10, 149, 443, 576
AD. *see* anaerobic digestion (AD)
adaptive neuro-fuzzy inference systems (ANFISs) 539
additive manufacturing (AM) 34, 219–220
ADs. *see* anaerobic digestor (ADs)
adsorption 307, 388–389, 433–435
advanced oxidation process (AOP) 2, 100, 106–107, 127–128, 177, 178, 435–436
advanced oxidation/reverse osmosis/membrane bioreactor (ARM) 242
advanced thermal methods 33
AEH. *see* Aravind Eye Hospital (AEH)
AEM. *see* anion-exchange membrane (AEM)
aerated lagoon (AL) 243
aerobic digestion 389
aerobic granular sludge (AGS) 439
aerobic membrane bioreactor 433
aerobic method 174–175
  activated sludge process 149
  biofilter 149–150
  nitrification–denitrification 148–149
  rotating biological contactors and advantages 150–151
agricultural waste 21
AGS. *see* aerobic granular sludge (AGS)
AI. *see* artificial intelligence (AI)
Air (Prevention and Control of Pollution) Act (1981) 79, 82, **197**
AL. *see* aerated lagoon (AL)
Al-Azraqain landfill 418–419
algae 274
algal biomass 395
algal biorefinery 525
AM. *see* additive manufacturing (AM)
AMD. *see* acid mine drainage (AMD)
ammonia stripping method 346
anaerobic baffled reactor (ABR) 10, 150, 323, *323*, 405, 413, 507
anaerobic digestion (AD) 243, 322, 326, 389–390
anaerobic digestor (ADs) 565
anaerobic membrane bioreactor (AnMBR) 243, 322
anaerobic method 176
anaerobic sludge blanket (UASB) 5–6
anaerobic stabilization reactor 507

anaerobic treatment 151
composting 152
membrane bioreactor 151
microbial fuel cell 151–152
ANFISs. *see* adaptive neuro-fuzzy inference systems (ANFISs)
anion-exchange membrane (AEM) 474
AnMBR. *see* anaerobic membrane bioreactor (AnMBR)
ANN. *see* artificial neural network (ANN)
AOP. *see* advanced oxidation process (AOP)
aquaculture 394
aquaponics systems 395
Aravind Eye Hospital (AEH) 507
decentralized wastewater treatment at 413, 507–508
ARM. *see* advanced oxidation/reverse osmosis/membrane bioreactor (ARM)
arsenic contamination 23
artificial intelligence (AI) 442, 524, 533–534
applications in WWT 540–544, *540*
biological methods 535–536
challenges and future perspectives 549–550
challenges in current WWT processes 536
chemical methods 535
deep learning 539–540
economic challenges 537
machine learning 538–539
physical methods 534–535
for resource recovery in WWT 544–549
social challenges 536
technical challenges 536–537
traditional processes in WWT 534–537
in WWT 537–540, *537*
artificial neural network (ANN) 437, 539
AS. *see* activated sludge (AS)
ASP. *see* activated sludge process (ASP)
Atal Mission for Rejuvenation and Urban Transformation 582
augmented epsilon-constraint approach (AUGMECON2) 54, **55**

b

Bacillus licheniformis 396
Bacillus subtilis 396
BAFs. *see* biological aerated filters (BAFs)
BC. *see* biochar (BC)
BES. *see* bioelectrochemical systems (BES)
Bhopal Gas Disaster (1984) 20
biochar (BC) 391, 521
biochemical oxygen demand (BOD) 3, 120, 145, 173, 222, 270, 314, 439
biodegradable polymers 395
biodiversity 578–579
bioelectrochemical systems (BES) 407–408, *407*, 566
bioenergy 374
biofilm 444–445
biofilter 149–150
biofuel 500–501
biodiesel 258, *259*
bio-oil 257–258
biological aerated filters (BAFs) 100
biological filtration system 9–10
biological method 350
constructed wetland and vermifilter treatment 350
micro-algal treatment 350–351
for urine and fecal treatment 405–406
in WWT 535–536
biological nutrient removal (BNR) 100, 256, 440
biological oxygen demand (BOD) 541, 563
biological technology
in circular economy 377–378
used for resource recovery 374, **375–376**
biological treatment methods 5–6
biological wastewater treatment, importance of 147–148
Bio-Medical Waste Management Rules (2016) **198**
bio-precipitation 309
biosorption 308–309
bipolar membrane electrodialysis (BMED) 348
blockchain technology 34
BMED. *see* bipolar membrane electrodialysis (BMED)
BNR. *see* biological nutrient removal (BNR)
BOD. *see* biochemical oxygen demand (BOD); biological oxygen demand (BOD)
BP. *see* British Petroleum (BP)
British Petroleum (BP) 254
broad slow feature neural network (BSFNN) 544

BSFNN. *see* broad slow feature neural network (BSFNN)
Bugolobi WWTP 416–417
business models 508
by-product management 541–542

C

CAGR. *see* Compound Annual Growth Rate (CAGR)
capacitive deionization (CDI) 348
CapEx. *see* capital expenditure (CapEx)
capital cost (CC) 153, 241
capital expenditure (CapEx) 332
carbon-based nanomaterials 464–465
carbon nanotubes (CNTs) 437, 463–464, 519
  with horizontal alignment 470–471
  membranes 481
  with mixed matrixes 470
  technology 466
carbon recovery 192
CAS. *see* conventional activated sludge (CAS)
cation-exchange membrane (CEM) 474
CBA. *see* cost-benefit analysis (CBA)
CBT. *see* conventional biological treatment (CBT)
CC. *see* capital cost (CC)
CCD. *see* Central Composite Design (CCD)
CDD. *see* Consortium for DEWATS Dissemination India (CDD)
CDI. *see* capacitive deionization (CDI)
CE. *see* circular economy (CE)
CEC. *see* contaminants of emerging concern (CEC)
CEEW. *see* Council on Energy, Environment and Water (CEEW)
CEM. *see* cation-exchange membrane (CEM)
cementation 547
Central Composite Design (CCD) 246
centralized approaches 411–413
  costs of wastewater reuse 353
  distribution for reuse 354–355
  sustainability 355, **356–357**
  technology and operation 353–354
  for utilization of wastewater 352–357
centralized sanitation concept 500
centralized sewage treatment 404
centralized systems 3–4, *4*, 353–354
  comparative analysis of decentralized systems and 408–410, *409*, **410**
  comparative economics of decentralized systems *vs.* 11–12
  history of 3
  limitations and effects or impacts of 7–8
  preliminary/primary treatment 4–5
  secondary treatment 5–6
  tertiary treatment 6–7
centralized treatment system for urine and fecal matter 403–404, 408–410
centralized wastewater treatment system (CWTS) 76–77, 328
Central Pollution Control Board (CPCB) 78, 79, 84, 577, 589
Central Public Health and Environmental Engineering Organisation (CPHEEO) 497
CETPs. *see* common effluent treatment plants (CETPs)
CFBBR. *see* circulating fluidized bed bioreactor (CFBBR)
cheese whey wastewater (CWW) 285
chemical and toxic substances 98
chemical consumption 133
chemical energy stored in sewage water 255–256
chemical industry 191
chemical methods 126
  advanced oxidation 127–128
  chemical precipitation 126
  electrochemical separation 129–130
  ion exchange 127
  ozonation 128–129
  in WWT 535
chemical oxygen demand (COD) 5, 46, 57, 130, 145, 171, 245–246, 270, 296, 314, 396, 403, 430, 541, 563
chemical precipitation 307–308, 547
chemical recycling 32
chemicals recovery 192
Chennai Metropolitan Water Supply and Sewerage Board (CMWSSB) 506
Chlorella vulgaris 396
chlorpyrifos (CPS) 543
CIP. *see* clean-in-place (CIP)
circular bioeconomy 305, 377

circular economy (CE) 24–25, 214, 270, 369, 373, 385
  benefits 34–35
  biological technology for resource recovery 374, **375–376**, 377–378
  case studies on conventional technologies 390–394, *394*
  energy generation 275–276
  environment, climate and economics 276–277
  implementation challenges 35
  implementation strategies 30–34
  modern technologies for wastewater treatment 394–396, *395*
  strategies 29–30
  waste treatment principles 30
  wastewater and 372–373, *372*
circulating fluidized bed bioreactor (CFBBR) 432
citywide inclusive urban sanitation (CWIS) services 568
clay minerals 434
cleaner production (CP) 275
  energy generation 275–276
  environment, climate and economics 276–277
  gold and iron ore waste management 279–281
  integrated approach 278–279
clean-in-place (CIP) 216
climate-resilient designs 111
closed-loop system 373
CM. *see* conventional manufacturing (CM)
CMWSSB. *see* Chennai Metropolitan Water Supply and Sewerage Board (CMWSSB)
CNNs. *see* convolutional neural networks (CNNs)
CNTs. *see* carbon nanotubes (CNTs)
coagulant chemicals 387
coagulation 173, 386–387, 437–438, *438*
Coastal Regulation Zone (CRZ) Notification (2019) 589
COD. *see* chemical oxygen demand (COD)
common effluent treatment plants (CETPs) 84
community
  engagement 111
  involvement 111–112
  well-being 579
comparative analysis of centralized and decentralized systems 408–410, *409*, **410**
complex salt-*Luffa cylindrica* seed extract (CS-LCSE) 543
composting 152
  toilets 10–11, 320
Compound Annual Growth Rate (CAGR) 41–42
compression 122
Consortium for DEWATS Dissemination India (CDD) 507
constructed wetlands (CWs) 9, 45–46, 350, 439, 565
Construction and Demolition (C&D) waste recovery 200
construction sector 190
Consultant Capacity Development (ConCaD) project 568
contaminants of emerging concern (CEC) 123, 125, 172
continuous stirred tank reactor (CSTR) 432–433
controlled waste 37
conventional activated sludge (CAS) 274
conventional biological treatment (CBT) 145–146, 148
  aerobic treatment 148–151
  anaerobic treatment 151–152
  challenges and future directions 156–157
  economic implications of 153–156
  importance of 147–148
  technologies 148–152
  wastewater treatment needs 146–147
conventional manufacturing (CM) 219
conventional polymers 522
conventional technologies
  adopting CE 390–394, *394*
  adsorption 388–389
  aerobic digestion 389
  anaerobic digestion 389–390
  coagulation 386–387
  filtration 387–388
  flocculation 387
  limitations of 179–180, *180*
  for wastewater treatment 386
convolutional neural networks (CNNs) 539
coronavirus (COVID-19) 54

Corporate Social Responsibility (CSR) 280
cost-benefit analysis (CBA) 135
Council on Energy, Environment and Water (CEEW) 509
coupled redox fuel cells (CRFCs) 548
COVID-19. *see* coronavirus (COVID-19)
CP. *see* cleaner production (CP)
CPCB. *see* Central Pollution Control Board (CPCB)
CPHEEO. *see* Central Public Health and Environmental Engineering Organisation (CPHEEO)
CPS. *see* chlorpyrifos (CPS)
CRFCs. *see* coupled redox fuel cells (CRFCs)
critical distance 353
CS-LCSE. *see* complex salt–*Luffa cylindrica* seed extract (CS-LCSE)
CSR. *see* Corporate Social Responsibility (CSR)
CSTR. *see* continuous stirred tank reactor (CSTR)
cutting-edge biotechnology 439
CWs. *see* constructed wetlands (CWs)
CWTS. *see* centralized wastewater treatment system (CWTS)
CWW. *see* cheese whey wastewater (CWW)

d

DAF. *see* dissolved air floatation (DAF)
dairy industry 244–245
  electrochemical treatment 247–248
  electrocoagulation process 245–246
  sequencing batch reactor 246–247
decentralized approaches 411–413
  costs of wastewater reuse 353
  distribution for reuse 354–355
  sustainability 355, **356–357**
  technology and operation 353–354
  for utilization of wastewater 352–357
decentralized onsite systems 315
decentralized project planning 316–327
decentralized sewage treatment in New Delhi 415
decentralized systems 11–12
  case study 327–330
  comparative analysis of centralized systems and 408–410, *409*, **410**
  decentralized project planning and resource recovery 316–327
  economic feasibility 330–332
  future prospects 332–333
  need of 315
  in rural areas 82
decentralized treatment
  DEWATS 507–508
  fecal sludge management (FSM) 506–508
  systems 103, 107
  of urine and fecal matter 408–410
  of wastewater, case studies on 506
decentralized wastewater treatment 413
decentralized wastewater treatment and reuse (DESAR) 47
Decentralized Wastewater Treatment Solutions 316, 319
decentralized wastewater treatment system (DEWATS) 76, 77, 81, 86, 314, 328–330, 566, 568
decentralized water regulatory authority 85–86
deep learning (DL) 539–540
deep tunnel sewerage system (DTSS) 102
Department of Science and Technology (DST) 319
Department of Biotechnology (DBT) 319
DESAR. *see* decentralized wastewater treatment and reuse (DESAR)
DEWATS. *see* decentralized wastewater treatment system (DEWATS)
digital innovations 34
discrete settling 122
dissolved air floatation (DAF) 283
dissolved oxygen (DO) 3, 314, 444, 540
DL. *see* deep learning (DL)
DO. *see* dissolved oxygen (DO)
domestic wastewater management 27, 339
drying 405
DTSS. *see* deep tunnel sewerage system (DTSS)
Dunaliella 501
DWWTS. *see* decentralized wastewater treatment system (DWWTS)
dyes 170–171

e

EAOPs. *see* electrochemical advanced oxidation processes (EAOPs)
EC. *see* electrocoagulation (EC)

EcoDrain 546
ecological health 578–579
ECOlogical SANitation (EcoSan) toilets 319
economic analysis
  Al-Azraqain landfill 418–419
  case study for 411
  of centralized and decentralized treatment 408–410
  decentralized wastewater treatment at Aravind Eye Hospital 413
  Devanahalli, fecal sludge management in 414
  Kampala, fecal sludge management in 417–418
  Leh, fecal sludge management in 414–415
  PPP for decentralized sewage treatment in New Delhi 415
  sewage treatment systems in Kampala 416–417
  solid waste management in Rural Maharashtra 415–416
  of source separation 508–511, **510**
  techno-economic analysis of nutrient recovery from urine 411–413
economic burden of industrial activity and strategies 222
economic consequences 104
economic costs 104
economic implications 153
  assessment of economic value 155–156
  capital costs 153
  externalities and social costs 155
  operational costs 153–154
  revenue generation 154–155
economic strategies for water body cleaning 584–587
economic value 41–47
  assessment 155–156
  economic analysis 47
  existing challenges and future prospects of nanotechnology 41–42
  status and practices to CE 43–46
economic viability strategies
  implementation challenges 38–39
  policy and regulation role 39–40
  potential for recovered resources 38
  public–private partnerships 40
  traditional *vs.* circular approaches, cost analysis of 36–37
EcoSan Systems 319–320
eco-toilet system (ETS) 505
ECP. *see* electrochemical peroxidation (ECP)
ECs. *see* emerging contaminants (ECs); environmental contaminants (ECs)
ED. *see* electrodialysis (ED)
EDCs. *see* endocrine-disrupting chemicals (EDCs)
EEE. *see* electronic and electrical equipment (EEE)
EF. *see* Electro Fenton (EF)
effluent treatment plants (ETPs) 83–84, 567
EGSB. *see* expanded granular sludge bed (EGSB)
EIA. *see* Environmental Impact Assessment (EIA)
electricity 258–260, *260*
electrochemical advanced oxidation processes (EAOPs) 447
electrochemical-Fenton (EF) 282
electrochemical oxidation (EO) 348
electrochemical peroxidation (ECP) 447
electrochemical reduction (ER) 348
electrochemical techniques 178, 348
electrocoagulation (EC) 129–130, 244, 283, 344
electrodialysis (ED) 129, 257, 344, 352, 474–475, **475**
Electro Fenton (EF) 246, 447
electronic and electrical equipment (EEE) 200
electronic waste (e-waste) 20, 21, 25–26
ELT. *see* emergency locator transmitters (ELT)
emergency locator transmitters (ELT) 408
emerging contaminants (ECs) 427
EMPs. *see* environmental management plans (EMPs)
EMS. *see* environmental management system (EMS)
endocrine-disrupting chemicals (EDCs) 148, 172
end-of-waste (EoW) criteria 503
energy
  consumption 132
  recovery 32–33, 191, 525, 546–547
  resources 304–305
  stored 271–272

entrepreneurship 223
The Environment Act (1986) 588
Environment Protection Act (EPA) (1986) 20, 82, 83, **197**
environmental contaminants (ECs) 147
environmental degradation 104
environmental impact 218–220, *219*
Environmental Impact Assessment (EIA) 78, 80–81, 237
environmental laws 587–589
  Coastal Regulation Zone Notification (2019) 589
  The Environment Act (1986) 588
  The National Green Tribunal Act (2010) 588–589
  The Water Act (1974) 588
environmentally sustainable financing 587
environmental management plans (EMPs) 80, 81
environmental management system (EMS) 80, 81, 279
environmental protection 17
Environmental Protection Act 196
Environmental Quality Act 6
EO. *see* electrochemical oxidation (EO)
EPR. *see* Extended Producer Responsibility (EPR)
EPS. *see* extracellular polymeric substances (EPS)
ER. *see* electrochemical reduction (ER)
Escherichia coli 464
ETPs. *see* effluent treatment plants (ETPs)
ETS. *see* eco-toilet system (ETS)
EU. *see* European Union (EU)
European Circular Economy Action Plan (2015) 28
European Union (EU) 221
evaporation 404
e-waste. *see* electronic waste (e-waste)
EWMA. *see* exponentially weighted moving average (EWMA)
Excel-based model 499
expanded granular sludge bed (EGSB) 322
exponentially weighted moving average (EWMA) 544
Extended Producer Responsibility (EPR) 199, **199**
extracellular polymeric substances (EPS) 296, 443–444

f

f-CNTs. *see* functionalized CNTs (f-CNTs)
fecal sludge management (FSM) 506–507
  in Devanahalli 414
  in Kampala 417–418
  in Leh 414–415, 508
  services 508
  systems 416
Fecal Sludge Treatment Plant (FSTP) 414, 506
fertilizer
  decentralized and centralized approaches for 352–357
  forms of nutrients applied as 340, 343
FIE. *see* food industry effluent (FIE)
filtration 306, 387–388
financial support mechanisms for water body cleaning 584–587
FISs. *see* fuzzy inference systems (FISs)
fixed film bioreactor 432
floc 444–445
flocculant settling 122
flocculation 387, 437–438, *438*
fluidized-bed bioreactor 432
FO. *see* forward osmosis (FO)
food industry effluent (FIE) 282
food processing industry 281–285
forward osmosis (FO) 257, 296, 322, 344, 352, 456, 463
fossil fuels 500
free water surface (FWS) 439
FSM. *see* fecal sludge management (FSM)
FSTP. *see* Fecal Sludge Treatment Plant (FSTP)
functionalized CNTs (f-CNTs) 466
fuzzy inference systems (FISs) 539, 542
FWS. *see* free water surface (FWS)

g

GA. *see* genetic algorithm (GA)
Ganga Action Plan (GAP) 581
Ganga river pollution 103–104
GAP. *see* Ganga Action Plan (GAP)
GDP. *see* gross domestic product (GDP)
GEMS. *see* Global Environmental Monitoring System (GEMS)
genetic algorithm (GA) 542
GGBFS. *see* Ground Granulated Blast-Furnace Slag (GGBFS)
GHG. *see* greenhouse gas (GHG)

GHSD. *see* Greenhouse Solar Drier Roofs (GHSD)
GIDC. *see* Gujarat Industrial Development Corporation (GIDC)
Global Environment and Health Management Standards 196
Global Environmental Monitoring System (GEMS) 84
GPCB. *see* Gujarat Pollution Control Board (GPCB)
granular sequencing batch reactors (gSBRs) 439
green building materials 260–261
greenhouse gas (GHG) 254, 272, 403
Greenhouse Solar Drier Roofs (GHSD) 507
green infrastructure 108
green reverse logistics network 53–54
  managerial insights 56
  proposed model 54
  result and discussion 54–55, **55**, *55*
  validation of model 54
green sanitation technology 327
grit
  chambers 172–173
  removal 4–5
gross domestic product (GDP) 223
Ground Granulated Blast-Furnace Slag (GGBFS) 31
groundwater 1
gSBRs. *see* granular sequencing batch reactors (gSBRs)
Gujarat Industrial Development Corporation (GIDC) 505
Gujarat Pollution Control Board (GPCB) 505

h

Hazardous Waste Management Rules **198**
Hazardous Waste Rules (1989) 20
healthcare 191
heat recovery 260, *261*
heavy metals 169–170, *170*, 192, 503–504
HHV. *see* High Heating Value (HHV)
High Heating Value (HHV) 45
high rate activated sludge (HRAS) 243
hindered settling 122
HLR. *see* hydraulic loading rate (HLR)
horizontally aligned CNT (HACNT) membranes 466
horizontal subsurface flow constructed wetland (HSFCW) system 448
hospital waste management 28
HRAS. *see* high rate activated sludge (HRAS)
HRT. *see* hydraulic retention time (HRT)
HTC. *see* hydrothermal carbonization (HTC)
HTL. *see* hydrothermal liquefaction (HTL)
human health 579
human urine 44
hybrid systems 438–441, *440*, 526, *527*
hydraulic loading rate (HLR) 346
hydraulic pressure 473
hydraulic retention time (HRT) 246, 274, 322, 346, 431, 443, 518
hydrochar/biochar 521
hydrothermal carbonization (HTC) 405
hydrothermal liquefaction (HTL) 405

i

independent regulatory agencies (IRAs) 78
India
  Central Pollution Control Board 589
  challenges in economic support and implementation 591
  economic strategies and financial support mechanisms 584–587
  environmental laws governing water body protection 587–589
  Ministry of Jal Shakti 590
  national-level initiatives for water conservation 580–582
  National Mission for Clean Ganga 590
  regulatory bodies and roles 589
  regulatory frameworks and enforcement 587
  State Pollution Control Boards 590
  state-level initiatives for water conservation 580–584
  sustainable development 591–592
  waste management laws 20
  water body pollution in 575–577
  water pollution and impact on India's water bodies 575–580
Indian regulations and guidelines 196
indirect potable reuse (IPR) projects 548
industrial activity generation 190–191
industrialization 222–223
industrial process water usage 214–218

Industrial Revolution 19, 24–25
Industrial symbiosis (IS) 33–34, 217–218
industrial waste 21, 31–32, 189
  recovery 191–193
industrial waste management
  regulations 199
  cleaner production technologies 200
  cost savings through resource efficiency 200–201
  enforcement and monitoring 203
  financial constraints 202–203
  global and cross-border challenges 204–205
  impacts of 199–201
  implementation challenges 202–205
  infrastructure 202
  legal and regulatory gaps 203
  legal liabilities and penalties 201
  pollution reduction 199–200
  promotion of innovation 201
  reduced landfill use 200
  socio-cultural and political challenges 204
industrial wastewater composition 168, 297
  agro-industry 299
  batteries-based industry 297–298
  choice of treatment strategy 178–179, **179**
  dairy industry 298–299
  dyes 170–171
  food industry 300
  heavy metals 169–170, *170*
  leather industry 299
  manufacturing and mining industries 300–301
  nuclear power facilities 298
  oil and grease 171–172
  oil extraction 299
  organic compounds 172
  petroleum and petrochemical sectors 298
  pharmaceutical industry 299
  pulp and paper industries 300
  steel and iron industry 298
  techniques 301
  textile industry 300
industrial wastewater economics 301–303
  ecological carbon source 302
  energy and electricity production 302–303
  health impact 301
  new substances production 302
  reducing dependence on fossil carbon 301–302
  sustainable agriculture 302
industrial wastewater, generation of 221–222, *221*
industrial wastewater treatment methods 172–173, 213–214, *215*
  economic burden 222–224
  environmental impact of 218–220, *219*
  generation of 221–222, *221*
  modified methods for 180–182
  primary treatment 173
  reuse and reduce approach 224–226
  secondary treatment 173–177
  tertiary treatment 177–178
  use of water 214–218
industrial wastewater utilization 295–297, *297*
  composition of 297–301
  economics of respective industry 301–303
  resource recovery methods 306–309
  resource waste during industrial usage 303–305
Industrial Water Footprint (IWF) technique 225
Industry 4.0 technologies 537, 539
inorganic matter 98
integrated treatment system 103
integrated water resource management (IWRM) 277
international regulations and guidelines 193, **194–195,** 196
Internet of Things (IoT) 537, 543
  IoT-based waste management 34
investment
  incentives 111
  infrastructure and technology 111, 112
ion exchange methods 307, 347, 547
ion flotation 548
IoT. *see* Internet of Things (IoT)
IRAs. *see* independent regulatory agencies (IRAs)
IS. *see* Industrial symbiosis (IS)
IWRM. *See* integrated water resource management (IWRM)

j

Jal Shakti Abhiyan (JSA) 582
Jalyukt Shivar Abhiyan 583, 585

Jatropha curcas L. 500
joint financing operation 111
JSA. *see* Jal Shakti Abhiyan (JSA)

k

KBDS. *see* Knowledge-Based Design System (KBDS)
Kingdom of Saudi Arabia (KSA) 329
Knowledge-Based Design System (KBDS) 570
KSA. *see* Kingdom of Saudi Arabia (KSA)

l

lab-scale treatment of wastewater 429
Ladakh Autonomous Hill Development Council (LAHDC) 508
Lagos Lagoon pollution 104
LAHDC. *see* Ladakh Autonomous Hill Development Council (LAHDC)
layered double hydroxide (LDH) 434
LCA. *see* life cycle assessment (LCA)
LCC. *see* life cycle cost (LCC)
LCCA. *see* life cycle cost analysis/assessment (LCCA)
LCFAs. *see* long-chain fatty acids (LCFAs)
LDH. *see* layered double hydroxide (LDH)
legal framework 189–190
  existing regulations and guidelines 193–199
  industrial activity and waste generation 190–191
  regulation impacts 199–201
  regulation implementation challenges 202–205
  resources, recovery of 191–193
life cycle assessment (LCA) 147, 200, 237, 355, 481, 501
life cycle cost (LCC) 134, 241–242
life cycle cost analysis/assessment (LCCA) 11, 136
long short-term memory (LSTM) networks 539
long-chain fatty acids (LCFAs) 245
low-pressure membrane (LPM)-adsorption coupling 518
low-pressure reverse osmosis/forward osmosis (LPRO/FO) 242
LPRO/FO. *see* low-pressure reverse osmosis/forward osmosis (LPRO/FO)
Lubigi WWTP 416–417

m

MABRs. *See* membrane-adsorption bioreactors (MABRs)
machine learning (ML) 524, 538
  models 442
  reinforcement learning 539
  supervised learning 538–539
  unsupervised learning 539
Mahindra & Mahindra 325
MBR. *see* membrane bioreactor (MBR)
MCA. *see* Multi-Criteria Analysis (MCA)
MCDA. *see* multi-criteria decision analysis (MCDA)
MCL. *see* Municipal Committee of Leh (MCL)
MD. *see* membrane distillation (MD)
MEC. *see* microbial electrolysis (MEC)
mechanical recycling 32
membrane-adsorption bioreactors (MABRs) 518
membrane-based methods 436–437, 456, 518
membrane bioreactor (MBR) 9, 76, 100, 106, 145, 151, 181, 243, 569
  system 322–323, 415
  treatments 548
membrane capacitive deionization (MCDI) systems 348
membrane distillation (MD) 257, 296, 322, 351–352
membrane filtration techniques 177–178, 306
membrane separation technology 351–352
membrane technology *457*, **458**, 476, 577
  application in wastewater treatment 472–475, *472*
  challenges and opportunities 481
  critical assessment of developments 481
  electrodialysis 474–475
  future research directions 482
  knowledge gaps 482
  membrane mechanisms driven by pressure 473–474, **473**, **475**
  nanostructure-based 463–472
  prospects 480–481
  role in resource recovery from wastewater 475–478, *476*, **477–478**
  techno-economic analysis 479–480
  types of membranes for wastewater treatment 459–463

metal organic framework (MOF) 390, 434
metal oxides 471
metal recovery 547–548
Metropolitan Water Reclamation District of Greater Chicago (MWRDGC) Calumet 547
MF. *see* microfiltration (MF)
MFC. *see* microbial fuel cell (MFC)
Miasmatic Theory 19
MICP. *see* microbe-induced calcium carbonate precipitation (MICP)
microalgae 274, 526
  microalgae-assisted aquaculture 395
  microalgae-based treatment 406
micro-algal treatment 350–351
microbe-induced calcium carbonate precipitation (MICP) 502
microbial-aided biotechnology 377
microbial-assisted remediation 378
microbial electrochemical technologies 344, 348–350
microbial electrolysis (MEC) 243, 348–349
  septic tank 407
microbial fuel cell (MFC) 151–152, 239, 247, 258–259, 273, 284, 326–327, 344, 348–349, 407, 436, 522–523, 548
microbial recycling cells (MRCs) 348–349
microfiltration (MF) 123, 177, 306, 344, *459*, 459–460
microorganisms 370
micropollutants (MPs) 439
  removal 132
Million Liter per Day (MLD) sewage 73
MINARS. *see* Monitoring of Indian National Aquatic Resources System (MINARS)
mining sector 190
Ministry of Environment, Forest and Climate Change (MoEF&CC) 17, 78, 80, 83, 85, 509
Ministry of Housing and Urban Affairs (MoHUA) 85–86, 497, 509
Ministry of Jal Shakti 590
Ministry of Power 509
MIP. *see* molecularly imprinted polymers (MIP)
Mission Kakatiya 584, 586
mixed liquor suspended solids (MLSS) 440
mixed matrix membranes (MMMs) 466
ML. *see* machine learning (ML)
MLR. *see* multiple linear regression (MLR)
MLSS. *see* mixed liquor suspended solids (MLSS)
MMMs. *see* mixed matrix membranes (MMMs)
MoEF&CC. *see* Ministry of Environment, Forest and Climate Change (MoEF&CC)
MOF. *see* metal organic framework (MOF)
MoHUA. *see* Ministry of Housing and Urban Affairs (MoHUA)
molecularly imprinted polymers (MIP) 521–522
Monitoring of Indian National Aquatic Resources System (MINARS) 84
moving bed biofilm reactor (MBBR) system 320–322, *321*
MPs. *see* micropollutants (MPs)
MRCs. *see* microbial recycling cells (MRCs)
MSW. *see* Municipal Solid Waste (MSW); municipal waste management (MSW)
Mucor hiemalis-based biosorbent 392
Mukhya Mantri Jal Swavlamban Abhiyan 583, 586
multi-criteria analysis (MCA) 24
multi-criteria decision analysis (MCDA) 136, 569
multidisciplinary approaches 132
  for wastewater treatment 523–524
multiple linear regression (MLR) 547
multi-step scale up
  interdependency of multi-step system 566–567
  multi-step wastewater treatment system 563–564
  sustainable approaches for scale-up technology 567–570, *568*
  technologies in treatment facility 564–566
multi-step wastewater treatment system 563–564
Municipal Committee of Leh (MCL) 508
Municipal Solid Trash Rules (2000) 20
Municipal Solid Waste (MSW) 17–18
municipal waste 20–21
municipal waste management (MSW) 28, 36, 47–56
MXenes 434

n

Namami Gange Programme 580–581, 585, 590
nanofiltration (NF) 123, 178, 306, 322, 344, 461, *461*
nanomaterials 458
nanosorbents 125
nanostructure-based membrane technology 463
  carbon-based nanomaterials 464–465
  CNT membranes with mixed matrixes 470
  CNTs with horizontal alignment 470–471
  fundamental CNT membranes technology 466, **467–468**
  metal oxides 471
  vertically aligned CNT membranes 466, 469, *469*
  zeolites 471–472
nanotechnology 456, 519–520
NAPCC. *see* National Action Plan on Climate Change (NAPCC)
Nasturtium officinale (watercress) 395
National Action Plan on Climate Change (NAPCC) 80
National Committee on Environmental Planning and Coordination (NCEPC) 83
National Conservation Strategy 80
National Forest Policy 80
The National Green Tribunal (NGT) Act (2010) 588–589, **197**
national industrial strategy 222–224
National Institution for Transforming India (NITI Aayog) 576
National Mission for Clean Ganga (NMCG) 590
National River Conservation Plan (NRCP) 581–582
national strategies 39–40
National Waste Policy—Action Plan (2018) 196
National Water Policy 78
National Water Quality Monitoring Programme 79
national-level budgets and allocations 584
  Atal Bhujal Yojana 585
  Namami Gange Programme 585
  National River Conservation Plan 585
national-level initiatives for water conservation 580–582
natural systems 30
NBV. *see* net benefit value (NBV)
NCEPC. *see* National Committee on Environmental Planning and Coordination (NCEPC)
NDMC. *see* New Delhi Municipal Corporation (NDMC)
Neeru–Chettu program 583
net benefit value (NBV) 47
net present value (NPV) 412
net-zero energy 255, 257
NEWater 102
New Delhi Municipal Corporation (NDMC) 415
New Delhi, PPP for decentralized sewage treatment in 415
NEW Factory concept 275
NEXheater 546
NF. *see* nanofiltration (NF)
NGT. *see* National Green Tribunal (NGT)
NITI Aayog. *see* National Institution for Transforming India (NITI Aayog)
nitrification–denitrification 148–149
nitrogen (N) 337–338
nitrogen dioxide (NO_2) 419
nitrogen mineralization 262
NMCG. *see* National Mission for Clean Ganga (NMCG)
NMPC. *see* nonlinear model predictive control (NMPC)
nonlinear model predictive control (NMPC) 569
NPV. *see* net present value (NPV)
NRCP. *see* National River Conservation Plan (NRCP)
nutrients 98
  by aqua-species 45
  dilution 408
  recovery 43–44, 192, 256–257, 545–546
  recycling 103

o

OC. *see* operational cost (OC)
odor issues 133
Oedegonium hatei-based biosorbent 392
O&G. *see* oil and grease (O&G)

oil and grease (O&G) 171–172
OLR. *see* organic loading rate (OLR)
O&M. *see* operation and maintenance (O&M)
on-site sanitation systems (OSS) 502, 506
on-site wastewater treatment system (OWTS) 76, 77–78
operating expenditure (OpEx) 332
operational cost (OC) 133, 134, 153–154, 242
operation and maintenance (O&M) 236
OpEx. *see* operating expenditure (OpEx)
organic compounds 172
organic loading rate (OLR) 349, 431, 444
organic matter 97–98
organic waste 31
OSS. *see* on-site sanitation systems (OSS)
OWTS. *See* on-site wastewater system (OWTS)

p

PAHs. *See* polycyclic aromatic hydrocarbons (PAHs)
partial nitrification 406
pathogens 98
PCBs. *see* polychlorinated biphenyls (PCBs)
PCPs. *see* pentachlorophenol (PCPs)
PDMS. *see* polydimethyl-siloxane (PDMS)
PDS. *see* peroxydisulfate (PDS)
PEADs. *see* photo-electrochemical anaerobic digestion system (PEADs)
pentachlorophenol (PCPs) 148–149
peroxydisulfate (PDS) 504
persistent organic pollutants (POPs) 172, 189
petroleum-based polyolefins 395
petroleum industries 190
PFCs. *see* photoelectrochemical fuel cells (PFCs)
PGF. *see* planted gravel filter (PGF)
PHA. *see* polyhydroxyalkanoates (PHA)
pharmaceuticals 191, 503–504
phosphorus (P) 43–44, 337–338
photocatalyst 128
photocatalytic membrane reactors (PMRs) 518
photocatalytic nanofiltration reactor (PCNFR) prototype 440
photocatalytic SBRs 431
photo-electrochemical anaerobic digestion system (PEADs) 273
photoelectrochemical fuel cells (PFCs) 548
phototrophic microorganisms 445
Phragmites australis shoot-based biosorbents 391
physical methods 121
  adsorption 125
  filtration 123–124
  flotation 124–125
  membrane-based technology 122–123
  screening 121
  sedimentation 121–122
  for urine and fecal treatment 404–405
  in WWT 534–535
physicochemical technology 344
  adsorption 346–347
  ammonia stripping 346
  ion exchange 347
  struvite precipitation 347–348
phytoremediation 107
"plant availability" 499
planted gravel filter (PGF) 507
plastic waste management 26–27
PMRs. *see* photocatalytic membrane reactors (PMRs)
pollution
  management 220
  prevention 109, 110
polychlorinated biphenyls (PCBs) 172
polycyclic aromatic hydrocarbons (PAHs) 172
polydimethyl-siloxane (PDMS) 472
polyhydroxyalkanoates (PHA) 260, 296, 395
polyvinylidene fluoride (PVDF)-proanthocyanidins-FeOOH membrane 391
pond treatment methods 176–177
POPs. *see* persistent organic pollutants (POPs)
"population expansion" 313
population growth impact 110
potassium (K) 337–338
PPP. *see* public–private partnership (PPP)
preliminary/primary treatment 4–5
pressure, membrane mechanisms driven by 473–474, **473**, **475**
primary sedimentation tank (PST) 4
primary treatment 99–100, 173
process scaling 436
products recovery 305

proportional-integral-derivative (PID) control technique 543
PST. *see* primary sedimentation tank (PST)
public awareness and campaign 112
public health crisis 104
public health interventions 220
public policy 135
public–private partnership (PPP) 111–112, 414–415, 508, 587
  for decentralized sewage treatment in New Delhi 415
pyrolysis 405

r

rare earth elements (REEs) 277
RAS. *see* recirculating aquaculture systems (RAS)
RBCs. *see* rotating biological contactors (RBCs)
RCRA. *see* Resource Conservation and Recovery Act (RCRA)
recirculating aquaculture systems (RAS) 394
recurrent neural networks (RNNs) 539
recycling technologies 32, 108
reduce, reuse and recycle (3Rs model) 370, 377
REEs. *see* rare earth elements (REEs)
regional strategic frameworks 40
regulatory authority of India
  central pollution control board (CPCB) 79
  compliance and monitoring 82
  for construction of treatment facility 78–82
  decentralized water regulatory authority 85–86
  Environmental Impact Assessment (EIA) 80–81
  Ministry of Environment, Forest, and Climate Change (MoEFCC) 80
  parameters of water quality 83–84
  for quality and maintenance evaluation treatment facility 83–86
  state pollution control board (SPCBs) 79–80
  surveillance of water quality 84
  technological standards 81–82
reinforcement learning (RL) 539
research strategy 220
Resource Conservation and Recovery Act (RCRA) 196
resource recovery 316, 344, **345**
  AI for resource recovery in WWT 544–549
  biological processes 308–309
  biological technology used for 374, **375–376**
  chemical processes 307–308
  decentralized project planning and 316–327
  incentives 111
  integration of treatments to enhance 524–526
  methods 306
  physical procedures 306
  planning infrastructure 316–319, *317–318*
  probable 324–327
  regulations promoting 196, **197–199**, 199
  technologies for simultaneous wastewater treatment and 344–352, **345**
  technology advancement 319–324
  urine and fecal matter as material for 402–403
  from wastewater 269–270, *271*
Restriction of Hazardous Substances (RoHS) 200
revenue generation 154–155
reverse osmosis (RO) 100, 123, 242, 296, 322, 344, 445, 456, 462, *462*, 499, 535, 548
RL. *see* reinforcement learning (RL)
RNNs. *see* recurrent neural networks (RNNs)
RO. *see* reverse osmosis (RO)
RoHS. *see* Restriction of Hazardous Substances (RoHS)
rotating biological contactors (RBCs) 145, 150–151, 175, 430

s

Saraswati 2.0 project 319
SBM. *see* Swachh Bharat Mission (SBM)
SBRs. *see* sequencing batch reactor (SBRs)
SBT. *see* soil biotechnology (SBT)
scale-up technology, sustainable approaches for 567–570, *568*
scaling up wastewater treatment systems 447
Scenedesmus biochar 396
screeners 172

screening 4–5, 306
SDGs. *see* sustainable development goals (SDGs)
SDG 6. *see* Sustainable Development Goal 6 (SDG 6)
secondary sedimentation tank (SST) 5
secondary treatment 5–6, 100, 173–174
  aerobic method 174–175
  anaerobic method 176
  pond treatment methods 176–177
sedimentation 173, 306
septic tank 8, 320
sequencing batch reactor (SBRs) 10, 76, 246, 430–431
settling process 122
sewage
  artificial intelligence and IoT 107–108
  chemical energy stored 255–256
  composition of 97–98, 254–255
  definition of 96–97
  energy, agriculture, and construction applications 44–45
  energy source 256–262
  generation 109–110
  prospects and challenges 262–263
  sources *96*, 254–255
  wasting resource 253–254
  water 253
  water quality 135
sewage treatment 99–100, 106, 119–120
  advanced treatment technologies 106–107
  chemical methods 126–130
  coagulation/flocculation 130–131
  current state of 99–100
  effect of economic analysis 136–139
  factors affecting economics 134–136
  future projections 109–110
  global sewage treatment statistics 101–102, **101**
  India 103–104
  innovations in sewage management 107–108
  limitations of 131–133
  mechanism of *99*
  Nigeria 104
  physical methods 121–125
  physicochemical processes methods 130–131
  potential solutions 109–112
  recommendations 112
  reducing environmental impact 108–109
  Singapore 102–103
  Sweden 103
  systems in Kampala 416–417
sewage treatment plants (STPs) 73, 403, 415, 496, 562
sewage wastewater
  electricity 258–260, *260*
  green building materials 260–261
  heat recovery 260, *261*
  nutrients recovery 256–257
  recovery of vermicompost sludge in agriculture 261–262, *262*
silver nanotubes (SNTs) 465
single-wall nanotube (SWNT) 464
skimming tank 173
slow sand filters (SSF) 388
sludge disposal issues 133
sludge retention time (SRT) 246
sludge volume index (SVI) 246
small and medium-sized enterprises (SMEs) 35, 82
SMC. *see* Surat Municipal Corporation (SMC)
SMEs. *see* small and medium-sized enterprises (SMEs)
SmF. *see* submerged fermentation (SmF)
SNTs. *see* silver nanotubes (SNTs)
soil biotechnology (SBT) 415
soil fertilizers 43–44
solid retention time (SRT) 322, 443
solid-state fermentation (SSF) 52
Solid Waste Management (SWM) 18, **198**, 418
  in Rural Maharashtra 415–416
source control 109
source-separated urine 504
source separation *496*
  biofuel 500–501
  case studies 505
  centralized treatment of wastewater systems, case studies on 505–506
  challenges of source separation systems and solutions 502
  concerns toward source-separated urine 504
  decentralized treatment of wastewater, case studies on 506–508

source separation (*Cont.*)
economic analysis of 508–511, **510**
energy savings 499–500
inconsistent regulations on waste-derived fertilizers 502–503
Koyambedu treatment plant of Chennai 506
lack of monitoring systems and construction spaces 502
nutrient recovery 498–499
perception 504–505
pharmaceuticals, heavy metals, and compounds 503–504
reuse of treated sewage by Surat's industries 505–506
urine-derived Products 501–502
wastewater treatment, important for 497–502, *498*
source variation 98
space requirement 133
SPCBs. *see* state pollution control board (SPCBs)
SRT. *see* sludge retention time (SRT); solid retention time (SRT)
SSF. *see* slow sand filters (SSF); solid-state fermentation (SSF)
SST. *see* secondary sedimentation tank (SST)
state-level budgets and allocations 585
Gujarat's Sujalam Sufalam Jal Abhiyan 586
Maharashtra's Jalyukt Shivar Abhiyan 585
Rajasthan's Mukhya Mantri Jal Swavlamban Abhiyan 586
Telangana's Mission Kakatiya 586
state-level initiatives for water conservation 583–584
state pollution control board (SPCBs) 78–80, 589, 590
stormwater detention tanks (SWDTs) 496
stormwater management 108, 496
STPs. *see* sewage treatment plants (STPs)
structural information 236–237
circular economy 239–240
energy conservation and management 238
energy recovery 238–239
environmental impact assessment and life cycle assessment 237
factors 240
identification of unit process and unit operations 238
influents/effluents characteristics 237
materials recovery 239
sewer segregation efforts 238
struvite ($NH_4MgPO_4{\cdot}6H_2O$) 499
precipitation 347–348, 404
submerged fermentation (SmF) 52
Sujalam Sufalam Jal Abhiyan 583, 586
sulfur dioxide (SO_2) 419
supervised learning 538–539
Supervisory Control and Data Acquisition (SCADA) system 238, 569
supply chain redesign 35
Surat Municipal Corporation (SMC) 505
surveillance of water quality 84
suspended solids 98
sustainability 355, **356–357**
of multi-step systems 562
sustainable approaches for scale-up technology 567–570, *568*
sustainable bioenergy generation network 47
concerns, sources and MSW composition 47–48, *49*
future directions 52–53
technologies for turning MSW into energy 48, **49–51**
waste management for circular bioeconomy energy generation plan 52
sustainable development 591–592
Sustainable Development Goal 6 (SDG 6) 217
sustainable development goals (SDGs) 12, 18, 256, 328, 378
sustainable resource management 102–103
SVI. *see* sludge volume index (SVI)
Swachh Bharat Mission (SBM) 416, 582
SWDTs. *see* stormwater detention tanks (SWDTs)
SWM. *see* Solid Waste Management (SWM)
SWNT. *see* single-wall nanotube (SWNT)
synthetic polymers 521

t

"take-make-use-dispose" approach 386
take-make-waste system 269
Tata Steel 325

TDS. *see* total dissolved solids (TDS)
TEA. *see* techno-economic analysis (TEA)
tech-driven industries 222–224
techno-economic analysis (TEA) 479
  cost calculation 479
  developing and validating model 480
  operational costs 480
  process design 479, *479*
  technical and economic assessment of effluent processing methods 480
  technical aspect 479
  of nutrient recovery from urine 411–413
technology adoption 82, 111
temperature 445
10Rs concept 30, *31*
tertiary treatment 6–7, 100, 177–178
tertiary treatment reverse osmosis (TTRO) 506
tetrahedron MFC (TP MFC) 408
textile manufacturing 191
TF. *see* trickling filter (TF)
The Vellore Institute of Technology (VIT) 326
thermal treatment 352
thermochemical valorization techniques 45
TMC. *see* Town Municipal Corporation (TMC)
TN. *see* total nitrogen (TN)
TOC. *see* total organic carbon (TOC)
total dissolved solids (TDS) 168, 563
total nitrogen (TN) 439, 563
total organic carbon (TOC) 274, 392, 471
total phosphorous (TP) 563
total suspended solids (TSS) 168, 323, 392
Town Municipal Corporation (TMC) 507
TP. *see* total phosphorous (TP)
TP MFC. *see* tetrahedron MFC (TP MFC)
trade-oriented industries 222–224
traditional techniques (TT) 243
transformer models 539
treated sewage effluents (TSE) 501
treated wastewater (TWW) 509
treatment system 75–76, 135
  centralized wastewater treatment systems (CWTS) 76–77
  decentralized wastewater treatment system (DEWATS) 77
  on-site wastewater system (OWTS) 77–78
trickling filter (TF) 2
Triticum aestivum (wheat) 395
TSE. *see* treated sewage effluents (TSE)
TSS. *see* total suspended solids (TSS)
TT. *see* traditional techniques (TT)
TTRO. *see* tertiary treatment reverse osmosis (TTRO)
22 WWTPs 56–57
TWW. *see* treated wastewater (TWW)

u

UASB. *see* anaerobic sludge blanket (UASB)
UASBR. *see* up-flow anaerobic sludge bed reactor (UASBR)
UDT. *see* urine-diverting toilet (UDT)
UF. *see* ultrafiltration (UF)
UFASFF. *see* up-flow anaerobic sludge-fixed-film reactor (UFASFF)
ultrafiltration (UF) 123, 178, 306, 351, 460–461, *460*, 505
ultrasonic (US) irradiation 435
ultraviolet (UV) light 100
UN. *See* United Nations (UN)
uncontrolled waste 37
UNEP. *see* United Nations Environment Programme (UNEP)
UNEP 2024 report 37
UNIDO. *see* United Nations Industrial Development Organization (UNIDO)
United Nations (UN) 29
United Nations Environment Programme (UNEP) 196
United Nations Industrial Development Organization (UNIDO) 196
United States Environmental Protection Agency (USEPA) 11, 28, 389
unsupervised learning 539
UN World Water Development Report (2023) 1, 218
upcycling technologies 32
up-flow anaerobic sludge bed reactor (UASBR) 431
up-flow anaerobic sludge blanket (UASB) reactor 5, 9, 273, 322, 324, 405, 525
up-flow anaerobic sludge-fixed-film reactor (UFASFF) 432
urbanization 110

urine and fecal matter
bioelectrochemical systems 407–408, *407*
biological methods 405–406
case study for economic analysis 411–419
comparative analysis of centralized and decentralized systems 408–410, *409*, **410**
economic analysis of centralized and decentralized treatment of 408–410
limitations of centralized treatment system for 403–404
as material for resource recovery 402–403
methods for urine and fecal treatment 404–408
physical methods 404–405
techno-economic analysis of nutrient recovery from 411–413
urine-derived Products 501–502
urine-diverting toilet (UDT) 500
USEPA. *see* United States Environmental Protection Agency (USEPA)

V

ventilated improved pit (VIP) 418
vermicompost sewage sludge 261–262, *262*
vermifilter treatment 350
vertically aligned CNT (VACNT) membranes 466, 469, *469*
VFAs. *see* volatile fatty acids (VFAs)
VIP. *see* ventilated improved pit (VIP)
VIT. *see* The Vellore Institute of Technology (VIT)
volatile fatty acids (VFAs) 305

W

waste 386
fertilizers 502–503
generation rates 18, 190–191
valorization 31–32
Waste Framework Directive (WFD) 196
waste management 29
economic perspectives 22–23
importance of 17–19
Waste Management and Public Cleansing Law (1970) 196
waste stabilization ponds (WSP) 134, 324
Waste-to-Energy (WTE) 23, 32–33, 52–53, 196
waste treatment
bridging the gap 23–24
case studies 47–58
circular economy strategies 29–35
economic perspectives 22–23
economic viability of strategies 36–40
future based on economic value 41–47
global waste generation trends and challenges 21–22
historical perspective of 19–20
importance of 17–19, *18, 19*
linear economy strategies 24–29
types and needs 20–21
wastewaters (WWs) 21, 32, 323, 337
and circular economy 372–373, *372*
construction of treatment facility 78–82
crop growth, improvement in 343
decentralized and centralized approaches for utilization of 352–357
energy generation 275–276
energy stored, forms of 271–272
energy, methods to derive 273–274
environment, climate and economics 276–277
food processing industry 281–285
management system 73–75
mining sector 278–281
nutrient composition of 340, **341–342**
nutrients applied as fertilizers, forms of 340, 343
quality and maintenance evaluation treatment facility 83–86
resource recovery from 269–270, *271*
source of nutrients in 339–340
as source of resources 338–343, *339*
treatment system 75–78
Wastewater Resource Recovery Facilities (WRRF) 275
wastewater treatment (WWT) 1–2, 377, 495
AI for resource recovery in 544
applications of AI technologies in 540
by-product management 541–542
centralized systems 3–7, *4* (*see also* centralized systems)
decentralized *vs.* centralized economics 11–12
energy recovery 546–547
fault detection 544

improving process performance 543–544
improving water accessibility 542–543
metal recovery 547–548
monitoring and control 540–541
needs 146–147
nutrient recovery 545–546
process design and optimization 542
simulating contaminant removal 543
strategies for 8–10
systems 533
water reuse 548–549
activated carbon 520–521
conventional technologies for 386–390
economic effects of 105–106
effects on communities 105
effects on industries 105
forward osmosis 463
government and municipal expenditures 106
hydrochar/biochar 521
membrane technology's application in 472–475, *472*
microbial fuel cells 522–523
microfiltration *459*, 459–460
modern technologies incorporating CE for 394–396, *395*
molecularly imprinted polymers 521–522
nanofiltration 461, *461*
nanotechnology 519–520
reverse osmosis 462, *462*
types of membranes for 459
ultrafiltration 460–461, *460*
usage of modern technologies for 519–523
wastewater treatment plant (WWTP) 42, 134, 147, 214, 235–236, 256, 271, 314, 330, 370–373, 416, 429, 499, 524
capacity and methods, dairy industry 244–248
construction, maintenance and working 241–243
structural information of 236–240
technology requirement for scaling up 441–442
wastewater treatment systems (WWTS) 314
wastewater treatment technologies (WWTTs) 427
adsorption 433–435
advanced oxidation process 435–436
biofilm and floc 444–445
biological method 350–351
biological wastewater treatment processes 430–433
case studies 447–448
dissolved oxygen 444
electrochemical technologies 348
extracellular polymeric substance 443–444
flocculation and coagulation 437–438, *438*
hybrid systems 438–441, *440*
hydraulic retention time 443
membrane-based processes 436–437
membrane separation technology 351–352
microbial electrochemical technologies 348–350
microbial fuel cells 436
organic loading rate 444
other parameters 445–446
parameters affecting scale-up of treatment technology 443
physicochemical technology 344, 346–348
and resource recovery 344–352, **345**
solid retention time 443
stepwise usage of multiple technologies for field implementation 446–447
technologies used for lab-scale treatment of wastewater 429–441
technology requirement for scaling up WWTPs 441–442
temperature 445
thermal treatment 352
water 385, 533
communities 517
consumption 225
contamination 235
management 524
molecules 465
national-level initiatives for water conservation 580–584
state-level initiatives for water conservation 580–584
recovery 191
recycling 226
reuse 548–549
reuse and reduce approach 224–226
scarcity 221
stress 1–2

Water Act (1974) 20, 79, 82, 83, **197**, 588
water body cleaning
  economic strategies and financial support mechanisms for 584–587
  environmentally sustainable financing 587
  government budgets and allocations 584
  international aids 586–587
  national-level budgets and allocations 584–585
  public–private partnerships 587
  state-level budgets and allocations 585
water footprint (WF) 217, 225
water pollution
  climate regulation and resilience 579–580
  ecological health and biodiversity 578–579
  human health and community well-being 579
  and impact on India's water bodies 575
  impact on public health, environment, and economy 577–578
  importance of clean water bodies 578
  water body pollution in India, overview of 575–577
water quality, parameters of 83–84
water resource recovery facilities (WRRFs) 413
Water Stress Index (WSI) 217
WCED. *see* World Commission on Environment and Development (WCED)
WF. *see* water footprint (WF)
WFD. *see* Waste Framework Directive (WFD)
WHO. *see* World Health Organization (WHO)
World Bank Group 22
World Commission on Environment and Development (WCED) 278
World Health Organization (WHO) 28, 313–314
World Trade Organization (WTO) 196
World Water Assessment Programme (WWAP) 221
WRRF. *see* Wastewater Resource Recovery Facilities (WRRF)
WRRFs. *see* water resource recovery facilities (WRRFs)
WSI. *see* Water Stress Index (WSI)
WSP. *see* waste stabilization ponds (WSP)
WTE. *see* Waste-to-Energy (WTE)
WTO. *see* World Trade Organization (WTO)
WWAP. *see* United Nations World Water Assessment Programme (WWAP)
WWs. *see* wastewaters (WWs)
WWT. *see* wastewater treatment (WWT)
WWTP. *see* wastewater treatment plant (WWTP)
WWTS. *see* wastewater treatment systems (WWTS)
WWTTs. *see* wastewater treatment technologies (WWTTs)

y

Yamuna Action Plan (YAP) 84, 581
YAP. *see* Yamuna Action Plan (YAP)

z

zeolites 471–472
zinc oxide (ZnO) nanoparticles 471
Zirconium dioxide (ZrO_2) nanoparticles 471
zone settling 122

Printed and bound by CPI Group (UK) Ltd, Croydon, CR0 4YY

17/09/2026

14964330-0005